Modern Approaches to
the Study of Crustacea

Modern Approaches to the Study of Crustacea

Edited by

Elva Escobar-Briones

Instituto de Ciencias del Mar y Limnología
Universidad Nacional Autónoma de México
México D.F., México

and

Fernando Alvarez

Instituto del Biología
Universidad Nacional Autónoma de México
México D.F., México

Kluwer Academic / Plenum Publishers
New York, Boston, Dordrecht, London, Moscow

Proceedings of The Crustacean Society 2000 Summer Meeting, held June 26–30, 2000, in Puerto Vallarta, Mexico

ISBN: 0-306-47366-6

©2002 Kluwer Academic / Plenum Publishers, New York
233 Spring Street, New York, New York 10013

http://www.wkap.nl/

10 9 8 7 6 5 4 3 2 1

A C.I.P. record for this book is available from the Library of Congress

CONTRIBUTORS

Daniela Triguerinho Alarcon   University of Taubaté, Department of Biology, Pca. Marcelino Monteiro 63, Taubaté, Sao Paulo, Brazil.

Javier Alcocer   Laboratorio de Limnología UIICSE, FES Iztacala, Universidad Nacional Autónoma de México, Av. de los Barrios s/n, Los Reyes Iztacala, Tlalnepantla 54090, Estado de México, México

Fernando Alvarez   Departamento de Zoología, Instituto de Biología, Universidad Nacional Autónoma de México, Apartado Postal 70-153, México 04510, D.F., México.

Judith Arciniega-Flores   Centro de Ecología Costera, Universidad de Guadalajara, Gómez Farías 82, San Patricio Melaque, Jalisco, México.

Hugo Aréchiga   División de Estudios de Posgrado e Investigación, Facultad de Medicina, Universidad Nacional Autónoma de México, México 04510, D.F., México.

Ian A. E. Bayly   501 Killiekrankie Rd., Flinders Island, Tasmania 7255, Australia.

Giovana Bertini   Departamento de Zoologia, Instituto de Biociencias, C.P. 510, Universidade Estadual Paulista 18618-000, Botucatú, Sao Paulo, Brazil.

José Luis Bortolini   Laboratorio de Invertebrados, Facultad de Ciencias, Apartado Postal 70-371, México 04510, D.F., México.

Enrique E. Boschi   Facultad de Ciencias Exactas y Naturales, Universidad de Buenos Aires, Paseo Victoria Ocampo 1, Mar del Plata B7605AVN, Argentina.

David Brailovsky   Facultad de Ciencias, Universidad Nacional Autónoma de México, México 04510, D.F., México.

Patricia Briones-Fourzán   Unidad Académica Puerto Morelos, Instituto de Ciencias del Mar y Limnología, Apartado Postal 1152, Cancún, Quintana Roo 77500, México.

Theodore H. Bullock   Neurobiology Unit, Scripps Institution of Oceanography and Department of Neurosciences, University of California, SanDiego, La Jolla, California 92093, USA.

Héctor Cabanillas-Beltrán   Instituto Tecnológico de Tepic, Tepic, Nayarit, México.

Gabina Calderón-Rosete   División de Estudios de Posgrado e Investigación, Facultad de Medicina, Universidad Nacional Autónoma de México, México 04510, D.F., México.

Ronald L. Carter   Loma Linda University, Department of Natural Sciences, Loma Linda, California 92350, USA.

Kathryn A. Coates   Bermuda Biological Station for Research, 17 Biological Lane, St. George's, Bermuda GE01.

Valter José Cobo   University of Taubaté, Department of Biology, Pca. Marcelino Monteiro 63, Taubaté, Sao Paulo, Brazil.

Pablo A. Collins   Instituto Nacional de Limnología, José Macía 1933, Santo Tomé 3016, Santa Fé, Argentina.

Rogerio C. Costa   Departamento de Zoologia, Instituto de Biociencias, C.P. 510, Universidade Estadual Paulista 18618-000, Botucatú, Sao Paulo, Brazil.

David L. Cowles   Loma Linda University, Department of Natural Sciences, Loma Linda, California 92350, USA.

Stacie E. Crowe   Nova Southeastern University Oceanographic Center, 8000 N. Ocean Drive, Dania, Florida 33004, USA

Javier Cruz-Hernández   Departamento El Hombre y su Ambiente, Universidad Autónoma Metropolitana – Xochimilco, México, D.F., México.

Elena Cuartas   Facultad de Ciencias Exactas y Naturales, Universidad Nacional de Mar del Plata, Funes 3350, 7600 Mar del Plata, Argentina.

Ana Cristina Díaz   Facultad de Ciencias Exactas y Naturales, Universidad Nacional de Mar del Plata, Funes 3350, 7600 Mar del Plata, Argentina.

Beth A. Dillon   Biology Department, University of Pittsburgh at Johnstown, Pennsylvania 15904, USA.

David W. Dunham   Department of Zoology, University of Toronto, 25 Harbord Street, Toronto, Ontario, Canada M5S 3G5.

Erich Eder   Institute of Zoology, University of Vienna, Althanstrasse 14, A-1090 Vienna, Austria.

Elva Escobar-Briones   Unidad Académica en Sistemas Oceanográficos y Costeros, Instituto de Ciencias del Mar y Limnología, UNAM, México 04510, D.F., México

María del Cármen Espinosa-Pérez   Unidad Académica Mazatlán, Instituto de Ciencias del Mar y Limnología, Universidad Nacional Autónoma de México, Apartado Postal 811, Mazatlán, Sinaloa 82000, México.

Brian Farrell   Museum of Comparative Zoology, Harvard University, 26 Oxford Street, Cambridge, Massachusetts 02138, USA.

Jorge L. Fenucci   Departamento de Ciencias Marinas, Facultad de Ciencias Exactas y Naturales, Universidad Nacional de Mar del Plata, Funes 3350, B7602AYL, Mar del Plata, Argentina.

Analia Verónica Fernández   Departamento de Ciencias Marinas, Facultad de Ciencias Exactas y Naturales, Universidad Nacional de Mar del Plata, Funes 3350, B7602AYL, Mar del Plata, Argentina.

Luisa E. Fiorito   Laboratorio de Histología Animal, Departamento de Ciencias Biológicas, Facultad de Ciencias Exactas y Naturales, Universidad de Buenos Aires, Pabellón II, Ciudad Universitaria, 1428 Buenos Aires, Argentina.

Adilson Fransozo   Departamento de Zoologia, Instituto de Biociencias, C.P. 510, Universidade Estadual Paulista 18618-000, Botucatú, Sao Paulo, Brazil.

José B. García   Universidad Central de Venezuela, Instituto de Zoología Tropical, Apartado 47058, Caracas 1041 – A, Venezuela.

Renata B. Garcia   Departamento de Biologia, FFCLRP, Av. Bandeirantes 3900, Riberao Preto, São Paulo, Brazil.

Henrik Glenner   Department of Evolution, Zoological Institute, University of Copenhagen, Universitetsparken 15, DK-2100, Copenhagen, Denmark.

Fernanda Jordão Guimarães   Departamento de Zoologia, Instituto de Biociencias, C.P. 510, Universidade Estadual Paulista 18618-000, Botucatú, Sao Paulo, Brazil.

Julia Halperin   Laboratorio de Histología Animal, Departamento de Ciencias Biológicas, Facultad de Ciencias Exactas y Naturales, Universidad de Buenos Aires, Pabellón II, Ciudad Universitaria, 1428 Buenos Aires, Argentina.

Michel E. Hendrickx   Unidad Académica Mazatlán, Instituto de Ciencias del Mar y Limnología, Universidad Nacional Autónoma de México, Apartado Postal 811, Mazatlán, Sinaloa 82000, México

Gladys N. Hermida   Laboratorio de Histología Animal, Departamento de Ciencias Biológicas, Facultad de Ciencias Exactas y Naturales, Universidad de Buenos Aires, Pabellón II, Ciudad Universitaria, 1428 Buenos Aires, Argentina.

Jesús Hernández-Falcón   Departamento de Fisiología, Facultad de Medicina, Universidad Nacional Autónoma de México, México 04510, D.F., México.

Walter Hödl Institute of Zoology, University of Vienna, Althanstrasse 14, A-1090 Vienna, Austria.

Jens T. Høeg Department of Zoomorphology, Zoological Institute, University of Copenhagen, DK-2100, Copenhagen, Denmark.

Jolande de Jonge Riza, Postbus 17, 8200 AA, Lelystad, The Netherlands.

Víctor Landa-Jaime Centro de Ecología Costera, Universidad de Guadalajara, Gómez Farías 82, San Patricio Melaque, Jalisco, México.

Karen T. Lee Biology Department, University of Pittsburgh at Johnstown, Pennsylvania 15904, USA.

Maria Helena de Arruda Leme University of Taubaté, Department of Biology, Pca. Marcelino Monteiro 63, Taubaté, Sao Paulo, Brazil.

Gilles LeMoulac Ifremer, Aquacop, Centre Océanologique du Pacifique, BP 7004, Taravao, Tahiti, Polinésie Français.

Enrique Lozano-Alvarez Unidad Académica Puerto Morelos, Instituto de Ciencias del Mar y Limnología, Apartado Postal 1152, Cancún, Quintana Roo 77500, México.

Carlos M. Luquet Laboratorio de Histología Animal, Departamento de Ciencias Biológicas, Facultad de Ciencias Exactas y Naturales, Universidad de Buenos Aires, Pabellón II, Ciudad Universitaria, 1428 Buenos Aires, Argentina.

Fernando L. M. Mantelatto Departamento de Biologia, FFCLRP, Av. Bandeirantes 3900, Riberao Preto, São Paulo, Brazil.

John C. Markham Arch Cape Marine Laboratory, Arch Cape, Oregon 97102-0105, USA.

Carlos A. Martínez-Palacios Instituto de Investigaciones sobre Recursos Naturales, Morelia, Michoacán, México.

Charles L. McKenney, Jr. US Environmental Protection Agency, National Health and Environmental Effects Research Laboratory, Gulf Breeze, Florida 32561, USA.

Luis M. Mejía-Ortiz Departamento El Hombre y su Ambiente, Universidad Autónoma Metropolitana – Xochimilco, México, D.F., México.

Liubov Nagorskaya Institute of Zoology NAS B, 27, Academitcheskaya st. 220072, Minsk, Belarus.

Sergio F. Nates National Research Council Postdoctoral Research Associate, USA.

Maria Lúcia Negreiros-Fransozo Departamento de Zoologia, Instituto de Biociencias, C.P. 510, Universidade Estadual Paulista 18618-000, Botucatú, Sao Paulo, Brazil.

Cristina Pascual Grupo de Biología Marina Experimental, Facultad de Ciencias, Universidad Nacional Autónoma de México, Apartado Postal 69, Ciudad del Carmen, Campeche, México.

Gladys N. Pellerano Laboratorio de Histología Animal, Departamento de Ciencias Biológicas, Facultad de Ciencias Exactas y Naturales, Universidad de Buenos Aires, Pabellón II, Ciudad Universitaria, 1428 Buenos Aires, Argentina.

Laura Peralta FES-Iztacala, Universidad Nacional Autónoma de México, Los Reyes Iztacala, Tlalnepantla 54090, Edo. de México, México.

Guido Pereira Universidad Central de Venezuela, Instituto de Zoología Tropical, Apartado 47058, Caracas 1041 – A, Venezuela.

Giuseppe L. Pesce Dipartimento di Scienze Ambientali, Universitá de L'Aquila, Via Vetoio, Localita Copito, 67100 L'Aquila, Italia.

Ana María Petriella Facultad de Ciencias Exactas y Naturales, Universidad Nacional de Mar del Plata, Funes 3350, 7600 Mar del Plata, Argentina.

Marcelo A. A. Pinheiro   Departamento de Biologia Aplicada, FACV, Universidade Estadual Paulista, 14870-000, Jaboticabal, São Paulo, Brazil.

Jesús T. Ponce-Palafox   Universidad Autónoma de Nayarit, CUVEDES, Tepic, Nayarit, México.

Fidel Ramón   División de Estudios de Posgrado e Investigación, Facultad de Medicina, Universidad Nacional Autónoma de México, México 04510, D.F., México.

Nancy A. Rayner   Zoology Department, University of Durban-Westville, P. Bag X54001, Durban 4000, South Africa.

Marjorie L. Reaka-Kudla   Department of Biology, University of Maryland, College Park, Maryland 20742, USA.

Y. Ranja Reddy   Department of Zoology, Nagarjuna University, Nagarjunanagar 522510, India.

Janet W. Reid   Department of Invertebrate Zoology, National Museum of Natural History, Smithsonian Institution, Washington, D.C. 20560-0163, USA.

Alvaro Luiz Diogo Reigada   Universidade Estadual Paulista, Campus do Litoral Paulista, Praça Infante D. Henrique s/n, CEP 11330-205, São Vicente, São Paulo, Brazil.

Carlos E. F. Rocha   Departamento de Zoologia, Universidade de São Paulo, C.P. 11461, São Paulo 05422-970, Brazil.

Lisa J. Rodrigues   Department of Zoology, University of Toronto, 25 Harbord Street, Toronto, Ontario, Canada M5S 3G5.

Leonardo Rodríguez-Sosa   División de Estudios de Posgrado e Investigación, Facultad de Medicina, Universidad Nacional Autónoma de México, México 04510, D.F., México.

Yolanda Rojas   Departamento de Zoología, Instituto de Biología, Universidad Nacional Autónoma de México, Apartado Postal 70-153, México 04510, D.F., México

Rosangela Romanos   Departamento de Ciencias Marinas, Facultad de Ciencias Exactas y Naturales, Universidad Nacional de Mar del Plata, Funes 3350, B7602AYL, Mar del Plata, Argentina.

Carlos Rosas   Grupo de Biología Marina Experimental, Facultad de Ciencias, Universidad Nacional Autónoma de México, Apartado Postal 69, Ciudad del Carmen, Campeche, México

José Salgado-Barragán   Unidad Académica Mazatlán, Instituto de Ciencias del Mar y Limnología, Universidad Nacional Autónoma de México, Apartado Postal 811, Mazatlán, Sinaloa 82000, México

Adolfo Sánchez   Grupo de Biología Marina Experimental, Facultad de Ciencias, Universidad Nacional Autónoma de México, Apartado Postal 69, Ciudad del Carmen, Campeche, México

Ariadna Sánchez   Grupo de Biología Marina Experimental, Facultad de Ciencias, Universidad Nacional Autónoma de México, Apartado Postal 69, Ciudad del Carmen, Campeche, México

Sandro Santos   Departamento de Biologia, Universidade Federal de Santa Maria, 97119-900, Santa Maria, RS, Brazil.

John R. Scarbrough   Loma Linda University, Department of Natural Sciences, Loma Linda, California 92350, USA.

Giséle Signoret   Departamento El Hombre y su Ambiente, Universidad Autónoma Metropolitana – Xochimilco, México, D.F., México.

Martha Signoret   Departamento El Hombre y su Ambiente, Universidad Autónoma Metropolitana – Xochimilco, México, D.F., México.

Liliana G. Sousa   Departamento de Ciencias Marinas, Facultad de Ciencias Exactas y Naturales, Universidad Nacional de Mar del Plata, Funes 3350, B7602AYL, Mar del Plata, Argentina.

Joseph Staton   Museum of Comparative Zoology, Harvard University, 26 Oxford Street, Cambridge, Massachusetts 02138, USA.

Hilda de Stefano Universidad Nacional Experimental Simón Rodríguez, Instituto de Estudios Científicos y Tecnológicos, Centro de Estudios Biomédicos y Veterinarios, Apartado 47925, Caracas 1041–A, Venezuela.

Eduardo Suárez-Morales El Colegio de la Frontera Sur – Unidad Chetumal, Apartado Postal 424, Chetumal, Quintana Roo 77000, México.

James D. Thomas National Coral Reef Institute, 8000 N. Ocean Drive, Dania, Florida 33004, USA

María del Pilar Torres-García Laboratorio de Invertebrados, Facultad de Ciencias, Apartado Postal 70-371, México 04510, D.F., México.

Shea R. Tuberty Center for Environmental Diagnostics and Bioremediation, University of West Florida, Pensacola, Florida 32514, USA.

Hiroshi Ueda Center for Marine Environmental Studies, Ehime University, Bunkyo-cho 3, Matsuyama, Ehime 790-8577, Japan.

Francisco Vargas-Albores Centro de Investigación en Alimentación y Desarrollo, A. C., Apartado Postal 1735, Hermosillo , Sonora 83000, México.

Eduardo J. Vernon-Carter Universidad Autónoma Metropolitana – Iztapalapa, México D.F., México.

José A. Viccon-Pale Departamento El Hombre y su Ambiente, Universidad Autónoma Metropolitana – Xochimilco, México, D.F., México.

José Luis Villalobos Departamento de Zoología, Instituto de Biología, Universidad Nacional Autónoma de México, Apartado Postal 70-153, México 04510, D.F., México.

D. Dudley Williams Division of Life Sciences, University of Toronto at Scarborough, 1265 Military Trail, Scarborough, Ontario, Canada M1C 1A4.

Verónica Williner Instituto Nacional de Limnología, José Macía 1933, Santo Tomé 3016, Santa Fé, Argentina.

PREFACE

When I teach my undergraduate course in Crustacea at the University of Copenhagen I always tell my students: " Now, if on reading through the chapter on Crustacea you wonder, what could possibly be said in common about these animals, you have got it exactly right! Indeed, there is next to nothing that unite them". Crustaceans range in size from much below millimeter size to the giant king crabs; in morphology from the almost stereotyped segment repetition of the Remipedia to animals with a highly diverse array of appendages such as the Decapoda; in life form from suspension feeding anostracans to the incredibly specialized rhizocephalan parasites; in importance from the all dominating calanoid copepods to the handful of meiobenthic cephalocarid species. But despite this almost incredible range in morphology, life form and species number the details of larval ontogeny and the marvelously preserved fossils of the Cambrian 'Orsten' fauna tells us that these animals belong together, and we now learn that they may even encompass the speciose insects. With or without insects the Crustacea are therefore unquestionably one of the most important animal groups. Is this why we study them? I think not. Our fascination with Crustacea is even deeper rooted and not easily analyzed. But one thing is important to me in this context. As I browse through all the papers in this book it brings up the memory of the highly rewarding Summer Meeting of *The Crustacean Society* held in 2000 in Puerto Vallarta, Mexico, where I had the privilege to serve as its president. We owe it to the editors, professors Elva Escobar and Fernando Alvarez to have compiled this book on Crustacea filled with contributions from authors all over the world. While each and every one of them bears witness to the fascination we share in Crustacea, the camaraderie across national barriers that is an inherent part of our science is perhaps our must precious legacy to the future.

Jens T. Hoeg
University of Copenhagen
President, the Crustacean Society, 2000-2001

In June of 2000, more than 200 scientists assembled in Puerto Vallarta, Mexico, to celebrate the Summer Meeting of the Crustacean Society. It was with the view of Bahía de Banderas and the Pacific Ocean that over 170 contributions were presented covering an important array of topics, ranging from neurophysiology to population biology, from conservation biology to immunology, from parasitology to evolution.

The present volume, organized in four sections, physiology, ecology, conservation and biodiversity, and systematics and evolution, is composed of 46 chapters, written by 100 authors from 17 countries, reflecting the truly international nature of our society. Our intention as editors was to bring together a volume representative of the diversity of topics and approaches presented in the Summer Meeting. As we prepared the final version we realized that, although what is included in the volume is but a small sample of what crustacean biologists do around the world, the resulting combination of chapters promises to call on the interest of a wide audience.

We wish to express our gratitude to the 40 colleagues who kindly accepted to review the submitted papers, to Mario Lara and Sergio Lara who composed the high quality final version of the volume and overcame all the technical difficulties involved, and to our home institutions, Instituto de Ciencias del Mar y Limnología and Instituto de Biología, Universidad Nacional Autónoma de México, which generously supported this project.

Elva Escobar-Briones
Fernando Alvarez

REVIEWERS

The editors thank the following people who reviewed the papers included in this volume.

Guillermina Alcaraz
Universidad Nacional Autónoma de México
México, D.F., México

Javier Alcocer
Universidad Nacional Autónoma de México
Iztacala, México, México

Patricia Briones Fourzán
Universidad Nacional Autónoma de México
Puerto Morelos, Quintana Roo, México

Richard Brusca
Arizona-Sonora Desert Museum
Tucson, Arizona, USA

Ernesto Campos
Universidad Autónoma de Baja California
Ensenada, Baja California, México

Xavier Chiappa
Universidad Nacional Autónoma de México
México, D.F., México

Keith A. Crandall
Brigham Young University
Provo, Utah, USA

Neil Cumberlidge
Northern Michigan University
Marquette, Michigan, USA

Manuel Elías
Colegio de la Frontera Sur
Chetumal, Quintana Roo, México

Alfonso Escobar Izquierdo
Universidad Nacional Autónoma de México
México, D.F., México

Carolina Escobar de Maurer
Universidad Nacional Autónoma de México
México, D.F., México

Ana Luisa Fanjul
Universidad Nacional Autónoma de México
México, D.F., México

Darryl L. Felder
University of Louisiana – Lafayette
Lafayette, Louisiana, USA

Joseph F. Fitzpatrick, Jr.
Tulane University Museum of Natural History
Belle Chasse, Louisiana, USA

Gabriela Gaxiola
Universidad Nacional Autónoma de México
Ciudad del Carmen, Campeche, México

Samuel Gómez Noguera
Universidad Nacional Autónoma de México
Mazatlán, Sinaloa, México

Adolfo Gracia
Universidad Nacional Autónoma de México
México, D.F., México

Pedro Joaquín Gutiérrez Yurrita
Universidad de Querétaro
Querétaro, Querétaro, México

Michel Hendrickx
Universidad Nacional Autónoma de México
Mazatlán, Sinaloa, México

William B. Jeffries
Dickinson College
Carlisle, Pennsylvania, USA

Rafael Lemaitre
Smithsonian Institution
Washington, D.C., USA

Virginia León
Universidad Nacional Autónoma de México
México, D.F., México

María Luisa Machain
Universidad Nacional Autónoma de México
México, D.F., México

Alejandro Maeda
Centro de Investigación Biológica del Noroeste
La Paz, Baja California Sur, México

Celio Magalhaes
Instituto Nacional de Pesquisas da Amazonia
Manaus, Brazil

John Markham
Arch Cape Marine Laboratory
Arch Cape, Oregon, USA

Jack O'Brien
University of South Alabama
Mobile, Alabama, USA

Gabino Rodríguez
Universidad Autónoma de Nuevo León
Monterrey, Nuevo León, México

Carlos Rosas
Universidad Nacional Autónoma de México
Ciudad del Carmen, Campeche, México

Miguel Rubio
University of Bristol
Bristol, United Kingdom

Nandini Sarma
Universidad Nacional Autónoma de México
Iztacala, México, México

Marylin Schotte
Smithsonian Institution
Washington, D.C., USA

Eduardo Suárez Morales
Colegio de la Frontera Sur
Chetumal, Quintana Roo, México

Alvaro de Tomás
Universidad Nacional Autónoma de México
México, D.F., México

Cecilia Vanegas
Universidad Nacional Autónoma de México
México, D.F., México

Ana Rosa Vázquez
Universidad Nacional Autónoma de México
México, D.F., México

Gerardo Vázquez Nin
Universidad Nacional Autónoma de México
México, D.F., México

Harold K.Voris
Field Museum of Natural History
Chicago, Illinois, USA

José Luis Villalobos
Universidad Nacional Autónoma de México
México, D.F., México

Mary K. Wicksten
Texas A & M University
College Station, Texas, USA

CONTENTS

ECOLOGY

CONSERVATION AND BIODIVERSITY

SYSTEMATICS AND EVOLUTION

Modern Approaches to the Study of Crustacea

SEASONAL RHYTHM OF SEROTONIN CONTENT IN THE CRAYFISH EYESTALK

Gabina Calderón-Rosete, Leonardo Rodríguez-Sosa
and Hugo Aréchiga

División de Estudios de Posgrado e Investigación, Facultad de Medicina, UNAM

arechiga@servidor.unam.mx

I. INTRODUCTION

5-Hydroxytryptamine (5-HT) is a common neurotransmitter and modulator in crustaceans. It has been shown to participate in a wide variety of functions, such as a) the facilitation of neuromuscular transmission (Glusman and Kravitz 1982, Dixon and Atwood 1985), b) the enhancement of visual input (Aréchiga et al. 1990) and of mechanoreception (El Manira et al. 1991, Rossi-Durand 1993, Pasztor and Golas 1993), c) the induction of the complete behavioral pattern of aggressiveness and social dominance (Kravitz 2000) and the modulation of the escape behavior (Glanzman and Krasne 1986, Yeh et al. 1996), d) the regulation of heart activity (Battelle and Kravitz 1978, Listerman et al. 2000), e) neurohormone release, facilitating that of the crustacean hyperglycemic hormone (Keller and Beyer 1968, Lee et al. 2000), and the molt-inhibiting hormone (Mattson and Spaziani 1985) and f) modulating the electrical activity of neurosecretory cells (Sáenz et al. 1997, Glowik et al. 1997, Alvarado-Álvarez et al. 2000), and of neurons at various levels of the crustacean central nervous system (Zhang and Harris-Warrick 1994).

All these effects have in common a facilitatory action of the level of alertness, and sensory and motor performance; thus, an overall image of 5-HT as an activity-promoting agent in crustaceans is emerging. Given the prolonged duration of 5-HT effects, its possible role in modulating long-term neuronal patterns is particularly interesting. In this regard, it is worth noticing that the motor and sensory activity levels in crustaceans show ample variations along the 24-hour cycle (Aréchiga et al. 1993, Aréchiga and Rodríguez-Sosa 1997); and correlative diurnal variations in 5-HT content have been documented in the nervous system of crustaceans, being its levels higher during the night phase (Fingerman and Fingerman 1977).

From field observations, crustaceans are known to display a pronounced seasonal rhythm of activity. For various species, it has been shown that individuals are very active from spring to autumn, with a reproductive season around springtime, and retract to their burrows in winter, while adopting a quiescent mode (see Naylor 1962, Aréchiga and Atkinson 1974, De Coursey 1983). In adults, molting is also linked to seasonal clues, with profound neuroendrocrine correlations (Aiken 1969). Seasonal rhythmicity has been documented in a crustacean neurohormone, the Red Pigment Concentrating Hormone (RPCH) (Rodríguez-Sosa et al. 1997, a).

However, no information is available on seasonal rhythmicity of the neurotransmitters which presumably underlie these behavioral and neurobiological rhythms. It was therefore interesting to explore whether the 5-HT content varied in the crayfish nervous system along the year. The results of a survey of 5-HT content, as well as that of its precursors L-Tryptophan (L-TRP) and 5-Hydroxy-tryptophan (5-OH-TRP) and its metabolite 5-Hydroxy-tryptophol (5-HTPH) are presented in this paper.

⚜ Kluwer Academic/Plenum Publishers

II. MATERIALS AND METHODS

The experiments were carried out in adult specimens of the crayfish *Procambarus clarkii* (Girard), with carapace lengths of 9 to 11 cm, corresponding to animals of one to two years of age, of either sex and in intermolt at the time of the experiments. They were collected from a location near Rio Conchos, Chihuahua, México, and kept for at least one week prior to the experiments under laboratory conditions of 12:12 light: dark cycles and ambient temperature of 20-24°C along the year.

For 5-HT determinations, the eyestalks were excised from animals anaesthetized under iced water. Usually 2 eyestalks were processed in each batch. The nervous tissue was removed by micro-dissection and homogenized in $400 \, \mu l$ of $HC10_4$ (100 mmol 1^{-1}) containing variable amounts of 3,4-dihydroxybenzylamine (DHBA) and centrifuged at 1600 g for 20 min at 4°C using a Sorvall centrifuge. The supernatant was filtered through a nylon sieve with a pore diameter of 0.22 um. The samples were stored at 0-5°C, for further analysis, usually carried out the same day. The protein content of the homogenized tissue was determined by a modification of the Lowry method, using small samples of crustacean nervous tissue (Cerbón and Aréchiga, 1986).

5-HT and its precursors and metabolite were determined by High Performance Liquid Chromatography, following the procedure reported by Rodríguez-Sosa et al (1997 b). The samples were injected into a guard column of C-18, connected to a reverse-phase analytical column of C-18 (25cm x 4.6 mm) packed with a particle size of 5 mm (LDC-Analytical). The amines were detected by means of a fluorescence detector. The HPLC system with fluorescence detector consisted of a programmable solvent delivery module (Waters, model 590) a universal LC injector (model U6K) and a scanning fluorescence detector (model 470). The signals were recorded on a data module (model 730). Fluorescence was detected at an excitation wavelength of 280 nm and an emission wavelength of 350.

Two mobile phases were used. One of them, mobile phase A was described by Leung and Tsao (1992); its composition was as follows: sodium acetate (40mmol.1^{-1}), citric acid (10 mmol.1^{-1}), disodium EDTA (130 μmol.1^{-1}), octane-1-sulphonic acid (420 μmol.1^{-1}), sodium chloride (13 mmol.1^{-1}) and methanol 10%

(v/v), pH 4.68. Mobile phase B only differed from phase A in the proportion of methanol, which was 20% (v/v). It was used for the detection of 5-HT and NA-5-HT. The mobile phases were degassed before use and recycled. The flow rate was variable for fluorescence detection.

The amine solutions were freshly prepared for each experiment and dissolved in perchloric acid (100 mmoll^{-1}). The system was calibrated to define the ranges of linearity for the following substances (all from Sigma Chemicals): L-Tryptophan (L-TRP), 5-hydroxytryptophan (5-OH-TRP), 5-hydroxytryptamine (5-HT), 5-hydroxy-tryptophol (5-HTPH). N-acetilserotonin (NA-5-HT), 5-hydroxy-indole-3-acetic acid (5-HIAA) and 3,4-dihydroxybenzylamine (DHBA), which was used as an internal standard. The amounts of the amines in the samples were calculated from the peak heights. The corresponding peaks in the chromatogram were confirmed by determining the increase in amplitude of the peak after adding pure substances to the column. The loss of substances during chromatography was estimated as 10-15%, with DHBA as the internal standard.

The results obtained were analyzed statistically with the non-paired t-Student's test, or with unpaired t test with Welch's correction.

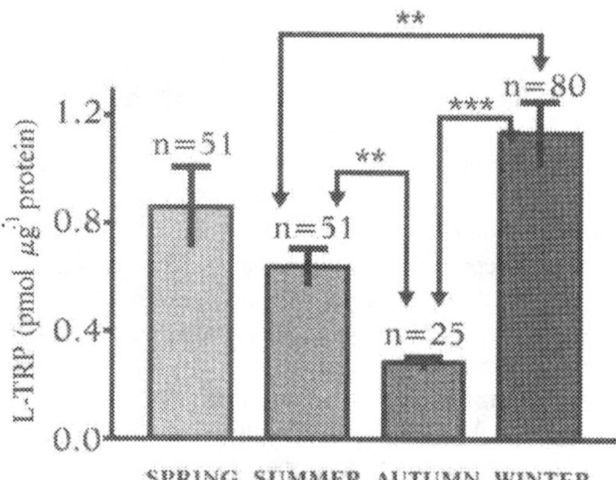

Figure 1. *Content of L-Tryptophan (L-TRP) in the crayfish eyestalk during spring, summer, autumn and winter. Bars indicate the content in each season and the values represent mean ± s.e. The number of experiments is indicated above the bars. The double asterisks indicate a significant difference (P < 0.01 student's t-test). Three asterisks indicate a significant difference (P < 0.001 student t-test). Values compared are shown by a two-arrow line.*

III. RESULTS

All four substances determined, showed seasonal variations in their content in the crayfish eyestalk. However, interesting differences were apparent. Whereas the earliest precursor in the chain, the aminoacid L-Tryptophan, and the product of 5-HT degradation by Monoamine Oxidase, 5-Hydroxytryptophol, attained their highest values during winter, both the immediate precursor of 5-HT (5-Hydroxytryptophan) and 5-HT itself were more abundant during summer. As seen in Fig. 1, L-Tryptophan content varied from a low value of 0.28 ± 0.023 pmol μg^{-1} protein in autumn, to a higher value of 1.13 ± 0.12 pmol μg^{-1} protein in winter, with intermediate values during spring and summer. The differences between summer and winter are very small, barely significant at $P < 0.01$.

The seasonal profiles of 5-HT and its immediate precursor, 5-Hidroxy-tryptophan are quite different (Figs. 2 and 3). Both are in phase, with a highest level during summer (48.79 ± 5.5 fmol μg^{-1} protein and 10.83 ± 1.88 pmol μg^{-1} protein respectively). While for 5-OH-TRP, the values obtained during the other three seasons are equally low. (4.26 ± 1.09; 4.08 ± 0.38; 4.25 ± 0.98 pmol μg^{-1} protein) respectively for spring, autumn and winter (Fig. 2), for 5-HT, the winter values are the lowest, at 14.95 ± 0.78 fmol μg^{-1} protein, with the spring and autumn levels being slightly higher, with 25.97 ± 2.75 fmol μg^{-1} protein and 31.09 ± 1.18 fmol μg^{-1} protein respectively (Fig. 3).

A striking difference was found in the seasonal distribution of the 5-Hydroxy-tryptophol values. As seen in Fig. 4, the winter values (76.34 ± 20.46 fmol μg^{-1} protein) are higher than the combined values obtained in the other three seasons (19.7 ± 5.94; 4.8 ± 1.21 and, 1.69 ± 1.29 fmol μg^{-1} protein) for spring, summer and autumn, respectively. Even considering a high variability for the winter values, the significance of the difference is quite clear.

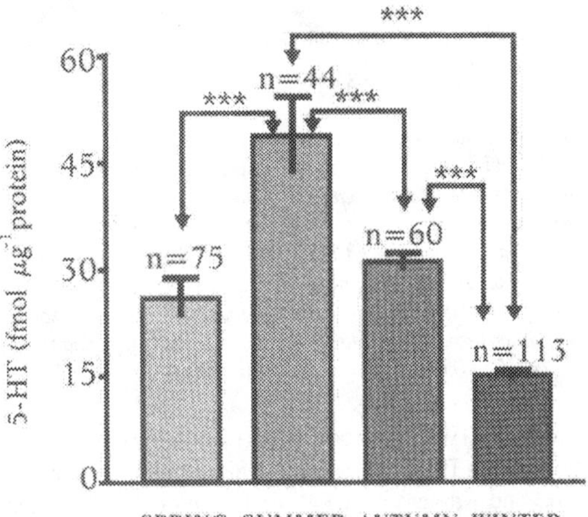

Figure 3. *Content of 5-Hydroxytryptamine (5-HT) in the crayfish eyestalk during spring, summer, autumn and winter. Bars indicate the content in each season. Values represent mean ± s.e. Above the bars is indicated the number of experiments in each season. Three asterisks indicate a significant difference (P < 0.001 student's t-test). Values compared are shown by a two-arrow line.*

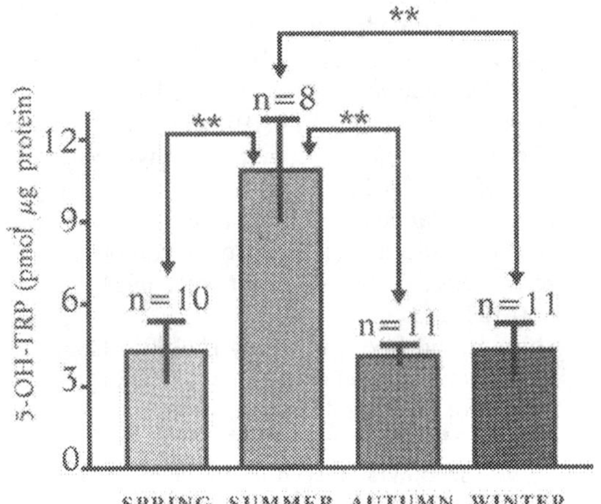

Figure 2. *Content of 5-Hydroxytryptophan (5-OH-TRP immediate precursor of 5-HT, in the crayfish eyestalk during spring, summer, autumn and winter. Bars indicate the content in each season and the values represent mean ± s.e. The number of experiments is indicated above the bars. The double asterisks indicate a significant difference (P < 0.01 student's t-test). Values compared are shown by a two-arrow line.*

IV. DISCUSSION AND CONCLUSIONS

The seasonal distribution of 5-HT content in the crayfish eyestalk is well correlated with the patterns detected in other physiological functions, and in a number of crustacean species, which attain higher levels during summer. Such is the case of the patterns of locomotor activity (Naylor 1962, Aréchiga and Atkinson 1974, De Coursey 1983), molting (Aiken 1969, von Gliscynski and

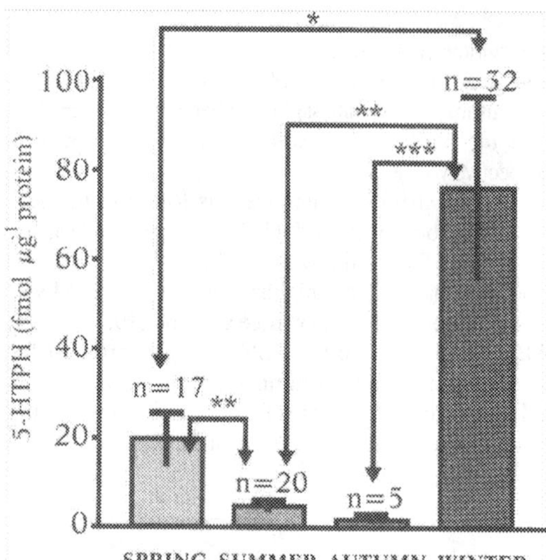

Figure 4. *Content of 5-Hydroxytryptophol (5-HTPH), metabolite of 5-HT in the crayfish eyestalk during spring, summer, autumn and winter. Bars indicate the content in each season. Values represent mean ± s.e. The asterisk indicates a significant difference (P < 0.05 student's t test) compared spring versus winter. The double asterisks indicate a significant difference (P < 0.01 student's t-test) compared with the content in summer, versus spring and winter, respectively. Three asterisk indicate a significant difference (P < 0.001 with Welch's-corrected t).*

Sedlemeier 1993), reproduction (Black 1966, Carpenter and de Roos 1970, Belleli et al. 1988), metabolic activities (Ono and Kamemoto 1969, Armitage et al. 1972) and sensory input (Aréchiga and Atkinson 1974, Hariyama and Tsukahara 1988). In all these instances, the values obtained at summer are much higher than those at winter, thus supporting the notion of a phase of high metabolic and behavioral activity, around summer, and another phase of quiescence in winter. Some neurobiological correlates support this contention. For instance, the size and responsiveness of crayfish neuromuscular junctions is greater in summer than in winter (Lnenicka and Zhao 1991, Lnenicka 1993) and the secretion of neurohormones, such as the Red Pigment Concentrating Hormone, is more than two-fold higher in summer than in winter (Rodríguez-Sosa et al. 1997a).

In this context, it is noteworthy that the seasonal distribution of 5-HTPH is in antiphase with that of 5-HT and 5-OH-TRP, suggesting that during winter, while the biosynthesis of 5-HT is at its lowest, its oxidation by mono-amine-oxidase proceeds at full extent. There is no available information on seasonal changes in other catabolites, and the issue deserves further analysis. The figures obtained for the content of L-TRP, which turned out to be higher in winter, can be ascribed to the fact that the supply of nutrients in the laboratory was kept constant all year round, while the biosynthesis of 5-HT was lower on winter time.

It will be interesting to establish in future, the proper correlation of the values obtained in this study, with the activity of the enzymes responsible for the biosynthesis and degradation of 5-HT.

ACKNOWLEDGEMENTS

This work was supported by grant 28089-N to Hugo Aréchiga from CONACyT.

REFERENCES

Aiken DE (1969) Photoperiod, endocrinology and the crustacean molt cycle. Science 164:149-155

Alvarado-Alvarez, R Aréchiga H and García U (2000) Serotonin activates a Ca^{2+}-dependent K^+ current in identified peptidergic neurons from the crayfish. J Exp Biol 203:715-723

Aréchiga H, Fernández Quiroz F, Fernández de Miguel F and Rodríguez-Sosa L (1993) The circadian system of crustaceans. Chronobiol Int 10:1-19

Aréchiga H and Atkinson RJA (1974) The eye and some effects of light on the locomotor activity in *Nephrops norvegicus*. Mar Biol 32:63-76

Aréchiga H, Bañuelos E, Frixione E, Picones A and Rodríguez-Sosa L (1990) Modulation of crayfish retinal sensitivity by 5-hydroxytryptamine. J Exp Biol 150:123-143

Aréchiga H and Rodríguez-Sosa L (1997) Coupling of environmental and endogenous factors in the control of rhythmic behaviour in decapod crustaceans. J Mar Biol Assoc 77:17-29

Armitage KB, Buikema AL Jr and Willems NJ (1972) Organic constituents in the annual cycle of the crayfish *Orconectes nais* (Faxon). Comp Biochem Physiol 41[a]:825-842

Battelle BA and Kravitz EA (1978) Targets of octopamine action in the lobster: cyclic nucleotide changes and physiological effects in haemolymph, heart and exoskeletal muscle. J Pharmacol Exp Ther 205:438-448

Bellelli A, Giardena B, Corda M Pellegrini MG, Cau A, Condo SG and Brunori M (1988) Sexual and seasonal variability of lobster haemocyanin M. Comp Biochem Physiol; 91[a]:445-449

Black JB (1966) Cyclic male reproductive activities in the dwarf crawfishes *Cambarellus shufeldtii* (Faxon) and *Cambarellus puer* (Hobbs). Trans Amer Microsc Soc 85:214-232

Carpenter MB and De Roos R (1970). Seasonal morphology and histology of the androgenic gland of the crayfish *Orconectes nais*. Gen Comp Endocrinol 15:143-57

Cerbón J and Aréchiga H (1986) Characterization of the Na$^+$, K$^+$-ATPase in a microsomal preparation from the abdominal ganglia of crayfish. Effect of neurodepressing hormone (NDH). Comp Biochem Physiol 85B:561-566

De Coursey P (1983) Biological timing. (pp 107-161) In: Vernberg J and Vernberg WB (eds) The Biology of Crustacea. New York, Academic Press

Dixon D and Atwood HL (1985) Crayfish motor nerve terminal's response to serotonin examined by intracellular microelectrode. J Neurobiol 16:409-424

El Manira A, Rossi-Durand C and Clarac F (1991) Serotonin and proctolin modulate the response of a stretch receptor in crayfish. Brain Res 541:157-162

Fingerman SW and Fingerman M (1977) Circadian variation in the levels of red pigment-dispersing hormone and 5-hydroxytryptamine in the eyestalk of the fiddler crab, *Uca pugilator*. Comp Biochem Physiol 56C:5-8

Glanzman DL and Krasne FB (1986) 5,7-Dihydroxytryptamine lesions of crayfish serotonin-containing neurons. Effect on the lateral giant escape reaction. J Neurosci 6:1560-1569

Glowik, R M, Golowash J, Keller R and Marder E (1997) D-Glucose-sensitive neurosecretory cells of the crab *Cancer borealis* and negative feedback regulation of blood glucose level. J Exp Biol 2000:1421-1431

Glusman S and Kravitz EA (1982) The action of serotonin on excitatory nerve terminals in lobster nerve-muscle preparations. J Physiol Lond 325:223-241

Hariyama T and Tsukahara Y (1988) Seasonal variation of spectral sensitivity in crayfish retinula cells. Comp Biochem Physiol; 81A:529-533

Keller R and Beyer J (1968) Zur hyperglykäemischen Wirkung von Serotonin und Augenstielextrakt beim Flusskrebs *Orconectes limosus*. Z Vergl Physiol 59:78-85

Kravitz EA (2000) Serotonin and aggression: insights gained from a lobster model system and speculations on the role of amine neurons in a complex behavior. J Comp Physiol A 186:221-238

Leung P Y and Tsao CS (1992) Preparation of an optimum mobile phase for the simultaneous determination of neurochemicals in mouse brain tissues by high-performance liquid chromatography with electrochemical detection. J Chromatogr 576:245-254

Listerman L R, Deskins J, Bradacs H and Cooper RL (2000) Heart rate within male crayfish: social interactions and effects of 5-HT. Comp Biochem Physiol 61C:229-237

Lee CY, Yau SM, Liau CS and Huang WJ (2000) Serotonergic regulation of blood glucose levels in the crayfish, *Procambarus clarkii*: Site of action and receptor characterization. J Exp Zool 286:596-605

Lnenicka GA (1993) Seasonal differences in motor terminals. Comp Biochem Physiol 104A:423-429

Lnenicka GA and Zhao Y (1991) Seasonal differences in the physiology and morphology of crayfish motor terminals. J Neurobiol 22:561-569

Mattson MP and Spaziani E (1985) 5-Hydroxytryptamine mediates release of molting-inhibiting hormone activity form isolated crab eyestalk ganglia. Biol Bull Mar Biol Lab, Woods Hole 169:246-255

Naylor E. (1962) Seasonal changes in a population of *Carcinus maenas* (L) in the littoral zone. J Anim Ecol 31:601-9

Ono JK and Kamemoto FI (1969) Annual and proecdysial variations in urine production in crayfish. Pac Sci 23:305-310

Pasztor UM and Golas L (1993) The modulatory effects of serotonin, Neuropeptide F$_1$, and Proctolin on the receptor muscles of the lobster abdominal stretch receptor and their exoskeletal muscle homologues J Exp Biol 174:363-374

Rodríguez-Sosa L, De la Vega T, Vergara P and Aréchiga H (1997a) Seasonal rhythm of red pigment concentrating hormone in the crayfish. Chronobiol Inter 14:639-645

Rodríguez-Sosa L, Picones A, Calderón-Rosete G, Islas S and Aréchiga H (1997b) Localization and release of 5-hydroxytryptamine in the crayfish eyestalk. J Exp Biol 200:3067-3077

Rossi-Durand C (1993) Peripheral proprioceptive modulation in crayfish walking leg by serotonin. Brain Res 632:1-15

Sáenz F, García U and Aréchiga H (1997) Modulation of electrical activity by 5-hydroxytryptamine in crayfish neurosecretory cells. J Exp Biol 200:3079-3090

Von Gliscynski U and Sedlemeier D (1993) Regulation of edcysteroid biosynthesis in crayfish Y-organs: II. Role of cyclic nucleotide-dependent protein kinases. J Exp Zool 265:454-458

Yeh SR, Fricke RA and Edwards DH (1996) The effect of social experience on serotonergic modulation of the escape circuit of crayfish. Science 271:366-369

Zhang B and Harris-Warrick RM (1994) Multiple receptors mediate the modulatory effects of serotonergic neurons in a small neural network. J Exp Biol 190 55-77

BRAIN ELECTRICAL SIGNALS IN UNRESTRAINED CRAYFISH

Fidel Ramón, Jesús Hernández-Falcón and Theodore H. Bullock

FR, División de Estudios de Posgrado e Investigación, Facultad de Medicina, UNAM, México City, México
JHF, Departamento de Fisiología, Facultad de Medicina, UNAM, México City, México
THB, Neurobiology Unit, Scripps Institution of Oceanography & Department of Neurosciences,
University of California, San Diego, La Jolla, CA 92093-0240, USA

I. INTRODUCTION

The electrical activity from the brain of vertebrates and some mollusks (octopus and cuttlefish; Bullock 1984) is dominated by slow components in the range 0 to 30 Hz and fast events such as spikes are seen by special effort. In contrast, electrical activity from the brain of several invertebrates such as crayfish, earthworm, slug and grasshopper, is dominated by spikes (Bullock 1945, Bullock and Horridge 1965, Bullock and Basar 1988), with relatively weak components of the power spectrum from around 5 Hz to 50 Hz. These differences in brain signals from vertebrates and invertebrates are not explained by brain size, different electrodes or other obvious factors, and they might depend on the different anatomical arrangements of neurons and their processes and neuroglia or on a relative weakness of synchrony in most invertebrates.

Because the spontaneous electrical activity of the brain of most invertebrates is comprised mainly of spikes, it is of interest to investigate whether spikes are also elicited in response to several stimuli and to evaluate their significance in relation to various behaviors. A type of electrical sign of neural activity is the evoked-potentials (EPs) which, although recorded from the brain, are more determined by the stimulus strength, rise-time, duration, modality and locus. These have been recorded in several invertebrates (Bullock and Basar 1986). Another type of brain electrical activity is the overlapping class of event-related potentials (ERPs), of more endogenous nature as they are more dependent on the state of the animal and brain. ERPs are almost unknown among invertebrates and are principally known from studies on humans, where they are considered to be signs of higher stages of processing and sometimes called 'cognitive waves' for their association with conscious experiences.

In these studies we recorded spontaneous as well as evoked and event-related potentials and investigated their characteristics and significance to specific animal behaviors. Our results show that some simple nervous mechanisms basic for behavior and cognition are present in crayfish, indicating that something has been conserved through evolution, even though the cellular mechanism may not be the same throughout.

II. MATERIALS AND METHODS

Experiments were performed on adult crayfish (*Procambarus clarkii*) 10-13 cm in length of either sex, obtained from local providers. Animals were kept for 2-3 months in well-aerated aquaria at room temperature (20-25 °C). Crayfish were chosen for their suitability for developing several experimental preparations.

For most experiments we used a chronically implanted crayfish with up to four electrodes on the brain (see Hernández et al. 1996). Animals were left to recover for a day before placing them in an appropriate experimental chamber for recording while in air or underwater.

Other experiments were performed with an isolated crayfish 'heart-brain' preparation (Serrato et al. 1995) in which we isolated the dorsal frontal part of the carapace with the heart, brain and vessels connecting them.

Kluwer Academic/Plenum Publishers

E. Escobar-Briones & F. Alvarez Eds.
MODERN APPROACHES TO THE STUDY OF CRUSTACEA
PP. 007-013

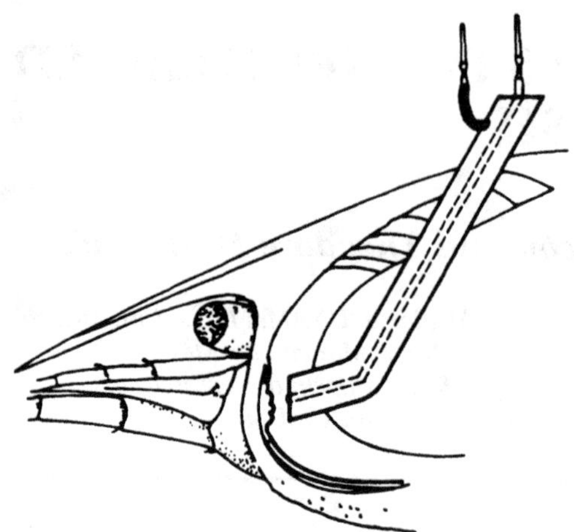

Figure 1. *Diagram of the position of the chronically implanted electrode. The brain surface is the wiggly line in front of the electrode tip, separated by an exaggerated space. An insulated central wire is surrounded by a grounded metal tube dental cemented to the carapace (modified from Aréchiga and Wiersma 1969).*

For stimulation we used light flashes isolated or in trains at various rates and durations, with the lamp at a distance of about 15 cm in front of the animal. We also used pulses of different odorant substances delivered directly to the antennulae. All records were sampled at a rate of 2-8 KHz and stored in a computer for off-line analysis.

III. & IV. RESULTS AND DISCUSSION

Crayfish chronically implanted with electrodes on the brain survive for up to 4-5 months while behaving normally, as ascertained visually. In fact, several complex behaviors such as eating, agonistic behavior and mating do not seem different in implanted as compared with intact animals. However, crayfish die while attempting to moult, apparently as a consequence of lesions while trying to remove body regions under the electrode glued to the carapace.

Spontaneous activity

The spontaneous brain electrical activity of crayfish is comprised of numerous spikes of various amplitudes, superimposed on a baseline where slow events are very small, usually within the input noise level. Two types of spikes are most common, fast (nar-

row) ones isolated and in bursts without pattern, and slow (wider) ones associated with movements or displacement of the animal. Burst activity is correlated with action potentials recorded simultaneously with an intracellular electrode nearby. Usually, synaptic-like activity precedes each burst. We could not find this kind of intra-extracellular correlation, for slow events (Serrato et al. 1995).

The spontaneous electrical activity is relatively invariant and no specific patterns have been observed (Fig. 2). In fact, spikes of different amplitude are continuously generated while the animal is engaged in all types of activity. However, during displacement the brief fast spikes are obscured by other larger and wider ones, most likely originating in muscles surrounding the brain, such as antennal, antennular and eyestalks muscles. This muscle activity subsides when crayfish stop and the frequent brief spikes reappear. The power spectrum of the spontaneous activity is very similar through 24 hour cycles (Hernández et al. 1996).

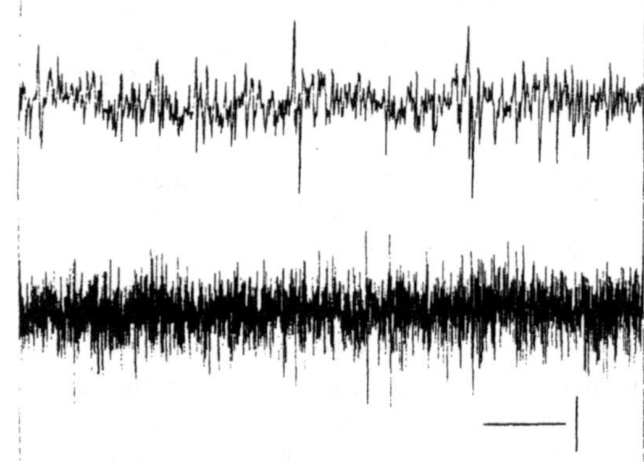

Figure 2. *Spontaneous electrical activity of the crayfish brain on two different scales. The horizontal bar corresponds to 26 msec (top) and 200 msec (bottom) and the vertical bar to 10 μV for both records (Hernández et al. 1996).*

Evoked activity

In response to a brief (10 μsec) flash of light, an electrode placed on the protocerebrum records a burst of spikes, the visually evoked potential (VEP. Fig. 3). This burst is comprised of 3-10 spikes that after one or two large ones, decrease in amplitude to reach the basal level in about 100 msec. Other stimuli, such as an odorant substance delivered directly to the antennulae, also elicit evoked potentials, but these spikes are smaller than those seen in response to light pulses.

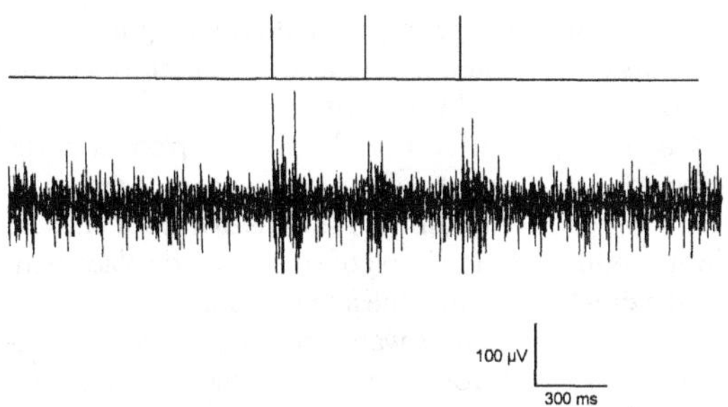

Figure 3. *Potentials evoked from the crayfish brain with each light pulse in a train of 3 pulses.*

We have used evoked potentials in response to light pulses as a tool to localize and study regions of integration in the crayfish nervous system (Fig. 4). For this purpose we delivered light pulses of wavelengths in the visible range (475-675 nm) while simultaneously recording the ERG from the cornea, the travelling burst of spikes on the optic nerve and the VEP from the brain (Serrato et al. 1996).

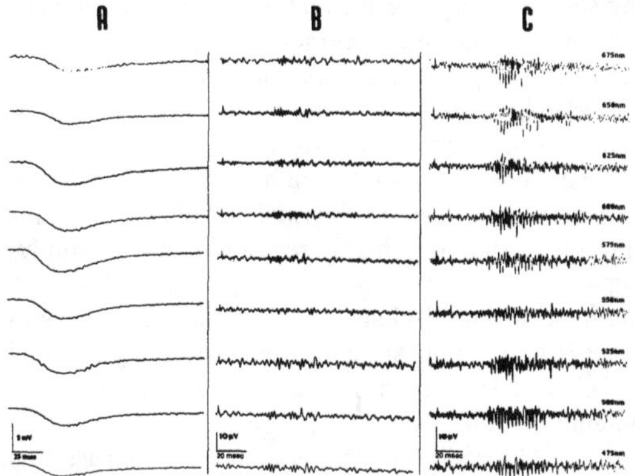

Figure 4. *Electrical activity recorded from three regions of the crayfish optic tract (A, retina; B, optic tract, and; C, brain) in response to light pulses of different wavelength. Records for each wavelength were obtained simultaneously from the three regions and are the average of 20 sweeps (Serrato et al. 1996).*

The recording of evoked potentials from the crayfish brain is not a surprise, as it is expected that in all animals peripheral sense organs communicate with the brain or with the nervous system engaged in integrating information from the external environment. In all experiments stimulation with light pulses evokes a brain activity comprised of spikes, but we have not been able to record slow events in this study. However, oc-casionally and under some conditions apparently difficult to reproduce, slow waves have been recorded from the brain of some invertebrates in response to light flashes (Adrian 1931, Jahn and Crescitelli 1938), suggesting that crayfish are also capable of producing them.

Spike bursts elicited at the crayfish eye travel via the optic tract to the brain. When recorded at the optic tract, the spikes elicited by pulses of different wavelength are evenly spaced and arrive at the recording electrode at regular intervals. However, the same spikes recorded at the brain appear at variable intervals that depend on the stimulus wavelength and these results can be better explained assuming that different signals are analyzed by different neuronal circuits, a hypothesis known as labeled-line (Bullock and Horridge 1965).

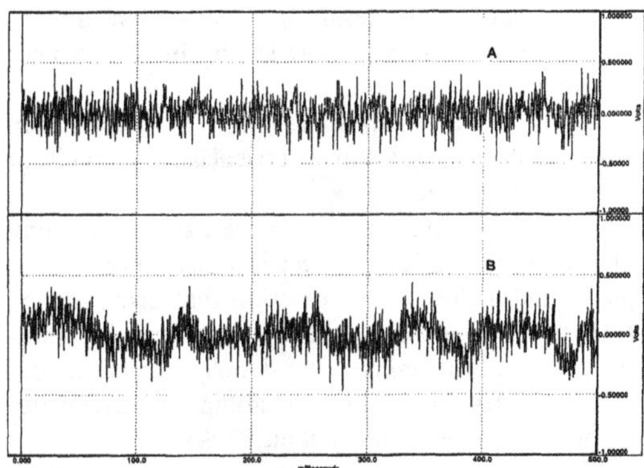

Figure 5. *Odor evoked potentials recorded from the crayfish brain in response to stimulation with a pulse of chicken liver extract. A= control record. B= 30 seconds after stimulation. Note that after the stimulus there is a mild increase in the number of spikes and slow waves. This latter characteristic is not seen after stimulation with light pulses.*

Contrary to results obtained stimulating with light flashes, odor stimulation in crayfish evokes a response composed of spikes and slow waves easily recorded from several regions of the brain, particularly from the deutocerebrum (Fig. 5). Slow waves have also been recorded from the olfactory system of other invertebrates such as slugs and snails (Gelperin and Tank 1990) as well as vertebrates. It has been suggested that in vertebrates and invertebrates the analyzing mechanism of odor, mechanical and gustatory evoked signals is also based on the labeled-line postulates (Derby et al. 1991, Surmeier et al. 1988, Azuma et al. 1984).

Although brief flashes of light induce visual evoked potentials at the crayfish brain, this type of stimulus is very intense and therefore somewhat unnatural, a reason which impelled us to seek more natural visual stimulation and to ascertain if it also produces VEP's, as well as the behavioral responses elicited by these signals. To this effect we performed experiments in which we used a special chamber to place chronically implanted animals underwater and stimulate them visually with objects that might be present in their natural habitat, as well as different odors.

Vision/olfaction-action correlations

We studied the correlation between potentials evoked at the brain due to vision of different objects or different odorant stimuli, and the resulting motor behaviors. We found three types of vision-action correlates:

a) vision of an (irrelevant?) object such as a moving black bar induces a short lasting burst of spikes without a motor reaction;

b) vision of a (mildly interesting?) motionless fish, evokes long lasting brain electrical activity without motor reaction, and;

c) vision of a (relevant?) relatively large swimming fish, evokes a long lasting electrical discharge of large spikes, and a slow motor reaction that displaces the crayfish in the direction of the fish. Large moving shadows on top of the animal occasionally induce the defense posture or the well known tail-flip escape response (Fig. 6, Hernández-Falcón et al. 1999).

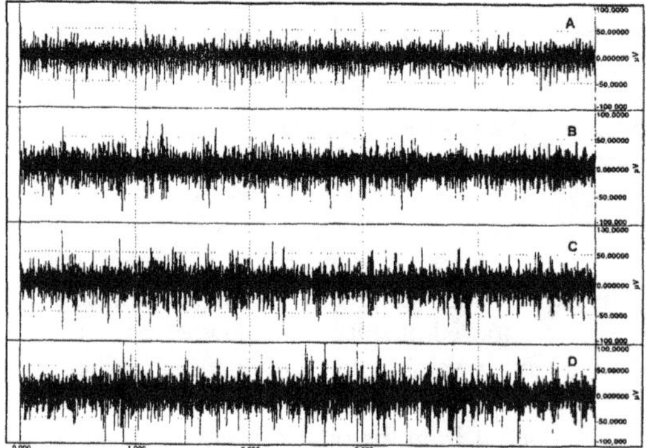

Figure 6. *Multiunit brain activity in response to several visual stimuli. A, control record; B, a motionless goldfish stimulates isolated but frequent spikes; C, same fish, swimming movements elicit bursts of spikes, and; D, bursts of large spikes are seen whenever the fish swims in front of crayfish (Hernández-Falcón et al. 1999). Stimuli are maintained throughout the record.*

Similar experiments, but searching for olfaction-vision correlations have also been performed and among results are the following,

a) a compound (irrelevant?) such as a piece of plastic tube induces a short burst of spikes without a motor reaction;

b) an odor (interesting?) such as extract of chicken liver produces a large burst of spikes and displacement of the crayfish towards the aliment, and;

c) a small amount of water containing putative compounds released from conspecifics elicits a long-lasting (10 min) burst comprised of small amplitude spikes, during which the crayfish sustains an 'attentive' position. This brain discharge is brief in reversibly blinded animals, suggesting some interaction in the processing of visual and olfactory information (Hernández-Falcón and Ramón 2000).

These results have been interpreted by assuming that some stimuli have certain 'relevance' for crayfish. 'Non-relevant' stimuli elicit a low level change in brain activity and 'highly-relevant' ones long-lasting discharges from various electrode placements, including some sensory-specific and others involved in general processing and integration. Thus, a light flash activates the photoreceptors and main pathways involved in light detection, while food activates chemical receptors and appropriate pathways that participate in nourishment. The presence of a second animal in the neighborhood is sensed by visual and olfactory organs and interpreted as a 'highly-relevant' stimulus, implying a great level of brain processing. This is shown by the long-lasting discharges and low frequency waves recorded, that suggest a concerted processing of visual and chemical information which determine the type and magnitude of behavioral responses.

Hopefully our 'natural' stimulus would allow us to study crayfish brain electrical and motion responses under conditions similar to those found in their natural habitat. We eliminated the olfactory component of the stimulus by placing crayfish and visual stimuli in two chambers isolated from each other, such that crayfish would be guided only by vision. Under these conditions we found that, although reactions were multiple and difficult to categorize, lumping them into three main groups based on the most obvious responses allows us to attribute them a significance. Thus, we artificially labeled them as, irrelevant, mildly interesting and very interesting, depending on the robustness of the evoked potential and the observed motion response. The weakest response was obtained by stimulating with artificial objects, such as a black bar moving at a constant speed in front of the crayfish, while more intense responses were obtained with a fish. In this case, when the fish is motionless the response is less robust, but when it swims the electrical response is intense and the crayfish also moves in the direction of the fish.

Similar patterns were obtained with olfactory stimulation, but in this case it was a correlation between the intensity of the brain electrical response and motion, suggesting that a component of the decision circuit is a threshold detecting circuit. We believe that further analysis of the electrical signals and their site and time of origin, will give us clues as to the brain regions involved in decision making in crayfish.

Omitted Stimulus Potentials (OSPs)

The fact that a relatively simple brain such as that of crayfish is capable of evoked activity similar to that recorded from the brain of vertebrates, suggested the possibility that other functions, labeled 'high' because of their presence in mammals could also be present in crayfish. Thus, to study this possibility we used a simple cognitive task used in mammals, including man and called expectation.

Expectation was studied stimulating with 1 sec trains of flashes at various rates, and searching for a spike discharge appearing at the end of the train, and which, as opposed to a regular off-response, is time-locked to the due-time of the stimulus omitted after the last one (Fig. 7). We found at the appropriate time a burst of spikes, the Omitted Stimulus Potential (OSP), with characteristics very similar to the 'fast' response obtained under similar conditions in vertebrates (Bullock et al. 1994, Karamürsel et al. 1994). Naturally, our recording of these signals does not imply that crayfish are capable of elaborate mental operations, but only that the neuronal circuit produces a similar response. In lower vertebrates a similar response seems to originate in the retina (Prechtl et al. 1994, Bullock et al. 1993), and several of our results suggest that the OSP in crayfish is also produced at the peripheral part of the visual pathway, modulated by the state of the brain.

A one second long train of pulses at rates between 3-15 Hz induces evoked potentials after each pulse, and a spike burst at the end of the train (Fig. 7). Usually after a long lasting stimulus there is an 'off' response and an OSP could be included in this classification. However, what places the OSP in a class by itself is that it has a latency that is constant when measured in due-time of the first omitted stimulus. This behavior implies that the system 'learns' the inter-stimulus interval and comes to ex-

pect it in order to deliver the appropriate response; however, at the end of a train the following stimulus does not arrive. Thus, the system behaves as though it has learned to expect the next stimulus on schedule and if it does not come it releases the OSP -a burst of spikes at a relatively fixed latency after the expected time, which is much longer than the latency of the flash-evoked bursts. Thus each flash not only evokes a visual potential but also acts to inhibit the later OSP.

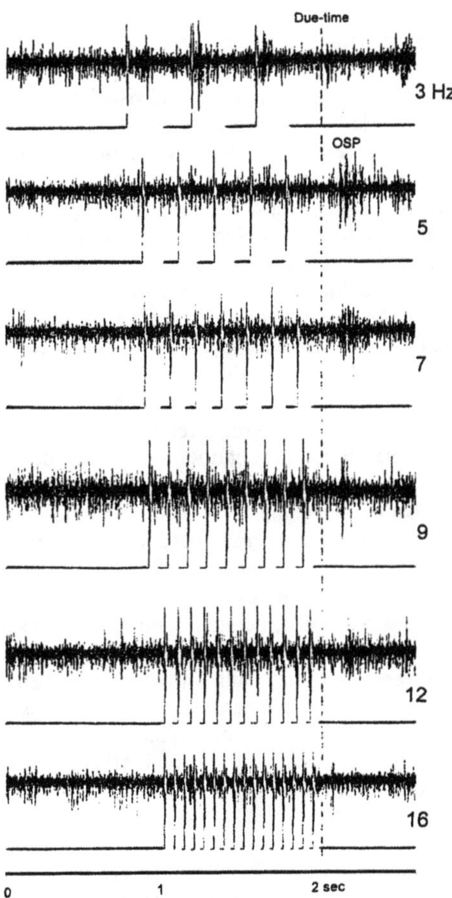

Figure 7. *Records from the crayfish brain showing responses after stimulation with trains of pulses at the frequency indicated at the right of each record. Records are lined-up by the 'due-time,' the time at which the next pulse of the train would have come. Records between 5 and 12 Hz show a burst of spikes (OSP) around 150 msec after due-time (Hernández et al. 1999)*

It is important to emphasize that the presence of an OSP does not necessarily mean that a higher level process, a 'thought,' 'mental' event or conscious experience has occurred. This is important, because these findings indicate a central state as though an expectation has resulted from the previous stimulation. However, we do not imply that this is equivalent to human expectations except operationally.

The presence of a response after a train of light flashes that closely resembles similar responses obtained from both lower and higher vertebrates, suggests that it originated in a very simple circuit. Some experimental results suggest that the generating mechanism is located in a peripheral part of the visual system, perhaps at the level of the retina. However, it can be influenced by the central state and other sensory input, such as produces arousal and darkness that can affect latency or suppress the OSP. Independently of the localization or complexity of the generating circuit, the fact that it can be recorded all the way from humans to crayfish allowes us to draw several important points.

As cognition (the neural mechanisms responsible for capturing and analyzing information collected by the senses, as well as those involved in deciding the appropriate course of action in response to that information) is in some degree a requirement for adaptation to the environment, it does not surprise us to find degrees of it in all sorts of animals. As the environment is probably more complex for many mammals than for crayfish, it would not surprise us either that more complex cognition mechanisms evolved in the former. However, a question arises whether or not cognitive mechanisms are all generated *de novo* for each animal species or group or are based on circuits invented in simple animals. Our results in these studies suggest that some cognitive mechanisms could be based on circuits developed early in evolution and conserved throughout many stages, in which other more complex circuits were added.

Summary

Our chronically implanted crayfish preparation has been very useful to study the event-related potentials generated at the brain during their normal behavior under semi-natural conditions, and has already allowed us to suggest mechanisms involved in the analysis of evoked-potentials in an unrestricted invertebrate, as well as to demonstrate correlation between those potentials and animal motion.

ACKNOWLEDGMENTS

Supported by NIH/NINDS, the Guggenheim Foundation and DEPI-357. The help of J.C. Gómez and C. Ramírez with illustrations and P. Ramírez with secretarial assistance is also appreciated.

REFERENCES

Adrian ED (1931) Potential changes in the isolated nervous system of *Dysticus marginalis*. J. Physiol Lond 72:132-151

Aréchiga H and Wiersma CAG (1969) The effect of motor activity on the reactivity of single visual units in the crayfish. J Neurobiol 1:53-69

Azuma S, Yamamoto T and Kawamura Y (1986) Studies on gustatory responses of amygdaloid neurons in rats. Exp Brain Res 56:12-22

Bullock TH and Horridge GA (1965) Stucture and Function in the Nervous Systems of Invertebrates. WH Freeman and Co San Francisco and London. Vol. I

Bullock TH (1984) Ongoing compound field potentials from octopus brain are labile and vertebratelike. Electroencephalogr Clin Neurophysiol 57:473-483

Bullock TH and Basar E (1986) A comparative analysis of EEG and evoked potentials of a mammal, an elasmobranch and a mollusc: the question of structure-function relationship. Int J Neurosci 29:163-164

Bullock TH and Basar E (1988) Comparison of ongoing compound field potentials in the brains of invertebrates and vertebrates Brain Res Rev 13:57-75

Bullock TH, Karamürsel S and Hofmann MH (1993) Interval-specific event related potentials to omitted stimuli in the electrosensory pathway in elasmobranchs: an elementary form of expectation. J Comp Physiol A 172:501-510

Bullock TH, Karamürsel S, Achimowicz JS, McClune MC and Basar-Eroglu C (1994) Dynamic properties of human visual evoked and omitted stimulus potentials. Electroencephalogr Clin Neurophysiol 91:42-53

Derby CD, Girardot MN and Daniel DC (1991) Responses of olfactory receptor cells of spiny lobsters to binary mixtures. II. Pattern mixture interaction. J Neurophysiol 66:131-139

Gelperin A and Tank DW (1990) Odour-modulated collective networkoscillations of olfactory interneurons in a terrestrial mollusk. Nature 345 (6274):437-440

Hernández OH, Serrato J and Ramón F (1996) Chronic recording of electrical activity from the brain of unrestrained crayfish: the basal, unstimulated activity. Comp Biochem Physiol 114A:219-226

Hernández OH, Ramón F and Bullock TH (1999) Expectation in invertebrates: crayfish have "omitted stimulus potentials." Proc 6th Joint Symp Neural Computation, Inst Neural Comput, UCSD, La Jolla, CA, USA 9:50-56

Hernández-Falcón J, Serrato J and Ramón F (1999) Evoked potentials elicited by natural stimuli in the brain of unanesthetized crayfish. Physiol Behav 66:397-407

Hernández-Falcón J and Ramón F (2000) Olfaction-action responses evoked in behaving crayfish. Soc Neurosc Abstr 26:

Jahn TL and Crescitelli F (1938) The electrical response of the grasshopper eye under conditions of light and dark adaptation. J Cell Comp Physiol 12:39-55

Karamürsel S and Bullock TH (1994) Dynamics of event-related potentials to trains of light and dark flashes: responses to missing and extra stimuli in rays. Electroencephalogr Clin Neurophysiol 90:461-471

Prechtl JC and Bullock TH (1994) Event-related potentials to omitted visual stimuli in a reptile. Electroencephalogr Clin Neurophysiol 91:54-66

Ramón F, Hernández OH and Bullock TH (1999) "Cognitive waves" from crayfish brains: Omitted stimulus potential. Soc Neurosci Abstr 25:1140

Serrato J, Hernández OH and Ramón F (1995) Multi- and unitary electrical activity in an isolated head-heart crayfish preparation. Soc Neurosci Abstr 21:407

Serrato J, Hernández OH and Ramón F (1996) Integration of visual signals in the crayfish brain: multiunit recordings in eyestalk and brain. Comp Biochem Physiol 114A:211-217

Surmeier DJ, Honda CN and Willis WD (1988) Natural groupings of primate spinothalamic neurons based on cutaneous stimulation. Physiological and anatomical features. J Neurophysiol 59:833-860

DIEL ACTIVITY RHYTHM IN *CALLINECTES ORNATUS* ORDWAY, 1863 AND *CALLINECTES DANAE* SMITH, 1869 (BRACHYURA, PORTUNIDAE) UNDER LABORATORY CONDITIONS

Alvaro Luiz Diogo Reigada

NEBECC, Group of Studies on Crustacean Biology, Ecology and Culture
UNESP, Câmpus do São Vicente, Praça Infante D. Henrique s/n., CEP: 11330-205,
São Vicente, Sao Paulo, Brazil

Email: areigada@uol.com.br.

ABSTRACT

The locomotory and feeding activity rhythms of *Callinectes ornatus* Ordway, 1863 and *Callinectes danae* Smith, 1869 were examined under laboratory conditions. Light significantly influenced the activity of these organisms. However, activity in both species was affected by the presence of food, independently of photoperiod regime.

I. INTRODUCTION

Exposure to light may affect many physiological processes in crustaceans, such as locomotion, feeding, mating, colour changes and metabolism (Meyer-Rochow 1994). One of the major goals in studies focusing on the feeding ecology of benthic predators is to determine diel feeding patterns (Fernández et al. 1991a, b). Laboratory observations have been used to address this specific issue (Hill 1976, Lipicius and Herrkind 1982, Lawton 1987).

Activity rhythms may be tightly connected with diel feeding patterns. Stomach analyses along 24 h cycles, including replenishment condition and food contents recordings, are often used to determine eventual feeding periods in brachyuran crabs (Mori 1986, Stevens et al. 1982, Wassemberg and Hill 1987, Freire et al. 1991a).

The purpose of this study was to examine the locomotory and feeding activity rhythms of *Callinectes ornatus* Ordway, 1863 and *Callinectes danae* Smith, 1869 under laboratory conditions.

II. MATERIALS AND METHODS

Crabs were placed in 38 l aquaria (25 X 32.5 X 46.5 cm) provided with a dry wet external filter in order to maintain water quality. Continuous aeration was also supplied. During the experiments, physical parameters of the water were monitored to keep captive conditions as close as possible to the natural environment. Sediments used were mainly constituted by fine and very fine sand, water temperature was kept at 25 ± 2°C and salinity maintained at 33 ± 2 $^o/_{oo}$, as measured with a hand refractometer. Only adult intermolt males of similar size were used in the experiments.

Individuals of both species were placed in isolated aquaria to test for locomotory activity. Forty *C. ornatus* and 20 *C. danae* were used. Before the experiments were run, crabs were acclimated for a 10 day period. Individuals were separated into two experimental groups. Both groups were kept in a 12 h dark: 12 h light photoperiod regime, but, while retaining the natural light-dark cycle in one of the groups, an inverted condition was used in the other experimental treatment.

N Kluwer Academic/Plenum Publishers

Observations were taken at 6 h intervals along daily cycles. Swimming crabs were fed every 24 h with shrimp muscle at 2:00 pm. The number of crabs emerged from the sand was counted both before and after feeding. The non-parametric Kruskal-Wallis test was used to compare swimming crabs activity. This test was complemented with the Student-Newman-Keuls procedure (Sokal and Rohlf 1979).

8:00 pm ($P < 0.05$) and at 2:00 pm, when they were fed (Fig. 1B). In the case of individuals maintained in the inverted photoperiod, activity differed significantly with a peak obtained at 8:00 am ($P < 0.05$, Fig. 2A). After feeding, a more intense activity was observed at 8:00 am and 2:00 am, with no significant differences between them. Lower activity levels were recorded at 8:00 pm, while no response was observed at 2:00 am (Fig. 2B).

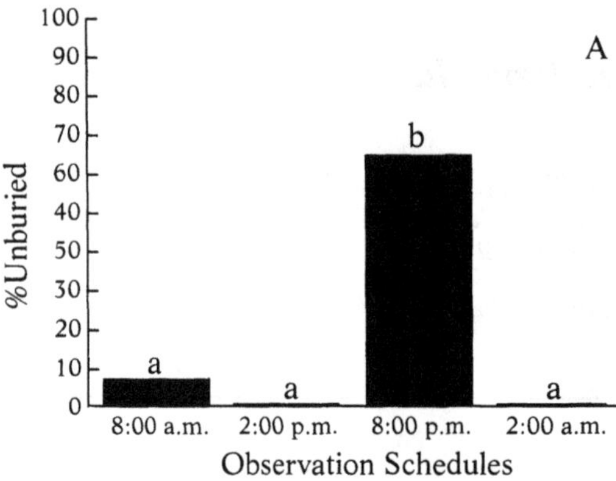

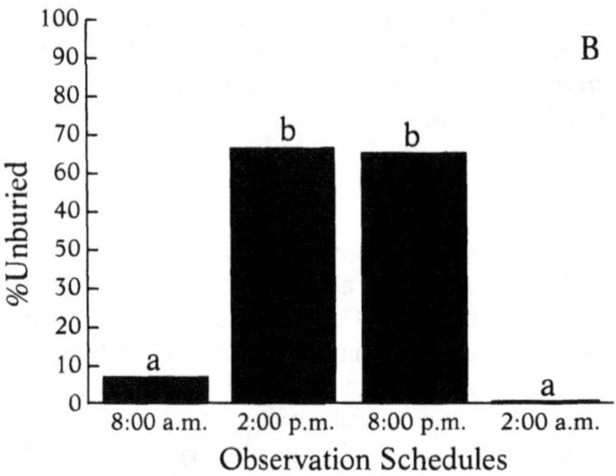

Figure 1. *Callinectes ornatus, bar graphs showing percentage of active crabs under natural photoperiod conditions; A) before feeding, B) after feeding (bars with same letters do not significantly differ).*

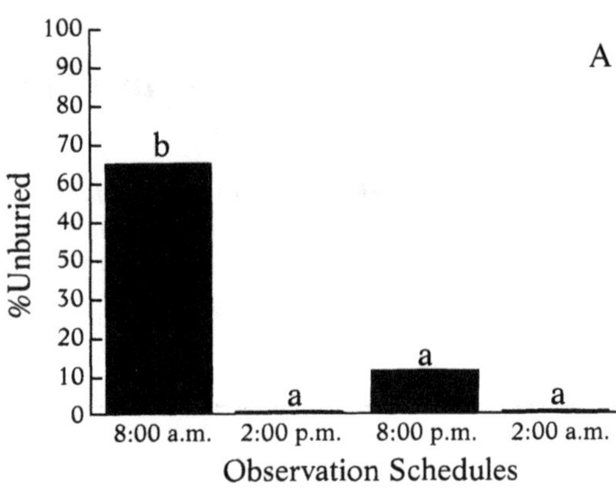

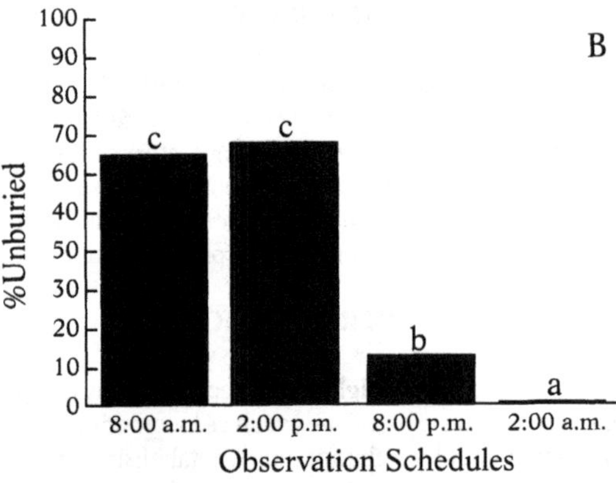

Figure 2. *Callinectes ornatus, bar graphs showing percentage of active crabs under inverted photoperiod conditions; A) before feeding, B) after feeding (bars with same letters do not significantly differ).*

III. RESULTS

Callinectes ornatus

The size of crabs used in this study varied from 58.6 to 74.4 mm of carapace width. The activity of *C. ornatus* kept in natural photoperiod conditions is presented in figure 1A. Individuals were more active at

Callinectes danae

Size of crabs used in this experiment ranged from 75.0 to 96.6 mm of carapace width. Activity of crabs kept in the natural photoperiod peaked at 8:00 pm compared to observations made at other times ($P < 0.05$). Percentage of buried females was however higher at 8:00 am than at 2:00 pm and 2:00 am (Fig.

3A). Figure 3B shows the same response after feeding. At this time, the activity of crabs increased at 2:00 pm. Organisms in the inverted photoperiod regime showed higher activity at 6:00 am, differing significantly from all the remaining observations (Figs. 4A, B). After feeding, periods of highest activity corresponded to the observations made at 8:00 am and 2:00 pm.

interval just after switching off the lights. The influence of light on the activity of these organisms is evident since results from trials using natural and inverted light regimes indicate the same trend. Captive crabs apparently respond spontaneously to the stimuli presented.

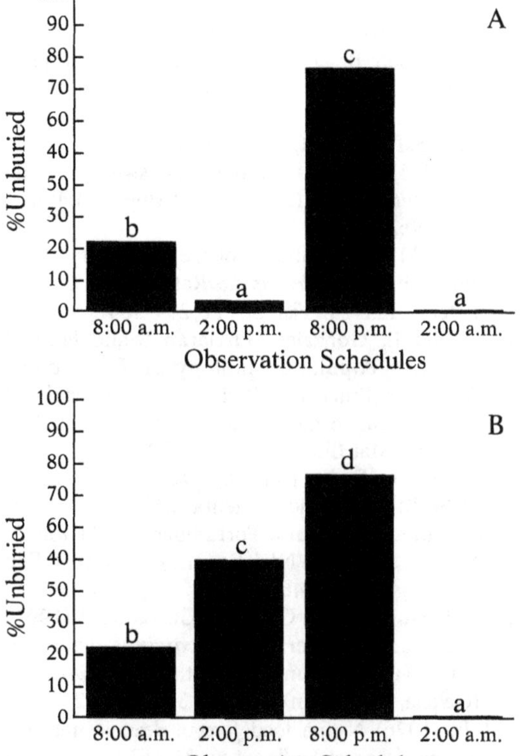

Figure 3. *Callinectes danae, bar graphs showing percentage of active crabs under natural photoperiod conditions; A) before feeding, B) after feeding (bars with same letters do not significantly differ).*

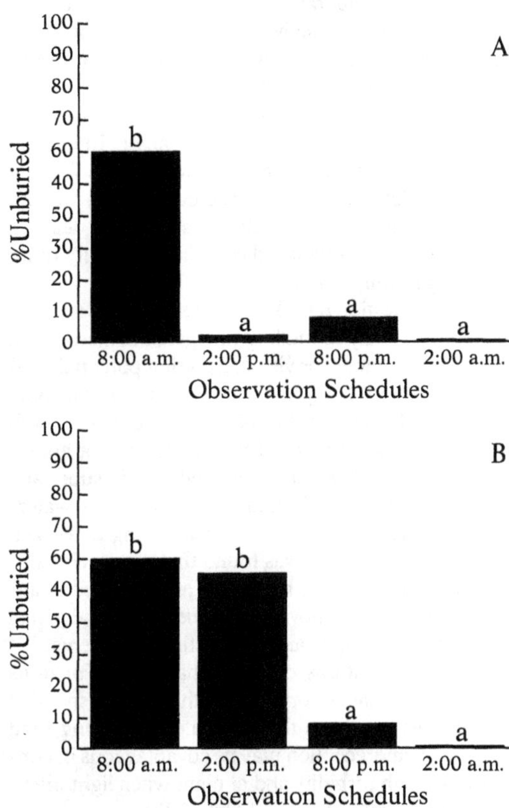

Figure 4. *Callinectes danae, bar graphs showing percentage of active crabs under inverted photoperiod conditions; A) before feeding, B) after feeding (bars with same letters do not significantly differ).*

IV. DISCUSSION

Present findings show that the swimming crabs *C. ornatus* and *C. danae* become more active and emerge from the sediment during periods of low light intensity. Under the experimental conditions used in this study, that period corresponded to a relatively short

Hill (1976), who studied the portunid *Scylla serrata* in the laboratory, verified that this species presents nocturnal activity restrained to 30 minutes during twilight, when unburied crabs remained motionless or moving over the sediment. Other studies on swimming crabs conducted under laboratory conditions, such as those carried out by Darnell (1959) with *Callinectes sapidus*, Caine (1974) on *Ovalipes guadalupensis* and Paul (1981) using *Callinectes arcuatus*, bring up similar results, indicating that nocturnal activity is a more general trend in portunids.

Activity in *C. ornatus* and *C. danae* was affected by the presence of food, independently of photoperiod regime. This is evident, since buried animals under dark conditions emerged from the sand in a number of occasions after stimulated by the presence of food. Still in the laboratory, the presence of food items causes an increase of activity in *Scylla serrata* by eliciting swimming behaviour of crabs near the surface of experimental tanks (Hill 1976).

According to Schembri (1980), the crab *Ebalia tuberosa* is usually active at night. In captive conditions, the species may shift to daytime feeding when stimulated by water-borne cues released by an injured animal. Therefore, the presence of a dead or near-dead animal might indicate the availability of an appealing feeding resource.

A question raised by the present study is which stimulus is actually controlling emergence of these crabs, i.e., chemical, tactile or visual. Spotting potential food items at night probably involves chemoreception (Barber 1961). Reabach et al (1990), who studied the crab *Cancer irroratus*, evidenced the importance of antennular sensitivity for detecting food in the substrate. Schembri (1981) had already observed that water-borne stimuli are responsible for inducing activity in *Ebalia tuberosa*. Yet, it was found that the species also engaged in locomotory behaviour and attacked inanimate objects when moving. So, mechanic and visual stimuli are also important in detecting food items.

Observations on the behavior of *Portunus pelagicus* in the laboratory showed that crabs responded to bait emerging from the substrate. Localizing food through chemoreception may be advantageous in conditions of high turbidity and at night when light intensity is low (Wassemberg and Hill 1987).

The blue crab *Callinectes sapidus* follows an endogenous activity rhythm in which locomotory behavior takes place mainly at night. This pattern may change in the presence of chemical stimuli (Zimmer-Faust et al. 1996). Cannicci et al. (1995), who studied the portunid *Thalamita crenata*, confirmed that visual cues play an important role in the locomotory activity of this diurnal species.

In the case of *C. ornatus* and *C. danae*, it is possible that the presence of food might have been the major factor triggering locomotory activity. It should be mentioned, however, that more detailed experiments are needed to determine what sort of cues, or what sort of interacting stimuli, are actually mediating such behavioural response.

ACKNOWLEDGEMENTS

CNPq and FAPESP (# 94/4878-4; # 95/3872-9) provided financial support. I am grateful to NEBECC colleagues for their aid during field and laboratory analyses.

REFERENCES

Caine EA (1974) Feeding of *Ovalipes guadalupensis* (Decapoda: Brachyura: Portunidae) and morphological adaptations to a burrowing existence. Biol Bull 174:550-559

Cannicci S, Dahdouh-Guebas F, Anyona D and Vannini M (1995) Homing in the mangrove swimming crab *Thalamita crenata* (Decapoda: Portunidae). Ethology 100:242-252

Darnell RM (1959) Studies of the life history of the blue crab (*Callinectes sapidus* Rathbun) in Lousiana waters. Trans Am Fish Soc 88:294-304

Fernandéz L, González-Gurriarán E and Freire J (1991A) Population biology of *Liocarcinus depurator* (Brachyura: Portunidae) in mussel raft culture areas in the Ría de Arousa (Galicia, NW Spain). J Mar Biol Ass 71:375-390

Fernandéz L, Freire J and González-Gurriarán E (1991B). Actividad alimentaria de *Liocarcinus arcuatus* (Brachyura, Portunidae) en la Ría de Arousa (Galícia, NW España). Bol Inst Esp Oceanogr 7:139-146

Freire J, Muiño R and González-Gurriarán E (1991) Diel feeding pattern of *Liocarcinus depurator* (Brachyura: Portunidae) in the Ría de Arousa (Galicia, NW, Spain). Ophelia 33:165-172

Hill BJ (1976) Natural food, foregut clearance-rate and activity of the crab *Scylla serrata*. Mar Biol 24:109-116

Lawton P (1987) Diel activity and foraging behavior of juvenile American lobsters *Homarus americanus*. Can J Fish Aquat Sci 44:1195-1205

Lipicius RN and Hernkind WF (1982) Molt cycle alterations in behavior, feeding and diel rhythms of a decapod crustacean, the spiny lobster *Panulirus argus*. Mar Biol 68:241-252

Meyer-Rochow VB (1994) Light-induced damage to photoreceptors of spiny lobsters and other crustaceans. Crustaceana 67:95-109

Mori M (1986) Alimentary rhythms in *Geryon longipes* A. Milne Edwards, 1881 (Crustacea, Decapoda: Brachyura). Quad lab Tecnol Pesca 3:169-172

Paul RKG (1981) Natural diet, feeding and predatory activity of the crabs *Callinectes arcuatus* and *C. toxotes* (Decapoda, Brachyura, Portunidae). Mar Ecol Prog Ser 6:91-99

Rebach S, French DP, von Staden FC, Wilber MB and Byrd VE (1990) Antennular sensitivity of the rock crab *Cancer irroratus* to food substances. J Crust Biol 10:213-217

Schembri PJ (1980) Aspects of the biology, behaviour and functional morphology of the crab *Ebalia tuberosa* (Pennant). PhD Thesis, University of Glasgow, Scotland, 194 p

Schembri PJ (1981) Feeding in *Ebalia tuberosa* (Pennant) (Crustacea: Decapoda: Leucosiidae). J Exp Mar Biol Ecol 53:1-10

Sokal RR and Rohlf FJ (1979) Biometría. Madrid, H. Blume Ediciones

Stevens BG, Armstrong DA and Cusimano R (1982) Feeding habits of the dungeness crab *Cancer magister* as determined by the index of relative importance. Mar Biol 72:135-145

Zimmer-Faust RK, O'Neill B and Schar DW (1996) The relationship between predator activity state and sensitivity to prey odor. Biol Bull 190:82-87

Wassemberg TJ and Hill BJ (1987) Feeding by the sand crab *Portunus pelagicus* on material discarded from prawn trawlers in Moreton Bay, Australia. Mar Biol 95:387-393

ULTRASTRUCTURAL AND BIOCHEMICAL STUDIES OF THE BRANCHIOSTEGITE OF THE BIMODAL BREATHING CRAB *CHASMAGNATHUS GRANULATA* DANA, 1851

Julia Halperin, Gladys N. Hermida, Luisa E. Fiorito, Gladys N. Pellerano and Carlos M. Luquet

Laboratorio de Histología Animal, Departamento de Ciencias Biológicas, Facultad de Ciencias Exactas y Naturales, Universidad de Buenos Aires, Pabellón II, Ciudad Universitaria, 1428 Buenos Aires, Argentina

halperin@bg.fcen.uba.ar

ABSTRACT

This study provides histologic and ultrastructural evidence of the capacity of *Chasmagnathus granulata* to perform aerial gas exchange across the lining of branchiostegites. Although the lungs of *C. granulata* seem to be suitable organs for aerial oxygen uptake, the low levels of carbonic anhydrase activity compared with the gills, indicate limited capacity to excrete carbon dioxide directly to air, resulting in a limiting factor for the crab to invade land.

I. INTRODUCTION

Terrestrial crabs possess the ability to exchange gases directly with air through the inner walls of the branchiostegites. While oxygen is readily available in air, favoring its diffusion through the respiratory wall, CO_2 excretion into air requires both elevated hemolymph CO_2 and the presence of carbonic anhydrase (CA) within the branchiostegites (see Henry 1994 for a review). This enzyme catalyzes the reversible CO_2 hydration/dehydration. The fraction of CA bound to the plasma membrane facing the hemolymph, participates in the conversion of bicarbonate to molecular CO_2 which diffuses across the cell to the medium (Henry 1988). Thus, the ability of the inner branchiostegal wall of terrestrial crabs to excrete CO_2 into air is directly related with the presence of membrane CA.

The active amphibious species, called bimodal breathing crabs, are also able to uptake O_2 from air through the branchiostegites. However, they rely on the gills to excrete carbon dioxide to a water reservoir kept within the branchial chambers. In bimodal breathing species such as *Cardisoma guanhumi*, membrane CA has been reported to be restricted to the gills (Henry 1991). The latter author has postulated that branchiostegal CA is a major achievement that confers terrestrial crabs their respiratory independence from water.

From a morphological viewpoint, the branchiostegites are lateral extensions of the thoracic body wall and comprise two integumental layers separated by spongy connective tissue. The inner integument is very thin and delimits the branchial chambers. In regard to vasculature, portal vessels interdigitate and branch as they descend towards the inner integument. In air breathing species, the minor vessels open in a system of flattened blood spaces in direct contact with the attenuated epithelium that lines the branchial chambers. This arrangement results in a minimal hemolymph-to-air distance, favoring the function of the inner branchiostegal integument as a gas exchange organ. Connective tissue is also characterized by possessing glycogen-packed cells, bundles of collagen-like fibers and chromatophores (Taylor and Greenaway 1979). The gas exchange surface area is increased by specializations like invaginations or evaginations of the inner wall (Díaz and Rodríguez 1977, Taylor and Taylor 1992) or by enlargement of the branchial chambers (Maitland 1990, Farrelly and Greenaway 1993).

E. Escobar-Briones & F. Alvarez Eds.
MODERN APPROACHES TO THE STUDY OF CRUSTACEA
PP. 021-027

Chasmagnathus granulata Dana, 1851, is an estuarine grapsid crab from Brazil, Uruguay and Argentina, which actively moves to supratidal areas even during daylight in the summer (Boschi 1964). In previous reports, this species has been considered as relying on branchial respiration during air exposure thanks to the maintenance and recirculation of branchial water reservoirs (Santos et al. 1987, Luquet and Ansaldo 1997, Luquet 1998). However, a recent study (Halperin et al. 2000) has shown that *C. granulata* is able to sustain a high rate of oxygen uptake and high venous oxygen pressure after extraction of the branchial chamber water, thus suggesting that the branchiostegites are involved in aerial oxygen uptake.

The aim of this study is to provide histologic and ultrastructural evidence of the capacity of this species to perform aerial gas exchange across the lining of branchiostegites. Branchiostegal carbonic anhydrase activity will also be studied in relation to the capacity of this species to excrete CO_2 into air.

II. MATERIALS AND METHODS

Maintenance of animals

Adult male crabs in intermoult stage C (Drach and Tchernigovtzeff 1967) were collected from a mud-sand beach near San Clemente del Tuyú, Province of Buenos Aires, Argentina. Crabs were kept in glass aquaria with artificial brackish water (12‰ salinity) and free access to air breathing. The animals were fed twice a week with rabbit food pellets and kept at room temperature ($20\pm1°C$) and a 12 h:12 h day/night photoperiod.

Transmission electron microscopy (TEM)

Small pieces of dorsal and lateral carapace were cut and fixed by immersion in 2.5% glutaraldehyde, buffered with sodium cacodylate at pH 7.4 and adjusted to 650 mOsml.l^{-1} with sucrose. Samples were postfixed with ice-cold 1% osmium tetroxide in the same buffer solution for 1h. The pieces were then rinsed in buffer, dehydrated in a graded ethanol series and embedded in Spurr resin. Thin sections (750-900 Å) were cut with a diamond knife in a Sorvall Porter-Blum MT2-B ultramicrotome, double stained with 1% aqueous uranyl acetate for 45 minutes at room temperature and lead citrate for 3 min at room temperature, and examined with a Zeiss EM T109 electron microscope operated at 80 kV.

Semithin ($1\mu m$) cross sections were mounted on glass slides and stained with 1% toluidine blue in 1% Na_2CO_3. Photomicrographs were taken with a Reichert Polivar III microscope with a M35 camera.

Light microscopy

Pieces of the branchiostegite were fixed in Bouin's mixture for 20 h. The pieces were decalcified using a 5% solution of nitric acid for 20 h and then suspended in a 5% solution of iron alum to avoid connective tissue swelling. Finally, the pieces were dehydrated through an ethanol series and embedded in Paraplast.

Serial sections 7 to 9 μm thick were stained with Masson's trichromic and hematoxylin-eosin. Histochemical techniques were performed to examine the presence of glycoconjugates. Photomicrographs were taken with a Zeiss Axioplan microscope.

Gas exchange surface area

The water stored in the branchial chambers of 10 *C. granulata* (18.76 ± 1.69 g) was emptied by placing a blotting paper over the exhalant openings for two minutes. After killing the animals, chelipeds and epipods were dissected and the gills were removed with a pair of pincers. After that, the branchial chambers were filled with latex through the Milne-Edwards openings using a 5 ml syringe.

After 1 h, the carapace was removed and the exposed surface of the cast was outlined with a felt-tip pen. Pieces of graph paper were cut to fit exactly the marked areas. Lung area was then determined by counting graph paper units (Farrelly and Greenaway 1993). Surface area was expressed as a function of live body mass.

Carbonic anhydrase activity

The third gill pair (respiratory gills), branchiostegite and muscle of chelipeds were gently removed from 6 individuals, homogenized and assayed potentiometrically for carbonic anhydrase activity. Data were analyzed by one way ANOVA. Differences were considered to be significant at $P < 0.05$.

III. RESULTS

The branchial chambers are delimited by an inner integument that is constituted of a single epithelium and a thin cuticle (Fig. 1). This epithelium is composed of flattened cells about 0.57 mm thick and a thin cuticle of about 0.78 mm thick, which is formed by epicuticle, exocuticle and endocuticle. This epithelium lays on a thick basal lamina with a fine fibrillar appearance (Fig. 2). Adjacent epithelial cells are joined by zonulae adherens, septate desmosomes and gap junctions. Few but deep lateral membrane interdigitations and sometimes apical infoldings are also seen (Fig. 3). The nucleus is irregular in contour, with peripheral heterochromatin and occupies great part of the cell volume. Almost no cell organelles are found within the cytoplasm, a few electron-dense bodies and elongated mitochondria can be observed in the perikarion. Some epithelial cells present cytoplasmic prolongations and conspicuous infoldings and interdigitations in the basal surface, dense microtubule arrangements occupy great part of their cytoplasm (Figs. 4A, B).

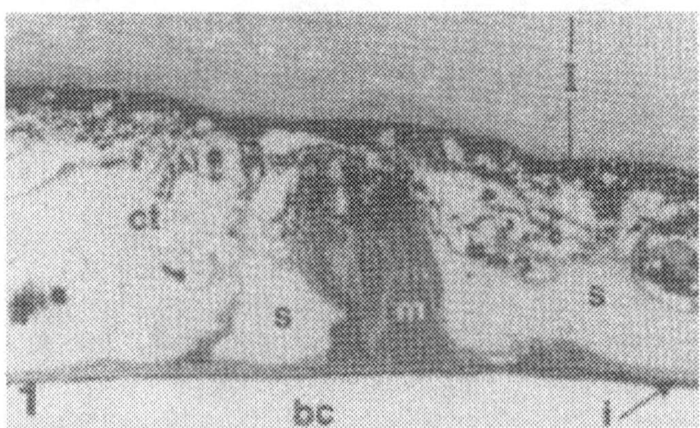

Figure 1. *Semithin cross section of the branchiostegite showing an outer integument (I) covering the outer surface, while the inner integument (i) is composed by a thin cuticle and an attenuated epithelium. Spongy connective tissue (ct) is situated between these two kinds of integuments. Periodic bundles of striated muscle (m) attached to the inner and outer integuments; bc, branchial chamber; s, sinus. A:250X.*

Periodic bundles of striated muscle associated with abundant collagen-like fibers attach by zigzag junctions to modified epithelial cells (Figs. 1, 5). This muscle is ultrastructurally different from the typical striated fibers of the dorso-ventral muscles since the striated structure is hardly evident (Fig. 6).

Fine sheets of connective tissue delimit many hemolymphatic spaces. The more flattened of these spaces are located beneath the respiratory epithelium and constitute the respiratory lacunae (Fig. 7). These lacunae are lined by a basal lamina continuous with the lining of the epithelium. Hemolymph contains hyaline and granular hemocytes, being the latter the most abundant. Granulocytes have a high nucleocytoplasmic ratio, being the nucleus irregular and central, with the

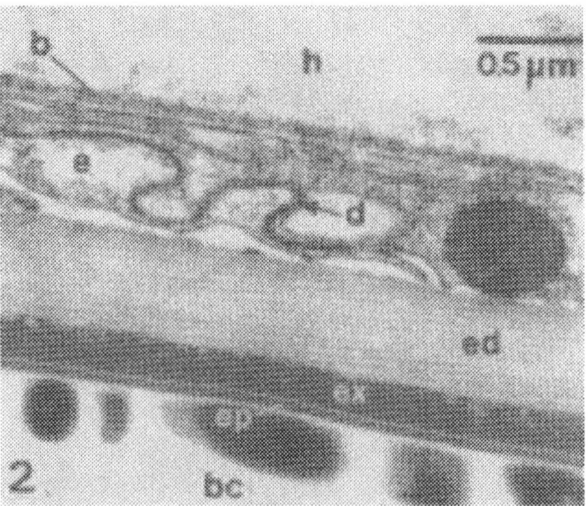

Figure 2. *TEM of a cross section of the branchiostegite showing the attenuated inner epithelium (e) and thin cuticle of the gas exchange surface. This respiratory epithelium lays on a thick basal lamina (b). Adjacent epithelial cells are joined by a septate desmosome (d); bc, branchial chamber; ep, epicuticle; ex, exocuticle; ed, endocuticle; h, hemolymph.*

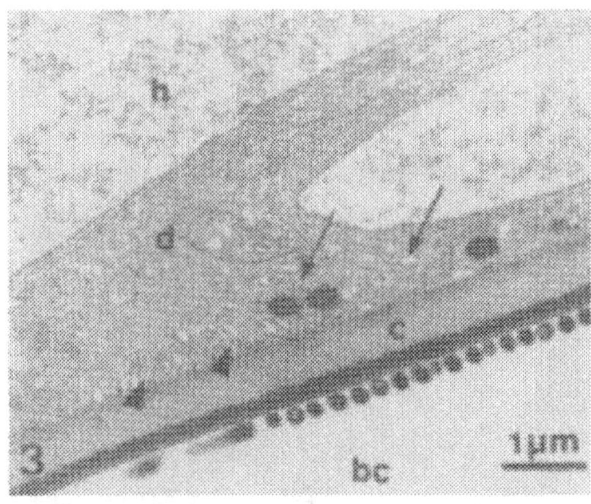

Figure 3. *TEM of a cross section of the branchiostegite showing deep membrane interdigitations (arrows) and apical infoldings (arrowheads) in the epithelium; bc, branchial chamber; c, cuticle; d, septate desmosome; h, hemolymph.*

chromatin condensed around the nuclear envelope. The cytoplasm contains abundant granules of varying size that characteristically emit numerous prolongations that contact other hemocytes or components of the extracelular matrix (Fig. 8).

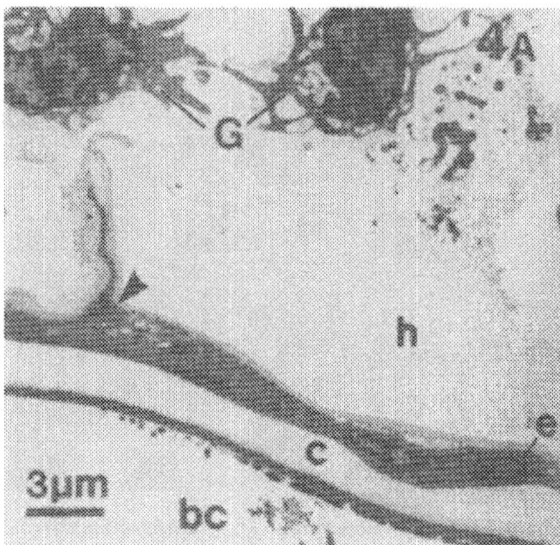

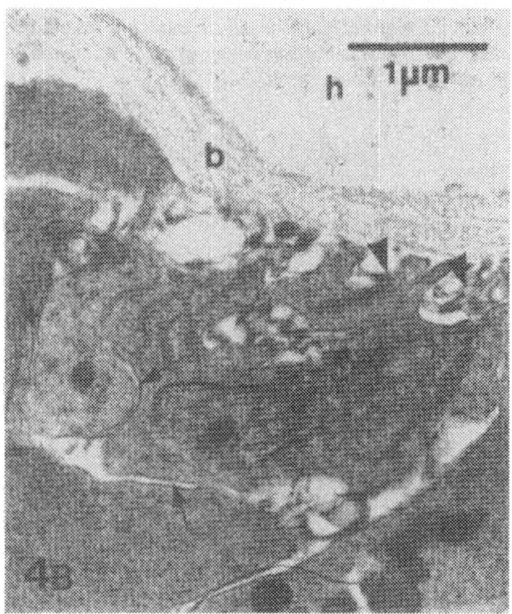

Figure 4. *A) TEM of a cross section of the branchiostegite showing modified epithelial cells (arrowhead); bc, branchial chamber; c, cuticle; e, epithelium; G, granulocytes; h, hemolymph. B) Detail of a modified cell showing dense microtubule arrangements occupying great part of their cytoplasm and conspicuous infoldings (arrowheads) and interdigitations (arrows) in the basal surface. b, basal lamina; h, hemolymph.*

Another feature of the connective tissue is the presence of large cells with lipid inclusions and thin cytoplasmic processes. Under histochemical study these cells show PAS and Alcian Blue (pH 3.5) positive staining (Fig. 9). Abundant chromatophores are also present near the external body wall.

Small arteries (6 mm lumen diameter) with stellate appearance in transverse section appear free in the hemocoel only supported by thin sheets of connective tissue (Fig. 10A). The arterial wall is composed,

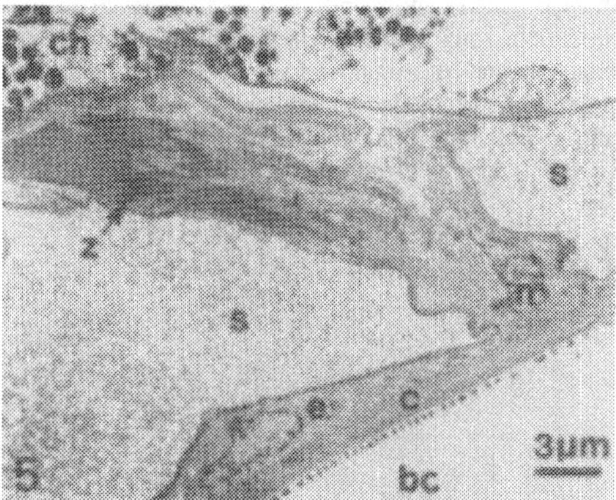

Figure 5. *TEM of a cross section of the branchiostegite showing bundles of striated muscle associated with abundant collagen-like fibers attached by zigzag junctions (z) to modified epithelial cells; bc, branchial chamber; c, cuticle; ch, chromatophores; e, epithelium; n, nucleus, s, sinus.*

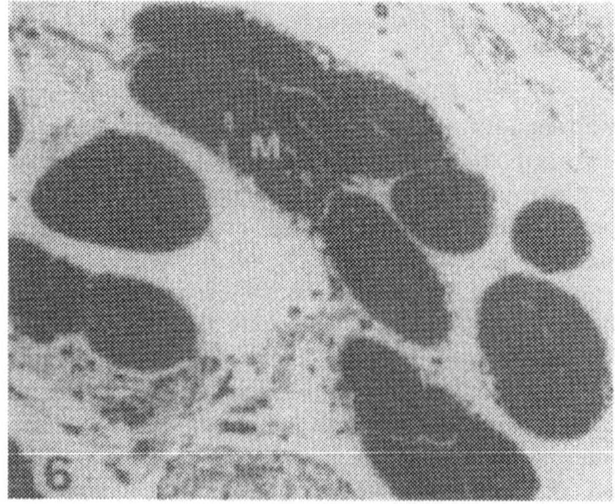

Figure 6. *Semithin cross sections of the branchiostegite showing striated fibers of the dorso-ventral muscles (M). A:320X.*

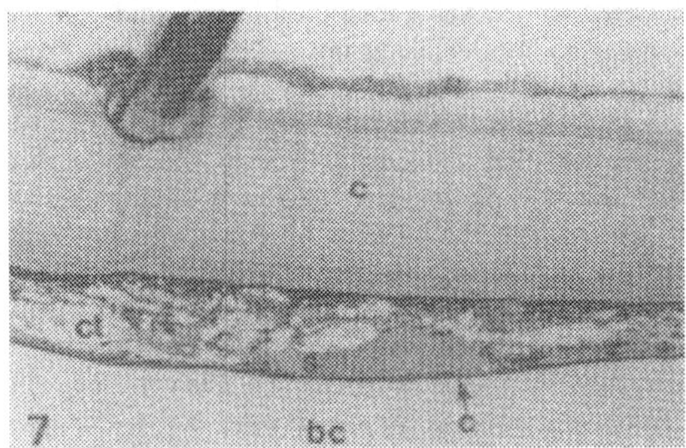

Figure 7. *Semithin cross sections of the branchiostegite showing a general view where delicate partitions of connective tissue delimit many hemolymphatic sinuses (s). Flattened spaces located beneath the respiratory epithelium constitute the respiratory lacunae; bc, branchial chamber; c, cuticle; ct, connective tissue. A:50X.*

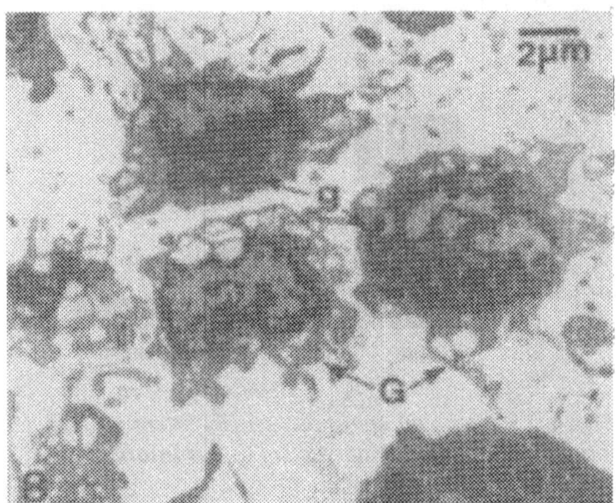

Figure 8. *TEM of a cross section of the branchiostegite showing some granulocytes (G). These cells have an irregular and central nucleus (n) and the cytoplasm contains abundant granules (g) of varying size and characteristically emits numerous prolongations which contact with other hemocytes or with components of the extracelular matrix.*

from the luminal side, by an intima with thick glycocalyx, an endothelial cell with numerous large membrane infoldings and an external basal lamina (Fig. 10B).

The putative respiratory surface area measured in the latex casts was 11.4 ± 0.3 mm^{2}g^{-1}. Branchiostegite carbonic anhydrase activity was significantly higher than that of muscle (considered as negative control) but significantly lower than the value obtained for respiratory gills (Fig. 11).

IV. DISCUSSION

Casts and serial light microscopy studies show that the branchial chambers of *Chasmagnathus granulata* are limited by a smooth integument, except for a reduced zone of the latero-dorsal angle, where simple folds and lobulated projections are evident. Morphometry and electron microscopy indicate that the respiratory surface is restricted to the dorsal and lateral sides, in the anterior region of the branchiostegites. This zone is believed to be ventilated with air when individuals of this species recirculate the water reservoir that they keep within the branchial chambers during emersion and, specially when this reservoir is being depleted by evaporation.

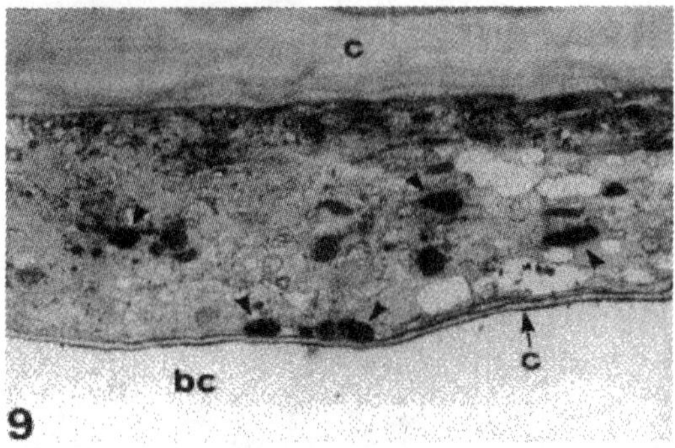

Figure 9. *Semithin cross sections of the branchiostegite showing the spongy connective tissue containing large cells with lipid inclusions (arrowheads); bc, branchial chamber; c, cuticle. A:125X.*

From an ultrastructural viewpoint, the branchiostegites of *C. granulata* are efficient gas exchange organs. The close contact between an attenuated epithelium and a system of branching, flattened blood sinuses are fundamental requirements. Moreover, the small arteries seen in branchiostegal preparations show endothelial cells with numerous plasma membrane infoldings. This suggests that these vessels possess the capacity to change in diameter in order to regulate the perfusion of the organ in response to changing respiratory conditions.

Modified epithelial cells with abundant microtubules and related with muscular fibers probably provide strength and support for preventing the separation of the inner and outer integuments by hemolymph pressure changes. These modified epithelial cells could be compared with the epidermal cells which attach to connective fimbriated cells in *Holthuisana_transversa*,

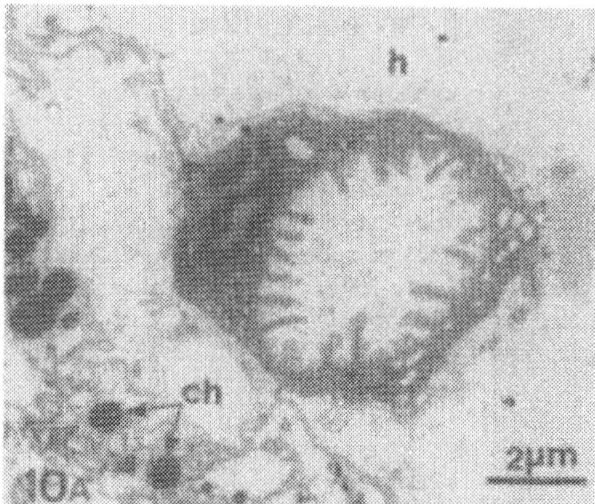

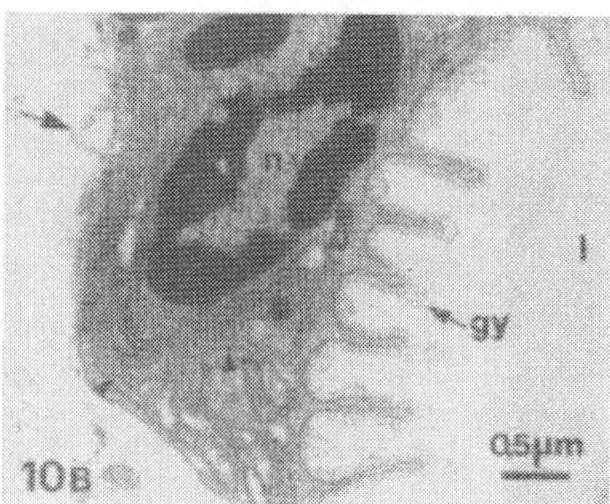

Figure 10. *A) TEM of a cross section of the branchiostegite showing a small artery in transverse section only supported by thin sheets of connective tissue. ch, chromatophores; h, hemolymph. B) The arterial wall is composed, from the luminal side, by an intima with thick glycocalyx (gy), an endotelial cell with numerous large membrane infoldings (arrowheads) and an external basal lamina (arrow); n, nucleus; l, lumen.*

Geograpsus grayi, G. crinipes and *Gecarcoidea natalis* (Taylor and Greenaway 1979, Farrelly and Greenaway 1993), since they also possess arrangements of microtubules and filamentous components of the cytoskeleton.

The lung surface area (11.4 ± 0.3 mm^{2}g^{-1}) of *C. granulata* measured herein is only a small fraction of the total respiratory gill surface area (278.5 mm^{2}g^{-1}) (Luquet et al. 2000). The diffusion distance reported by Luquet et al. (2000) ($3.74\,\mu$m) is almost three times

higher than the value measured in this study for the lung (1.35 mm). Taking into account the ratios between respiratory surface areas and diffusion distances, the gills could be about 8 times more efficient for oxygen uptake than the lungs. However, the capacitance and diffusibility of oxygen in air are much higher than in water. Thus, the lungs of *C. granulata* seem to be suitable organs for aerial oxygen uptake.

The low levels of carbonic anhydrase activity compared with the gills, indicate limited capacity to excrete carbon dioxide directly to air. In contrast, fully terrestrial crabs such as *Birgus latro* and *Gecarcinus lateralis* possess important branchiostegal carbonic anhydrase activity (Morris and Greenaway 1990, Henry 1991). Thus, this reduced capacity to excrete carbon dioxide to air in *C. granulata* results in a limiting factor for its ability to invade land.

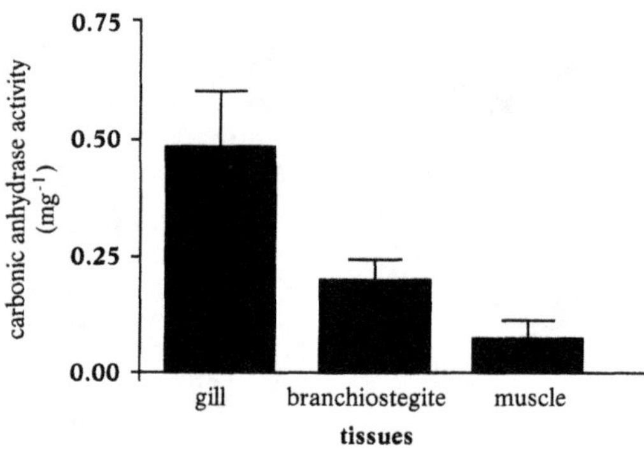

Figure 11. *Carbonic anhydrase activity in branchiostegite, muscle tissue and respiratory gill of **C. granulata**.*

ACKNOWLEDGEMENTS

This study was supported by a grant UBACyT TX07 from Universidad de Buenos Aires. We acknowledge Dr. Ines O'Farrel for revising the English translation.

REFERENCES

Boschi EE (1964) Los crustáceos decápodos brachyura del litoral bonaerense. Bol Inst Biol Mar 164:1-34

Díaz H and Rodriguez G (1977) The branchial chamber in terrestrial crabs: a comparative study. Biol Bull 153: 485-504

Drach P and Tchernigovtzeff C (1967) Sur la méthode de détermination des stades d'intermue et son application générale aux crustacés. Vie Milieu 18:597-607

Farrelly C and Greenaway P (1993) Land crabs with smooth lungs: Grapsidae, Gecarcinidae and Sundathelphusidae, ultrastructure and vasculature. J Morphol 215:245-260

Halperin J, Ansaldo M, Pellerano GN and Luquet CM (2001) Bimodal breathing in the estuarine crab *Chasmagnathus granulata*, Dana 1851. Physiological and morphological studies. Comp Biochem Physiol (in press)

Henry RP (1988) Multiple functions of carbonic anhydrase in the crustacean gill. J Exp Zool 248:19-24

Henry RP (1991) Branchial and branchiostegite carbonic anhydrase in decapod crustaceans: the aquatic to terrestrial transition. J Exp Zool 259:194-303

Henry RP (1994) Morphological, behavioral, and physiological characterization of bimodal breathing crustaceans. Am Zool 34:205-215

Luquet CM and Ansaldo M (1997) Acid-base balance and ionic regulation during emersion in the estuarine entertidal crab *Chasmagnathus granulata*, Dana 1851 (Decapoda Grapsidae). Comp Biochem Physiol 117A:407-410

Luquet CM, Cervino CO, Ansaldo M, Carrera Pereyra V, Kocmur S and Dezi RE (1998) Physiological response to emersion in the amphibious crab *Chasmagnatus granulata* Dana 1851 (Decapoda Grapsidae): biochemical and ventilatory adaptations. Comp Biochem Physiol 121A:385-393

Luquet CM, Rosa GA, Ferrari CC, Genovese G and Pellerano GN (2000) Gill morphology of the intertidal estuarine crab *Chasmagnathus granulata* Dana, 1851 (Decapoda, Grapsidae) in relation to habitat and respiratory habits. Crustaceana 73:53-67

Maitland DP (1990) Aerial respiration in the semaphore crab, *Heloecius cordiformis*, with or without branchial water. Comp Biochem Physiol 95A:267-274

Morris S and Greenaway P (1990) Adaptations to a terrestrial existence by the robber crab, *Birgus latro* L. V. The activity of carbonic anhydrase in gills and lungs. J Comp Physiol 160:217-221

Santos EA, Baldiseroto B, Bianchini A, Colares EP, Nery LEM and Manzoni GC (1987) Respiratory mechanism and metabolic adaptations of an intertidal crab, *Chasmagnathus granulata* (Dana 1851). Comp Biochem Physiol 88A:21-25

Taylor HH and Greenaway P (1979) The structure of the gills and lungs of the arid-zone crab, *Holthuisana* (*Astrothelphusa*) *transversa* (Brachyura: Sundathelphusidae) including observations on arterial vessels within the gills. J Zool 189:359-384

Taylor HH and Taylor EW (1992) Gills and lungs: the exchange of gases and ions. In: Harrison FW and Humes AG (eds) Microscopic Anatomy of Invertebrates. Decapod Crustacea. Vol. 10 (pp. 203-293) Wiley-Liss Inc, New York

POLYCLONAL ANTISERA AGAINST ESTUARINE CRUSTACEAN VITELLINS: A MOLECULAR APPROACH TO REPRODUCTIVE ENDOCRINOLOGY AND TOXICOLOGY

Shea R. Tuberty, Sergio F. Nates and Charles L. McKenney, Jr.

SRT, Center for Environmental Diagnostics and Bioremediation, University of West Florida, Pensacola, Florida 32514, USA

stuberty@uwf.edu

SFN, Zeigler Bros., Inc., Gardners, Pennsylvania 17324, USA
CLM, U.S Environmental Protection Agency, National Health and Environmental Effects Research Laboratory, Gulf Breeze, Florida 32561, USA

ABSTRACT

To fully elucidate the action of crustacean hormones, or their agonists, on vitellogenesis and reproduction, it has become increasingly important to develop sensitive assays that indicate a stimulatory or inhibitory effect on easily measured endpoints. Because of the relative abundance of vitellin in crustacean yolk and the ease with which it can be isolated, this protein makes an excellent model for studying the mechanisms that control and regulate reproduction. Vitellin was purified from eggs of adult female *Lepidophthalmus louisianensis, Palaemonetes pugio, Rhithropanopeus harrisii, Americamysis (Mysidopsis) bahia,* and *Uca panacea* which were collected from estuarine localities in Santa Rosa Sound, Gulf Breeze, Florida during late spring and summer of 1999. The purified proteins were used to immunize rabbits for polyclonal antibody production. Specificity of antisera was tested against respective vitellins by Western blotting. These antisera will be used to develop assays useful to the study of reproductive endocrinology, vitellogenesis, and toxicology.

I. INTRODUCTION

Recently there has been much attention focused on vitellogenin (Vg), the precursor to the egg yolk protein vitellin (Vt) in egg-laying vertebrates and invertebrates alike. This focus is a result of the growing economic importance of crustacean and fish mariculture and the use of vertebrate Vt assays as indicators of exposure to estrogenic xenobiotics. The importance of yolk proteins, in both invertebrate and vertebrate species, as the maternal input for embryo growth and development makes this area of research important to basic biology, aquaculture, and more recently, toxicology.

Vitellin, a high molecular weight lipo-glyco-carotenoprotein (Kerr 1969) ranging from 2 to 5 x 10^5 daltons, is the major yolk protein of mature crustacean eggs and is vital to the nutritional needs of a developing embryo (Lee 1991). Crustacean Vt, consisting of between 28 and 35% lipid (Fyffe and O'Connor 1974, Wallace *et al.* 1967), is the primary source of lipid for membrane formation and energy storage for the developing young. In some lecithotrophic marine decapods, which depend on long distance dispersal of embryos by ocean currents, the quantity and quality of yolk is of

E. Escobar-Briones & F. Alvarez Eds.
MODERN APPROACHES TO THE STUDY OF CRUSTACEA
PP. 029-037

utmost importance for survival. Because of the relative abundance of Vt in crustacean yolk and the ease with which it can be isolated, Vt makes an excellent subject for studying the mechanisms which control and regulate reproduction. The control and production of Vg, termed vitellogenesis, are being intensively studied because yolk is an excellent model for studying mechanisms of hormonal control at the cellular and molecular levels.

Growing interest in effects of xenobiotic compounds in the environment on organismal metabolism, development, and reproduction has opened the door to a new area of research for arthropod biologists. This is true because of recent advances in insect pesticide chemistry resulting in the creation of juvenile hormone agonists (e.g., methoprene, fenoxycarb, and pyriproxifen). In order to assess the gonadotropic action of newly purified hormones or identify adverse effects of xenobiotics on crustacean reproduction, it is important to accurately measure Vt and Vg in crustacean models. In the development of a quantitative enzyme-linked immunosorbent assay (ELISA), a significant amount of Vt must be purified from the species to use it as a standard in the ELISA method. In this study, we developed a two-step chromatographic method to purify Vt from six important crustacean species. The Vt was then partially characterized and subsequently used to produce polyclonal antiserum for use in immunocytochemistry and ELISA. Future applications of the assays are discussed.

II. MATERIALS AND METHODS

Animals

Ovigerous female *Lepidophthalmus louisianensis* with secondary vitellogenic ovaries were collected with yabby pumps from burrows (Manning 1975), and *Americamysis* (*Mysidopsis*) *bahia*, *Palaemonetes pugio*, *Uca panacea*, and *Rhithropanopeus harrisii* were collected with dipnets or by hand in shallow waters of Santa Rosa Island, Gulf Breeze, Florida in May of 1999. The red swamp crayfish, *Procambarus clarkii*, were provided to us by the University of Louisiana at Lafayette Crawfish Research Center. The animals were transported to the laboratory (US Environmental Protection Agency, National Health and Environmental Effects Research Laboratory, Gulf Ecology Division, Gulf Breeze, Florida, USA) in individual perforated 50 mL plastic tubes and maintained in aquaria with filtered (20 mm) seawater at 28 ± 1°C and 20 ± 2 ‰ S.

Source of vitellin

Ovaries or egg masses were dissected from the animals, weighed, rinsed with deionized water and homogenized in a Potter-Elvehjem tissue homogenizer (Kimble-Kontes Glass, Vineland, NJ, USA) attached to a Stir Pak laboratory stirrer (Cole-Parmer, Chicago, IL, USA) with 10 ml of crustacean extraction buffer (136 mM NaCl, 10 mM Na_2HPO_4, 2.7 mM KCl, 1.8 mM KH_2PO_4, 2% glycine ethyl ester, 0.03% EDTA, 1 mM PMSF, 0.2% aprotinin, 2 mM leupeptin, and 0.02% sodium azide; pH 7.3). Homogenates were first separated by centrifugation at 4°C in a Sorvall RC5-C refrigerated centrifuge (Dupont Co., Newtown, CT, USA) for 15 minutes at 10,000 x g. The lipid layer was removed from the surface and the supernatant was transferred to new tubes and centrifuged a second time. Crude homogenates were stored at 4°C for further purification steps.

Sample preparation and clarification: gel permeation chromatography

A modified version of a technique reported by Brion *et al.* (2000) for purification of fish Vt was used to isolate the crustacean yolk proteins. Samples of the crude homogenates were loaded onto an FPLC System (Amersham Pharmacia Biotech, Uppsala, Sweden) by injecting 3 ml from a 10 ml sample loop (Superloop 10 ml, Amersham Pharmacia Biotech) onto a HiLoad 16/60 200 prep grade Superdex column. The column was equilibrated with 20 mM TBS, 5 mM EDTA, pH 7.6 during two column volumes. Elution of the sample was performed with 3 column volumes at a flow rate of 1 ml/min. Eluates were monitored at 280 and 474 nm for concomitant maximum protein and carotenoid pigment absorption, respectively, and 2 ml fractions were automatically collected. The diluted fractions that eluted with concominant peaks at both 280 and 474 nm were pooled and concentrated under nitrogen in an ultrafiltration cell (Amicon, Beverly, MA, USA) with a 10,000 molecular weight cut-off membrane (PM 10, Amicon).

Capture and purification step: anion-exchange chromatography

The partially purified samples from the gel permeation step were loaded onto a HiLoad 16/10 Q Sepharose High Performance anion exchange column. The column was equilibrated with 20 mM TBS, 5 mM EDTA, pH 7.6 during two column volumes. Proteins were eluted isocratically in 3 column volumes of 130 mM NaCl, 20 mM TBS, and 5mM EDTA, pH 7.6 at a flow rate of 1 ml/min. Eluates were monitored at 280 and 474 nm for concomitant maximum protein and carotenoid pigment absorption, respectively, and 1 ml fractions were automatically collected. Samples were assayed for total protein by a modified method of (Bradford 1976), using Pierce Coomassie Protein Plus assay reagent (Pierce, Rockford, IL, USA).

Gel electrophoresis

Purity of the isolated Vt was analyzed by 4% polyacrylamide gel electrophoresis (PAGE) or 7.5% sodium dodecyl sulphate-polyacrylamide gel electrophoresis (SDS-PAGE) in the absence of ß-mercaptoethanol (Laemmli 1970) on mini gels (Mighty Small II SE250, Hoefer Scientific Instruments, San Francisco, CA, USA). The molecular weights of polypeptide subunits were determined by reducing 7.5% SDS-PAGE after diluting and heating the purified Vt in sample buffer (0.5 M Tris HCl, pH 6.8, 10% glycerol, 0.02% Bromophenol Blue, and 5% ß-mercaptoethanol; 95°C, 5 min) and comparing to high molecular weight protein standards (SigmaMarkers, Sigma, St. Louis, MO, USA). The protein bands on the gels were visualized by Coomassie brilliant blue or Silver Stain Plus (Bio-Rad, Hercules, CA, USA). The gels were also stained for glycoproteins and lipoproteins with GlycoPro (Sigma, St. Louis, MO, USA) and Sudan Black B, respectively.

Rabbit immunization and immunoglobulin preparation

Pre-immunized blood was collected from New Zealand white rabbits, centrifuged, and the serum stored at -70°C until needed. The primary immunization was injected intramuscularly with purified Vt in Complete Freund's Adjuvant, followed every three weeks by boosts of Vt and Incomplete Freund's Adjuvant. Test bleeds were collected and tested for anti-Vt antibodies by capture enzyme-linked immunosorbent assay (ELISA). Once a suitable titer of antibody was reached, the rabbits were bled weekly, the blood was then centrifuged, and the antisera pooled in vials and frozen at -70°C until used for Western blot analysis.

Western blotting

Proteins were transferred onto nitrocellulose membrane (0.45 mm) from SDS-PAGE gels using a TE Series Transphor Electrophoresis unit (Hoefer Scientific Instruments). Transfer was conducted at 90 V for 3 hours or 40 V overnight at 4°C in Towbin buffer, pH 8.3 (Towbin *et al.*, 1979). The membrane was blocked overnight with 5% nonfat dehydrated milk in 150 mM phosphate buffered saline (PBS), pH 7.2. The blot was then washed 3 x 10 min in PBS and 0.05% Tween20 (PBS-T) and incubated with anti-Vt polyclonal antisera (1:2000 in PBS-T) for 1 hour at room temperature (RT). The blot was washed 3 times with PBS-T and incubated 2 hours with secondary antibody (horseradish peroxidase-conjugated goat anti-rabbit IgG, Boehringer Mannheim, Indianapolis, IN, USA) diluted 1:5000 in PBS-T. After 3 washes the antigen-antibody complexes were identified by addition of color developing solution (20 mg 4-Cl-1-Napthol, 10 ml methanol, 30 ml PBS, and 200 μl hydrogen peroxide). Color development was allowed to proceed until bands were easily detected at which time the reaction was terminated by washing the blot in several changes of deionized water.

III. RESULTS

Sample preparation and purification of the Vt

Using the purification of Vt from *R. harrisii* as an example, the elution profile of crude ovarian homogenate from gel permeation chromatography on a HiLoad 16/60 Superdex 200 HR column is presented in Figure 1. During separation of the crustacean ovarian homogenates, two to three peaks eluted as a group at 45 minutes (45 mL of buffer volume). Due to the absorption spectra at both 474 nm and 280 nm, these peaks were assumed to be Vt because it is known to be associated with carotenoid pigments (e.g., astaxanthin, canthaxanthin). Several protein peaks eluted from the column at minute 100-140, although they do not show absorption at 474 nm.

An example of the elution profile from the Q Sepharose anion exchange column is shown in Figure 2. A double peak (Peak 1) eluted with the binding buffer at 15 minutes, and appears to be made of degraded Vt not bound to the column. After the application of the elution buffer a highly concentrated peak (Peak 2) containing the Vt was released. Only fractions contained in the main peak area (fractions 46-48) were collected in order to avoid contaminants in the final product. Samples were either used immediately or frozen and lyophilized on a Flexi-Dry MP (FTS Systems, Stone Ridge, NY, USA) and stored in sterile tubes in the dark at -70°C.

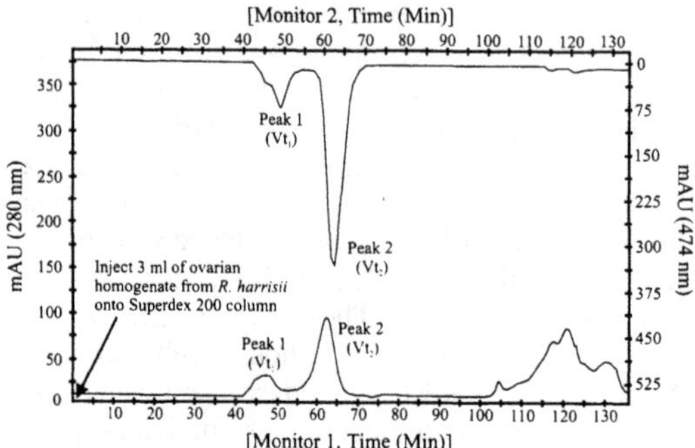

Figure 1. *Gel permeation chromatography of crude ovarian homogenate from **R. harrisii** on a HiLoad 16/60 Superdex 200 HR column with duel monitors (280 and 474 nm) in series. Eluent buffer: 0.02 M Tris-HCl, 5 mM EDTA, pH 7.6; flow rate 1 ml/min.*

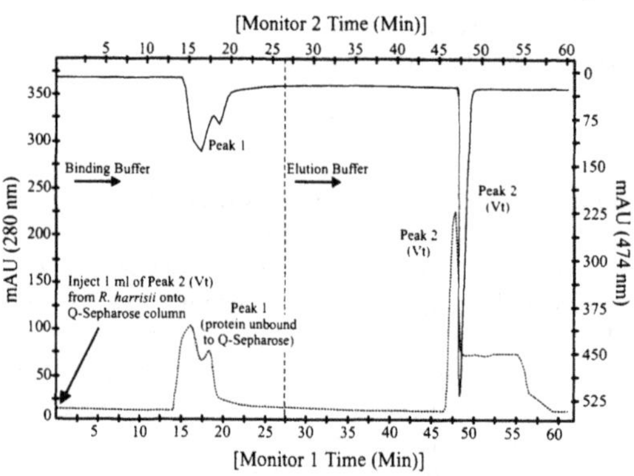

Figure 2. *Anion-exchange chromatography on HiLoad 16/10 Q Sepharose High Performance column: elution profile of ovarian peak 2 from **R. harrisii**. Eluates were monitored at 280 nm (monitor 1, dotted line) and 474 nm (monitor 2, continuous line). Binding buffer: 0.02 M Tris-HCl, 5 mM EDTA, pH 7.6; Elution buffer: 0.13 M NaCl, 0.02 M Tris-HCl, 5 mM EDTA, pH 7.6; flow rate 1 ml/min.*

Electrophoresis and Western blotting

Under native electrophoretic conditions, the six Vt solutions obtained after the two-step chromatography appeared as single protein bands ranging from 200-500 kDa in molecular weight. These bands were not seen in male hemolymph or homogenates used as negative controls for Vt (Fig.3). After freezing the samples, two to five protein bands per sample could be seen on native gels, indicating breakdown of the parent protein (not shown). Furthermore, GlycoPro and Sudan Black B staining indicated that all of the Vt proteins and their respective subunits are glycolipoproteins. Electrophoresis of purified Vt by SDS-PAGE under reducing and denaturing conditions produced the breakdown of the native forms to 6-12 subunits depending on the species (Fig.3). Western blots of reduced and denatured SDS-PAGE gels produced 2 to 6 bands, generally corresponding to the most concentrated of the protein subunits (Fig.3).

IV. DISCUSSION

This study provides the first report of polyclonal antisera produced against Vt of *Americamysis* (*Mysidopsis*) *bahia, Lepidophthalmus louisianensis, Palaemonetes pugio, Procambarus clarkii, Rhithropanopeus harrisii,* and *Uca panacea.* These particular genera were selected either for their importance in estuarine food webs of the Gulf of Mexico (Felder and Griffis 1994) or because they are easily reared in controlled laboratory settings and much of their physiology has been previously reported (Celestial and McKenney Jr. 1994, Eastman-Reks and Fingerman 1985, McKenney 1996, McKenney and Celestial 1993, Quackenbush and Keeley 1988).

Interests in Vt and its precursor, Vg, have ranged widely from carotenoid pigment content and characterization to lipid class composition, but most interests relate directly to the maternal transfer of necessary energy and resources to offspring. So it may not be surprising that isolation of Vt or Vg has been performed for many crustacean species (Table 1). The methods used for the isolation of Vt vary tremendously from ultracentrifugation to high performance liquid chromatography. The method described here has been adopted since it is economical, takes only one day, and allows purification of large volumes of crude homogenates.

Recently, molecular approaches to invertebrate reproductive endocrinology have utilized antisera produced against Vt in order to more fully elucidate the mechanisms of control and synthesis of egg yolk production. We plan to use these antisera as tools for immunocytochemical investigation of ovary and extra-ovarian tissues, which may be responsible for Vg synthesis. Although not reported here, enzyme-linked immunosorbent assays have been developed for these antisera and will be valuable in determining Vg concentrations in hemolymph

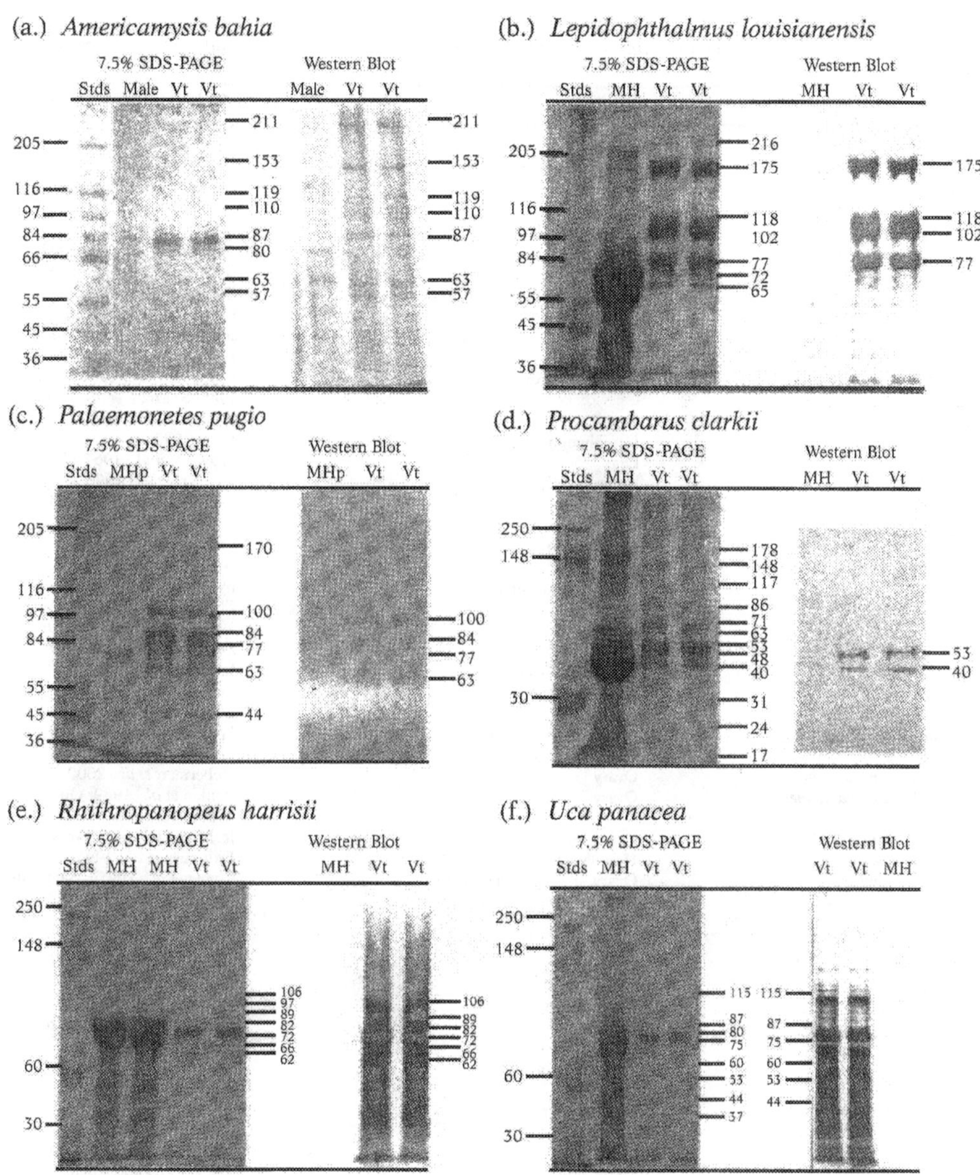

Figure 3. *Sodium dodecyl sulfate - polyacrylamide gel electrophoreses (SDS-PAGE) and corresponding Western blotting of vitellins purified by two-step chromatography and negative controls from six species of crustaceans. (a.) Americamysis bahia; (b.) Lepidophthalmus louisianensis; (c.) Palaemonetes pugio; (d.) Procambarus clarkii; (e.) Rhithropanopeus harrisii; (f.) Uca panacea. Each vitellin lane was loaded with 5 mg of protein, while lanes with male controls were loaded with between 5 and 20 mg. Gels were stained with Coomassie Blue R250, blots were probed with species specific polyclonal antisera and localized with 4-chloro-1-naphthol. Male; whole homogenized male; MHp, male hepatopancreas; MH, male hemolymph; Stds, molecular weight standards in kilodaltons (kDa); Vt, vitellin.*

Table 1. *Crustacean species from which vitellin, vitellogenin, or lipovitellin has been isolated and partially characterized.*

Species	Tissue	References
Acanthephyra sp.	Eggs	Herring and Morris, 1975
Americamysis (Mysidopsis) bahia	Whole female	Present study
Artemia salina	Yolk granules	de Chaffoy de Courcelles and Kondo, 1980
Astacus leptodactylus	Ovary, Hemolymph	Durliat, 1984
	Hemolymph, hepatopancreas	Durliat and Vranckx, 1982
Branchinecta packardi	Yolk platelets of eggs	Gilchrist and Zagalsky, 1983
Branchipus stagnalis	Yolk platelets of eggs	Zagalsky *et al.*, 1983
Callinectes sapidus	Hemolymph, ovary	Horn and Kerr, 1969; Kerr, 1969; Lee and Puppione, 1988; Lee and Walker, 1995
Cancer antennarius	Hemolymph, ovary	Puppione *et al.*, 1986; Spaziani *et al.*, 1986; Spaziani *et al.*, 1995
Cancer irroratus	Eggs	Wallace *et al.*, 1967
Cancer magister	Hemolymph	Allen, 1972
Cancer pagurus	Eggs	Zagalsky *et al.*, 1967
Cherax quadricarinatus	Hemolymph	Yehezkel *et al.*, 1998
Daphnia magna	Whole female	Oberdörster *et al.*, 2000b
Fenneropenaeus (Penaeus) chinensis	Ovary	Chang and Shih, 1995
	Hemolymph	Chang *et al.*, 1996
Homarus americanus	Eggs	Wallace *et al.*, 1967
	Ovary, hemolymph	Byard and Aiken, 1984
Homarus gammarus	Eggs	Cheesman *et al.*, 1967; Zagalsky *et al.*, 1967; Renstrom *et al.*, 1982; Zagalsky, 1985
Idothea granulosa	Eggs, hemolymph	Lee, 1966b
Idothea montereyensis	Eggs, hemolymph	Lee, 1966a
Lepas sp.	Eggs	Cheesman *et al.*, 1967
Lepidophthalmus louisianensis	Eggs	Present study
Libinia emarginata	Eggs	Wallace *et al.*, 1967
Litopenaeus (Penaeus) schmitti	Ovary	Huberman *et al.*, 2000
Litopenaeus (Penaeus) vannamei	Ovary	Rankin *et al.*, 1989; Quackenbush, 1989b; Tom *et al.*, 1992
Macrobrachium idella	Ovary, hemolymph	Joshi and Diwan, 1996
Macrobrachium nipponense	Hemolymph	Okumura *et al.*, 1992
Macrobrachium rosenbergii	Ovary, blood, hepatopancreas	Chang *et al.*, 1993b; Sagi *et al.*, 1995; Lee and Chang, 1997
Marsupenaeus (Penaeus) japonicus	Ovary	Vazquez-Boucard *et al.*, 1986
Metapenaeus ensis	Ovary	Qiu *et al.*, 1997
Pachygrapsus crassipes	Ovary	Lui and O'Connor, 1977
Pagurus pollicaris	Ovary	Wallace *et al.*, 1967
Palaemon paucidens	Hemolymph	Nakagawa *et al.*, 1982
Palaemontes pugio	Eggs	Oberdörster *et al.*, 2000b; Present study
Pandalus kessleri	Ovary, hemolymph	Quinitio *et al.*, 1989
Panulirus vulgaris	Eggs, ovary	Zagalsky, 1964
Panulirus interruptis	Hemolymph	Lee and Puppione, 1978
Parapenaeus longirostris	Ovary	Tom *et al.*, 1987
Penaeus monodon	Ovary	Chang *et al.*, 1993a; Chang *et al.*, 1994; Chen and Chen, 1994
	Hemolymph	
Penaeus semisulcatus	Ovary	Tom *et al.*, 1992
Plesionika edwardsi	Eggs	Zagalsky *et al.*, 1967
Procambarus sp.	Hemolymph	Fyffe and O'Connor, 1974
Procambarus clarkii	Eggs, Ovary	Tuberty, 1998
Rhithropanopeus harrisii	Eggs	Oberdörster *et al.*, 2000b, Present study
Sesarma reticulatum	Ovary, Eggs	Wallace *et al.*, 1967
Uca panacea	Eggs	Present study
Uca pugilator	Ovary	Wallace *et al.*, 1967; Eastman-Reks and Fingerman, 1985

and tissue homogenates at different stages of ovarian maturation. Quantitative ELISA could also be used in conjunction with *in vitro* studies for validation of gonadotropic effects of newly isolated hormones (e.g., gonad stimulating hormone, gonad inhibition hormone). Finally, these assays could be adopted for investigating effects of xenobiotics on crustacean vitellogenesis in lab or field studies (Oberdörster et al. 2000a, b).

ACKNOWLEDGEMENTS

Thanks to Dr. Jay Huner, University of Louisiana at Lafayette Crawfish Research Center, for graciously providing *Procambarus clarkii* specimens and to Dr. Peter Schoor, USEPA, for aid and guidance in the chromatographic procedures. The U.S. Environmental Protection Agency, through its Office of Research and Development, partially funded the research described here under cooperative agreement #CT826431 to the University of West Florida. This work has been subjected to the Agency's peer and administrative review and has been approved for publication as an EPA document. The views expressed herein are those of the authors and do not necessarily reflect the views of the supporting agencies, nor does mention of commercial products imply endorsement by the EPA. This is contribution No. 1114 from the Gulf Ecology Division, NHEERL, Gulf Breeze, FL.

REFERENCES

Allen WV (1972) Lipid transport in the dungeness crab, *Cancer magister* Dana. Comp Biochem Physiol B 43:193-207

Bradford MM (1976) A rapid and sensitive method for the quantitation of microgram quantities of protein utilizing the principle of protein-dye binding. Anal Biochem 72:248-254

Brion F, Rogerieux F, Noury P, Migeon B, Flammarion P, Thybaud E and Porcher JM (2000) Two-step purification method of vitellogenin from three teleost fish species: rainbow trout (*Oncorhynchus mykiss*), gudgeon (*Gobio gobio*) and chub (*Leuciscus cephalus*). J Chromatogr 737:3-12

Byard EH and Aiken DE (1984) The relationship between molting, reproduction, and a heamolymph female-specific protein in the lobster, *Homarus americanus*. Comp Biochem Physiol A 77:749-757

Celestial DM and McKenney Jr CL (1994) The influence of an insect growth regulator on the larval development of the mud crab *Rhithropanopeus harrisii*. Environ Pollut 85:169-173

Chang CF and Shih TW (1995) Reproductive cycle of ovarian development and vitellogenin profiles in the freshwater prawns, *Macrobrachium rosenbergii*. Invertebr Reprod Dev 27:11-20

Chang CF, Lee FY and Huang YS (1993a) Purification and characterization of vitellin from the mature ovaries of prawn, *Penaeus monodon*. Comp Biochem Physiol B 105:409-414

Chang CF, Shih TW and Hong HH (1993b) Purification and characterization of vitellin from the mature ovaries of prawn, *Macrobrachium rosenbergii*. Comp Biochem Physiol B 105:609-615

Chang CF, Lee FY, Huang YS and Hong TH (1994) Purification and characterization of the female-specific protein (vitellogenin) in mature female hemolymph of the prawn, *Penaeus monodon*. Invertebr Reprod Dev 25:185-192.

Chang CF, Jeng SR, Lin MN and Tin YY (1996) Purification and characterization of vitellin from the mature ovaries of prawn, *Penaeus chinensis*. Invertebr Reprod Dev 29:87-93

Cheesman DF, Lee WL and Zagalsky PF (1967) Carotenoproteins in invertebrates. Biol Rev 42:132-160

Chen CC and Chen SN (1994) Vitellogenesis in the giant tiger prawn, *Penaeus monodon* Fabricius, 1789. Comp Biochem Physiol B 107:453-460

de Chaffoy de Courcelles D and Kondo M (1980) Lipovitellin from the crustacean, *Artemia salina*. Biochemical analysis of lipovitellin complex from the yolk granules. J Biol Chem 255:6727-6733

Durliat M (1984) Occurrence of plasma proteins in ovary and egg extracts from *Astacus ptodactylus*. Comp Biochem Physiol B 78:745-753

Durliat M and Vranckx R (1982) Proteins of aqueous extracts from the hepatopancreas of *Astacus leptodactylus*- 2. Immunological identities of proteins from hepatopancreas and blood. Comp Biochem Physiol B 71:165-171

Eastman-Reks SB and Fingerman M (1985) *In vitro* synthesis of vitellin by the ovary of the fiddler crab, *Uca pugilator*. J Exp Zool 233:111-116

Felder DL and Griffis RB (1994) Dominant infaunal communities at risk in shoreline habitats: Burrowing thalassinid Crustacea. OCS Study # MMS 94-0007, U.S. Dept. of the Interior, Minerals Mgmt.

Service, Gulf of Mexico OCS Regional Office, New Orleans

Fyffe WE and O'Connor JD (1974) Characterization and quantification of a crustacean lipovitellin. Comp Biochem Physiol B 47: 851-867

Gilchrist BM and Zagalsky PF (1983) Isolation of a blue canthaxanthin-protein from connective tissue storage cells in *Branchinecta packardi* Pearse (Crustacea: Anostraca) and its possible role in vitellogenesis. Comp Biochem Physiol B 76:885-893

Herring PJ and Morris RJ (1975) Embryonic metabolism of carotenoid pigments and lipid in species of *Acanthephyra* (Crustacea: Decapoda). In: Barnes H (ed) Proceedings of the 9th European Marine Biology Symposium (pp 299-310) Aberdeen University Press, Aberdeen

Horn EC and Kerr MS (1969) The hemolymph proteins of the blue crab, *Callinectes sapidus* – I. Hemocyanins and certain other major protein constituents. Comp Biochem Physiol 29:493-508

Huberman A, Ramos L, and Mitre IB (2000) Ovarian lipovitellin from the Caribbean shrimp, *Penaeus (Litopenaeus) schmitti*. In: The Crustacean Society 2000 Summer Meeting Program, Puerto Vallarta, Mexico, p 29

Joshi VP and Diwan AD (1996) Biochemical changes in the tissues of female prawn *Macrobrachium idella* (Hilgendorf, 1898) during different breeding seasons. J Aquacult Trop 11:227-251

Kerr MS (1969) The hemolymph proteins of the blue crab, *Callinectes sapidus*. II. A lipoprotein serologically identical to oocyte lipovitellin. Dev Biol 20:1-17

Laemmli UK (1970) Cleavage of structural protein during the assembly of the bacteriophage T4. Nature 227:680-685

Lee FY and Chang CF (1997) The concentrations of vitellogenin (vitellin) and protein in hemolymph, ovary and hepatopancreas in different ovarian stages of the freshwater prawn, *Macrobrachium rosenbergii*. Comp Biochem Physiol A 117:433-439

Lee RF (1991) Lipoproteins from the hemolymph and ovaries of marine invertebrates In: Gilles R (ed) Advances in Comparative and Environmental Physiology (pp 187-207) Springer-Verlag, Berlin

Lee RF and Puppione DL (1978) Serum lipoproteins in the spiny lobster, *Panulirus Interuptus*. Comp Biochem Physiol 59:239-243

Lee RF and Puppione DL (1988) Lipoproteins I and II from the hemolymph of the blue crab *Callinectes sapidus*: Lipoprotein II associated with vitellogenesis. J Exp Zool 248:278-289

Lee RF and Walker A (1995) Lipovitellin and lipid droplet accumulation in oocytes during ovarian maturation in the blue crab, *Callinectes sapidus*. J Exp Zool 271:401-412

Lee WL (1966a) Pigmentation of the marine isopod *Idothea montereyensis* (Maloney). Comp Biochem Physiol 18:17-36

Lee WL (1966b) Pigmentation of the marine isopod *Idothea granulosa* (Rathke). Comp Biochem Physiol 19:13-29

Lui CW and O'Connor JD (1977) Biosynthesis of crustacean lipovitellin-III. The incorporation of labelled amino acids into the purified lipovitellin of the crab *Pachygrapsus crassipes*. J Exp Zool 199:105-108

Manning RB (1975) Two methods for collecting crustaceans in shallow water. Crustaceana 29:317-319

McKenney Jr CL (1996) The combined effects of salinity and temperature on various aspects of the reproductive biology of the estuarine mysid, *Mysidopsis bahia*. Invertebr Reprod Dev 29:9-18

McKenney Jr CL and Celestial DM (1993) Variations in larval growth and metabolism of an estuarine shrimp *Palaemonetes pugio* during toxicosis by an insect growth regulator. Comp Biochem Physiol C 105:239-245

Nakagawa H, Salam A and Kasahara S (1982) Female-specific lipoprotein level in hemolymph during egg formation in freshwater shrimp. Bull Jap Soc Sci Fish 48:1073-1080

Oberdörster E, Brouwer M, Hoexum-Brouwer T, Manning S and McLachlan JA (2000a) Long-term pyrene exposure of grass shrimp, *Palaemonetes pugio*, affects molting and reproduction of exposed males and offspring of exposed females. Environ Health Perspect 108:641-646

Oberdörster E, Rice CD and Irwin LK (2000b) Purification of vitellin from grass shrimp *Palaemonetes pugio*, generation of monoclonal antibodies and validation for the detection of lipovitellin in Crustacea. Comp Biochem Physiol, In Press.

Okumura T, Han CH, Suzuki Y, Aida K and Hanyu I (1992) Changes in hemolymph vitellogenin and ecdysteroid levels during the reproductive and non-reproductive molt cycles in the freshwater prawn *Macrobrachium nipponense*. Zool Sci 9:37-45

Puppione DL, Jensen DF and O'Conner JD (1986) Physiochemical study of rock crab lipoproteins. Biochem Biophys Acta 875:563-568

Qiu YW, Ng TB and Chu KH (1997) Purification and characterization of vitellin from the ovaries of the shrimp *Metapenaeus ensis* (Crustacea: Decapoda: Penaeidae). Invertebr Reprod Dev 31:217-223

Quackenbush LS (1989b) Yolk protein production in the marine shrimp *Penaeus vannamei*. J Crust Biol 9:509-516

Quackenbush LS and Keeley LL (1988) Regulation of vitellogenesis in the fiddler crab, *Uca pugilator*. Biol Bull 175:321-331

Quinitio ET, Hara A, Yamauchi K, Mizushima T and Fuji A (1989) Identification and characterization of vitellin in a hermaphrodite shrimp, *Pandalus kessleri*. Comp Biochem Physiol B 94:445-451

Rankin S, Bradfield J and Keeley LL (1989) Ovarian protein synthesis in the South American white shrimp, *Penaeus vannamei*, during the reproductive cycle. Invertebr Reprod Dev 15:27-33

Renstrom B, Ronneberg H, Borch G and Liaaen-Jensen S (1982) Animal carotenoids 27. Further studies on the carotenoproteins crustacyanin and ovoverdin. Comp Biochem Physiol B 71:249-252

Sagi A, Soroka Y, Snir E, Chomsky O, Calderon J and Milner Y (1995) Ovarian protein synthesis in the prawn *Macrobrachium rosenbergii*: Does ovarian vitellin synthesis exist? Invertebr Reprod Dev 27:41-47

Spaziani E, Havel RJ, Hamilton RL, Hardman DA, Stoudemire JB and Watson RO (1986) Properties of serum high-density lipoproteins in the crab, *Cancer antennarius* Stimpson. Comp Biochem Physiol B 85:307-314

Spaziani E, Wang WL and Novy LA (1995) Serum high-density lipoproteins in the crab *Cancer antennarius* - IV. Electrophoretic and immunological analyses of apolipoproteins and a question of female-specific lipoproteins. Comp Biochem Physiol B 111:265-276

Tom M, Fingerman M, Hayes TK, Johnson V, Kerner B and Lubzens E (1992) A comparative study of the ovarian proteins from the penaeid shrimps, *Penaeus semisulcatus* (De Haan) and *Penaeus vannamei* (Boone). Comp Biochem Physiol B 102:483-490

Tom M, Goren M and Ovadia M (1987) Purification and partial characterization of vitellin from the ovaries of *Parapenaeus longirostris* (Crustacea, Decapoda, Penaeidae). Comp Biochem Physiol B 87:17-23

Towbin H, Staehelin T and Gordon J (1979) Electrophoretic transfer of proteins from polyacrylamide gels to nitrocellulose sheets: Procedure and some applications. Proc Natl Acad Sci USA 76:4350-4354

Tuberty SR (1998) Vitellogenesis in the red swamp crayfish, *Procambarus clarkii*. Department of Ecology and Evolutionary Biology, Tulane University, New Orleans, Louisiana, USA. PhD thesis

Vazquez-Boucard C, Ceccaldi HJ, Benyamin Y and Roustan C (1986) Identification, purification, and characterization of the lipovitellin from a natantian decapod crustacean, *Penaeus japonicus* (Bate). Int J Invertebr Repr Dev 12:227-240

Wallace RA, Walker SL and Hauschka PV (1967) Crustacean lipovitellin. Isolation and characterization of the major high-density lipoprotein from the eggs of decapods. Biochemistry 6:1582-1590

Yehezkel G, Khalaila I, Abdu U and Soreanu S (1998) A secondary-vitellogenic specific lipoprotein from the hemolymph of *Cherax quadricarinatus*. Society for Integrative and Comparative Biology Annual Meeting, Denver, Colorado, USA, 89A

Zagalsky PF (1964) The association of carotenoids with proteins in certain invertebrates. University of London, London, England. PhD thesis

Zagalsky PF (1985) A study of the astaxanthin-lipovitellin, ovoverdin, isolated from the ovaries of the lobster, *Homarus gammarus* (L.). Comp Biochem Physiol B 80:589-597

Zagalsky PF, Cheesman DF and Ceccaldi HJ (1967) Studies on carotenoid-containing lipoproteins isolated from the eggs and ovaries of certain marine invertebrates. Comp Biochem Physiol 22:851-871

Zagalsky PF, Gilchrist BM, Clark RJH and Fairclough DP (1983) The canthaxanthin-lipovitellin of *Branchipus stagnalis* (L.) (Crustacea: Anostraca): a resonance Raman circular dichroism study. Comp Biochem Physiol B 75:163-167

CYTOARCHITECTURE OF THE HEPATOPANCREAS OF THREE SPECIES OF CRABS FROM MAR CHIQUITA LAGOON, ARGENTINA

Elena Cuartas and Ana María Petriella

*Facultad de Ciencias Exactas y Naturales, Universidad Nacional de Mar del Plata,
Funes 3350, B7602AYL, Mar del Plata, Argentina
Consejo Nacional de Investigaciones Científicas y Técnicas*

ecuartas@mdp.edu.ar

ABSTRACT

Based on the various functions of the crustacean hepatopancreas, this work analyzes the possible structural differences in brachyurans from different microhabitats in the Mar Chiquita lagoon "cangrejal" (37° 45' S, 57° 26' W). Adult males in intermolt (Stage C) of *Uca uruguayensis, Chasmagnathus granulata* and *Cyrtograpsus angulatus* were collected at the same time. The hepatopancreas was removed and processed by using standard histological techniques. The organ completely occupies the dorsal hemocoel in *C. granulata* and *C. angulatus*, while it is arranged in two dorsal, lateral masses in *U. uruguayensis*. The cytoarchitecture of the three species is similar to that observed in other brachyurans. Differences in the tubular structure are related to the presence of a peritrophic lamina in all its extension and a marked differentiation at the medial zone in *C. angulatus*. The tubular lumen in *C. angulatus* and *U. uruguayensis* is cross-shaped, though very branched in *C. granulata*. These characteristics indicate a different rhythm in the cycle of cellular activity. The tubular and cell dimensions are similar for the three species; it is outstanding the augmented intertubular tissue in *U. uruguayensis*.

I. INTRODUCTION

The midgut gland (hepatopancreas) is responsible for the synthesis and secretion of digestive enzymes, the final digestion of food and the absorption of nutrients (Al-Mohanna and Nott 1989, Vogt et al. 1989, Ceccaldi 1997). It is also involved in other processes, such as maintenance of salt and ion balance and vitellogenesis. It plays an immunological role in the removal of foreign bodies from the blood system, excretion of waste metabolites from the blood system, and detoxification of metals from foreign organic substances (Hryniewiecka-Szyfter and Babula 1997, Vogt 1987, Roldán and Shivers 1987).

Because of the importance of the hepatopancreas in the general body economy, physiological abnormalities caused by diseases and environmental changes are often visualized in the histological structure of this organ and its interstitial tissues (Vogt 1992).

Mar Chiquita lagoon (37° 45' S, 57° 26' W) lies parallel to the shore, and it is characterized by a wide northern part with freshwater contribution and a narrow southern part (the mouth) opening to the sea. It is inhabited by three species of brachyurans that belong to a community called "cangrejal" (Boschi 1964).

Uca uruguayensis is a semiterrestrial species found in muddy substrates (Boschi 1992), in the upper tidal mark. The water salinity and type of substrate are factors that influence the distribution of these semiterrestrial brachyurans (Boschi 1964). Crabs of the genus *Uca* feed on organic matter in the sediments of the estuaries and beaches where they live (Grahame 1983).

Kluwer Academic/Plenum Publishers

E. Escobar-Briones & F. Alvarez Eds.
MODERN APPROACHES TO THE STUDY OF CRUSTACEA
PP. 039-044

Chasmagnathus granulata inhabits caves in a zone that goes from the middle tidal to the upper tidal line. It burrows into flats and banks in sandy or muddy beaches (Boschi 1964). It also feeds on organic detritus in the sediment.

Cyrtogtrapsus angulatus lives in sandy or muddy beaches, it can be observed swimming in shallow waters or buried in the sediments from oceanic to oligohaline waters (Spivak 1999). *Cyrtogtrapsus angulatus* is mainly carnivorous and can sometimes be cannibalistic.

The inhabitants of the intertidal zone may also be subjected to important osmotic stress when confronted with changes in salinity (Gilles and Pequeux 1983); crustaceans, in particular, have developed a number of adaptive mechanisms to cope with these changing conditions (Vernberg 1983). The three crab species subject of this study, being animals that live in shallow waters and estuaries, should be able to tolerate fluctuations in salinity. In turn, these fluctuations lead to physiological adaptations that should be evident in the histological structure of the hepatopancreas.

Our previous experimental work on penaeids show that the hepatopancreas reacts to the osmotic stress with deep structural modifications that alter its functionality (Cuartas et al. unpublished). Based on the various functions of the crustacean midgut gland, this study describes the structure of this organ in three species of brachyurans from different microhabitats in Mar Chiquita lagoon.

II. MATERIALS AND METHODS

Uca uruguayensis,Chasmagnathus granulata and *Cyrtogtrapsus angulatus* adult males in intermolt (stage C), with 12 h starvation, and 1.0-1.4 cm, 1.5-2.0 cm and 2.0-2.5 cm carapace width, respectively, were collected by hand simultaneously. Samplings were done in November-December 1999; the salinity varied between 34.5 and 35 ups and water temperature was 22±2°C.

For the macroscopic observation of the hepatopancreas, the thoracic zone of 7-10 individuals of each species was fixed for 24 h in Davidson's fluid (ethanol, formol, acetic acid and water) (Bell and Lightner 1988).

The digestive gland was removed under a dissecting microscope. For the histological description, the hepatopancreas was dehydrated in increasing concentrations of ethanol, butyl alcohol (two changes of

24 hours), butyl-paraffin 50:50 (for 24 hours) and finally embedded in paraffin. Sections (3 μm) were stained with hematoxylin-eosin, Mallory's triple stain and periodic acid-Schiff (PAS) method.

Twenty-five epithelial cell heights of each cell type (F, B and R-cells) and 15 tubule diameters at the apical and medial zones were measured in five individuals. Measurements were made under a light microscope furnished with a micrometer eyepiece.

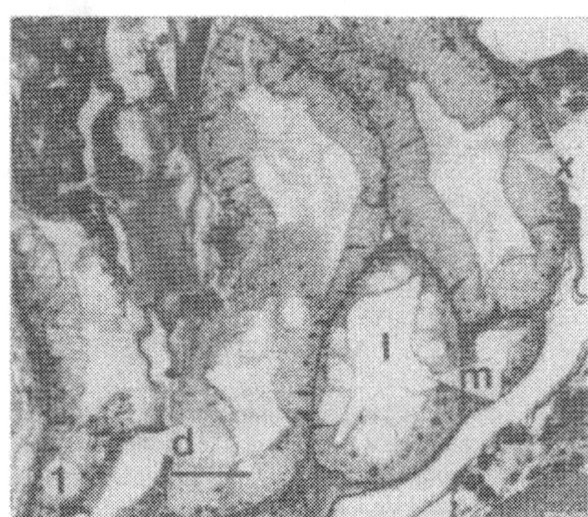

Figure 1. *Uca uruguayensis. Section of hepatopancreas, general view. H and E. d: distal zone. l: lumen. m: midtubulart zone. x: proximal zone. Scale bar= 50*μ*m.*

III. RESULTS

The hepatopancreas size and its position varies according to the species. In *C. granulata* it occupies the full dorsal hemocoel, lateral to the digestive tract and heart, and partially covering the branchial chambers; in *C. angulatus* it is placed at both sides of the digestive tract in the anterior portion of the hemocoel, occupying two thirds of it; and in *U. uruguayensis* it appears like two small dorsal lateral masses away from the branchial chambers.

The cytoarchitecture in the three species is similar to that observed in other brachyurans: the gland is composed of many blind tubules, each tubule is lined with a cubic or columnar epithelium and surrounded by connective tissue. The epithelial cells are differentiated into four types: E (embryonic), R (resorptive), B (blisterlike) and F (fibrillar). Their distribution along

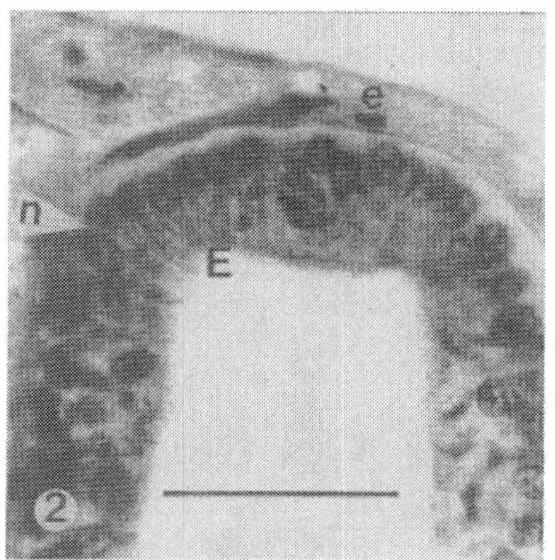

Figure 2. *Uca uruguayensis*. Distal tips of tubule. H and E. e: *mioepithelial nucleus. E: E cell. n: mitotic nucleus. Scale bar= 50μm.*

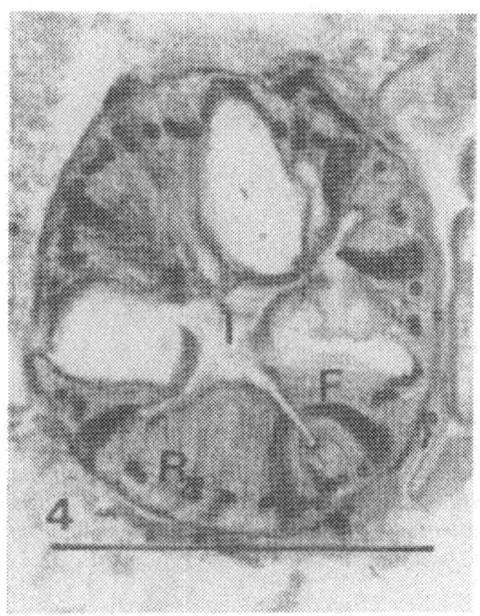

Figure 4. *Uca uruguayensis*. Cross section of tubule. H and E. R_2: binuclear R cell. Scale bar=100μm.

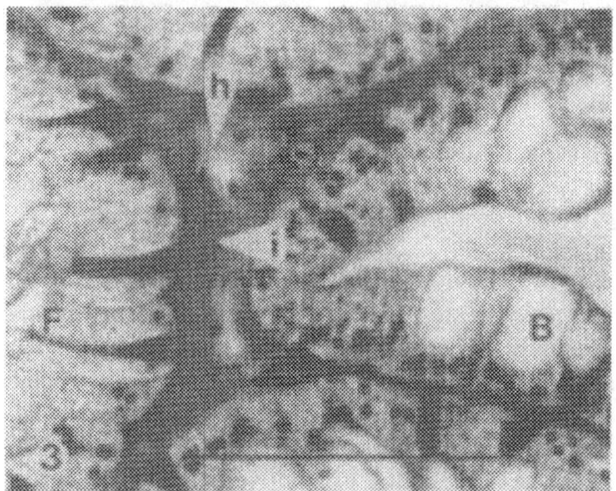

Figure 3. *Uca uruguayensis*. Cross section of medial zone tubles (rigth), proximal zone tubules (left) and intertubular space. H and E. B: B cell. h: hemal space. i: intertubular area. F: F cell. Scale bar=50 μm.

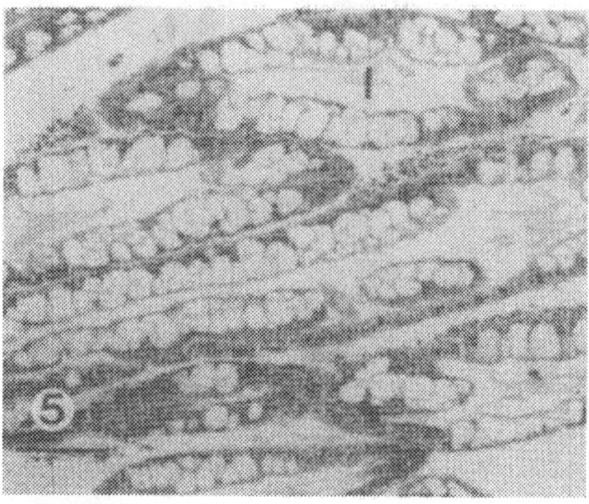

Figure 5. *Cyrtograpsus angulatus*. Longitudinal section of medial zone. H and E. Scale bar=50μm.

the tubules was originally thought to be E-cells in the distal tips, which often present mitotic figures, young R and F-cells in the adjacent cell differentiation zone, mature R, F and B-cells in the midtubular region and R and F-cells in the proximal region (Figs.1, 2). The entire tubule epithelium has a microvillous border, difficult to observe in the apical zone, although evident

with the PAS method. Although tubular diameter among the three species is variable, cell size is similar in *C. angulatus* and *U. uruguayensis*, while it is larger in *C. granulata* (Table 1).

The *U. uruguayensis* cell distribution along the tubule is in accordance to the general pattern described. The apical zone showing only E cells gradually turns

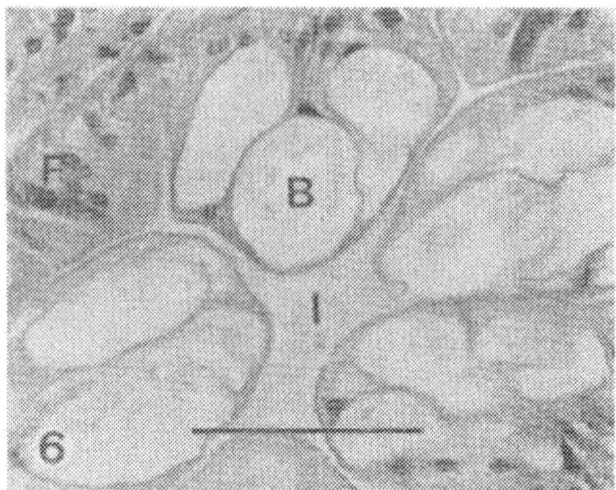

Figure 6. *Cyrtograpsus angulatus*. *Cross section of medial zone (assembled in cluster of B cell). H and E. .l: lumen. Scale bar=50μm.*

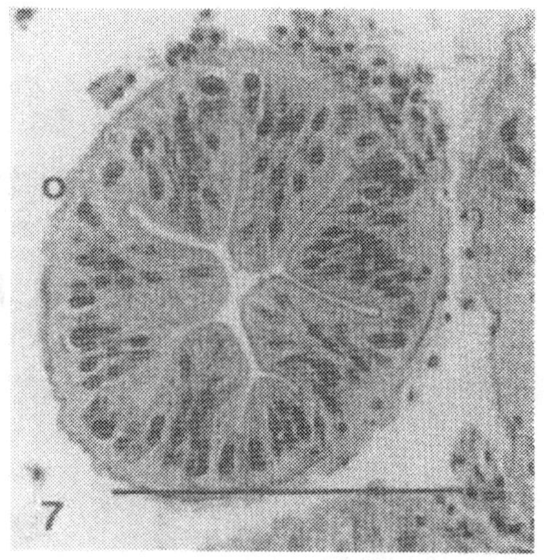

Figures 7. *Chasmagnathus granulata*. *7: Cross section of proximal area. H and E. o: adherent connective. Scale bar=100 μm.*

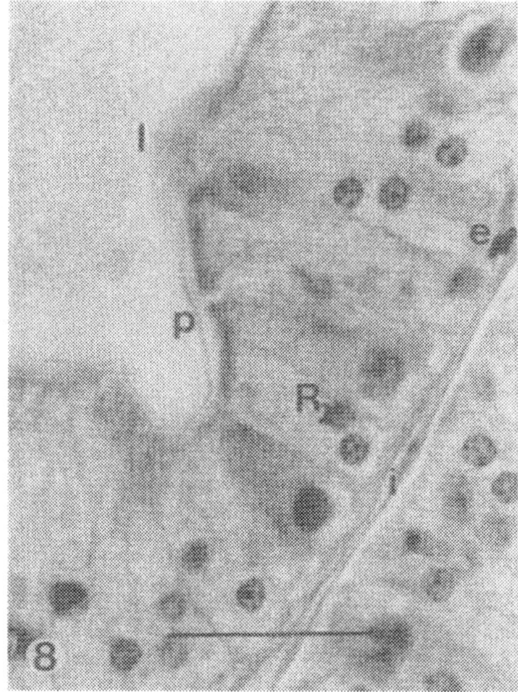

Figure 8. *Chasmagnathus granulata*. *Cross section showing thin intertubular area. H and E. e: mioepithelial nucleous. p: laminar peritrophic membrane. Scale bar= 50μm.*

into an intermediate zone where R and F cells alternate; towards the medial zone the B cells alternate among the R and F cells or differentiate into compact groups in some tubules (Figs. 3, 4). Even though the other two species have a spatial arrangement similar to *U. uruguayensis*, the transition between the apical and middle zones is not so gradual. The sharpest difference is seen in *C. angulatus* where the medial zone of the tubules is almost exclusively organized in B- cell (Figs. 5, 6).

The tubular luminae of *C. angulatus* and *U. uruguayensis* are cross-shaped, while in *C.granulata* are greatly branched (Fig. 7). Differences in the tubular structure are related to the presence of a peritrophic membrane all along the tubule: lamellar in *U. uruguayensis* and *C.granulata,* and fibrilar in the medial zone of the tubule in *C. angulatus* (Figs. 8, 9, 10).

Bands of longitudinal muscles and bundles of circular muscles are seen beneath the basal lamella. Capillaries and hemocytes occur in the hemal spaces between tubules; the hemal spaces are more evident and easier to visualize in *U. uruguayensis*.

IV. DISCUSSION

The smaller size of the *U. uruguayensis* hepatopancreas can be related to the semiterrestrial way of life of the species. Among the malacostracans, the terrestrial isopods are the other group where a sharp de-

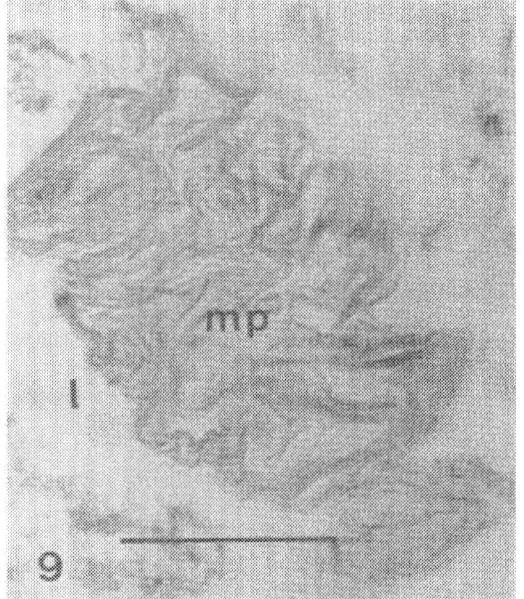

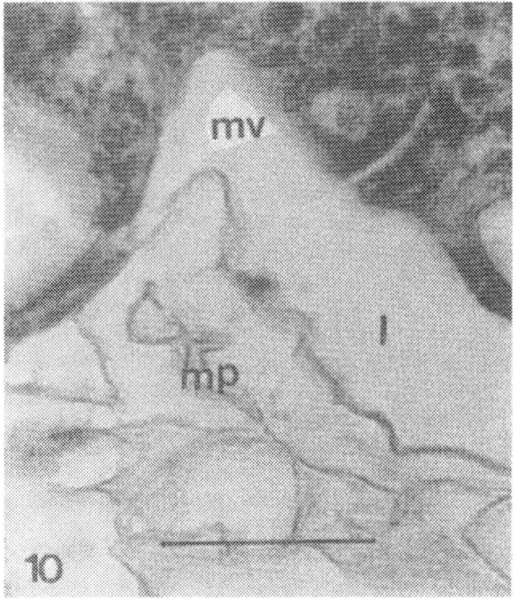

Figures 9 and 10. *C. angulatus* *multilayered peritrophic membrane. H and E. mp: multilayered peritrophic membrane. mv: microvillous border. Scale bar= 50μm.*

Table 1. *Cell and tubular dimensions*

	Tubular diameter (m)	Cellular height (m)			
		E	R	F	B
C. angulatus	204±63	25.9±6	52.2±7	41.4±6	62.9±9
U. uruguayensis	183±44	21.4±4.	59.7±12	46.6±9	68.5±10.6
C. granulata	216±84	41.2±2	92.4±3	72.6±2	75.3±2

crease in the number of tubules and cellular diversity are observed (Storch 1984).

Our observations indicate that these brachyurans possess a "frayed" looking midgut gland due to the delicacy of the connective tissue surrounding the tubules and to the lack of a limiting external wall of that same tissue. In general terms, this structure is not strictly the one described for the majority of decapods, particularly the genus *Homarus* (Farkas 1995).

Of the three studied species, the morphology of *C. granulata* is the one closest to *Callinectes sapidus* (Johnson 1980) with regard to the latter's generalized model. The tubules of some decapod hepatopancreas are zoned so that different kinds of cells are prominent at different levels in the tubule (Gibson and Barker 1979). In *C. granulata* and *U. uruguayensis* B cells mixed with R and F cells are observed in the middle zone just as Hopkin and Nott (1980) described for *Carcinus maenas.* In *C. angulatus* B cells are assembled into clusters in the same zone. Binuclear R and B cells are found in *U. uruguayensis* and in *C. angulatus* that are similar in appearance to those described by Johnson (1980) for *Callinectes sapidus.* The diverse forms of the tubular lumen suggest a differential rhythm in cellular activity and a different structure or spatial distribution of the three cellular types, specifically the size of the F cells.

The peritrophic membrane is commonly found in Crustacea; it is secreted by gut cells and serves to enclose material in the lumen, producing a membranous barrier between the enclosed material and the gut epithelium. Hopkin and Nott (1980) showed that a peritrophic membrane is also secreted within the tubules of the digestive gland of *Carcinus maenas*. The peritrophic membrane can be easily recognized in the three species studied herein, although with different morphological characteristics: it generally appears laminar in *C. granulata* and usually laminar and multilayered (cellophane-like) in the other two species.

In *C. angulatus* and *C. granulata* the interstitial tissue is a thin layer of adherent connective tissue which covers the tubules externally. However, in *U. uruguayensis,* the blood vessels and their associated tissues are suspended in a more abundant interstitial substance that could be the product of the outer connective tissue lamella that infiltrate into the tubular spaces. This image belongs to the inflammatory response type and it would be necessary to clear this point up with further studies.

It would be also necessary to initiate an experimental study in order to determine whether some of the observed differences have their cause on the various salinities to which these organism are subjected to.

ACKNOWLEDGEMENTS

Financial support was provided by a University of Mar del Plata Grant (EXA: 109/97).

REFERENCES

Al-Mohanna SY and Nott JA (1989) Functional cytology of the hepatopancreas of *Penaeus semisulcatus* (Crustacea: Decapoda) during the moult cycle. Mar Biol 101:535-544

Bell TA and Lightner DV (1988) A handbook of normal Penaeid shrimp histology. World Aq Soc Allen Press, Inc., USA

Boschi EE (1964) Los crustáceos decápodos Brachyura del litoral bonaerense (R. Argentina). Bol Inst Biol Mar, Mar del Plata 6:1-99

Boschi EE, Fischbach CE and Iorio MI (1992) Catálogo ilustrado de los Crustáceos Estomatópodos y Decápodos marinos de Argentina. Frente Marítimo 10, Sección A:7-94

Ceccalddi HJ (1997) Anatomy and Physiology of the Digestive System. Crustacean Nutrition. (pp 261-290) In: D' Abramo LR, Conklin DE and Akiyama DM (eds) Advances in World Aquaculture. World Aqu Soc, Baton Rouge, USA

Farkas RF (1995) The Digestive System. (pp 395-440) In: Farkas JR (ed) Biology of the Lobster *Homarus americanus* Academic Press, New York

Gibson O and Barker PL (1979) The decapod hepatopancreas. Oceanogr Mar Biol Annu Rev 17:285-346

Gilles R and Pequeux A (1983) Interactions of chemical and osmotic regulation with the environment. (pp 109-165) In: Bliss DE (ed) The Biology of Crustacea Vol. 8, Academic Press, New York

Grahame J (1983) Adaptive aspects of feeding mechanism. (pp 65-101) In: Bliss DE (ed) The Biology of Crustacea Vol. 8, Academic Press, New York

Hyrniewiecka-Szyfter Z and Babula A (1997) Ultrastructural changes in the hepatopancreas cells of *Saduria entomon* (Linnaeus, 1785) (Isopoda, Valvifera) from the Baltic infected with *Cryptococco laurentii* (Kufferath) Skinner. Crustaceana 70:822-830

Hopkin SP and Nott JA (1980) Studies on the digestive cycle of the shore crab *Carcinus maenas* (L) with special reference to the B-cells in the hepatopancreas. J Mar Biol Ass UK 60: 891-907

Johnson PT (1980) Histology of the Blue Crab *Callinectes sapidus*. A Model for the Decapoda. Praeger Publishers, New York

Roldan BM and Shivers RR (1987) The uptake and storage of iron lead in cells of the crayfish (*Orconectes propinquus*) hepatopancreas and antennal gland. Comp Biochem Physiol 86C:201-214

Spivak ED (1999) Effects of reduced salinity on juvenile growth of two co-occurring congeneric grapsid crabs. Mar Biol 134:249-257

Spivak E, Anger K, Luppi T, Bas C and Ismael D (1994) Distribution and habitat preferences of two grapsid crab species in Mar Chiquita Lagoon (Province of Buenos Aires, Argentina) Helgol Meeresunters 48:59-78

Storch V (1984) The influence of nutritional stress on the ultrastructure of the hepatopancreas of terrestrial isopods. Symp Zool Soc London 53:167-184

Vernberg FJ (1983) Respiratory Adaptations. (pp 1-42) In: Bliss DE (ed) The Biology of Crustacea Vol. 8, Academic Press, New York

Vogt G (1987) Monitoring of environmental pollutants such as pesticides in prawn aquaculture by histological diagnosis. Aquaculture 67:157-164

Vogt G (1992) Transformation of anterior midgut and hepatopancreas cells by monodon baculovirus (MBV) in *Penaeus monodon* postlarvae. Aquaculture 107:239-248

Vogt G, Stöcker W, Storch V and Zwilling R (1989) Biosynthesis of *Astacus* protease, a digestive enzyme from crayfish. Histochemistry 91:373-381

ACCLIMATION OF ADULT MALES OF *LITOPENAEUS SETIFERUS* EXPOSED AT 27 °C AND 31 °C: BIOENERGETIC BALANCE

Ariadna Sánchez, Cristina Pascual, Adolfo Sánchez, Francisco Vargas-Albores, Gilles LeMoullac and Carlos Rosas

AS, CP, AS, CR, Grupo de Biología Marina Experimental, Facultad de Ciencias, UNAM, Apdo. Post. 69, Cd. del Carmen, Campeche, México
FVA, Centro de Investigación en Alimentación y Desarrollo, A.C.
Apartado post. 1735, Hermosillo, Sonora 83000, México
GL, Ifremer, Aquacop, Centre Océnologique du Pacifique, BP 7004, Taravao, Tahiti, Polinésie Française

ABSTRACT

The reduction of reproductive performance of adult males of *Litopenaeus setiferus* in captivity has been limiting the massive nauplii production, mainly by lowering the spermatophore attachment success associated with the male reproductive tract degenerative and the male reproductive melanization syndromes. Both syndromes had been related to the captivity and management stress. In this study the bioenergetic alterations were measured after seven days in captivity through absorption efficiency (AE), absorption (A), routine respiratory rate (R_{ROUT}), apparent heat increment (R_{AHI}), ammonia excretion (U) and post-prandial nitrogen excretion ($PPNE$), as indexes of captivity stress in adult males of *L. setiferus* kept at 27 and 31 °C. All this parameters were integrated through the production equation (P_T) = A - (R_{AHI} + R_{ROUT} + U). At 31 °C, the equation values were higher than those observed at 27 °C, except for R_{ROUT} and ammonia excretion. The amount of energy directed to P_T at 31 °C was 1784.99 J g^{-1} afdw d^{-1}, in contrast with shrimp exposed at 27 °C, where it was 1269.22 J g^{-1} afdw d^{-1} ($P > 0.05$). The O:N ratio obtained was lower than 10, indicating the use of a proteic substrate ($P > 0.05$). The reduction of the metabolic responses at high temperature shows what the adaptation capability of this shrimp is to tolerate a wide range of temperatures, also reflecting the adaptation mechanisms associated with its distribution in shallow coastal waters.

I. INTRODUCTION

The development of the culture techniques for *L. setiferus* is increasing because it has high survival (95-99%), attractive production rates of more than 5000 kg/ha, final weights of 13.5 g/shrimp, and can be cultured in relatively high densities of 40 shrimps/m^2 (Sandifer et al. 1993). This species can be cultured for bait (Samocha et al. 1998) or to grow to 12 - 14 g when shrimps are stocked between 40 to 60 shrimps/m^2 (Hopkins et al. 1993). However, massive nauplii production in *L. setiferus* by means of natural mating has not been reported. The main reason is the low spermatophore attachment rate which has been associated to the male sterilization by the male reproductive tract degenerative syndrome (MRTDS) (Bray et al. 1985, Brown et al. 1979, Leung-Trujillo and Lawrence 1987). Characterized by the progressive reduction in the number of sperm cells and an increase in the percentage of abnormal and dead cells, MRTDS and male reproductive system melanization (MRSM) were separated as two distinct syndromes by Alfaro (1990). MRSM is caused by melanin production by the shrimp immune system after a chitinolitic bacterial infection (Alfaro 1993). Although the process that triggers MRTDS remains poorly understood, it has been confirmed that electrostimulation and temperature-induced stress cause spermatophore degeneration (Pascual et al. 1998, Rosas et al. 1993, Chamberlain et al. 1983, Sandifer et al. 1984). Both

Kluwer Academic/Plenum Publishers

E. Escobar-Briones & F. Alvarez Eds.
MODERN APPROACHES TO THE STUDY OF CRUSTACEA
PP. 045-052

syndromes have been related to the captivity and management stress that affect the physiological state and immune system (Pascual et al. 1998, Pascual 2000, Alfaro 1996).

Recent studies have demonstrated that the acclimation period to captivity is critical in the first days for adult male *L. setiferus*, when important biochemical and immunological changes occur (Pascual 2000, Sanchez et al. 2001). These changes, observed between days 10 and 15 depending on the temperature, are reflected in the loss of sperm quality (Pascual et al. 1998, Pascual 2000). The MRSM and MRTDS syndromes were induced 5 days after electrostimulation in adult males of *L. setiferus* maintained at 28 °C, showing that the stress of captivity and management could be the main variables associated to both syndromes (Rosas et al. 1993).

In this context, the energy production (P_T), an integrative response of the physiological state of organisms, can be used as a stress index because through this bionergetic parameter it is possible to quantify the energy that is channeled to growth or to reproduction. P_T results from the integration of physiological rates in a energy balance which can be expressed through the equation (Lucas 1996):

$$P_T = A - (R_{AHI} + R_{ROUT} + U)$$

where P_T is the energy production, A is the energy absorbed from the food, $R_{AHI} + R_{ROUT}$ is the energy used in respiratory metabolism including the energy used in the routine metabolism and the losses through mechanical and biochemical transformation of the ingested food (formally called apparent heat increment or AHI). Although some studies have been published on the influence of environmental factors on shrimp bioenergetics and related to dissolved oxygen (Rosas et al. 1998), salinity and temperature (Guerin and Stickle 1997, Vernberg and Piyatiratitivorakul 1998), there are no data in relation to the energetic alterations produced during the acclimation of adult males, just when the male reproductive syndromes start appearing (Sanchez et al. 2000).

Molecular and biochemical indicators have a good potential to evaluate stress, because stress typically leads to a rapid onset and a cascade of molecular and physiological responses. Hall and van Ham (1998) reported on the effect of a variety of putative stressors on blood glucose concentrations in an attempt to obtain a simple and reliable biological index of stress for shrimp. These authors showed a significant increase of blood glucose in *P. monodon* after a depletion of dissolved oxygen, and increased dissolved CO_2 levels. Racotta and Palacios (1998) showed that blood glucose is strongly and rapidly increased in response to repeated blood-sampling stress, with lactate levels more slowly affected, in *L. vannamei* juveniles.

Stress makes the animal more susceptible to microbial infections and disease, reducing its capacity to resist bacteria normally present in seawater (Lee 1992). Against a bacterial infection as observed in MRSM, cellular defenses should be activated. In crustaceans these mechanisms rely on hemocytes with several functions, such as coagulation, phagocytosis, encapsulation, and wound healing (Le Moullac et al. 1997, Smith and Ratcliffe 1980, Johansson and Söderhäll 1988). Certain types of hemocytes have been associated to several proteins, for example the prophenoloxidase (proPO) system which is involved in encapsulation and melanization and has a role as a non-self recognition system (Söderhäll et al. 1984, Perazzolo and Barraco 1997). Smith and Ratcliffe (1980) suggested the attaching proteins of the phenoloxidase cascade are strong non-self signals for the hemocytes, causing them to degranulate and release previously cell-bound recognition factors into the hemolymph, where they are free to trigger activation of adjacent hemocytes. All this responses use energy, probably modifying the energy balance of a shrimp during the acclimation period.

The present investigation was undertaken to study the effects of 7 days of acclimation at 27 °C and 31 °C, through the evaluation of the energy budget, as index of captivity stress in adult male *L. setiferus*.

II. MATERIALS AND METHODS

Capture and maintenance of the organisms

The shrimps used in the present study came from a natural population of *L. setiferus* caught over the continental shelf of the Gulf of Mexico, off Laguna de Términos, Campeche, Mexico. The sampling took place in October (27 °C experiments) and November (31 °C experiments), 1998. Once the shrimps were caught they were placed in plastic bags with cold seawater (24 °C) with a saturated oxygen atmosphere at a density of 10 shrimps/40 l, and transported to the laboratory. The shrimps were reared in oval dark ponds (10 m²) at a density of 40 shrimps/pond and provided with aerated, filtered (20 mm), flow-through seawater

(200% day) at 27 °C and 34‰ salinity. The shrimps were kept in the ponds for 24 h before any measurement was taken. For the first experiment the mean wet weight of the shrimps was 32.5 ± 0.7 g (n = 80); while for the second experiment it was 35.4 ± 0.87 g (n = 51). The shrimps were not fed during the first 24 h.

After 24 h, a group of shrimps were acclimated in a closed recirculating system with two 450 l tanks holding 9 organisms each. The temperature was maintained at 27 °C (first experiment) or 31 °C (second experiment) using a temperature-control system and a water exchange rate of 300% per day was established. Fresh squid was supplied as food, totaling 5% of the total biomass, twice a day. Every day the unconsumed food and feces were carefully removed with a siphon. A 10:14, light:dark photoperiod was used in the acclimation area.

Bioenergetic balance

After 7 days, 5 shrimps were separated from the rearing system from each temperature (27 and 31 °C) to evaluate the Total Production (P_T). It was calculated from the equation: $P_T = A - (R + U)$ (Lucas 1996), where A = energy absorbed from food consumed (A = energy ingested as food x absorption efficiency), R = energy lost as respiration, and U = energy lost as ammonia excretion. Absorption efficiency was calculated using the equation (Conover 1966):

$$AE = (F - E) \cdot ((1-E)F)^{-1} \cdot 100$$

where F = ratio of ash-free dry weigth (afdw) to dry weight in the food, and E = ratio of afdw to dry weight in the feces. The fecal production of individual shrimp was determined directly by collecting feces. The feces, squid, and entire shrimp were dried at 60 °C until constant weight was achieved. The samples were incinerated in a muffle (Sybron Thermolyne) at 600 °C for 4 h to obtained the ash free dry weight (afdw). Caloric values of the entire organism and squid were determined using a bomb calorimeter (Parr).

Routine metabolism was determined through the oxygen consumption of the shrimps. Five shrimps from each temperature were starved for 12 h to prevent handling stress. Each shrimp was placed individually in a 2.5 l respirometric chamber which was connected to a water recirculating system (Martínez-Otero and Díaz-Iglesia 1975). Oxygen consumption was determined as in a closed system flow respirometer (Rosas et al. 1998) and was calculated as: $VO_2 = O_{2E} - O_{2Ex}$ x Fr, where Vo_2 is the

oxygen consumption (mg O_2 h^{-1} animal^{-1}), O_{2E} is the oxygen concentration entering the chamber (mg l^{-1}), O_{2Ex} is the oxygen concentration exiting the chamber (mg l^{-1}), and Fr is the flow rate (ml hr^{-1}). Oxygen concentration was measured with a digital oxygen- meter (YSI 50B digital, USA) with a polarographic sensor (± 0.01 mg l^{-1}), previously calibrated with oxygen-satured seawater. The readings were corrected with respect to an empty control chamber. Oxygen consumption was measured every hour for a 6 h period, between 8 and 14 h. The first reading was considered as (R_{ROUT}) (mg g^{-1}h^{-1}) and was estimated from the VO_2 of the unfed shrimps. Afterwards, a pre-weighed ration of squid, representing 5 % of the shrimp's body weigth was added to each chamber, including the control. After 30 min, the uneaten food was removed and weighed to measure the food consumed. The apparent heat increase (R_{AHI}) was estimated from the difference between VO_2 of the unfed shrimps and the maximum value attained after feeding.

The ammonia excretion rate (N-NH$_3$ mg l^{-1}) was measured in conjunction with the oxygen consumption measurements by collecting water samples from the respirometric chambers. It was determined from the differences between the ammonia concentration at the entrance and the exit of each chamber and by multiplying it by the water flow. The ammonia concentration (total ammonia: NH$_4^+$ + NH$_3$) was measured using a flow injection-gas diffusion (Hunter and Uglow 1993). The postprandial nitrogen excretion (*PPNE*) was estimated from the difference between ammonia excretion of the unfed shrimp and the maximum value attained after feeding.

The VO_2 and ammonia excretion of unfed and fed shrimps was related to the afdw of the organism. A 14.3 and 20.5 conversion factors were used to transform the unfed and fed VO_2 and ammonia excretion, respectively, to J g^{-1} afdw d^{-1} (Brafield 1985, Lucas 1996).

The O:N atomic ratio was calculated by atoms of consumed oxygen and excreted nitrogen (Mayzaud and Conover 1988). The O:N ratio was obtained for the unfed and fed shrimps. The maximum values of consumed oxygen and excreted nitrogen were considered for the fed shrimps.

Energetic efficiencies were obtained of each element of the energetic balance equation, through: R/A x 100, U/A x 100, P_T/A x 100.

Statistical analysis

The differences between temperatures were obtained using a one way analysis of variance (ANOVA). The homogeneity of variances was verified. Means obtained during the treatment were compared using Duncan´s multiple range test (Zar 1974).

III. RESULTS

The temperature did not affect the energy balance values obtained ($P > 0.05$), except for R_{AHI} (Table 1). A reduction of energetic metabolism R_{ROUT} was observed in shrimps acclimated at 31°C. The absorption efficiency was similar between temperatures, 65.2 and 70.5 %, for 27 and 31 °C, respectively. Absorption energetic values on shrimps exposed at 27 °C were lower (1697.83 J g $^{-1}$ afdw d^{-1}) than those observed for shrimps acclimated at 31 °C (2174.19 J g^{-1} afdw d^{-1}) ($P > 0.05$). Routine metabolism was higher at 27 °C, (402.7 J g $^{-1}$ afdw d^{-1}) than that observed at 31 °C (225.5 J g^{-1} afdw d^{-1}) ($P > 0.05$). In contrast the R_{AHI} of shrimps acclimated at 27 °C was lower (25.84 J g^{-1} afdw d^{-1}) than that observed at 31 °C (73.97 J g^{-1} afdw d^{-1}) ($P < 0.05$). The ammonia excretion (86.11 and 78.80 J g $^{-1}$ afdw d^{-1}), and PPNE (7.70 and 10.86 J g^{-1} afdw d^{-1}), respectively, did not show significant differences. The energy directed to P_T showed a higher value in shrimps kept at 31 °C (1784.99 J g $^{-1}$ afdw d^{-1}) than in shrimps kept at 27 °C which only showed 1175.42 J g^{-1} afdw d^{-1} ($P > 0.05$). The O:N ratio was not significantly different in any case, and was always lower than 10 (Table 1).

Regarding the energetic efficiencies, the respiration (($R_{ROUT} + R_{AHI}$)$/A$) was the largest component of the bioenergetic balance with 25.2 % and 13.8 %, for 27 and 31 °C, respectively. The ammonia excretion ((N-NH_3 + $PPNE$)$/A$) represented 5.52 % and 4.12 %, for 27 and 31 °C, respectively (Fig. 1). The lower energy for growth and reproduction or net production efficiency (P_T/A) was obtained in shrimps exposed at 27°C (69.2 %) in comparison to that obtained for shrimps acclimated at 31°C (82.1 %) (Fig. 1).

IV. DISCUSSION

The captivity conditions affected the bionergetic responses of *L. setifers* adult males exposed at 27 and 31 °C. Temperature seems to have an effect on the metabolism, since growth rate, feeding rate and metabolic rate, typically increase as temperature increases (Fry 1947). In this study, although the results did not show any significant differences between the two experimental temperatures, captive shrimps kept at 31 °C showed the highest physiological values for the absorption efficiency, absorption, R_{AHI}, and PPNE. We can observe that *L. setiferus* adult males exhibited compensatory mechanisms to preserve energy regardless of the temperature effect. At 31 °C, the shrimps responded faster than those exposed at 27 °C, suggesting the action of compensatory mechanisms to acclimate faster.

The differences on the R_{ROUT} rates between both experimental temperatures could be associated to the compensatory mechanisms involved in the thermal

Table 1.- *Effect of temperature in the energy balance. Values correspond to J g^{-1} adfw d^{-1}. Mean values are followed ± SE. Means with the same letter are not significantly different.*

	27 °C	31 °C
Absorption Efficiency % (AE)	65.2 ± 7.38 [a]	70.5 ± 1.27 [a]
Absorption (A)	1697.83 ± 69.52 [a]	2174.19 ± 390.46 [a]
Respiratory rate (R)		
R_{ROUT}	402.77 ± 40.79 [a]	225.56 ± 8.03 [a]
R_{AHI}	25.84 ± 3.42 [b]	73.97 ± 15.80 [a]
Ammonia excretion (U)		
N-NH_3	86.11 ± 12.86 [a]	78.80 ± 10.91 [a]
PPNE	7.70 ± 2.03 [a]	10.86 ± 4.47 [a]
Total Production (P_T)	1175.42 ± 111.50 [a]	1784.99 ± 369.86 [a]
O:N ration		
O:N unfed	9.07 ± 1.01 [a]	7.01 ± 1.25 [a]
O:N fed	8.59 ± 1.14 [a]	9.69 ± 1.70 [a]

acclimation. Paterson (1993) demonstrated that the respiratory metabolism could be compensated through the decrease of oxygen consumption in high temperature when animals are acclimated at temperatures that are within their range of tolerance. Then, the anaerobic metabolism could be activated accumulating lactate, which is a final product of the anaerobic metabolism and its accumulation results when its concentration is higher than the capacity of lactate oxidation, producing an accumulation in the muscle. To avoid damage associated with lactate crystal accumulation in muscle, lactate is stored in the hemolymph, and is transported to the muscle to be metabolized reducing its hemolymph concentration (Spotts and Lutz 1981, Paterson 1993). This mechanism can explain the higher lactate level observed at 31 °C (Sanchez et al. 2001).

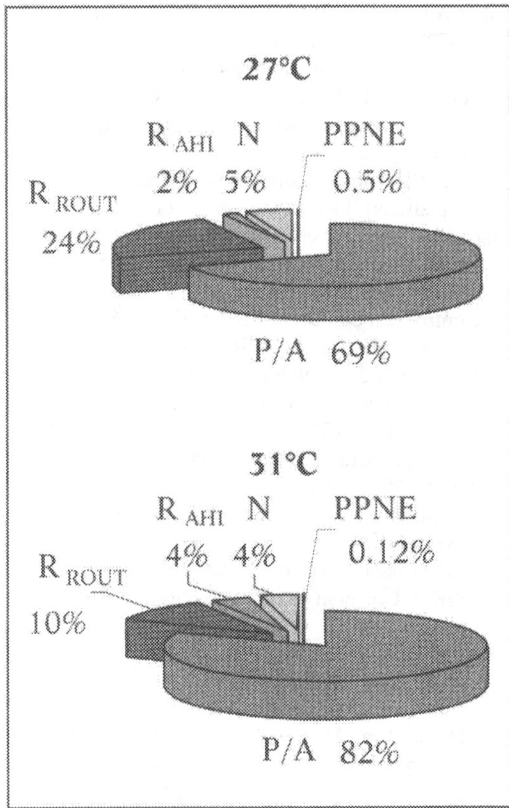

Figure 1. *Energetic efficiencies of **L. setiferus** adult males at 27°C and 31°C at captivity day seven.*

The absorption efficiency was 65.2 and 70.5 % at 27 and 31 °C, respectively (Table 1). We used the same diet for all organisms, considering that the absorption efficiency can be related to the proportion of proteins in the food and is maintained if the energetic substrate does not change (Mayzaud and Conover 1988). The values for the O:N ratio obtained in the present study demonstrated that *L. setiferus* adult males fed with squid used only proteins as a source of metabolic energy, independently of temperature. In a recent paper (Rosas et al. 2000) it has been demonstrated that shrimps are well adapted to use protein as a source of energy, because its carbohydrate and lipid metabolism are limited. It is an important adaptation for adult shrimps because at this age the animals will require more protein to produce the reproductive cells, than what will be chanelled to the spermatophore formation.

The R_{AHI} in crustaceans and fishes depends on the quantity, quality, and balance of the energy components of the diet (Beamish 1974, Du-Preez et al. 1992), and is a term that may be used as an index of the transformation efficiency of digestible energy (Chakraborty et al. 1992, Rosas et al. 1996). In this study, although we used the same kind of food, our results showed that the R_{AHI} was significantly different for both experimental temperatures (Table 1). Then, the R_{AHI} values could be related to the absorption values, lower values at 27 °C, in contrast with higher at 31 °C, for both parameters.

In the nitrogen metabolism of marine crustaceans, ammonia-nitrogen constitutes between 60 and 100 % of the final products of the protein metabolism (Regnault 1979, 1986). In this metabolism, the formation of glucose through aminoacidic gluconeogenesis is the most common way of forming ammonia (Campbell 1991). Beamish and Trippel (1990) and Rosas et al. (1996) demonstred a reduction in the transformation efficiency of digestible energy to growth as the protein catabolism increased. These mechanisms resulted in a higher value of ammonia excretion in shrimps maintained at 27 °C compared with the values obtained for shrimp kept at 31 °C (Table 1).

According to the energetic efficiencies (Fig. 1), the total energy expenditure $(R+U)$ had a different behavior between temperatures, mainly due to the higher percent of energy that was utilized for respiration $(R_{ROUT}+R_{AHI})$ at the lower temperature (25.2 %).

The proportion of absorbed energy directed to production was 69.2% for shrimps exposed at 27 °C and 82.1% for those maintained at 31 °C (Table 1, Fig. 1). Lucas and Beninger (1985) showed that the P_T/A ratio is a good physiological condition index that reflects the general health of shrimps exposed to varying experimental conditions. *Litopenaeus setiferus* is a eurythermal species and the adults are more sensitive to lower temperatures (Muncy 1984, Wiesepape 1975). Vernberg and Piyatiratitivorakul (1998) found for *Palaemonetes pugio* adults that energy allocated for reproduction at 17 °C was less than 10% of the energy ingested, but it increased with increasing temperature reaching the maximal value, 51.7%, at 32 °C.

In the present study it was shown that the energy targeted for reproduction increased with increasing temperatures, and this response was associated to different compensatory mechanisms. These results demonstrated that adult males of *L. setiferus* are well adapted to tolerate both temperatures, modulating the energy used for metabolism.

A positive energetic balance for growth and reproduction during acclimation was obtained for both experimental temperatures. As has been observed by Pascual et al. (1998) and Sanchez et al. (2001), there are severe metabolic changes that modify the immunological state of shrimps during the first seven days of captivity. The immunological and biochemical changes are compensated through the energetic metabolism, which in the end represents the shrimp homeostasis.

ACKNOWLEDGEMENTS

This project was partially supported by Dirección General de Asuntos de Personal Académico - UNAM (IN206199), and CONACYT (31137-B). Special thanks to Ramón Méndez Lanz, President of Fideicomiso Para Estudios y Proyectos from the Fisheries Secretariat of the State of Campeche. Thanks to Dr. Ellis Glazier for editing the English manuscript.

REFERENCES

Alfaro J (1990) A contribution to the understanding and control of the male reproductive system melanization disease of broodtsock *Penaeus setiferus*. College Station, Texas, USA, 65 p

Alfaro J (1993) Reproductive quality evaluation of male *Penaeus stylirostris* from a grow-out pond. J World Aquac Soc 24:6-11

Alfaro J (1996) Effect of 17 alpha -methyltestosterone and 17 alpha-hydroxyprogesterone on the quality of white shrimp *Penaeus vannamei* spermatophores. J World Aquac Soc 27:487-492

Beamish FWH and Trippel EA (1990) Heat increment: A static or dynamic dimension in bioenergetic models. Trans Amer Fish Soc 119:649-666

Brafield AE (1985) Laboratory studies of energy budgets. Crom Helm, London 251 p

Bray WA, Leung-Trujillo JR, Lawrence AL and Robertson SM (1985) Preliminary investigations of the effect of temperature, bacterial inoculation and EDTA on sperm quality in captive *Penaeus setiferus*. J World Maric Soc 16:250-257

Brown A Jr, McVey JP, Middledich BS and Lawrence AL (1979) Maturation of white shrimp *Penaeus setiferus* in captivity. Proc World Maric Soc 435:135-136

Campbell JW (1991) Excretory nitrogen metabolism. In: Prosser (ed) Comparative Animal Physilogy. 4th Ed, Wiley-Liss, New York

Conover RJ (1966) Assimilation of organic matter by zooplancton. Limnol Oceanogr 11:338-345

Chakraborty SC, Ross LG and Ross B (1992) The effect of dietary protein level and ration on excretion of ammonia in common carp *Cyprinius carpio*. Comp Biochem Physiol 103A:579-584

Chamberlain GW, Johnson SK and Lewis DH (1983) Swelling and melanization of the male reproductive system of captive adult peneid shrimp. J World Maric Soc 14:135-136

Du Prezz HH, Chen HY and Hsieh CS (1992) Apparent specific dynamic action of food in the grass shrimp *Penaeus monodon* Fabricus. Comp Biochem Physiol 103A:173-178

Fry FEJ (1947) Effect of the environment on animal activity. University of Toronto Studies, Biol Series No 55, 62 p

Guerin JL and Stickle WB (1997) Effect of salinity on survival and bioenergetics of juvenile lesser blue crabs, *Callinectes similis*. Mar Biol 129:63-69

Hall MR and van Ham EH (1998) The effects of different types of stress on blood glucose in the giant tiger prawn *Penaeus monodon*. J World Aquac Soc 29:290-299

Hopkins JS, Hamilton RD, Sandifer PA, Browdy CL and Stokes AD (1993) Effect of water exchange rate on production, water quality, effluent characteristics and nitrogen budgets of intensive shrimp ponds. J World Aquac Soc 24:304-320

Hunter DA and Uglow RF (1993) A technique for the measurement of total ammonia in small volumes of seawater and hemolymph. Ophelia 37:31-40

Johansson MW and Söderhäll K (1988) Isolation and purification of a cell adhesion factor from crayfish blood cells. J Cell Biol 106:1795-1803

Le Moullac G, Le Groumellec M, Ansquer D, Froissard S, Levy P and Aquacop (1997) Hematological and phenoloxidase activity changes in the shrimp *Penaeus stylirostris* in relation with the moult cycle: protection against vibriosis. Fish Shellfish Immunol 7:227-234

Lee DOC (1992) Crustacean farming. Blackwell Scientific Publications, Oxford, England, 335 p

Leung-Trujillo JR and Lawrence AL (1987) Observations on the decline in sperm quality of *Penaeus setiferus* under laboratory conditions. Aquaculture 65:363-370

Lucas A. (1996) Bioenergetics of Aquatic Animals. Taylor & Francis. Great Bretain, 169 p

Lucas A. and Beninger PG (1985) The use of physiological condition indices in marine bivalve aquaculture. Aquaculture 44:159-166

Martínez-Otero A and E. Diaz-Iglesia (1975) Instalación respirométrica para el estudio de la acción de diversos agentes presentes en el agua de mar. Rev Inv Mar 111:1-8

Mayzaud P and Conover RJ (1988) O:N ratio as a tool to describe zooplancton metabolism. Mar Ecol Prog Ser 45:289-302

Muncy RJ (1984) Species profiles: Life histories and environmental requeriments of coastal fishes and invertebrates (South Atlantic). White Shrimp. Fish and Wildlife Service-U.S. Department of the Interior and Coastal Ecology Group Waterways Experiment Station-U.S. Army Corps of Engineers. FWS/OBS-82/11.27, TR EL-8-4, September 1984, USA, 20 p

Pascual C, Valera E, Re-regis C, Gaxiola G, Sanchez A, Ramos L, Soto LA and Rosas C (1998) Effect of temperature on reproductive tract condition of *Penaeus setiferus* adult males. J World Aquac Soc 29:477-484

Pascual C (2000) Caracterización inmunologica y bioquímica de una población de machos adultos de *Penaeus setiferus*, línea de base y estrés térmico. Tesis de Licenciatura, Facultad de Ciencias, UNAM, 70 p

Paterson BD (1993) The rise in inosine monophosphate and L-lactate concentrations in muscle of live penaeid prawns (*Penaeus japonicus, Penaeus monodon*) stressed by storage out of water. Comp Biochem Physiol 106B:395-400

Perazzolo L and Barraco M (1997) The prophenoloxidase activating system of the shrimp *Penaeus paulensis* and associated factors. Dev Comp Immunol 21:385-395

Racotta IS and Palacios E (1998) Hemolymph metabolic variables in response to experimental manipulation stress and serotonin injection in *P. vannamei*. J World Aquac Soc 29:351-356

Regnault M (1979) Ammonia excretion of the sand shrimp *Crangon crangon* (L) during the moult cycle. J Comp Physiol 141:549-555

Regnault M (1986) Excretion d´azote chez les crustaces influence de l´etat physiologique. Cah Biol Mar 28:361-373

Rosas C, Sanchez A, Chimal ME, Saldaña G, Ramos L and Soto LA (1993) The effect of electrical stimulation on spermatophore regeneration in white shrimp *Penaeus setiferus*. Aq Liv Res 8:139-144

Rosas C, Sanchez A, Diaz-Iglesia E, Soto LA, Gaxiola G and Brito R (1996) Effect of dietary protein level on apparent heat increment and post-pandrial nitrogen excretion of *Penaeus setiferus, P. schmitti, P. duorarum*, and *P. notialis* postlarvae. J World Aquac Soc 27:92-102

Rosas C, Martinez E, Gaxiola G, Brito R, Diaz-Iglesia E and Soto LA (1998) Effect of dissolved oxygen on energy balance and survival of *Penaeus setiferus* juveniles. Mar Ecol Prog Ser 174:67-75

Rosas C, Cuzon G, Gaxiola G, Arena L, Lemaire P, Soyez C and Van Wormhoudt A (2001) Influence of dietary carbohydrate on the metabolism of juvenile *Litopenaeus stylirostris*. J Exp Mar Biol Ecol (in press)

Samocha TM, Burkott BJ, Lawrence AL, Juan YS, Jones ER and McKee DA (1998) Management strategies for production of the Atlantic white shrimp *Penaeus setiferus* as bait shrimp in outdoor ponds. J World Aquac Soc 29:211-220

Sanchez A, Pascual C, Sanchez A, Vargas-Albores F, Le Moullac G and Rosas C (2001) Hemolymph metabolic variables and immune response in *Litopenaeus setiferus* adult males: the effect of acclimation. Aquaculture 198:13-28

Sandifer PA, Lawrence AL, Harris SG, Chamberlain GW, Stokes AD and Bray WA (1984) Electrical stimulation of spermatophore expulsion in marine shrimp, *Penaeus* spp. Aquaculture 41:181-187

Sandifer PA, Hopkins JS, Stokes AD and Browdy CL (1993) Preliminary comparisons of the native *Penaeus setiferus* and Pacific *P. vannamei* white shrimp for pond culture in South Carolina, USA. J World Aquac Soc 24:295-303

Sandifer PA, Lawrence AL, Harris SG, Chamberlain GW, Stokes AD and Bray WA (1984) Electrical stimulation of spermatophore expulsion in marine shrimp, *Penaeus* spp. Aquaculture 41: 181-187

Smith V and Ratcliffe NA (1980) Host defense reactions of the shore crab, *Carcinus maenas* (L.): clearance and distribution of the injected test particles. J Mar Biol Ass UK 60:89-102

Smith VJ and Söderhäll K (1983) Induction of degranulation and lysis of haemocytes in the freshwater crayfish, *Astacus astacus* by components of the phenoloxidase activating system in vitro. Cell Tissue Res 233:295-303

Soderhall K, Vey A and Ramstedt M (1984) Hemocyte lysate enhancement of fungal spore encapsulation by crayfish hemocytes. Develop Comp Immunol 8:23-29

Vernberg FJ and Piyatiratitivorakul S (1998) Effects of salinity and temperature on the bioenergetics of adult stages of the grass shrimp (*Palaemonetes pugio* Holthuis) from the North Inlet Estuary, South Carolina. Estuaries 21:176-193

Wiesepape LM (1975) Thermal resistance and acclimation rate in young white and brown shrimp, *Penaeus setiferus* Linn and *Penaeus aztecus* Ives. Texas A & M University Sea Grant 76-202, USA, 196 p

Zar JH (1974) Biostatistical Analysis. Englewood Cliff: Prentice Hall. New Jersey, USA, 413 p

EVALUATION OF LEGUMINOUS SEED MEALS (*RYNCHOSIA MINIMA* AND *CAJANUS CAJAN*) AS PLANT PROTEIN SOURCES IN DIETS FOR JUVENILE *LITOPENAEUS VANNAMEI*

Héctor Cabanillas-Beltrán, Jesús T. Ponce-Palafox, Carlos A. Martínez-Palacios and Eduardo Jaime Vernon-Carter

HCB, Instituto Tecnológico de Tepic, Tepic, Nayarit, México

hcabanillas@netscape.net

JTPP, UAN-CUVEDES, Tepic, Nayarit and UAEM-CIB, Cuernavaca, Morelos, México
CAMP, Instituto de Investigaciones sobre Recursos Naturales, Morelia, Michoacán, México
EJVC, Universidad Autónoma Metropolitana-Iztapalapa, México, D.F., México

ABSTRACT

The potential use of the leguminous seed meals obtained from *Rynchosia minima* and *Cajanus cajan*, which grow in northwestern Mexico, as protein sources was evaluated in *Litopenaeus vannamei* postlarvae. These meals were included in practical diets, replacing fish meal as the protein source in the basal diet, in order to obtain isoproteic and isoenergetic diets. The leguminous seed meals were included at dietary levels of 10 and 18% of total protein. As a prior step, the meals were submitted to a thermal process (wet heat, 121°C) during 1, 45 and 90 min to reduce anti-nutritional factors and to evaluate its effect in the experimental organisms.

Litopenaeus vannamei postlarvae (mean initial weight 10.2±0.5 mg) were fed the practical diets for 38 days. Specific growth rates (SGR), feed conversion (FC) and survival rates (SR) were evaluated. The SGR, FC and SR of shrimp given *R. minima* or *C. cajan* meals were comparable to that of shrimp fed a control diet. Results suggest that it is possible to replace 18% of animal protein with *R. minima* or *C. cajan* protein.

I. INTRODUCTION

Various protein sources have been tested to partially or totally replace fish meal in practical shrimp diets. Soybean meal is widely used as a protein source for shrimp due to its high protein content, digestibility and amino acid composition (Dominy and Lim 1991, Boonyaratpalin et al. 1988, Cabanillas-Beltrán et al. 1997). Other legumes, specifically cowpea and mumgbean, are known for their relatively high protein and energy content (Eusebio and Coloso 1998). White cowpea meal can be used as a protein source in practical diets for *P. monodon* at 29% of the diet without affecting their growth (Eusebio 1991). Peñaflorida (1995) found that papaya leaf meal can partially replace 10% of the animal protein in shrimp diets and serve as a source of exogenous proteolytic enzyme. Inclusion of white cowpea meal or cassava leaf meal in diets at 8.6 to 8.8% of the total protein did not significantly affect the growth of the *Fenneropenaeus indicus* juveniles (Eusebio and Coloso 1998).

Inclusion levels of these feed ingredients in shrimp diets are often limited by their nutrient bioavailability (De Silva 1989). This could be due to

E. Escobar-Briones & F. Alvarez Eds.
MODERN APPROACHES TO THE STUDY OF CRUSTACEA
PP. 053-057

the presence of anti-nutritional factors and other toxic substances in legumes (Elias et al. 1979) and leaf meals (Peñaflorida 1995). The objective of this study was to assess the use of leguminous seed meals available in northwestern Mexico as protein sources in practical diets formulated for juvenile *L. vannamei*.

II. MATERIALS AND METHODS

Diets

Twelve diets were formulated using the leguminous seed meals from *Rynchosia minima* and *Cajanus cajan* in substitution of 10% and 18% of the dietary animal protein (Mexican brown fish meal). The leguminous seed meals were subjected to a thermal process (wet heat, 121°C) during 1, 45 and 90 minutes to reduce anti-nutritional factors and to evaluate its effect in the experimental organisms. Each diet was formulated to provide 35% protein and 420 Kcal contents. These diets were compared with a control diet in which fish meal was the sole protein source. The diets were prepared following the method described by Eusebio and Coloso (1998), using a meat mill to form crumbles that were then dried for 24 h at 37°C in a forced-air stove, and stored in a freezer until use. Samples were taken for proximate analysis using standard methods (AOAC 1990). Table 1 shows the formulation and proximate composition of the diets.

Experimental procedure

A 38-day experiment was conducted in a flow-through system with filtered aerated seawater. For the growth experiments, 500 *L. vannamei* PL15 postlarvae, commonlly known as the Pacific white shrimp, were purchased from Laboratorios Génesis (Puerto Peñasco, Sonora, México). Twenty *P. vannamei* postlarvae (mean initial weight 10.18±0.5 mg) were stocked into each of 16 oval plastic tanks containing 15 L seawater. All shrimp were maintained for five days on the control diet prior to feeding them the test diets. The organisms were fed *ad libitum* every three hours (from 09:00 to 18:00 h). The water quality was monitored throughout the 6-week experimental period. The following mean (±SE) values were recorded: temperature = 26±2°C; salinity = 31±1°/∞; pH = 7.8±0.5; dissolved oxygen concentration = 6.0±0.10 mg/L; ammonia = 0.003±0.001 mg/L and nitrate = 0.0012±0.0007 mg/L. Temperature, salinity and oxygen were measured daily, and pH and N products were recorded weekly. The shrimp

in each tank were weighed every ten-days. Mortality was recorded every day and dead shrimp were removed.

Table 1. *Formulation and proximate composition of experimental diets (average of two replicates).*

Ingredients	Control Diet	Diet 1 R. minima	Diet 2 R. minima	Diet 3 C. cajan
		10%	18%	10%
[1]Fish meal	49.55	44.59	40.63	44.59
R..minima meal	0.00	18.90	34.02	0.00
C. cajan meal	0.00	0.00	0.00	22.49
Fish oil	2.90	5.72	7.71	4.92
Soybean oil	1.42	2.09	2.56	1.96
Starch	35.78	18.35	2.73	15.79
[2]Mineral mix	2.00	2.00	2.00	2.00
[3]Vitamin mix	3.00	3.00	3.00	3.00
Vitamin C	0.60	0.60	0.60	0.60
Lecithin	0.25	0.25	0.25	0.25
Binder	4.00	4.00	4.00	4.00
Gelatin	0.00	0.00	2.00	0.00
Cholesterol	0.50	0.50	0.50	0.50
Proximal analysis				
Protein (%)	35.00	35.00	35.00	35.00
Lipid (%)	7.10	10.45	12.80	9.80
Energy (kcal)	419.97	420.07	420.15	420.19

[1]Mexican brown fish meal, [2]Arredondo et al. (2000), [3]Vernon-Carter et al. (1996).

Analytical methods

The individual protein diets were analyzed for crude protein (N x 6.25) and lipid using standard methods (AOAC 1990). Gross energy in diets was estimated based on the following conversion factors: protein = 5.65 kcaVg, fat = 9.5 kcal/g and carbohydrate (as NFE) = 4.1 kcal/g.

Trypsin, lectin, phytate and glycoside cyanogen inhibitory activities were the anti-nutritional factors evaluated. Trypsin inhibitory activity was determined by putting an aqueous extract of the sample in a standard solution of trypsin (40 mg/10 ml). Afterwards the residual proteolytic activity is determined using a synthetic substrate (benzoyl-arginine-p-nitroaniline, BAPNA), and reading the absorbance at 410 nm in a spectrophotometer. Phytic acid inhibitory activity was determined in accordance to the technique of Tangkongchirt et al. (1981), in which phytic acid is obtained and phosphorous content is evaluated following Pereira' s (1988) technique. Lectin inhibitory activity was determined by a semi-quantitative micro-titration technique. Glycoside cyanogen inhibitory activity was measured taking advantage of the sensitive and specific Guignard reaction, also used in qualitative HCN tests. For quantifying the total HCN that may be potentially liberated, an enzymatic hydrolysis (by means of a B-glucosidase) of the corresponding glycoside cyanogen is carried out. The liberated HCN reacts with picric acid, giving a positive response if a brown-redish color is developed.

Statistical methods

The results for growth, specific growth rate, and ingestion feed conversion ratio were analyzed statistically using a factorial analysis of variance. The data were analyzed by means of the SAS software package (SAS 1988). Factorial analysis of variance was also performed on the arcsin square root transformation of survival data. Treatment means were compared by Duncan's new multiple range test. Differences were significant at $P < 0.05$.

III. RESULTS

The anti-nutritional factors content of the leguminous seed meals and the time of thermal treatment are shown in Table 2. The *R. minima* seed meal had a higher lectin and trypsin and a lower acid phytic inhibitory activity than the *C. cajan* seed meal. Glycoside cyanogen inhibitory activity was not detected in either. Longer thermal treatment times had a stronger inhibitory effect on the anti-nutritional factors.

Growth experiment

Table 3 shows the growth response and survival rate of the juvenile *L. vannamei*. The diets with *R. minima* were well accepted, although a food rejection

trend existed as the thermal treatment was shorter. Survival was variable among treatments and exhibited a non diet-related pattern (Table 3). The highest survival (50%) was obtained with the shrimps fed the control, R-90-10 and R-90-18 diets, while a low shrimp survival of 30% was recorded with the R-O 1-18, R-45-10, C-45-10 and C-45-18 diets. The survival differences were attributed to handling, because no deleterious effect related to the meal seed content was observed.

Table 2. *Content of anti-nutrient in leguminous seed meal as function of thermal treatment times.*

Protein source	Thermal treatment time	Trypsin inhibitory activity[1]	Phytic acid inhibitory activity[2]	Lectin inhibitory activity[3]
R. minima	1	43.38	<0.92	6.0
R. minima	45	28.34	<0.92	4.0
R. minima	90	12.51	<0.92	0.0
C. cajan	1	9.28	1.14	0.0
C. cajan	45	5.04	1.40	0.0
C. cajan	90	0.98	0.92	0.0

[1]TIU/mg sample (units of inhibited trypsin/ mg sample), [2]mg phytic acid/g sample, [3]agglutination title.

Key variables such as final body weight, specific growth rate and feed intake showed the highest values in shrimps fed with C-90-10 and C-90-18 diets. Longer thermal treatment times of the seed meals resulted in an enhanced shrimp growth. The growth of the shrimp fed the *R.minima* and *C. cajan* diets was comparable with that of the shrimp fed the control diet. The highest FCR values were obtained with the R-01-18, C-01-10 and C-Ol-18 diets and the lowest with R-90-10, C-45-10 and control diets.

Table 3. *The mean growth and survival of **Litopenaeus vannamei** fed diets containing protein from locally available leguminous seed meals.*

	Control	R0110	R0118	R4510	R4518	R9010	R9018	C0110	C0118	C4510	C4518	C9010	C9018	±SE[4]
Survival (%)	50.0[a]	30.0[a]	25.0[a]	25.0[a]	30.0[a]	50.0[a]	50.0[a]	30.0[a]	30.0[a]	25.0[a]	25.0[a]	30.0[a]	30.0[a]	2.09
Initial weight (mg)	10.0[a]	11.0[a]	10.0[a]	9.5[a]	11.0[a]	10.0[a]	10.0[a]	10.0[a]	11.0[a]	10.5[a]	10.0[a]	9.5[a]	11.0[a]	1.26
Final weight (mg)	89.0[b]	63.3[b]	62.5[b]	68.0[b]	56.6[c]	92.0[b]	56.0[c]	66.7[b]	66.7[b]	86.0[b]	60.0[b]	96.7[b]	121.7[a]	5.53
Feed intake (mg/d)	116.0[b]	85.3[b]	116.4[b]	106.1[b]	71.3[c]	106.2[b]	86.8[b]	115.9[b]	114.8[b]	99.9[b]	79.6[b]	146.4[a]	154.6[a]	8.41
[2]SGR (% day)	2.61[a]	2.0[a,b]	2.25[a]	2.09[a]	1.87[b]	2.54[a]	1.97[a,b]	2.17[a]	2.06[a]	2.40[a]	2.05[a]	2.65[a]	2.75[a]	0.09
[3]FCR	1.48[b]	1.63[b]	2.22[a]	1.81[a]	1.56[b]	1.29[b]	1.89[a]	2.04[a]	2.06[a]	1.32[b]	1.59[b]	1.68[b]	1.39[b]	0.10

Values followed by the same superscript letters are not significantly different ($P < 0.05$), [2]SGR = Specific Growth Rate, [3]FCR = Feed Convertion Ration, [4]SE = standard error, calculated from the mean-square for error of the ANOVA.

IV. DISCUSSION

The present study shows that *R. minima* and *C. cajan* possess an adequate nutritional value for juvenile shrimp at low inclusion levels, making possible substitution levels up to 18% of the dietary animal protein without adverse effects on growth and feeding efficiency. The results obtained in the present study were similar to those found in previous experiments with *F. indicus* (Eusebio and Coloso 1998) fed with leguminous seed meals. The lower survival obtained in the present study may be due to an imbalance in the amino acid component of the protein sources, the diets may have been limited in arginine and methionine, or to the stress encountered by the shrimp during moulting.

Trypsin inhibitory activity in *R. minima* (12.51-43.38 TIU/mg protein) had a more marked effect than *C. cajan* (0.98-9.28 TIU/mg protein) upon shrimp growth. This result is in agreement with that found by Noor et al. (1980) who observed a relatively high trypsin inhibitory activity in cooked mungbean (24.1-27.7 TIU/mg protein), that was attributed to tannin content which is heat resistant. No published data for *R. minima* anti-nutrients was found, but the seeds used in this work had lectin, phytic acid and trypsin inhibitory activity, that were decreased and inactivated with the thermal treatment.

Studies by Newman (1997), Dominy and Lim (1991) and Akiyama (1991b) including soybean meal in the diets showed that good growth, FC and survival results were obtained in penaeids, coinciding with those found in this study. The present study demonstrates the feasibility of using *R. minima* and *C. cajan* as an alternative protein source in practical diets, providing a strong argument seeking an increase in the availability of this product for use as substitute of animal proteins in shrimp diets.

In summary, the results suggest that the *R. minima* seed meal (R-90-10) or *C. cajan* seed meal (C-90-18) diets can partially replace the animal protein in shrimp diets at inclusion levels of 10 and 18%, respectively.

ACKNOWLEDGEMENTS

The authors wish to thank Ana Ocampo and Leonor Sánchez for helping with the bioassay. M.Sc. Mario Ortiz for the statistical analysis of the data. The authors wish to acknowledge the financial support of the Sistema de Investigación del Mar de Cortés (SIMAC) through project 970106027, the Consejo Nacional de Ciencia y Tecnología (CONACyT) through project 25153-B and the Government of the State of Nayarit, México. The authors wish to thank Blanca T. González Rodríguez and Irma Eugenia Martínez Rodríguez (CIAD-Mazatlán) for helping with the diet analysis.

REFERENCES

Akiyama DM (1991) Soybean meal utilization by marine shrimp. (pp 207-225) In: Proceedings of the aquaculture feed processing and nutrition workshop, 2da. Edition. American Soybean Association, Thailand and Indonesia, September 19-25

AOAC (1990) Official Methods of Analysis. K. Helrich (ed) 15th Edition AOAC. Arlington, Virginia, 1298 p

Arredondo-Figueroa JL, Ponce-Palafox JT and Vernon-Carter EJ (2000) Dose response to Aztec marigold (*Tagetes erecta*) pigment of white shrimp (*Penaeus vannamei*) fed various dietary carotenoid concentrations. Crustacean Issues XI:134-141

Boonyaratpalin M, Phromkunthong W and Sumapataya K (1988) Substitution of soybean meal for fish meal in *Penaeus merguiensis* feed. J Sci Tech 10:45

Cabanillas-Beltrán H, Martinez CA and Ponce-Palafox JT (1997) Evaluación de la proteina de soya en dietas prácticas a diferentes temperaturas y salinidades en el camarón blanco, *Penaeus vannamei* Boone, 1931. (pp 227-232) Memorias del IV Congreso Ecuatoriano de Acuicultura, Guayaquil, Ecuador, 550 p

De Silva SS (1989) Digestibility evaluation of natural and artificial diets. (pp 36-45) In: De Silva SS (ed) Fish Nutrition Network Meeting. Asian Fish Soc Spec Publ 4, Manila

Dominy W and Lim C (1991) Evaluation of soybean meal extruded with wet squid viscera as a source of protein in shrimp feeds. (pp 116-120) In: Proceedings of the Aquaculture Feed Processing and Nutrition Workshop, 2nd Edition, American Soybean Association, Thailand and Indonesia

Elias LG, De Fernandez DH and Bressani R (1979) Posible effects of seed coat pholyphenolic on the nutritional quality of bean protein. J Food Sci 44:524-527

Eusebio PS (1991) Effect of dehulling on the nutritive value of some leguminous seed as protein source for the tiger prawn *Penaeus monodon* juveniles. Aquaculture 99:297-308

Eusebio PS and Coloso RM (1998) Evaluation of leguminous seed meals and leaf meals as plant protein source in diet for juvenile *Penaeus indicus*. The Journal of Aquaculture- Bamidgeh 50:47-54

Newman M (1997) El efecto de la harina de soya en el desenvolvimiento del crecimiento de *Penaeus vannamei* en tanque de concreto con sustrato y luz natural. (pp 30-31) In: Resúmenes IV Congreso Ecuatoriano de Acuicultura Guayaquil, Ecuador

Peñaflorida VD (1995) Growth and survival of juvenile tiger shrimp fed diets where fish meal is partially replaced with papaya (*Carica papaya* L.) or camote (*Ipomea batata*.v IJam) leaf meal. Israel J Aquacult 47:25-33

Pereira F (1988) Manual de análisis de los alimentos. Ed. UAY, Mérida, México 230 p

SAS (1985) Statiscal Analysis System Institute Inc, Cary NC, USA

Vernon-Carter E, Ponce-Palafox JT and Pedroza RI (1996) Pigmentation of Pacific white shrimp (*Penaeus vannamei*) using aztec marigold (*Tagetes erecta*) extracts as the carotenoid source. Arch Latin Nutr 46:243-246

HISTOLOGICAL ALTERATIONS IN HEPATOPANCREAS OF *FARFANTEPENAEUS AZTECUS* CAUSED BY THE CESTODE *GILQUINIA* SP. IN TAMIAHUA LAGOON, VERACRUZ, MEXICO

José Luis Bortolini-Rosales and María del Pilar Torres-García

Laboratorio de Invertebrados, Facultad de Ciencias, A. P. 70-371, México 04510, D. F., México, Universidad Nacional Autónoma de México

jlbr@hp.fciencias.unam.mx

ABSTRACT

The histological alterations of the hepatopancreas of *Farfantepenaeus aztecus*, caused by the cysts of the cestode *Guilquinia* are presented. The damage found in the hepatopancreatic tissue includes an encapsulation produced by leukocytes, fibroblasts and the deposition of collagen like fibers. The characteristic tubules of the hepatopancreas appear eroded, reduced in size, stain more intensely basophilic than normal, and are frequently incorporated into the cyst wall and destroyed. This represents the first report for *Gilquinia* parasitizing the brown shrimp in the Gulf of Mexico.

I. INTRODUCTION

The species of the genus *Farfantepenaeus* (Decapoda: Penaeidae) from the Gulf of Mexico, are organisms of great economic importance. However, it has been observed that they are frequently infected by a diverse group of parasites, including virus, bacteria, fungi, protozoa and helminths. Among this last group of parasites, a reduced number of cases of cestodes (Order Trypanorhyncha) have been observed in the hepatopancreas and other organs of penaeid shrimps. The traditional systematics of the trypanorhynchs has been based upon scolex and proglottid anatomy, type of tentacular armature and structure of the plerocercus larvae. The key character that distinguishes the cestodes is the structure of the scolex, which is relatively large, with 2 or 4 bothridia anteriorly and 4 (less frequently) or 2 very contractile and retractable proboscids (Grabda 1991).

Cestode larvae occasionally infect penaeid shrimps (Brock and Main 1994). Penaeid shrimps are used as intermediate hosts by the larval stages of the parasite; however, they typically require a vertebrate as a definitive host, which in this case could be a fish or a bird. In the Gulf of Mexico, cestodes can use sharks and rays as their final host. In some cases, penaeid shrimps and fishes of the Gulf of Mexico, have been found to carry thousands of cestodes encysted in the hepatopancreas, intestine, abdominal muscle, gills, and heart (Overstreet 1973, 1978, Ward 1962, Villella et al. 1970).

The first reports of the presence of cestodes in shrimps, began with the study of Kruse in 1959 (Conroy and Conroy 1990), in which he reported the presence of *Prochristianella penaei* in *Farfantepenaeus aztecus, F. duorarum* and *Litopenaeus setiferus*, species all from the Gulf of Mexico, finding prevalences of 91%, 97% and 94%, respectively. *Parachristianella monomegacantha* and *P. dimegacantha* were described as parasites of *F. aztecus* from the Gulf of Mexico by Corkern (1970). In 1975, Felgenbaum (Conroy and Conroy 1990) published the larval description of *Prochristianella heteromegacanthus, P. monomegacantha, P. dimegacantha* and *Renibulbus penaeus*, and suggested *F. aztecus* could be the normal intermediate host.

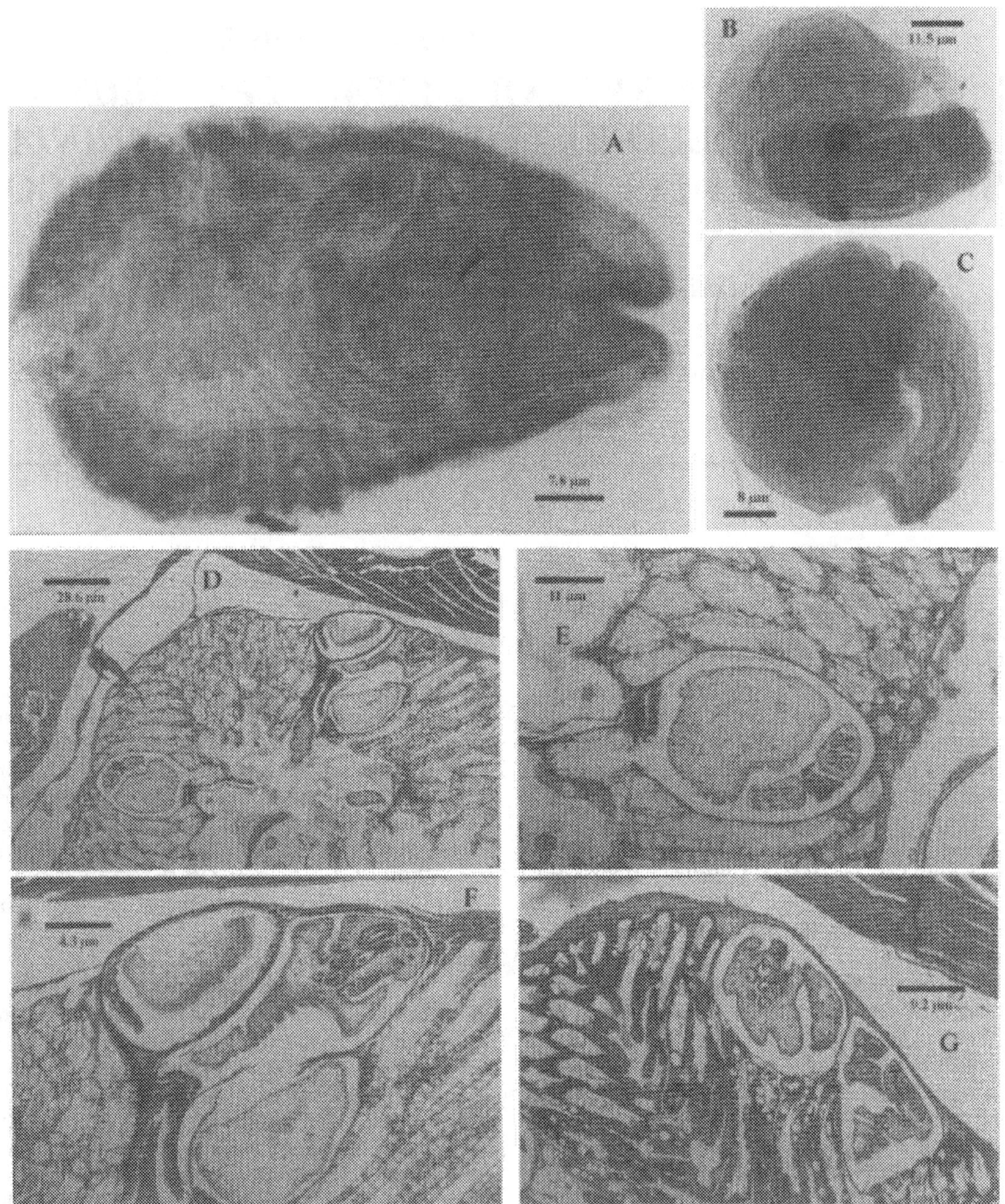

Figure 1. *A-C, cysts of cestodes of the genus* **Gilquinia**: *A, early larval stage; B, intermediate larval stage; C, final larval stage; D, view of the hepatopancreas of* **Farfantepenaeus aztecus** *with the alterations caused by the cestodes (fibroblast increase); E, cestode in the hepatopancreatic tubules; F, detail of the cestode in the hepatopancreas; G, view of the hepatopancreas of* **Farfantepenaeus aztecus** *parasitized with cestodes of the genus* **Gilquinia**.

Penaeid shrimps spawn offshore and enter the bays as postlarvae, averaging about 12 mm in length, then migrate into low salinity nursery grounds where they become juveniles and grow rapidly (Sparks 1985). Prevalence of *Prochristianella penaei* increases with size of the shrimp, demonstrating infection is acquired in the estuary (Aldrich 1965). Toft et al. (1991) concluded that penaeid shrimps are strictly intermediate hosts of cestodes, appearing mainly in the digestive gland or hepatopancreas and intestine. Intensity of infection (number of worms per host) increases with increasing size in smaller shrimp, reaching a peak in brown shrimp at 13-15 mm carapace length and in white shrimp at about 18 mm. Brown shrimp in the Galveston Bay complex are generally more heavily parasitized than white shrimp (Sparks 1985). Sparks and Fontaine (1973) reported that the hemocoel adjacent to the hepatopancreas is a more favorable site for survival until the intermediate host is eaten by the final host, the southern sting ray, *Dasyatis sabina*. Larval stages of the genus *Gilquinia* occur in teleosts and crustaceans (Wardle and McLeod 1952, Yamaguti 1959). Adult organisms of *Gilquinia squali* were found in the eyes of the North Sea whiting *Odontogadus merlangus* (MacKenzie 1965), and of *Merlangius merlangus* (MacKenzie 1975). In this last report he found 1 to 18 larvae per fish of 24,000 observed fishes and there was apparently no seasonality. Infection with trypanorynch larvae sometimes shows marked local differences, Hislop and MacKenzie (1976) distinguished stocks of *Merlangius merlangus* in the northern North Sea by tagging experiments as well as by the presence of larval *Gilquinia squalli*.

In this paper, we report the presence of cestodes parasitizing postlarval and juvenile brown shrimp, *Farfantepenaeus aztecus*, in Tamiahua Lagoon, Veracruz, Mexico. The cysts are elliptical and milky white in color, the larval stages are slender and elongated, present separated oval and patelliform bothridia, and the hooks on the proboscis are roughly arranged in oblique rows with varying numbers.

II. MATERIALS AND METHODS

A total of 254 shrimps, ranging from postlarvae to early juveniles of approximately 35 mm of total length, were captured in Tamiahua Lagoon. The samples were fixed with Davidson´s solution (Bell and Lightner 1988) and transported to the Invertebrates Laboratory of the School of Sciences of the National Autonomous University of Mexico for histological processing. The hepatopancreas was dissected, separating the cysts present with the help of a stereoscopic microscope and using Gallego´s staining technique in order to make the internal structures more evident. Permanent slides in synthetic resin were obtained from this dissections. For the histological analysis of host and parasite, the tissues were embedded in paraffin, sections 7-8 μm thick were obtained, and these were stained using Hematoxylin-Eosin of Gil and Masson´s trichromic techniques. The resulting slides were classified and photographed using a compound microscope.

III. RESULTS

From the total sample of 254 shrimps, 187 (73.6%) were found to carry cestode cysts. All shrimp sizes analyzed carried cestodes, and all were found in the hepatopancreas exclusively (Figs. 1A-C). A total of 253 cestodes were found, individual shrimps carried from 0 to 17 cestodes.

The cestodes in all cases were identified as *Gilquinia* sp., this genus have a modified plerocercoid, in which the rest of the body can withdraw. These plerocercoids are encapsulated in a thin cellular cysts, which progressively thickens by migration of leukocytes and fibroblasts and deposition of collagen like fibers. Hepatopancreatic tubules in the vicinity of an encapsulated plerocercoid are eroded, reduced in size, stain more intensely basophilic than normal, and are frequently incorporated into the cyst wall and destroyed.

The damage caused by the cestodes to the hepatopancreas is evident in several ways. First, there is a cellular atrophy, and in extreme cases a lysis of the hepatopancreatic tubules, due to the implantation of the cysts of *Gilquinia*. There is also an increase in vascularization and an increase in the amount of fibroblasts within the hepatopancreatic tissue in comparison to healthy hepatopancreas.

No members of the genus *Gilquinia* have ever been reported as parasite of shrimps of the genus *Farfantepenaeus*, although other cestodes have been found in *F. aztecus*, like *Prochristianella penaei* in the northern Gulf of Mexico (Couch 1978, Conroy and Conroy 1990).

IV. DISCUSSION

In this study cestodes of the genus *Gilquinia* in several larval stages, were found in the hepatopancreas of *Farfantepenaeus aztecus* (Figs. 1 D,G). As it has already been established, penaeid shrimps serve as intermediate hosts of trypanorynch cestodes (Toft et al. 1991, Feingenbaun 1975). Similar to the species of the genus *Prochristianella*, *Gilquinia* sp. appears to be using penaeid shrimps as intermediate hosts, while the final hosts are fishes. No previous records exist on the presence of *Gilquinia* in the Gulf of Mexico.

The damage caused to the hepatopancreas was not reflected in the general condition of the host shrimp. When caught, parasitized individuals appeared completely normal. The histological alterations of the hepatopancreatic tissues are limited to the contact zone with the cestode cyst, suggesting that no severe damage will be visible unless there is a heavy infection.

REFERENCES

Aldrich DV (1965) Observations on the ecology and life cycle of *Prochristiniella penaei* Kruse (Cestoda: Trypanorhyncha). J Parasitol 51:370-376

Bell TA and Lightner DV (1988) A handbook of normal penaeid shrimp histology. World Aquaculture Society. Aquaculture Development Program, State of Hawaii, USA

Brock JA and Main KL (1994) A guide to the common problems and diseases of cultured *Penaeus vannamei*. World Aquaculture Society, Baton Rouge, Louisiana, USA

Conroy DA and Conroy G (1990) Manual de patología de los camarones peneidos. 2a. Ed, Maracay, Venezuela

Corkern CC (1970) Investigations on helminths from the hepatopancreas of the brown shrimp, *Penaeus aztecus* Ives, from Galveston Bay, Texas. MA thesis, Texas A&M University, College Station, Texas, USA

Couch JA (1978) Diseases, parasites, and toxic responses of commercial penaeid shrimps of the Gulf of Mexico and south Atlantic coasts of North America. Fish Bull 76:1-44

Feingenbaun DL (1975) Parasites of the commercial shrimp *Penaeus vannamei* and *Penaeus brasiliensis* Latraille. Bull Mar Sci 25:491-514

Grabda J (1991) Marine fish parasitology. An outline, Poland. Polish Scientific Publishers 306 p

Hislop JR and MacKenzie K (1976) Population studies of the whiting *Merlangius merlangus* (L.) of the northern North Sea. J Cons Int Explor Mer 37:98-111

MacKenzie K (1965) The plerocercoid of *Gilquinia squali* Fabricius, 1794. Parasitology 55

MacKenzie K (1975) Some aspects of he biology of the plerocercoid of *Gilquinia squali* Fabricius 1794 (Cestoda: Trypanorhyncha). J Fish Biol 7:321-327

Overstreet RM (1973) Parasites of some penaeid shrimps with emphasis on reared hosts. Aquaculture 2:105-140

Overstreet RM (1978) Marine Maladies? Worms, Germs and other Symbionts from the Northern Gulf of Mexico. Mississippi-Alabama Sea Grant Consortium, Blossman Printing Inc., USA

Sparks AK (1985) A synopsis of Invertebrate Pathology. Elsevier Science Publisher, The Netherlands 423 p

Toft C, Aeschlimann A and Bolis L (1991) Parasite-Host Associations. Oxford University Press, USA

Villela JB, Iversen ES and Sindermann CJ (1970) Comparasion of the parasites of pond-reared and wild pink shrimp (*Penaeus duorarum* Burkenroad) in South Florida. Trans Am Fish Bull 84: 197-200

Ward JW (1962) Helminth parasites of some marine animals, with special reference to those from the yellow-fin tuna, *Thunnus albacares* (Bonnaterre). J Parasitol 48:155-157

Wardle RA and McLeod JA (1952) Zoology of tapeworms. University of Minnesota Press, Minneapolis

Yamaguti S (1959) Systema helminthum Vol. II. The cestodes of vertebrates. Interscience Publishers. New York

HISTOLOGICAL ALTERATIONS IN *MACROBRACHIUM PANAMENSIS* CAUSED BY *PROBOPYRUS* SP.

María del Pilar Torres-García and José Luis Bortolini-Rosales

Laboratorio de Invertebrados, Facultad de Ciencias,
Universidad Nacional Autónoma de México,
Apartado Postal 70-371, México 04510, D. F., México

mptg@hp.fciencias.unam.mx

ABSTRACT

The histological damage caused by *Probopyrus* sp. to the gills of *Macrobrachium panamensis* is presented. A cellular lysis in the branchial branches was found in all the cases examined, suggesting that the bopyrids feed directly from the hosts tissues and hemolymph.

I. INTRODUCTION

The genus *Macrobrachium* includes a number of species with commercial interest, which appear infected not only by microorganisms including virus, bacteria, fungi and protozoa, but also by other crustaceans such as isopods. This phenomenon occurs more frequently when water levels are low and the area of reservoirs decrease, changing the habits of the hosts which become more sedentary. Under these circumstances the interaction among all the components of the ecosystem is more intense and prevalences increase.

The degree of damage that bopyrid isopods inflict to their hosts is variable across species. Hiraiwa and Sato (1939) found the gonads of individuals of parasitized *Penaeopsis akayeby* reduced or, in some males, completely atrophied by some parasitic isopoda. The presence of the branchial parasite, *Bopyrus squillarum*, on the shrimp, *Leander serrifer*, causes suppression of the ovaries and the breeding characteristics

of the pleopods (Yoshida 1952). Large protuberances are commonly seen in the shrimp *Hippolysmata wurdemanni*, in Biscayne Bay, Florida, where 50% of the population was reported to be parasitized and deformed by the isopod *Probopyrus* sp. (van Arman and Smith 1970). Other studies indicate that cymothoid isopods feed on the blood and/or tissue of the host, the diet depending on the site of infestation (O´Connor 1979). O´Connor observed the effects of *Irona melanostricta*, parasitic on the branchial chamber, which included pressure atrophy of the gill filaments, damage to the gill rakers, and damage to the epithelial tissue within the gill chamber.

In spite of the evidence of damage and possible loss of viability of the host, the overall effects of cymothoid isopods are not usually severe under favourable conditions, but may influence host survival under adverse conditions. The presence of *Probopyrus* sp. caused atrophy of the underlying gills and musculature of *Hippolysmata wurdemanni*, in addition to the exoeskeletal abnormality. Another species of *Probopyrus*, *P. pandalicola*, a branchial parasite of numerous species of palaemonid shrimps along the Atlantic coast of North America, produced the atrophy of the gonads of female *Palaemonetes paludosus*, by preventing ovarian maturation, while no effects on male gonads were found (Beck 1980). The secondary sex characteristics of the females were not affected, but those of the males were.

E. Escobar-Briones & F. Alvarez Eds.
MODERN APPROACHES TO THE STUDY OF CRUSTACEA
PP. 063-065

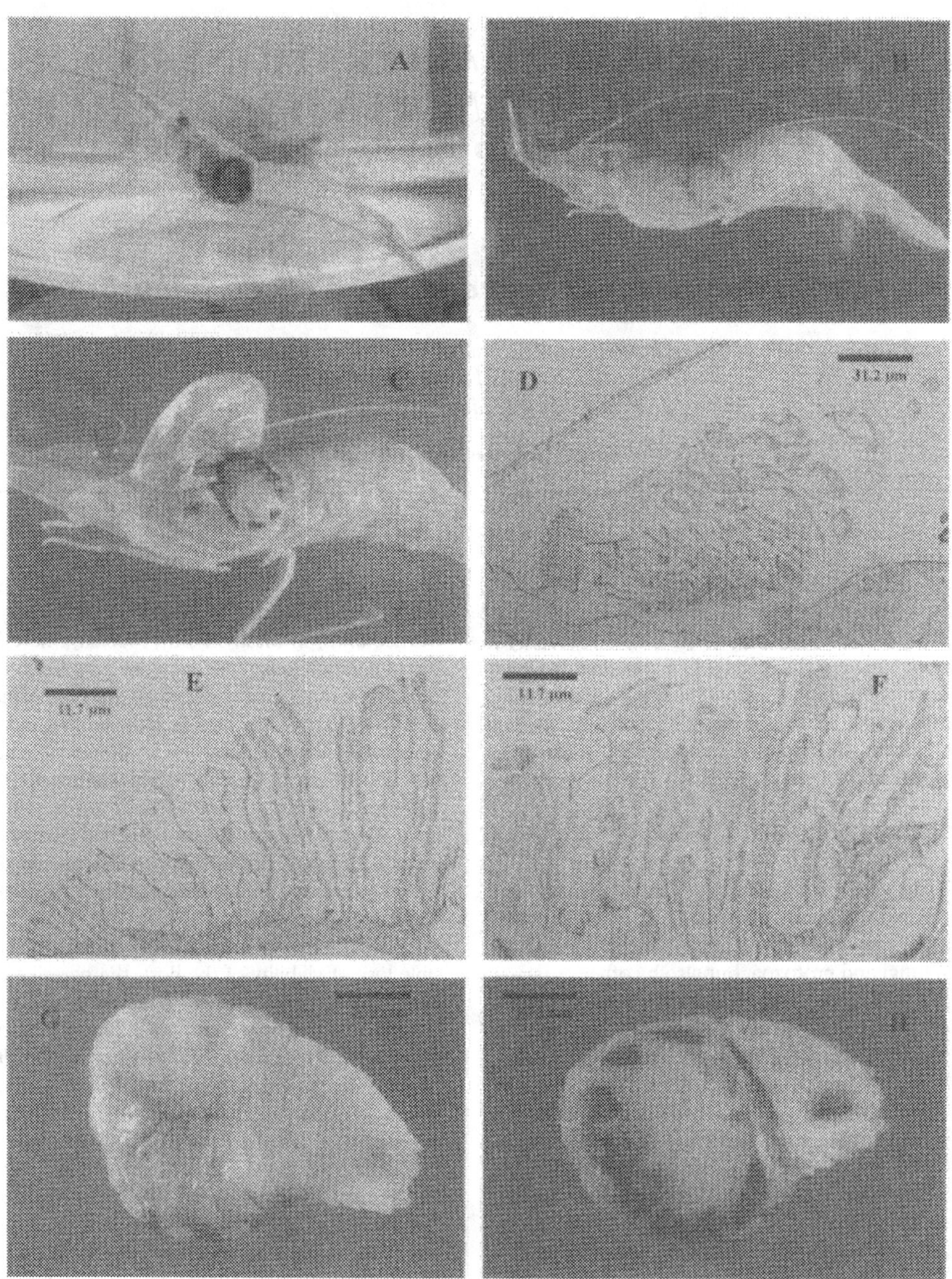

Figure 1. A, *Macrobrachium panamensis* parasitized by **Probopyrus** sp.; B-C, female of **Probopyrus** sp. in the branchial chamber of *M. panamensis*; D-E, view of branchial branches with abnormal histology of **M. panamensis**; F, detail of lysis on branchial branches; G, dorsal view of **Probopyrus** sp. female; H, ventral view of **Probopyrus** sp. female with eggs between appendages.

In this study we describe for the first time at the histological level, the damage caused by an undescribed species of *Probopyrus* to the gill tissue of *Macrobrachium panamensis*, from Ecuador.

II. MATERIALS AND METHODS

Thirteen *Macrobrachium panamensis* presenting a dark protuberance in the branchial chamber (Fig. 1, A-C) were collected near the Guayas River, south from the City of Guayaquil, Ecuador, at depths ranging between 1.5 - 2.5 m with a dragnet. These organisms were fixed in RNA friendly (R-F) fixative solution (Hasson et al. 1997), preserved in 70% alcohol (Bell and Lightner 1988), and transported to the Invertebrate Laboratory of the School of Sciences of the National Autonomous University from Mexico (UNAM). The organisms were dissected and the parasites separated for their identification. The cephalothorax of each prawn was embedded in paraffin, sections 5-7 μm thick were obtained and stained with the Hematoxylin-Eosin technique. The sections were observed and photographed with a compound microscope.

III. RESULTS AND DISCUSSION

At the morphological level, large protuberances on the branchial chamber region were observed in the hosts. In all cases, at the histological level, a cellular lysis in the branchial branches caused by the attachment organs of the parasite was observed (Fig. 1, D-F). The damage could be caused by the pressure exerted by the appendages of the parasite on the branchial branches, as well as by the direct consumption of the peripheral tissue of the branchial branches, which can be found in the oral region of the parasite. All the 13 female parasites examined were carrying eggs (Fig. 1, G-H). The presence of males was not observed in any of the females. This evidence suggests that *Probopyrus* sp. may be feeding directly on the gill tissue of *M. panamensis*, as well as on its hemolymph.

REFERENCES

Bell TA and Lightner DV (1988) A Handbook of Normal Penaeid Shrimp Histology. World Aquaculture Society. Aquaculture Development Program. State of Hawaii

Beck JT (1980) The effects of an isopod castrator, *Probopyrus pandalicola*, on the sex characters of one of its caridean shrimp hosts, *Palaemonetes paludosus*. Biol Bull 158:1-15

Hasson WK, Hasson J, Aubert H, Redman R and Lightner DV (1997) A new RNA-friendly fixative for the preservation of penaeid shrimp samples for virological detection using cDNA genomic probes. J Virol 66:227-236

Hiraiwa YK and Sato M (1939) On the effect of parasitic isopoda on a prawn, *Penaeopsis akayeby* Rathbun, with a consideration of the effect of parasitization on the higher Crustacea in general. J Sci Hiroshima Univer, Ser B7:105-124

O´Connor PF (1979) The biology, life-cycle, and host-effect of isopods (Cymothoidae) found as parasites of the Luderick, *Girella tricuspidata*. PhD Thesis, University of Sydney, Sydney, Australia

van Arman JA and Smith AC (1970) The pathobiology of an epibranchial bopyrid isopod in a shrimp, *Hippolysmata wurdemanni*. J Invertebr Pathol 15:133-135

Yoshida M (1952) On the breeding character of the shrimp, *Leander serrifer*, parasitized by bopyrids. Annot Zool Jpn 25:362-365

HEPATOPANCREAS STRUCTURE OF *PALAEMONETES ARGENTINUS* (DECAPODA, CARIDEA) FED DIFFERENT LEVELS OF DIETARY CHOLESTEROL

Ana Cristina Díaz, Liliana G. Sousa and Ana María Petriella

Departamento de Ciencias Marinas, Universidad Nacional de Mar del Plata,
Funes 3350. B7602AYL Mar del Plata, Argentina
CIC Pcia. Buenos Aires CONICET

ABSTRACT

Cholesterol is not synthesized "de novo" in crustaceans and it constitutes an essential nutrient for growth and survival, with an optimum dietary amount. In this study the effect of dietary cholesterol on hepatopancreas structure of *Palaemonetes argentinus* was evaluated. Individuals from Los Padres Lagoon (Mar del Plata, Argentina; 37°57'S, 57°44'W) were maintained in aquaria with a diet containing three cholesterol levels (0.4, 0.8, and 1.2%). After 60 days, the hepatopancreas from individuals in intermolt, premolt and postmolt were removed and processed by using standard histological techniques. No differences were found in molting rate and growth. The most important alterations were observed in postmolt prawns fed the 0.4% cholesterol diet. Besides the typical desquamated degeneration in postmolt, haemocytic infiltration, cellular dysplasia and necrosis were present. In all stages, R-cells showed hypervacuolization in prawns fed 0.8 and 1.2% cholesterol diets. The tissular changes related to the molt cycle are similar to those observed in wild prawns.

I. INTRODUCTION

Palaemonetes argentinus is distributed throughout the littoral region of Argentina and southern Brazil (Boschi 1981). It is a species of great ecological interest because it is part of the diet of fishes and birds in freshwater ponds and lagoons (Destefanis and Freyre 1972). This caridean prawn has been used as an experimental species showing a good ability to adapt to captivity.

Cholesterol has been found to be the most important sterol in crustaceans and is present in nearly all their tissues (Whitney 1969, Guary et al. 1976); however, crustaceans have been shown to be incapable of synthesizing cholesterol "de novo" from acetate (Teshima and Kanazawa 1971). Because it serves as a precursor of a number of compounds such as the molting hormone, dietary cholesterol is essential for growth and survival (Teshima 1997). This compound is also needed to maintain the membrane structure and other cellular organelles (Teshima 1981). The hepatopancreas or midgut gland is the main site for lipid storage in crustaceans. The quantitative and qualitative composition of lipids found in this organ are the result of absorption, secretion and catabolism; this reserve of lipids has its principal function in molting processes (Dall 1981, Chang and O'Connor 1983). Considering that the hepatopancreas plays an important function in the digestion and suffers histological alterations during molting, it is assumed that the level of dietary cholesterol could alter its structure affecting its metabolic role.

The hepatopancreas of *P. argentinus* is a bilobed gland, which is composed by a mass of blind tubules connected to the pyloric stomach by two primary ducts. Each tubule is lined by a simple columnar epithelium and surrounded by connective fibers and myoepithelial cells. The typical four cellular types were found along the tubular epithelium, which represent E (embryonic), F (fibrillar), R (resorptive) and B (blisterlike). There is an important desquamation in the medial and proximal zones of the tubules, particularly of B-cells undergoing holocrine secretion and degenerating R-cells. The general structure in intermolt (stage C) is coincident with that described for other decapod species (Sousa and Petriella 2000).

There are many studies about the effect of dietary cholesterol on growth, survival and molt (Kanazawa et al. 1971, Chen and Jenn 1991, Petriella 1996, Samuel et al. 1997), but hardly any work has been done on the relationship between dietary cholesterol and the hepatopancreas structure. The aim of this work was to evaluate the effect of dietary cholesterol on functional morphology of the hepatopancreas of *P. argentinus*.

II. MATERIALS AND METHODS

Experimental conditions

Adult individuals, at sexual rest, were collected from Los Padres Lagoon (Mar del Plata, Argentina; 37°57'S, 57°44'W) with a hand net. The prawns were maintained for 60 days in six aquaria (25 in each one) with freshwater and filter beds of shell and sand. The water was gently aerated at all times. They were acclimatized during a week at $20 \pm 2°C$ and a 13h:11h light/dark photoperiod.

The experiment involved feeding diets with three different cholesterol levels: 0.4, 0.8 and 1.2% ($E_{0.4}$, $E_{0.8}$, $E_{1.2}$). The basal diet (45% protein, 17% lipid, 7% moisture, 7% ash) was prepared in the laboratory using a cold extrusion method (Fenucci et al. 1981). The used animal meals were not defattened. Diet composition is shown in table 1. Each treatment was tested in duplicate.

The prawns were fed once daily, recording the presence of exuviae and dead prawns. The individuals were weighed at the beginning of the experiment and every fifteen days to the nearest 0.001 g. The molting rate was determined according to the number of molts recorded at the end of the experiment/initial number of individuals (Guary et al. 1976).

Table 1. *Percentage composition of the diet fed to Palaemonetes argentinus.*

Ingredients	Diet E
Fish meal (65.6% crude protein)	20.0
Squid meal (78% crude protein)	15.0
Mussel meal (73% crude protein)	30.0
Soybean meal (42.8% crude protein)	5.0
Wheat bran	22.0
Sodium alginate	2.0
Sodium hexametaphosphate	1.0
Cholesterol	0 - 0.5 - 1.0
Fish oil	1.0
Fish soluble	2.0
Vitamins*	2.0

Expressed as g kg⁻¹ dry matter in whole wheat meal: cholecalciferol 1.8; thiamin 8.2; riboflavin 7.8; pyridoxine 10.7; calcium panthothenate 12.5; biotin 12.5; niacin 25.0; folic acid 1.3; B_{12} HCl 1.0; ascorbic acid Rovimix Stay C 39.1; menadione 1.7; inositol 0.3; choline chloride 0.2; a tocopherol acetate 75; vitamin A acetate 5.

Chemical analysis

The content of cholesterol in the diets was analyzed by GLC (Gas Liquid Chromatography) using a Hewlett Packard 5840A chromatograph equipped with a 12 m x 0.2 mm fused silica capillary column coated with metil silicone fluid. The cholesterol was chromatographed between 200 and 280°C at a rate of 10°C/min, with nitrogen as the carrier gas. Quantification was made by automatic integration of the peaks before and after addition of a known amount of 5a-colestan used as an internal standard.

Statistical analysis

The data of mean weight, survival and molting rate were compared using Cochran, t-Student and χ^2 tests (Sokal and Rohlf 1995).

Histology

Twenty wild prawns (at different molt stages) were used as controls. At the end of the trial, a sample of ten individuals per aquarium were randomly collected and stocked according to the molt stage, which was determined by the microscopic observation of the setae on the external ramus of the uropod (Díaz et al. 1998). The hepatopancreas of prawns in intermolt, premolt and postmolt were removed and fixed in Davidson fluid (ethanol, acetic acid, formol, and distilled water) (Bell and Lightner 1988). They were dehydrated in progressive series of ethanol, butyl alcohol, butyl-paraffin and embedded in paraffin. Sections 3 μm thick were stained with hematoxylin-eosin. Histopathological effects (Meyers and Hendricks 1985) of dietary cholesterol were studied on treated specimens in comparison to the untreated ones.

III. RESULTS

Survival in the three cholesterol levels was relatively high (73, 90 and 80% with $E_{0.4}$, $E_{0.8}$ and $E_{1.2}$ diets, respectively). Statistical differences in survival were found between the 0.4 and 0.8% cholesterol diets (χ^2 test, $P < 0.05$), but there were no differences in molting rate and weight increment among the treatments (t-Student, $P > 0.05$) (Table 2). The percentage of animals in the different molt stages at the end of the experiment is presented in Table 3.

Table 2. *Wet weight (g) and molting rate of **Palaemonetes argentinus**.*

Diet	Initial mean weight		Final mean weight		MR
	x	± s	x	± s	
E0.4	0.110	0.035	0.119	0.006	1.59±0.177
E0.8	0.107	0.011	0.117	0.011	1.39±0.156
E1.2	0.110	0.007	0.122	0.003	1.49±0.219

x: mean weight of two aquaria, s: standard deviation, MR: mean molting rate of two aquaria.

Table 3. *Percentage of individuals at the different molt stages at the end of the trial.*

Stage	E0.4	E0.8	E1.2
A-B (postmolt)	4%	16.6%	9.1%
C (intermolt)	36%	33.3%	59%
D_0-D_2 (premolt)	60%	50%	31.8%

Regarding the hepatopancreas, some alterations were found according to the molt stage. In wild animals the hepatopancreas in intermolt presents the four typical cells: E, F, R, and B. E-cells undergo mitosis during all the molt cycle. There is degenerative desquamation of R and B-cells in the proximal zone of the tubules (Fig. 1), which increases towards the end of premolt and in postmolt (Fig. 2). R-cells show a higher content of lipidic vacuoles in premolt (Figure 3). Some of the tubules present a folded basal lamina at the end of premolt and during postmolt (Sousa, pers. obs.).

0.4% cholesterol

The hepatopancreas from animals in premolt and intermolt stages present similar characteristics to those of wild animals. In postmolt, important haemocytic infiltration, cellular dysplasia and necrosis of B-cells in the proximal zone are observed, and the basal lamina presents infoldings in some of the tubules (Fig. 4).

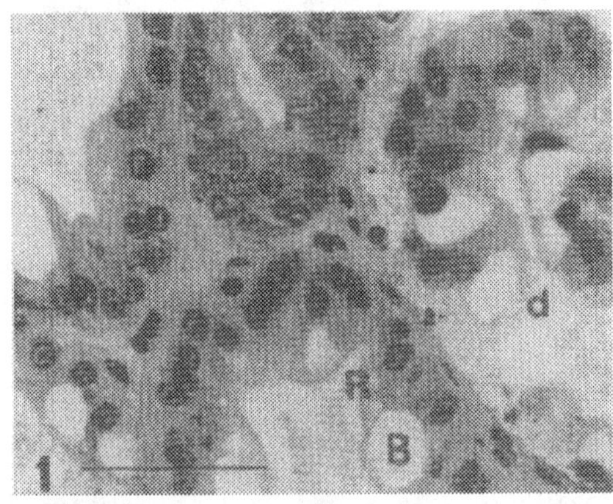

Figure 1. *Hepatopancreas of wild **Palaemonetes argentinus** in intermolt. B: B-cell; d: degenerative desquamation; F: F-cell; R: R-cell. Scale bar: 50µm.*

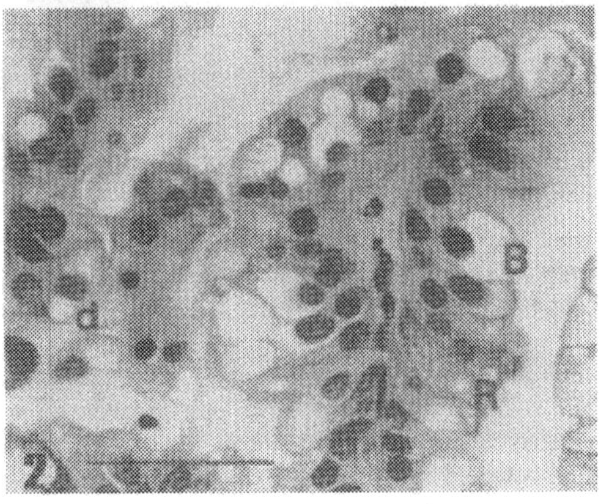

Figure 2. *Hepatopancreas of wild **Palaemonetes argentinus** in postmolt. B: B-cell; d: degenerative desquamation; R: R-cell. Scale bar: 50µm.*

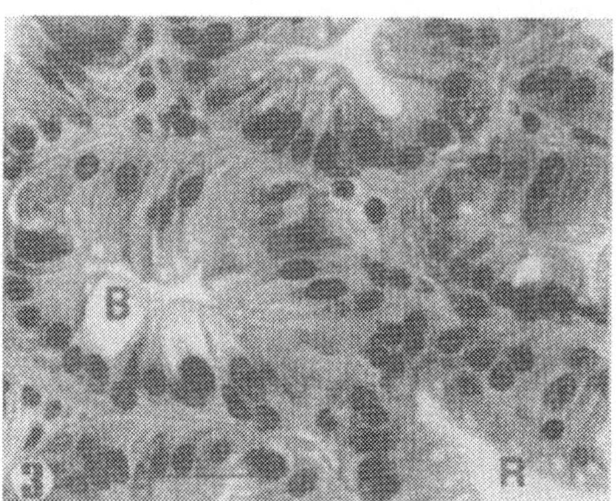

Figure 3. *Hepatopancreas of wild **Palaemonetes argentinus** in premolt. B: B-cell; F: F-cell; R: R-cell. Scale bar: 50μm.*

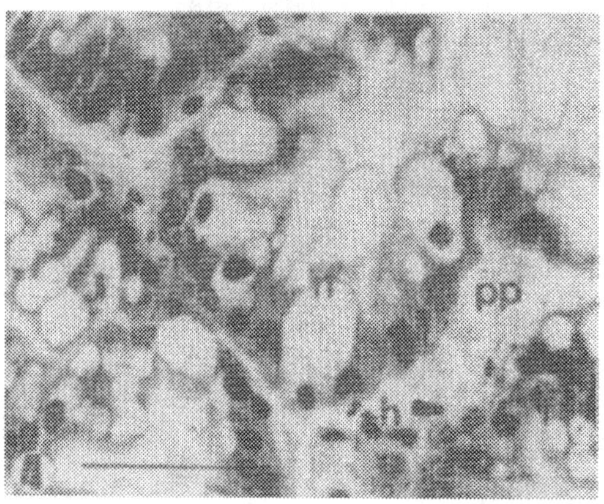

Figure 4. *Hepatopancreas in postmolt of **Palaemonetes argentinus** fed 0.4% cholesterol. h: haemocytic infiltration; n: necrotic B-cell; pp: protein precipitation. Scale bar: 50μm.*

0.8% cholesterol

The most conspicuous alterations were found in intermolt and premolt. In intermolt the basal lamina appears folded in some of the tubules and the R-cells show an intensive vacuolization (Fig. 5). In premolt, R-cells are hipervacuolated and the basal lamina is interrupted in some tubules. Haemocytic infiltration and protein precipitation in the intertubular space are also observed.

1.2% cholesterol

In intermolt, the epithelium is hyperthophied and there is an important vacuolization of R-cells (Fig. 6). The basal lamina is well conserved and there is a protein precipitation in the intertubular space. During premolt and postmolt the tubules are very well conserved, the general appearance is similar to the gland of wild animals.

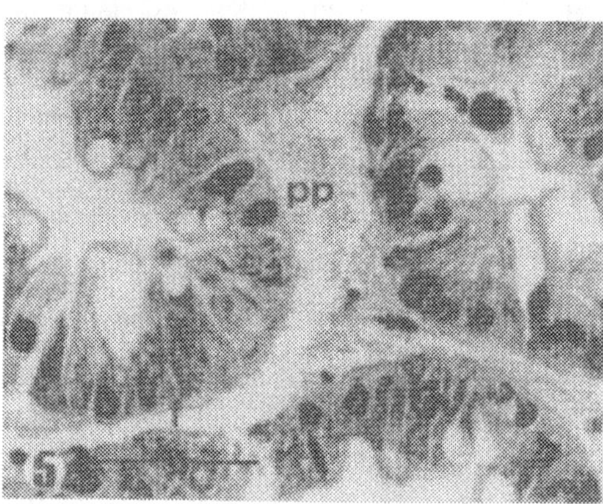

Figure 5. *Hepatopancreas in intermolt of **Palaemonetes argentinus** fed 0.8% cholesterol. f: folded basal lamina; pp: protein precipitation. Scale bar: 50μm.*

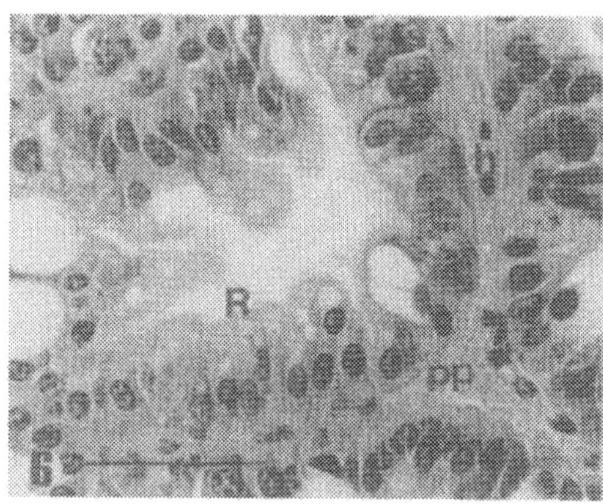

Figure 6. *Hepatopancreas in intermolt of **Palaemonetes argentinus** fed 1.2% cholesterol. h: haemocytic infiltration; pp: protein precipitation; R: multivacuolated R-cell. Scale bar: 50μm.*

IV. DISCUSSION

Several researchers have demonstrated the necessity of dietary cholesterol for optimal growth and survival in decapods (Kanazawa et al. 1971, Castell et al. 1975, D'Abramo et al. 1984). The optimum levels of dietary cholesterol for different shrimp species ranged from 0.12 to 2.0% of the diets. There is no general agreement about the exact cholesterol requirement for crustaceans because the physiological responses to nutrients are, as a rule, graded (Sheen et al. 1994).

The three cholesterol levels used in the present study produced normal growth and high survival of *P. argentinus*. However the lowest concentration (0.4%) reduced the survival but did not affect growth. In *Macrobrachium rosenbergii* a supplement of 0.5 to 1% cholesterol improved growth and survival, meanwhile a semi-purified diet containing 0.12% did not improve growth (Teshima 1997). Petriella (1990) found that survival decreased in individuals of the Argentine prawn *Artemesia longinaris* fed with low cholesterol diets, however survival increased when cholesterol in the diet increased to 2.1%.

Regarding the molting rate, Guary et al. (1976) determined that a 1 % total lipid diet increased the molting rate in *Penaeus japonicus*. Castell et al. (1975), working with juvenile *Homarus americanus*, reported the greatest molting rate with a 1% cholesterol diet. Petriella (1996) determined that in *A. longinaris*, the higher the percentage of cholesterol, the higher the molting frequency. In the present study, the molting rate was not affected significantly by the diet; maybe diets containing more than 1.2% cholesterol may affect molting frequency.

Several parameters are used in determining the effectiveness of a diet. Growth and survival are the most common parameters, but the effectiveness of the diet is not recognized until after several weeks of feeding. In contrast, histological changes in the liver, hepatopancreas, gills and other internal organs due to the nutritional stress are detected relatively early (Storch et al. 1984). The midgut gland of decapods plays an important role in lipid metabolism; however, the midgut gland lipids, although not important during periods of starvation, may have a central role during molting, with perhaps a secondary role in oogenesis (Dall 1981).

The hepatopancreas of *P. argentinus* in intermolt is similar in structure to the rest of decapods (Sousa and Petriella 2000). In this study the tissular changes related to the molt cycle were similar to those observed in wild prawns. The hepatopancreas from individuals fed diet $E_{0.4}$ showed important alterations in postmolt. The characteristics of these hepatopancreas, such as dysplasia, necrosis of B-cells, and the presence of infoldings in the basal lamina of the tubules, were observed by several authors in different decapod species after a long period of starvation (Storch and Anger 1983, Strus 1987). In wild *P. argentinus* we never observed dysplasia and necrosis of B-cells, so these are apparently diet effects.

Other authors noted infiltration of haemocytes in the hepatopancreas and observed a correlation between the nutritional state of the animals and haemocyte counts (Rodriguez Souza et al. 1996). Haemocytes appear to have a role in several physiological functions such as haemostasis, ecdysis, storage of glycogen, production of haemocyanin, and defense reactions. Haemocytes are also involved in coagulation of haemolymph, agglutinating and producing substances, which precipitate the haemolymph proteins (Martin and Hose 1992).

In most crustaceans there is a premolt increase in lipid content, probably due to an increased capacity for esterification of ingested fatty acids (Chang and O'Connor 1983). In *Orconectes virilis* and *Gecarcinus lateralis*, the incorporation of radiolabeled acetate into hepatopancreatic fatty acids was markedly elevated in premolt when compared to tissue obtained from intermolt organisms (O'Connor and Gilbert 1968). In contrast, hypervacuolization of R-cells was observed in the hepatopancreas of *P. argentinus* in intermolt when fed with $E_{0.8}$ and $E_{1.2}$, this is doubtlessly the result of an increment in lipid reserves and the hypertrophy of the epithelium reflects an increased cellular activity. Other authors have reported that dietary cholesterol increased the protein and lipid contents of tissues (Chen and Jenn 1991, Ramesh and Kathiresan 1992, Samuel et al. 1997).

From the results of this study it is possible to say that the three cholesterol levels (0.4, 0.8 and 1.2%) promote growth and survival during 60 days, although diet $E_{0.4}$ produces detrimental effects in the hepatopancreas structure. Maybe the important dynamic and the high cellular turnover rate (Sousa and Petriella 2000) allows this gland to maintain its functions during relatively long periods of time, considering that the intermolt period in this species is about 20 days under similar experimental conditions (Díaz et al. 1998). Although further investigations are necessary to determine the optimum level of cholesterol for *P. argentinus*, the present study shows that a diet containing 0.4% of cholesterol seems to be insufficient to complete the molt cycle in a good nutritional condition. A low content of cholesterol affects the histological structure of the hepatopancreas and, in the long term, its function, with reduction of the metabolic activity. Histological variation between diets is detected much earlier than differences in growth or survival, suggesting that histological criteria constitute a practical means for the preliminary assessment of the acceptability and nutritional value of diets (Rodriguez Souza et al. 1996). This phenomenon was also proved by Vogt et al. (1986) for the hepatopancreas of *Penaeus monodon* and in the present work for *P. argentinus*.

REFERENCES

Bell TA and Lightner DV (1988) A handbook of Normal Penaeid Shrimp Histology. The World Aquaculture Society, Baton Rouge

Boschi EE (1981) Decapoda Natantia. (pp 1-61) In: Fauna de agua dulce de la República Argentina. Vol. 26, FECIC, Buenos Aires

Castell JD, Mason EG and Covey JF (1975) Cholesterol requirements of juvenile American lobster (*Homarus americanus*). J Fish Res Bd Can 32:1431 1435

Chang ES and O'Connor JD (1983) Metabolism and transport of carbohydrates and lipids. (pp 263-287) In: Bliss D (ed.) The Biology of Crustacea. Vol. 5 Academic Press, New York

Chen HY and Jenn JS (1991) Combined effects of dietary phosphatidylcholine and cholesterol on the growth, survival and body lipid composition of marine shrimp, *Penaeus penicillatus*. Aquaculture 96:167-178

D'Abramo LR, Bordner CE, Conklin DE and Baum NA (1984) Sterol requirement of juvenile lobsters, *Homarus* sp. Aquaculture 42:13-25

Dall W (1981) Lipid absorption and utilization in the Norwegian lobster, *Nephrops norvergicus* (L.). J Exp Mar Biol Ecol 50:33-45

Destefanis S and Freyre L (1972) Relaciones tróficas de los peces de la laguna de Chascomús con un intento de referenciación ecológica y tratamiento bioestadístico del espectro trófico. Acta Zool Lilloana 29:17-53

Díaz AC, Sousa LG and Petriella AM (1998) Setogenesis and growth of the freshwater prawn *Palaemonetes argentinus* (Decapoda, Caridea, Palaemonidae). Iheringia, Sér. Zoologia 85:59-65

Fenucci JL, Müller MI and Petriella AM (1981) Efecto de la alimentación natural y artificial en el crecimiento del camarón *Artemesia longinaris* Bate. Rev Lat Acui 10:10-17

Guary JC, Kayama M, Murakami Y and Ceccaldi H (1976) The effects of a fat-free diets and compounded diets supplemented with various oils on molt, growth and fatty acid composition of prawn *Penaeus japonicus* Bate. Aquaculture 7:245-254

Johnston MA, Elder HY and Davies PS (1973) Cytology of *Carcinus* haemocytes and their function in carbohydrates metabolism. Comp Biochem Physiol 46:569-583

Kanazawa A, Tanaka N, Teshima S and Kashiwada K (1971) Nutritional requirements of prawn. II. Requirement for sterols. Bull Jap Soc Sci Fish 37:211 215

Martin GG and Hose JE (1992) Vascular elements and blood (hemolymph). (pp 117-149) In: Harrison FW and Humes AG (eds) Microscopic Anatomy of Invertebrates, Vol. 10 Wiley-Liss, New York

Meyers TR and Hendricks JD (1985) Histopathology. (pp 283-331) In: Rand GM and Petrocelli SR (eds) Fundamentals of Aquatic Toxicology. Taylor & Francis, USA

O'Connor JD and Gilbert LI (1968) Aspects of lipid metabolism in crustaceans. Am Zool 8: 529-539

Petriella AM (1990) Study of the molting cycle of the Argentine prawn *Artemesia longinaris* Bate. III. Influence of cholesterol. J Aqua Trop 5:77-85

Petriella AM (1996) Effect of dietary cholesterol upon setogenesis and molting frequency in the Argentine prawn *Artemesia longinaris* Bate (Crustacea, Decapoda, Penaeidae). J Aqua Trop 11:167-174

Ramesh MX and Kathiresan K (1992) Mangrove cholesterol in the diet of penaeid prawn *Penaeus indicus*. Indian J Mar Sci 21:164-166

Rodríguez Souza JC, Sekine S, Suzuki S, Shima Y, Strüssmann CA and Takashima F (1996) Usefulness of histological criteria for assessing the adequacy of diets for *Panulirus japonicus* phyllosoma larvae. Aquacult Nutr 2:133-140

Samuel MJ, Soundarapandian P and Kannupandi T (1997) Impact of dietary cholesterol on the growth and conversion efficiency of the freshwater prawn *Macrobrachium malcolmsonii* (H. Milne Edwards). Isr J Aquacult 49:3-11

Sheen SS, Liu PC, Chen SN and Chen JC (1994) Cholesterol requirement for juvenile tiger shrimp (*Penaeus monodon*). Aquaculture 125:131-137

Sokal R. and Rohlf J (1995) Biometry. WH Freeman, New York

Sousa GL and Petriella AM (2000) Histology of the hepatopancreas of the freshwater prawn *Palaemonetes argentinus* (Crustacea, Caridea). Biocell 24:189-195

Storch V and Anger K (1983) Influence of starvation and feeding on the hepatopancreas of larval *Hyas araneus* (Decapoda, Majidae). Helgol Meeresunters 36:67-75

Storch V, Juario JV and Pascual FP (1984) Early effect of nutritional stress on the liver of milk fish *Chanos chanos* (Forsskal) and on the hepatopancreas of the tiger prawn *Penaeus monodon* (Fabricius). Aquaculture 36:229-236

Strus J (1987) The effects of starvation on the structure and function of the hepatopancreas in the isopod *Ligia italica*. Inv Pesq 51:505-514

Teshima S (1981) Sterol metabolism. (pp 205-216) In: Pruder GD, Landgon C and Conklin D (eds) Proceedings of the second international conference on aquaculture nutrition: biochemical and physiological approaches to shellfish nutrition. Special Publication N° 2 Louisiana State University Division, Baton Rouge

Teshima S (1997) Phospholipids and sterols. Advances in World Aquaculture 6:85-107

Teshima S and Kanazawa A (1971) Biosynthesis of sterols in the lobster *Panulirus japonica*, the prawn *Penaeus japonicus* and the crab *Portunus tuberculatus*. Comp Biochem Physiol 38B:597-602

Whitney JO (1969) Sterol, fatty acids and sterol content in eggs and hepatopancreas of the blue crab *Callinectes sapidus* (Rathbun). Acta Embriol Exp 1969:111-121

COMPARATIVE EVALUATION OF DIFFERENT ANIMAL PROTEIN SOURCE IN JUVENILES OF *PLEOTICUS MUELLERI* (CRUSTACEA, PENAEOIDEA)

Ana Cristina Díaz and Jorge L. Fenucci

*Departamento de Ciencias Marinas, Universidad Nacional de Mar del Plata, Funes 3350, B7602AYL
Mar del Plata, Argentina.CIC Pcia. Buenos Aires
CONICET*

acdiaz@mdp.edu.ar

ABSTRACT

A study was conducted to evaluate alternative protein supplements that could be used to reduce the cost of formulated diets for shrimp. Marine proteins, such as fish meal, squid meal, and bivalve meal, are considerably more expensive than byproduct meals such as meat and bone meal. The objective of this study was to compare the growth, survival, and body composition of juvenile *Pleoticus muelleri* fed diets containing different levels of meat and bone meal as a substitute of the marine fish meal. Feeding trials were carried out on early juveniles (0.70 ± 0.05 g initial weight) held in 150 l glass aquaria (33‰ salinity, 20°C, 12:12 h photoperiod, 20 shrimp/m² density). Each diet (control, PL1, PL2, PL3, PL4) was tested in three replicate groups of 10 shrimps during 50 days. The juveniles were obtained from hatchery-raised postlarvae (wild broodstock from Mar del Plata, Argentina) at Nagera Station, dependent of the Marine Science Department (Mar del Plata National University, Argentina). Four isoproteic and isolipidic diets (34% crude protein) were prepared to contain 0, 11, 15 and 23% meat and bone meal in substitution of fish meal. The control group was fed with fresh squid mantle. Percentage increment in mean weight varied between 81 and 103%. Survival rates ranged from 50 to 76%. No significant differences were detected in final weight gain or survival among dietary treatments ($P < 0.05$). Whole body moisture, ash, crude protein, and lipid content of shrimps were not affected by diet ($P < 0.05$). The feeding experiment suggests that meat and bone meal can be utilized by *P. muelleri* as a suitable replacement for fish meal in a formulated diet.

I. INTRODUCTION

The design of inexpensive high quality diets is extremely important as feeding costs represent more than 50% of the production costs for most aquaculture enterprises (Lovell 1989). The high level of protein required in shrimp diets increases the cost and makes of the feed a major expense in shrimp production. Therefore, it is very important to identify low-cost protein-rich ingredients in order to lower shrimp feed production cost.

Variations in protein requirements are attributed mainly to different sources of proteins used in diets. Cost reduction is possible in several major areas including: better definition of nutrient requirements and ingredient characteristics, more precise formulation methods, and improved processing procedures (Chamberlain 1995). Selection of an appropriate protein source for cost reduction is important. This study was conducted to evaluate alternative protein supplements that could be used to reduce the cost of shrimps feed. Many agriculture and livestock activities generate byproducts, some of which have been used as animal feed

![Kluwer logo] Kluwer Academic/Plenum Publishers

E. Escobar-Briones & F. Alvarez Eds.
MODERN APPROACHES TO THE STUDY OF CRUSTACEA
PP. 075-078

ingredients. Meat meals are the principal by-products of animal slaughterhouse; including meat scraps and trimmings. The quality of meat meal as a protein supplement depends on its production process as well as on the raw material used (Tacon and Akiyama 1997). Argentina produces meat meal on a commercial scale and its use as part of shrimp diets has been shown to be important.

The aim of this study was to compare the growth, survival, and body composition of juvenile Argentine red shrimp *Pleoticus muelleri* fed diets containing different levels of meat and bone meal as a substitute of the marine fish meal.

II. MATERIALS AND METHODS

Feeding trials were carried out with 150 early juveniles (0.70 ± 0.05 g initial weight) held in 150 l glass aquaria with under gravel filter and a sand and crushed shell bed. The juveniles were reared from hatchery-raised postlarvae (wild broodstock from Mar del Plata, Argentina) at Nagera Station, of the Marine Science Department, Mar del Plata National University, Argentina.

The treatments consisted of four diets (34% crude protein) that were prepared to contain 0, 11, 15 and 23% meat and bone meal in substitution of fish meal. Percent composition of the experimental diets is given in Table 1. Formulations were made according to the chemical composition results of the by-product meal in order to obtain isoproteic and isolipidic diets. The chemical composition of the diets was confirmed through proximate analysis (Table 2) according to AOAC (1990). All ingredients were mixed and cold pelleted (<50°C) by extrusion (Fenucci and Zein-Eldin 1976) to obtain 3 mm diameter pellets. The pellets were oven-dried for 24 h at 50°C.

The control group was fed with fresh squid mantle *(Illex argentinus)*. The animals were fed *ad libitum* once a day. Feeding rate was adjusted daily in each tank in order to maintain feed waste at a minimum. Diets (control, 1, 2, 3, 4) were tested in three replicate groups of 10 shrimps randomly chosen, during 50 days. The experimental conditions were 33‰ salinity, 20±1°C temperature and 12:12 h photoperiod. Shrimps were stocked at a density of 20m². Feed intake, mortality and presence of exuviae were recorded daily. Individual shrimp weights were determined at the beginning of the experiment and after 50-days. At the end of the experiment, animals were anesthetized on ice and were pooled by treatment for the analysis of growth and whole body composition.

Growth performance and survival were measured in terms of final individual weight, percentage of gain weight [(final mean weight–initial mean weight)/ initial mean weight] x100, and percentage survival.

Whole body composition was analyzed for dry matter (by oven drying to constant weight at 100°C), crude protein (Kjeldahl, nitrogen x 6.25), total lipid (Soxhlet extraction with petroleum ether bp 35-60°C for 6 h) and ash (residue muffle furnace ignition at 550°C).

To determine differences among diets data were analyzed with analysis of variance (ANOVA). A χ^2 test was used to assess survival. Statistical significance was accepted at the $P < 0.05$ level (Sokal and Rohlf 1995).

Table 1. *Percent composition of experimental diets.*

Ingredient (g/100g dry diet)	Diets			
	1	2	3	4
Fish meal (65% crude protein)	35.0	27.0	39.0	48.0
Meat and bone meal (61% crude protein)	15.0	23.0	11.0	-
Soybean meal (42% crude protein)	17.0	17.0	17.0	17.0
Manioc starch	20.0	20.0	20.0	20.0
Wheat	7.5	7.5	7.5	9.5
Fish oil	2.0	2.0	2.0	2.0
Fish soluble	2.0	2.0	2.0	2.0
Lecithin	0.5	0.5	0.5	0.5
Cholesterol	0.5	0.5	0.5	0.5
Vitamin supplement*	0.5	0.5	0.5	0.5

* Values are g/kg: cholecalciferol, 1.8; thiamin, 8.2; riboflavin, 7.8; pyridoxine, 10.7; calcium panthothenate, 12.5; biotin, 12.5; niacin, 25.0; folic acid, 1.3; B12 HCl, 1.0; ascorbic acid Rovimix Stay C, 39.1; menadione, 1.7; inositol, 0.3; choline chloride, 0.2; a-tocopherol acetate, 75; vitamin A acetate, 5.

Table 2. *Proximate composition of the experimental diets.*

	Diet				
	1	2	3	4	Squid
Moisture	7.02	7.30	7.11	7.05	82.53
Crude protein*	36.88	33.38	34.05	30.28	78.04
Total lipid*	6.48	5.48	4.92	5.66	14.96
Ash	6.38	5.45	4.74	6.92	1.93

* Dry matter.

III. RESULTS

Neither a clear tendency nor a significant difference was obtained for any of the studied parameters. Shrimps fed diets containing 0, 11, 15, and 23% of meat and bone meal in substitution of fish meal exhibited similar growth patterns throughout the 50-day period. Percentage increment in mean weight varied between 81 and 103%. No significant differences in growth rates among dietary treatments were obtained (ANOVA, $P > 0.05$). Weight gain and survival data for shrimps fed the various diets are presented in the Table 3.

Survival in all treatments was high, ranging between 50 and 76%. However, mean survival was higher with the formulated diets (67-77%) than with the control diet (50%).

Whole body moisture, ash, crude protein, and lipid content were not significantly affected by the diet ($P < 0.05$). Final body composition data are presented in Table 4, with protein percentage going from 74 to 85%.

Table 3. *Weight gain and survival of juveniles fed different diets.*

Diet	n_i	Mean weight (g)*		n_f	%Δw	% S
		Initial	Final			
PL1	30	0.71 ± 0.051	1.33 ± 0.140	22	85	73.3
PL2	30	0.70 ± 0.012	1.26 ± 0.095	20	81	66.7
PL3	30	0.70 ± 0.005	1.34 ± 0.102	22	90	73.3
PL4	30	0.69 ± 0.041	1.26 ± 0.190	23	83	76.7
Squid	30	0.73 ± 0.035	1.48 ± 0.108	15	103	50.0

*Values are means ± standard error of triplicates. n_i and n_f: initial and final number of individuals; %Δw: percentage in mean weight; % S: survival.

IV. DISCUSSION

A number of studies have tried to determine the suitability of various protein sources and optimum protein levels for supporting crustacean growth (Piedad-Pascual et al. 1990, Cruz-Suárez et al. 1993, Tidwell et al. 1993b, Reigh and Ellis 1994, Lim 1996). In addition, other factors such as amino acid composition, mineral content and digestibility of various protein sources may also be important. The diets used in this study were calculated to be isoproteic and isolipidic; therefore it is assumed that quality of the protein is due to the substitutions made in the diet with various levels of bone and meat meal. Meat and bone meal protein has variable quality and usually is deficient in methionine and tyrosine. Amino acids considered essential for

Table 4. *Proximate composition of shrimps fed different diets.*

Diet	Moisture	Crude protein*	Total lipid*	Ash
1	80.3	83.02	1.76	7.94
2	80.5	85.05	1.34	7.88
3	80.2	76.38	2.14	8.07
4	81.2	74.12	2.43	10.35
Squid	81.4	76.39	1.82	8.53

* Dry matter.

shrimp are methionine, arginine, threonine, tryptophan, histidine, isoleucine, leucine, valine and phenylalanine (Cowey and Forster 1971, Shewbart et al. 1972, Kanazawa and Teshima 1981). Tyrosine is believed to be synthesized from phenylalanine (Akiyama 1992). In the present study the level of this compound in the fish meal probably satisfies the methionine requirement in the Argentine red shrimp.

While there is no doubt of the high nutritional value of fish meal and other ingredients used in diets for cultured shrimps, they represent an expensive and finite commodity which is also competitively utilized for livestock feeds or human consumption. The use of compound feeds is rapidly increasing; Wijkstrom and New (1989) reported the potential resource limitation that fish meal availability would cause on further expansion of the aquaculture industry. The world supply of fish meal will likely remain stable, so there will be an increasing need to reduce the level of fish meal in aqua feeds (Chamberlain 1993). Meat and bone meal has been used as an alternative protein supply due to its low price compared to fish meal (Watanabe et al. 1993). Tidwell et al. (1993a) working with *Macrobrachium rosenbergii* used diets containing 0 and 8% meat and bone meal, and did not find differences in mean weight increment and survival. Results of this study show that the substitution of almost 21% of fish meal for bone and meat meal does not affect the growth and survival of *P. muelleri* juveniles.

Despite the fact that decapods, both in nature and in the laboratory, seem to prefer other crustaceans as food, economical and logistical reasons have made mollusks, such as bivalves and cephalopods, the commonest natural food item used in crustacean grow-out operations. Apparently, this practice is more related to the availability and cost of these products than to the real preferences of shrimps. However, in *P. muelleri* the best results in weight gain were obtained feeding squid mantle. From the commercial point of view the use of squid is not feasible due to its high price and the poor survival of the shrimps.

Terrestrial animal by-products are not generally used in shrimp feeds because the saturated lipid contents of these ingredients are believed to reduce the reproductive performance of shrimps (Akiyama et al. 1991, Jory 1995). In contrast, the present feeding experiment suggests that meat and bone meal can be utilized by *P. muelleri* as a suitable replacement for fish meal in formulated diet at levels of 23% of the total formulation.

REFERENCES

AOAC (1990) Official methods of analysis of the Association of Official Chemists. 15th edition, AOAC Inc., Arlington, Virginia, USA

Akiyama DM (1992) Future considerations for shrimp nutrition and the aquaculture feed industry. (pp 198-205) In: Wyban J (ed) Proceedings of the Special Session on Shrimp Farming. World Aquaculture Society, Baton Rouge

Akiyama DM, Dominy GW and Lawrence AL (1991) Penaeid shrimp nutrition for the commercial feed industry: Revised. (pp 80-98) In: Akiyama DM and Tan RKH (eds) Proceedings of the Aquaculture Feed Processing and Nutrition Workshop, Singapore

Chamberlain GW (1993) Aquaculture trends and feed projections. World Aquaculture 24:19-29

Chamberlain GW (1995) Frontiers in shrimp nutrition research. (pp 108-117) In: Browdy CL and Hopkins JS (eds) Proceedings of the Special Session on Shrimp Farming. World Aquaculture Society

Cowey CB and Forster JRM (1971) The essential amino acid requirements of the prawn, *Palaemon serratus*. Mar Biol 10:77-81

Cruz-Suárez LE, Ricque-Marie D, Martínez-Vega JA and Wesche-Ebeling P (1993) Evaluation of two shrimp by-product meals as protein sources in diets for *Penaeus vannamei*. Aquaculture 115:53-62

Fenucci JL and Zein-Eldin ZP (1976) Evaluation of squid mantle meal as a protein source in penaeid nutrition. FAO Technical Conference on Aquaculture, Kyoto, Japan, 76/E.36, 9 p

Jory DE (1995) Feed management practices for a healthy pond environment. (pp 118-143) In: Browdy CL and Hopkins JS (eds) Proceeding of the Special Session on Shrimp Farming World Aquaculture Society

Kanazawa A and Teshima S (1981) Essential amino acids of the prawn. Nippon Suisan Gakkaishi 47:1375-1377

Lim C (1996) Substitution of cottonseed meal for marine animal protein in diets for *Penaeus vannamei*. J World Aquaculture Soc 27:402-409

Lovell T (1989) Nutrition and feeding of fish. Ed. Van Nostrand Reinhold, New York

Piedad-Pascual F, Cruz EM and Sumalangcay Jr A (1990) Supplemental feeding of *Penaeus monodon* juveniles with diets containing various levels of defatted soybean meal. Aquaculture 89:183-191

Reigh RC and Ellis SC (1994) Utilization of animal-protein and plant-protein supplements by red swamp crayfish *Procambarus clarki* fed formulated diets. J World Aquaculture Soc 25:541-552

Shewbart KL, Meis WL and Ludwig PD (1972) Identification and quantitative analysis of the amino acids present in protein of the brown shrimp, *Penaeus aztecus*. Mar Biol 16:64-67

Sokal R and Rohlf J (1995) Biometry. WH Freeman, New York

Tacon AGJ and Akiyama DM (1997) Feed ingredients. Advances in World Aquaculture 6:411-472.

Tidwell JH, Webster CD, Clark JA and D'Abramo LR (1993a) Evaluation of distillers dried grains with solubles as an ingredient in diets for pond culture of the freshwater prawn *Macrobrachium rosenbergii*. J World Aquaculture Soc 24:66-70

Tidwell JH, Webster CD, Yancey DH and D'Abramo LR (1993b) Partial and total replacement of fish meal with soybean meal and distillers' by-products in diets for pond culture of the freshwater prawn (*Macrobrachium rosenbergii*). Aquaculture 118:119-130

Watanabe T, Pongmaneerat J, Sato S and Takeuchi T (1993) Replacement of fish meal by alternative protein sources in rainbow trout diets. Nippon Suisan Gakkaishi Bulletin 59:1573-1579

Wijkstrom UN and New MB (1989) Fish for feed: a help or a hindrance to aquaculture in 2000? INFOFISH International 6/89:48-52

EFFECT OF DIFFERENT DIETARY ARGININE AND LYSINE LEVELS FOR ARGENTINE PRAWN *ARTEMESIA LONGINARIS* BATE (CRUSTACEA, DECAPODA, PENAEIDEA)

Rosangela Romanos Mangialardo and Jorge L. Fenucci

Departamento de Ciencias Marinas, Facultad de Ciencias Exactas y Naturales,
Universidad Nacional de Mar del Plata, Funes 3350, (B7602AYL),
Mar del Plata, Argentina
Becaria FONCYT, CONICET

ABSTRACT

Amino acids are essential for synthesis of tissular proteins and many other important compounds. It has been determined a lysine - arginine dietary relationship and it is believed that the ratio of these compounds in diets should be maintained between 1:1 and 1:1.1. Three experiments were conducted using *Artemesia longinaris* to determine the effect of different arginine:lysine ratios in diets. The experimental diets contained 1:4, 1:1, 1:2 and 2:1 arginine to lysine ratios. Prawns exhibited no significant differences in growth and survival among treatments.

I. INTRODUCTION

The Argentine prawn *Artemesia longinaris*, distributed in the southwestern Atlantic from Espirito Santo, Brazil to Chubut, Argentina, is one of the most valuable species in the markets of Brazil, Uruguay and Argentina. Because the availability of the species undergoes yearly fluctuations, it is important to establish the feasibility of culturing *A. longinaris* in order to maintain a continuous supply to the market.

The formulation of a nutritionally adequate and economical diet is essential for the success of a commercial culture. Determinations of essential amino acid requirements are considered to be the highest priority area in shrimp nutrition research (Akiyama 1986). Although amino acids have a central role in the cell metabolism, are essential for the synthesis of tissular proteins and many important compounds (i.e., adrenaline, thyroxin, melanin, histamine, porphyrins-hemoglobin) and can function as a metabolic source of energy (Tacon 1987), information concerning their role in penaeid nutrition is still limited.

According to Cowey and Forster (1971), Shewbart et al. (1972) and Kanazawa and Teshima (1981), ten amino acids are considered to be essential for fish and shrimp: threonine, leucine, methionine, lysine, arginine, valine, isoleucine, tryptophan, histidine and phenylalanine. Early works (Shewbart et al. 1972, Faranda et al. 1984, Teshima et al. 1986, Gallardo et al. 1989) stated that the amino acid profile of diets should reflect the amino acid composition of the shrimps; however, Kanazawa (1990) pointed out that crustacean meal has not proven to be an ideal source of protein for crustaceans. Studies to quantify the essential amino acid requirements using diets supplemented with crystalline amino acids were generally unsuccessful due to poor shrimp growth and survival (Deshimaru and Kuroki 1974, Deshimaru 1981, Teshima et al. 1986, Pascual 1990). More recently, quantification of requirements of shrimp species, using microencapsulated amino acids had been reported (Chen et al. 1992, Milamena et al. 1998).

Kluwer Academic/Plenum Publishers

E. Escobar-Briones & F. Alvarez Eds.
MODERN APPROACHES TO THE STUDY OF CRUSTACEA
PP. 079-084

It is well known that there is a lysine and arginine dietary relationship. Kaushik and Fauconneau (1984) and Kim et al. (1992) presented evidence that a lysine-arginine antagonism may exist in rainbow trout, although this relationship has not been found in shrimp. According to Akiyama et al. (1991) it is believed that the lysine:arginine ratio should be maintained at 1:1. The objective of this study was to investigate the effects of different ratios of arginine:lysine in semi-purified diets on the growth and survival of the Argentine prawn *Artemesia longinaris*.

II. MATERIALS AND METHODS

Three feeding trials were conducted to determine the action of diets with different levels of arginine and lysine. Based on a standard diet (A) with a known essential amino acid composition and an arginine:lysine ratio of 1:4, a series of pelletized feeds were designed by adding different amounts of arginine to obtain 1:1, 1:2 and 2:1 ratios (diets B, C and I, respectively). The ingredient composition and proximate analysis of standard diet A are shown in Table 1.

Table 1. *Ingredient composition (%) and proximate analysis (%) of diet A.*

Ingredient composition (%)			
Casein[1]	40	fish soluble	0.20
Cholesterol[2]	1.5	vitamin premix[3]	0.11
Cellulose	18	wheat bran	10
Fish oil	3.0	starch	21.75
Dextrin	5.0	choline	0.4
Inositol	0.04	arginine[4]	0
Proximate analysis (%)			
Moisture	7.5	Total protein	44.4
Ash	1.9	Total lipids	8.8

[1] Research Organics Inc., USA. [2] Berna, Argentina. [3] Vitamin premix (mg/kg diet): cholecalciferol 35; thiamin 163; rivoflavin 156; pyridoxine 213; calcium pantothenate 250; biotin 250; niacin 500; folic acid 25; $B_{12}HCl$ 20; ascorbic acid Rovimix STAYC 781; menadione 34; inositol 300; choline chloride 200; vitamin A acetate 100 (Leydi, Argentina). [4] Encapsulated L-arginine HCL 70% active (CAP-SHURE, Balchem Corporation).

High Performance Liquid Chromatography (HPLC) using a o-phthaldialdehyde mercaptoethanol analyzed amino acid composition of diet A; the samples were treated according to Perham (1978). HPLC was performed with a Kauer HPLC. The amino acid composition of diet A (1:4) is shown in Table 2.

The experiments were conducted with juvenile *A. longinaris*, using duplicate tanks for the first and second experiments and with four replicates in the third one. The diets tested per trial were: A, B and C, in the first one; A, B and I, in the second one; and A, B, C and I, in the third one.

The specimens were obtained from commercial fishermen in the coastal waters of Mar del Plata. They were individually weighed to the nearest 0.01 g with a Mettler P1210 scale and stocked in 100 l glass aquaria fitted with under-gravel filters. The filter bed consisted of crushed oyster shells and sand; salinity was kept at 34‰ and pH 6.

Prawns were fed daily; at the beginning of the trial feeding was 8% of the total shrimp biomass in each tank. Uneaten food, exuviae and dead animals were recorded and removed daily. Density was 20 prawns/m². Experiment one ran for 29 days, experiment two for 30 days, and experiment three for 33 days. Water temperature was monitored every day, temperature was 23.6°C ±1.8, 20°C ± 1.7 and 17.8°C ±1.2, in the first, second and third experiments, respectively. Prawns were weighed every 15 days and at the end of each experiment; they were sacrificed for posterior analyses. Individual molting rates were calculated according to Guary et al. (1976): $Mr = m/n$, where m = total number of molts and n = number of living prawns.

Total protein in diets was determined by the micro Kjeldahl method, according to Barnes (1959); total lipids were measured by the Soxhlet method (AOAC 1990). Ash and moisture were determined following Moreno and Aizpun (1969) technique.

The following statistical tests were performed: Cochran or Bartlet test for homocedasticity of variances; Student's t test for increase in mean weight; and significance in survival was determined with a x^2 test (Sokal and Rohlf 1969). In all cases, the level of significance was set for ($P < 0.05$).

III. RESULTS

In the first trial, after 29 days, the growth rate varied between −3.3 and 2.8% and survival varied from 55 to 65%. Molting rate varied from 2.2 to 2.5. These results were lower than those obtained in the other two experiments (Table 3).

In the second experiment diets A, B and I were tested. After 30 days the mean weight gain ranged from 15 to 20%. Feeds I and B resulted in survival rates from

Table 2. *Amino acid composition of diet A.*

Amino Acid	Composition (g/100g diet A)	Amino Acid	Composition (g/100g diet A)
Essential amino acids		*Non-essential amino acids*	
Leucine	9.27	Tyrosine	1.96
Lysine	11.15	Alanine	4.91
Isoleucine	4.96	Glycine	7.22
Phenylalanine	5.02	Serine	8.28
Methionine	NA[1]	Glutamic acid	20.6
Valine	5.54	Aspartic acid	7.27
Arginine	2.79	Tryptophan	NA[1]
Threonine + Histidine	11.29		

[1] Data not available (Not Analyzed)

Table 3. *Mean initial and final weight, percent weight gain, survival and molting rates of A. **longinaris** fed different diets.*

	Experiment one			Experiment two			Experiment three			
Diets	**A**	**B**	**C**	**A**	**B**	**I**	**A**	**B**	**C**	**I**
arginine:lysine	1:4	1:1	1:2	1:4	1:1	2:1	1:4	1:1	1:2	2:1
w. initial (g)[1]	1.80	1.85	1.84	2.09	2.02	2.09	0.93	0.93	0.90	0.92
± s.deviation[2]	0.191	0.206	0.194	0.425	0.430	0.417	0.318	0.353	0.317	0.366
w. final (g)[3]	1.85	1.85	1.78	2.41	2.40	2.51	1.19	1.26	1.18	1.15
± s.deviation[2]	0.273	0.261	0.215	0.297	0.390	0.469	0.439	0.489	0364	0.301
increase[4]	2.8	0	-3.3	15	19	20	28	35	31	21
survival (%)	55	65	55	94	81	75	75	61	67	56
M. rate (%)[5]	2.5	2.2	2.3	1.00	1.36	1.36	0.67	0.62	0.68	0.65

[1] Mean weight initial (g). [2] ± standard deviation. [3] Mean weight final (g). [4] Increase (% weigh gain). [5] Molting rate (%).

75 to 81%. Animals fed diet A exhibited the best results in survival (94%), however no significant differences were determined among treatments. Molting ratios ranged from 1.0 to 1.36 (Table 3).

In the third experiment, survival varied from 56 to 75%; however, no significant differences were found among treatments. The highest percentage in mean weight gain (35%) was obtained with diet B and the lowest (21%) with diet I, but no significant differences were found. Molting rates varied from 0.62 to 0.67 (Table 3).

IV. DISCUSSION

Amino acids are used in tissues to synthesize new protein. Thus, animals do not necessarily require protein, but do have a dietary requirement for amino acids. Information about this requirement could con-siderably reduce feed costs with a reduction in protein content of feeds or by using less expensive protein supplements (Akiyama 1992).

The requirements of essential amino acids have been reported for a number of shrimp and fish species (*P. monodon*, Chen et al. 1992; *Chanos chanos*, Borlongan and Coloso 1993; fingerling of sea bass *Dicentrarchus labrax*, Tibaldi et al. 1994; *Salmon salar*, Lall et al. 1994; *Litopenaeus vannamei*, Fox et al. 1992, 1995; hybrid striped bass, Keembiyehetty and Gatlin III 1992; *Paralichthys olivaceus* and *Pargus major*, Forster and Ogata 1998; *M. japonicus*, Hew and Cuzon 1982; *Sarotherodon mossambicus*, Jackson and Capper 1982; *Oncorhyncus mykiss*, Kim et al. 1992).

Lysine and arginine are found in high quantities in the muscular protein of penaeid shrimps and are considered as the most limiting amino acids

(Akiyama 1986, Pascual 1990). Determination of requirements of essential amino acids are necessary to validate whether the variability in protein requirements is real; casein supplement with encapsulated amino acids or protein enriched through covalent bonding of some amino acids might be the best experimental or reference protein source for such studies (Guillaume 1997). Based on these ideas, semi purified diets, based on casein and supplemented with different percentage of microencapsulated lysine and arginine were tested.

The effects of feeding with different levels of lysine and arginine have been evaluated in few species and the results obtained so far differ considerably. Tibaldi et al. (1994) found that sea bass fingerlings are not sensitive to moderate dietary excess of arginine and lysine. Colvin and Brand (1977) estimated the lysine requirements of *L. vannamei* and *L. stylirostris* from a growth plateau, but concluded that lysine was probably not a limiting factor. Working with *M. japonicus*, Hew and Cuzon (1982) found an improvement of casein as protein source by adding both lysine and arginine, which indicate that adding lysine to a diet has more positive growth-promoting effects than arginine.

In this study, the lowest growth levels were obtained in the first experiment, in which the mean temperature was the highest. Lopez and Fenucci (1988) and Haran et al. (1992) found that at temperatures ranging from 24 to 27°C the growth of *A. longinaris* is reduced and mortality is high. However, the effects of diet and temperature can not be assessed separately with the current results.

It is well known that there is a lysine and arginine dietary relationship known as the lysine:arginine antagonism. Although this relationship has not been found in shrimp, it is believed that the lysine:arginine ratio should be maintained at 1:1 to 1:1,1 (Akiyama et al., 1991). There is no ready explanation for the different responses of shrimp and fish species to disproportionate amounts of dietary arginine and lysine nor for the lysine:arginine antagonism. Evidence for a definite lysine:arginine antagonism at the level of ureagenesis has been shown in 35 g rainbow trout by Kaushik and Fauconneau (1984). Robinson et al. (1981) found no apparent antagonism between lysine and arginine in their study with channel catfish on the basis of growth response and serum free amino acid levels. While this antagonism was not observed in this study for *A. longinaris*, it can not be ruled out because there are many factors that can affect amino acid requirements:

differences among species, differences in amounts and quality of proteins used in the diets, other dietary and environmental factors, experimental conditions, as well as differences in the techniques used to calculate the requirement.

In conclusion, for the levels tested in this study, the different ratios of dietary lysine:arginine do not affect the growth and survival of the Argentine prawn *Artemesia longinaris*. Furthermore, more research is necessary to assess the amino acid requirement of this species.

ACKNOWLEDGEMENTS

The authors wish to thank Dr. Gustavo Daleo and Lic. Alba Majoral for of amino acid analysis. This work was funded by PICT 97 n° 07-00000-01414, from Agencia Nacional de Promoción Científica y Tecnológica, Argentina.

REFERENCES

Akiyama DM (1986) The development of purified diet and nutritional requirement of lysine in penaeid shrimp. PhD Dissertation, Texas A&M University, College Station, 78 p

Akiyama DM (1991) Penaeid shrimp nutrition for the commercial feed industry: Revised*. (pp 80-98) Proceedings of the Aquaculture Feed Processing and Nutrition. Workshop, Thailand and Indonesia, September 19-25

Akiyama DM (1992) Future considerations for shrimp nutrition and the aquaculture feed industry. (pp 198-205) Proceedings of the special session on shrimp farming. World Aquaculture Society, LA, USA

AOAC (Association of Official Analytical Chemists) (1990) Official Methods of Analysis. 15th edition. Association of Official Analytical Chemists, Arlington, VA, 1141 p

Barnes H (1959) Apparatus and Methods of Oceanography. George Allen & Unwin Ltd.,NY, 339 p

Borlongan IG and Coloso RM (1993) Requirements of juvenile milkfish *Chanos chanos* for essential amino acid. J Nutr 123:125-132

Chen HY, Len YT and Roelants I (1992) Quantification of arginine requirements of juveniles marine shrimp *Penaeus monodon* using microencapsulated arginine. Mar Biol 114:229-233

Colvin LB. and Brand C.W (1977) The protein requirement of penaeid shrimp at various life cycle stages controlled environment system. Proceedings of the World Mariculture Society 8:821-840

Cowey CB and Forster JRM (1971) The essential amino acid requirements of the prawn *Palaemon serratus*. The growth of prawns on diets containing proteins of different amino acid composition. Int J Life Oceans Coast Water 10:77

Deshimaru O (1981) Protein and amino acid nutrition of prawn *Penaeus japonicus*. (pp 106-123) In: Pruder GD, Langdon C and Conklin D (eds) Proceedings of the 2nd International Conference on Aquaculture Nutrition, Biochemical and Physiological Approaches to Shellfish Nutrition. Rehoboth Beach, DE, USA

Deshimaru O and Kuroki K (1974) Studies on a purified diet for prawn: III. A feeding experiment with amino acid test diets. Bull Jpn Soc Sci Fish 41:101-103

Faranda F, Salleo A, Lo Paro G and Manganaro A (1984) Quantitative requirement of *Penaeus kerathurus* for natural unprocessed diet. Aquaculture 37:125-131

Forster I and Ogata HY (1998) Lysine requirement of juvenile Japanese flounder *Paralichthys olivaceus* and juvenile red sea bream *Pargus major*. Aquaculture 161:131-142

Fox J, Lawrence AL and Li-Chan E (1992) Apparent lysine requirement of *Penaeus vannamei* (Boone) using covalent and crystalline lysine supplementation. (pp 96-97). Abstracts of Aquaculture'92. Marriott's Orlando World Center, Orlando, FL, May 21-25

Fox JM, Lawrence AL and Li-Chan E (1995) Dietary requirement for lysine by juvenile *Penaeus vannamei* using intact and free amino acid sources. Aquaculture 131:279-290

Gallardo N, González R, Carrillo O, Valdés O and Forrellat A (1989) Una aproximación a los requirimientos de aminoácidos esenciales de *Penaeus schmitti*. Rev Inv Mar 10:259-266

Guary JC, Kayama M, Murakami Y and Ceccaldi H (1976) The effects of a fat-free diet and compounded diets supplemented with various oils on molt, grown and fatty acid composition of prawn, *Penaeus japonicus* Bate. Aquaculture 7:245-54

Guillaume J (1997) Protein and Amino Acids. Crustacean Nutrition. Advances in World Aquaculture 6:26-50

Haran NS, Fenucci JL and Diaz AC (1992) Efectos de la temperatura y la salinidad sobre el crecimiento y la supervivencia del camarón (*Artemesia longinaris*) y del langostino (*Pleoticus mulleri*). Frente Maritimo 11:79-83

Hew M and Cuzon G (1982) Effects of dietary lysine and arginine levels, and their ratio, on the growth of *Penaeus japonicus* juveniles. J. World Maricul Soc 13:154-156

Jackson AJ and Capper BS (1982) Investigations on the requirements of the tilapia *Sarotherodon mossambicus* for dietary methionine, lysine and arginine in semi-synthetic diets. Aquaculture 29:289-297

Kanazawa A (1990) Protein requirements of penaeid shrimp. (pp 261-270) Advances in Tropical Aquaculture, Actes de Colloques 9, IFREMER, Plouzané, France

Kanazawa A and Teshima S (1981) Essential amino acids of the prawn. Bull Japan Soc Sci Fish 47:1375

Kaushik SJ and Fauconneau B (1984) Effect of lysine administration on plasma arginine and some nitrogenous catabolites in rainbow trout. Comp Biochem Physiol 79A: 459-462

Keembiyehetty CN and Gatlin III DM (1992) Dietary lysine requirement of juvenile hybrid striped bass (*Morone chrysops* x *M. saxatilis*). Aquaculture 104:271-277

Kim Kyu-II, Kayes TB and Amundson CH (1992) Requirements for lysine and arginine by rainbow trout (*Oncorhynchus mykiss*). Aquaculture 106:333-344

Lall SP, Kaushik SJ, Le Bail PY, Keith R, Anderson JS and Plisetskaya E (1994) Quantitative arginine requirement of Atlantic salmon (*Salmon salar*) reared in sea water. Aquaculture 124:13-25

Lopez AV and Fenucci JL (1988) Acción de la temperatura y algunos contaminantes en el crecimiento del camarón *Artemesia longinaris* Bate. Rev Latinoam Acuicult 38:109-120

Millamena OM, Bautista-Teruel MN, Reyes OS and Kanazawa A (1998) Requirements of juveniles marine shrimp, *Penaeus monodon* (Fabricius) for lysine and arginine. Aquaculture 164:95-104

Moreno VJ and Aizpun JE (1969) Manual de métodos de análisis de harina de pescado. I. Métodos Químicos. CARPAS/Oc. Documentos Ocasionales, Rio de Janeiro, 12:1-25

Pascual FP (1990) Terminal report submitted to SEAFDEC Aquaculture Department. Tigbauan, Iloilo, Philippines, 84 p

Perham RN (1978) Techniques for determining the amino-acid composition and sequence of proteins. Techniques in Protein and Enzyme Biochemistry 110:1-39

Robinson EH, Wilson RP and Poe WE (1981) Arginine requirements and apparent absence of a lysine-arginine antagonism in fingerling channel catfish. J Nutr 111:46-51

Shewbart KL, Meis WL and Ludwig PD (1972) Identification and quantitative analysis of the amino acids present in protein of the brown shrimp, *Penaeus aztecus*. Mar Biol 16:64

Sokal RR and Rohlf FJ (1969) Biometry, WH Freeman and Co, San Francisco, CA, 776 p Tacon AGJ (1987) The nutrition and feeding of farmed fish and shrimp – A training manual -1. The essential nutrients. FAO, Field document No 2

Teshima S, Kanazawa A and Yamashita M (1986) Dietary value of several proteins and supplemental amino acids for larvae of the prawn *Penaeus japonicus*. Aquaculture 51:225-235

Tibaldi E, Tulli F and Lanari D (1994) Arginine requirements and effects of different dietary arginine and lysine levels for fingerling sea bass (*Dicentrarchus labrax*). Aquaculture 127:207-218

VITAMIN E REQUIREMENT OF THE PRAWN *ARTEMESIA LONGINARIS* (DECAPODA, PENAEIDAE)

Analia Verónica Fernández Gimenez and Jorge L. Fenucci

Departamento de Ciencias Marinas, Facultad de Ciencias Exactas y Naturales, Universidad Nacional de Mar del Plata, CONICET. Funes 3350, B7602AYL, Mar del Plata, Argentina

ABSTRACT

Two trials were conducted to evaluate the growth and survival of Argentine prawn *Artemesia longinaris* fed with different levels of vitamin E and BHT (butylated hydroxytoluene) in a semi purified diet. The feeding trial consisted of 0, 1500 and 3000 mg vitamin E/kg diet, and 0.5 mg BHT/kg diet (trial 1) and 500, 1000, 1500 and 2000 mg vitamin E/kg diet (trial 2). After 40 days (trial 1), survival of prawns fed diets with 3000 mg vitamin E/kg diet was significantly ($P < 0.05$) lower (59%) than the other treatments. Prawns fed diets supplemented with 1500 mg/kg diet exhibited high weight gain (71.3%), however no significant differences were found among treatments. After 32 days (trial 2), prawns fed a diet containing 1500 mg vitamin E, exhibited higher survival (72%), than the other treatments (53-64%), and the percent weight gain ranged from 38.1 to 61.5% with no significant differences among treatments. The results of this study indicate that the dietary vitamin E requirement of *A. longinaris*, under the particular experimental conditions, is around 1500 mg/kg diet.

I. INTRODUCTION

One of the main problems of growing penaeid shrimps is that related to the inadequate knowledge of their nutritional requirements. Vitamins are a miscellaneous group of organic compounds required in trace amounts by shrimp and fish for the maintenance of their life processes. Early work on vitamin nutrition in crustaceans showed that dietary vitamin E is required by *Daphnia magna* (Viehoever and Cohen 1938) and *Moina macrocopa* (Conklin and Provasoli 1977). Vitamin requirements of shrimps have yet to be completely defined, but progress since 1990 has greatly accelerated in response to the needs of penaeid shrimp culture.

There is increasing evidence that a number of vitamins such as C and E have a vital antioxidant role in protecting lipid tissues of aquatic animals (Conklin 1989). Vitamin E may play a significant role in shrimp nutrition as an antioxidant, preventing polyunsaturated fatty acid oxidation in feeds as well as in shrimp tissues (Kanazawa 1985). Only algae synthesize the vitamin E required by crustaceans in nature (Conklin 1997). Synthetic antioxidants, such as butylated hydroxytoluene (BHT), butylated hydroxyanisole (BHA), and ethoxyquin, are used in animal feeds to reduce oxidative rancidity (He and Lawrence 1993).

Nutritional requirements for growth or reproduction are different in penaeid shrimp (Cahu et al. 1991). A previous study found that addition of vitamin E to diets resulted in improvement of survival of larval *Marsupenaeus japonicus* (Kanazawa 1985). He et al. (1992) also reported that *Litopenaeus vannamei* showed significantly lower survival and less weight gain when fed a vitamin E free diet for 8 weeks. Alava et al. (1993) reported that dietary vitamins A, E and C en-

E. Escobar-Briones & F. Alvarez Eds.
MODERN APPROACHES TO THE STUDY OF CRUSTACEA
PP. 085-089

hanced ovarian development in *M. japonicus*. The a-tocopherol is one of the components affecting the quality of penaeid eggs, in terms of hatching rate and larval viability (Cahu et al. 1991, Fakhfakh et al. 1991).

The Argentine prawn *A. longinaris* is distributed along the South American coast, from 23° S to 43° S (Boschi and Scelzo 1968). In Argentina, the regions of greatest abundance of this species are Mar del Plata and Rawson, and the best catches are obtained from October to January. The annual catch of this species is highly variable; it is therefore important to determine if the culturing on a commercial basis of this species is feasible.

Although *A. longinaris* has been studied for several decades (Boschi 1963, Boschi 1969 a,b, Boschi and Scelzo 1976, Fenucci and Petriella 1981, Fenucci et al. 1983, Petriella et al. 1984, Aquino 1996), no information on its vitamin E requirement exists. This study provides data on the dietary vitamin E requirement based on growth and survival of *A. longinaris*.

II. MATERIALS AND METHODS

Two experiments of 40 and 32 days, were conducted to determine the dietary vitamin E requirement of *A. longinaris*. The experimental system consisted of twelve glass aquaria (150 l capacity) with under-gravel filter as described by Boschi (1972). The adult prawns, obtained from a commercial fisherman in the coastal waters of Mar del Plata, Argentina, were kept in the aquarium and fed squid mantle for seven days prior to initiation of the feeding trial. Triplicate groups of 20 prawns/m², were stocked in each aquarium ranging from 1.01 to 1.11 g (experiment one) and from 0.84 and 0.94 g (experiment two).

Different levels of vitamin E and BHT were added to a semi purified diet for experiment one: 0, 1500 and 3000 mg vitamin E/kg diet and 0.5mg BHT/kg diet. Following the results obtained in experiment one, in the second trial the levels of dietary vitamin E (500, 1000, 1500 and 2000 mg vitamin E/kg diet) were modified. Dietary ingredients (Table 1) were mixed and diets were prepared by using the cold extrusion process and then were dried in an oven at 45°C for 24 h (Fenucci and Zein-Eldin 1976). Initial feeding was 5% of the biomass in each aquarium, the amount of food was adjusted daily and exuviae and dead individuals were recorded daily.

Water temperature ranged from 17 to 22°C, salinity was 34‰ and pH 6. Every twenty days and at the end of the experiments, prawns were counted and weighed. Diets were analyzed for ash (550°C) and dry matter (60°C). Total protein was estimated by the micro-Kjeldahl method (Barnes 1959) and total lipids were determined using Soxhlet method (IRAM 1985). For the statistical analysis the following tests were performed: Cochran or Bartlett, analysis of variance (ANOVA), Student's t test and χ^2 (Sokal and Rolhf 1979).

Table 1. *Ingredient composition (g/kg diet) of semi purified diets.*

Ingredient	g/kg diet	Ingredient	g/kg diet
Casein, vitamin-free[1]	300	sodium alginate	36.6
squid protein[2]	100	cholesterol[3]	20
manioc starch	220	lecithin	10
gelatin	120	vitamin premix[4]	20
cellulose	100	mineral premix[5]	3.4
free fatty acids	70		
Proximal composition	%	**Proximal composition**	%
total protein	48.09	moisture	6.42
total lipids	10.08	ash	8.41

[1] Research Organics Inc. USA; [2] Prepared from squid mantle; [3] Berna, Argentina; [4] vitamin premix (mg/kg diet): cholecalciferol 35, thiamin 163, rivoflavin 156, pyridoxine 213, calcium pantothenate 250, biotin 250, niacin 500, folic acid 25, B_{12} HCL 20, ascorbic acid Rovimix STAY C 781, menadione 34, inositol 300, choline chloride 200, vitamin A acetate 100, wheat semolina csp (Leydi, Argentina) cellulose was replaced by appropriate amounts of vitamin E in the premix to give different levels of vitamin E; [5] mineral premix: calcium 1000mg, magnesium 500mg, potassium 99mg, zinc 30mg, iron 10mg, copper 2mg, iodine 150 mcg, selenium 200mcg, chromium 200 mcg, molybdenum 500mcg (Twin laboratories, Inc., USA).

III. RESULTS

For experiment one, weight gain, survival and molting rates of prawns fed diets with different levels of vitamin E are presented in Table 2. After 40 days, survival of prawns fed diets containing 3000 mg vitamin E/kg diet was significantly ($P < 0.05$) lower than that for prawns fed diets with 0, 1500 mg vitamin E/kg diet and BHT. Prawns fed diets supplemented with 1500 mg/kg exhibited high weight gain (71.3%), however no significant differences were found among treatments. Molting rates were higher for prawns fed diets supplemented with 0, 1500 and 3000 mg/kg than for those in other treatments (0.5 mg/kg BHT), however, there were no significant differences ($P < 0.05$).

In the second experiment, levels of dietary vitamin E between 500 and 2000 mg/kg diet were tested in order to more accurately determine the vitamin E requirement of *A. longinaris*. After 32 days, prawns fed the diet with 1500 mg vitamin E, exhibited higher survival (72%), than the other treatments (53-64%), but the differences were not significant. There was no significant difference in weight gain at the end of the second trial among treatments; values ranged from 38.10 to 61.50% (Table 3). Molting rates ranged from 0.13 to 0.51, without differences, these values were lower than those observed in experiment one.

Table 2. *Mean weight initial and final, percent weight gain, survival and molting rates of prawn fed different diets for 40 days (experiment one).*

vitamin E	mean weight (g±s)*		weight gain	survival	molting
(mg/kg diet)	initial	final	(%)	(%)	rates
0	1.11±0.388	1.63±0.492	46.8	85	0.82
1500	1.01±0.407	1.73±0.646	71.3	71	0.81
3000	1.05±0.396	1.62±0.566	54.3	59	1.08
0.5 (BHT)	1.07±0.408	1.68±0.612	57.0	85	0.51

*Values are means ± standard deviations of triplicates.

Table 3. *Mean weight initial and final, percent weight gain, survival and molting rates of shrimp fed different diets for 32 days (experiment two).*

vitamin E	mean weight (g±s)*		weight gain	survival	molting
(mg/kg diet)	initial	final	(%)	(%)	rates
500	0.91±0.905	1.47±0.567	61.5	56	0.34
1000	0.84±0.312	1.16±0.339	38.1	64	0.13
1500	0.94±0.329	1.42±0.364	51.1	72	0.51
2000	0.89±0.286	1.43±0.331	60.7	53	0.51

*Values are means ± standard deviations of triplicates.

IV. DISCUSSION

Predicting the need of fat-soluble vitamins A, D, E and K in shrimps is difficult (Conklin 1989). Poor growth has resulted when vitamins A, D and E, rather than vitamin K, were each removed from a diet prepared for *L. vannamei* (He et al. 1992, Conklin 1997). A quantitative requirement of 99 mg vitamin E/kg diet for juvenile *L. vannamei* was determined based on growth rate analysis (He and Lawrence 1993). This is significantly less than the level of 300 mg/kg previously recommended for commercial formulations (Akiyama et al. 1992). However in this study the best results for growth and survival were obtained with a diet supplemented with 1500 mg/kg diet of vitamin E.

Growth reduction resulting from feeding vitamin E deficient diets in experiment one resembles the observation of a previous study on *L. vannamei* (He et al. 1992, He and Lawrence 1993). These studies also reported other vitamin E deficient signs, such as the darkening of the hepatopancreas (He et al. 1992); this symptom, however, was not observed in this study. Previous experiments used shrimp larvae (Kanazawa 1985) or juvenile shrimp (He et al. 1992, He and Lawrence 1993), while adult individuals were utilized in this study.

The growth and survival results obtained with the different experimental diets used are related to histological characteristics. Signs of malnourishment were observed in the histology of the hepatopancreas in animals fed diets with the lower levels of vitamin E (Fernandez et al., pers obs).

The synthetic compound BHT is commonly used as antioxidant in animal feeds (Rumsey 1978). The use of good quality oils and the incorporation of synthetic antioxidants into formulations of crustacean diets may decrease the need for dietary vitamin E as was found for trout (*Salmo gairdneri*) (Hung et al. 1981) and catfish (*Ictalurus punctatus*) (Murai and Andrews 1974). In the present study, supplying BHT in the diet could prevent low survival of *A. longinaris*; however, these individuals exhibited a reduction in growth with respect to treatments with 1500 mg vitamin E/kg diet. He and Lawrence (1993) determined that dietary BHT at 16 mg/kg diet prevented growth reduction of *L. vannamei*. This synthetic antioxidant was highly effective in inhibiting dietary lipid peroxidation, but was less active in tissue protection and had no protective effect on membrane lipid peroxidation.

The results of this study indicate that the vitamin E requirement of *Artemesia longinaris*, under the experimental conditions used is approximately 1500 mg/kg diet, and this level is higher than levels previously recommended for *L. vannamei* in commercial formulations.

ACKNOWLEDGEMENTS

This research was funded under PICT 97 N° 07-00000-01414 from Agencia Nacional de Promoción Científica y Tecnológica and Universidad Nacional de Mar del Plata, Argentina.

REFERENCES

Akiyama DM, Dominy WG and Lawrence AL (1992) Penaeid shrimp nutrition. (pp 535-568) In: Fast E W and Lester L J (eds) Marine shrimp culture: principles and practices. Elsevier Science Publishers, Amsterdam, The Netherlands

Alava VR, Kanazawa A, Teshima S and Koshio S (1993) Effect of dietary vitamins A, E and C on ovarian development of *Penaeus japonicus*. Nippon Suisan Gakkaishi 59:1235-1241

Aquino JB (1996) Variaciones cuali-cuantitativas de la dieta natural del camarón *Artemesia longinaris* Bate (Crustacea, Decapoda, Penaeidae) durante el ciclo de muda. Tesis de Licenciatura en Ciencias Biológicas, Facultad de Ciencias Exactas y Naturales, Universidad Nacional de Mar del Plata, Argentina, 31 p

Barnes H (1959) Apparatus and methods of oceanography. George Allen and Unwin Ltd., New York, 339 p

Boschi EE (1963) Los camarones comerciales de la Familia Penaeidae de la costa Atlántica de América del Sur. Bol Inst Biol Mar 3, 39 p

Boschi EE (1969a) Estudio biológico pesquero del camarón *Artemesia longinaris* Bate de Mar el Plata. Bol Inst Biol Mar 19, 47 p

Boschi EE (1969b) Crecimiento, migración y ecología del camarón comercial *Artemesia longinaris* Bate. FAO Fish Rep 57

Boschi EE (1972) El acuario de agua salada. Contrib Inst Biol Mar 220, 23 p

Boschi EE and Scelzo MA (1968) Nuevas campañas exploratorias camaroneras en el litoral
argentino, 1967/68. CARPAS 4 Documento técnico, Mar del Plata, Argentina. 9:1-6

Boschi EE and Scelzo MA (1976) El cultivo de camarones peneidos en la Argentina y la posibilidad de su producción en mayor escala. FAO Technical Conference on Aquaculture, Japan, 3 p

Cahu C, Fakhfakh M and Quazuguelet P (1991) Effect of dietary a-tocopherol level on reproduction of *Penaeus indicus*. (pp 242-244) In: Lavens P, Sorgeloos P, Jaspers E and Ollevier F (eds) Larvi' 91- Fish & Crustacean Larviculture Symposium. Special Publication N°15 European Aquaculture Society, Gent, Belgium

Conklin DE (1989) Vitamin requirement of juvenile Penaeid shrimp. Advances in Tropical Aquaculture. Actes de Colloque 9:287-308

Conklin DE (1997) Vitamins. (pp 123-149) In: D'Abramo LR, Conklin DR and Akiyama DM (eds) Advances in World Aquaculture Crustacean Nutrition. Vol. 6 Copyright, Baton Rouge, Louisiana, USA

Conklin DE and Provasoli L (1977) Nutritional requirements of the water flea Moina macrocopa. Biol Bull 152:337-350

Fakhfakh M, Villette M et Cahu Ch (1991) L'action de l'a-tocopherol dans le developpment embryonnaire de *Penaeus indicus*: role d'antioydant ou role specifique? ICES Mariculture Committee F:35

Fenucci JL and Zein-Eldin ZP (1976) Evaluation of squid mantle meal as a protein source in penaeid nutrition. FAO Technical Conference on Aquaculture, 76/E36, 9 p

Fenucci JL and Petriella AM (1981) Efectos de la alimentación natural y artificial en el crecimiento del camarón *Artemesia longinaris*. Rev Lat Acui 10:10-17

Fenucci JL, Petriella AM and Müller MI (1983) Estudios sobre el crecimiento del camarón *Artemesia longinaris* Bate alimentado con dietas preparadas. Contribución Instituto Nacional de Investigación y Desarrollo Pesquero, Mar del Plata, Argentina, 424, 13 p

He H and Lawrence AL (1993) Vitamin E requirement of *Penaeus vannamei*. Aquaculture 118:245-255

He H, Lawrence AL and Liu R (1992) Evaluation of dietary essentiality of fat-soluble vitamins, A, D, E and K for penaeid shrimp (Penaeus vannamei). Aquaculture 103:177-185

Hung SSO, Cho CY and Slinger SJ (1981) Effect of oxidized fish oil, dl-a-tocopherol acetate and ethoxyquin supplementation on the vitamin E nutrition of rainbow trout (*Salmo gairdneri*) fed practical diets. J Nutr 111:648-657

IRAM (1985) Método de determinación de la materia grasa por la técnica de extracción en un aparato tipo Soxhlet o Twiselmann. Instituto Argentino de Racionalización de Materiales, Norma IRAM 15040-1, 7 p

Kanazawa A (1985) Nutrition of penaeid prawn and shrimp. (pp 123-130) In: Taki Y, Primavera LH and Lobrera JA (eds) Proceedings of the First International Conference on Culture of Penaeid Prawn/Shrimps. Aquaculture Department, Southeast Asian Fisheries Development Center, Iloilo, Philippines

Murai T and Andrews JW (1974) Interaction of dietary a- tocopherol, oxidized menhaden oil and ethoxiquin on channel catfish (*Ictalurus punctatus*). J Nutr 104:1416-1431

Petriella AM, Müller MI, Fenucci JL and Saez MB (1984) Influence of dietary fatty acids and cholesterol on the growth and survival of the argentine prawn *Artemesia longinaris* Bate. Aquaculture 37:11-20

Rumsey GL (1978) Antioxidants in compounded feeds. (pp 177-182) In: Fish Feed Technology, FAO/UNDP Training Course. FAO/UNDP, Rome

Sokal R and Rohlf J (1979) Biometria. H. Blume, Madrid, 832 p

Viehoever A and Cohen I (1938) The responses of *Daphnia* to vitamin E. Am J Pharm 110:297-315

DISTRIBUTION AND ABUNDANCE OF *DIACYCLOPS* SP. (CRUSTACEA: COPEPODA) IN GABRIEL CAVE, OAXACA, MEXICO

*Javier Cruz-Hernández, Luis M. Mejía-Ortíz,
Martha Signoret-Poillon and José A. Viccon-Pale*

*Departamento El Hombre y su Ambiente, Universidad Autónoma Metropolitana-Xochimilco,
Apartado Postal 70-458, México 04510 D.F., México*

jcruz@cueyatl.uam.mx

ABSTRACT

The distribution and abundance of the stygobiont copepod *Diacyclops* sp., as well as values of some environmental variables (depth of the pools, temperature, dissolved oxygen, pH, percent saturated oxygen and organic matter content) in Gabriel Cave are reported for the first time. The copepods were collected in 12 pools using traps and plankton nets with a 10 μm mesh opening. The highest abundance of *Diacyclops* sp. was observed in the central zone of the cave and it was related to the organic matter content. The depth of the pools ranged from 10 to 100 cm, temperature from 20.1 to 24 °C, dissolved oxygen from 0.1 to 7 ml/l, pH ranged 6.3 to 8.3 and organic matter content ranged from 0.001 to 21.3 %. Water quality as well as the amount of available food determine the distribution and abundance of this planktonic stygobiont cyclopoid in the cave.

I. INTRODUCTION

Cyclopoida represents about 60% of the freshwater free-living Copepoda and are the most successful copepods in limnetic environments (Suárez-Morales et al. 1996). In Mexico there are few available data on the hypogean Copepoda. Osorio-Tafall´s papers (1941a, 1941b, 1942a, 1942b, 1943) yielded the first knowledge of the Mexican cave copepods. Reddell (1981) reported nine hypogean species of cyclopoids in some Mexican states. Regarding the genus *Diacyclops* Kiefer, 1927, only three hypogean species of this genus have been reported from Mexico until now: *D. chakan, D. pucc* Fiers et al., 1969 and *D. bernardi* Reid, 1993. These species are all benthic and have been recorded in karstic environments (cenotes) of the Yucatan Peninsula (Suárez-Morales and Reid 1998).

In this study, the distribution and abundance of an undescribed species of *Diacyclops* from Gabriel Cave, Oaxaca, Mexico, a karstic hypogean system explored by Mejía-Ortíz et al. (1997), is presented.

II. MATERIALS AND METHODS

Study Area

Gabriel Cave is located in the northern portion of the State of Oaxaca, at an altitude of 110 m (18° 27' 25" N, 96° 40' 34" W) (Fig. 1). It is a horizontal karstic cave with a 20 m wide entrance which continues as a long and irregular tunnel. The floor goes from bare rock to mud, with stalactite and stalagmite formations. The floor to ceiling distance ranges from 5 to 15 m; in some places there are higher chambers more than 15 m high and about 50 m wide. So far, it has been possible to explore approximately 1.5 km of galleries inside the cave. There is a junction at 1150 m from the entrance, dividing the cave into two branches. Inside, the air temperature is 20°C during the dry season.

Kluwer Academic/Plenum Publishers

E. Escobar-Briones & F. Alvarez Eds.
MODERN APPROACHES TO THE STUDY OF CRUSTACEA
P.P. 091-094

The cave has many irregular pools of variable depth, shape, extension and type of sediment. The water level depends mostly on the input of an underground river.

The 14 sampling sites constitute a mosaic of microhabitats, isolated or not from the main stream each one has its own characteristics. The first station represented the deepest pool (2 m), while the rest of the stations were shallower, except for station 11. Only the first two stations had poor light conditions, while the rest were in complete darkness. The sediments varied from mud, to sand and gravel rich in calcium carbonate. The water was clean and transparent. The volume as well as the speed of the water flow inside the cave depends mostly on the seasonal rain pattern in the region and the topography of the cave.

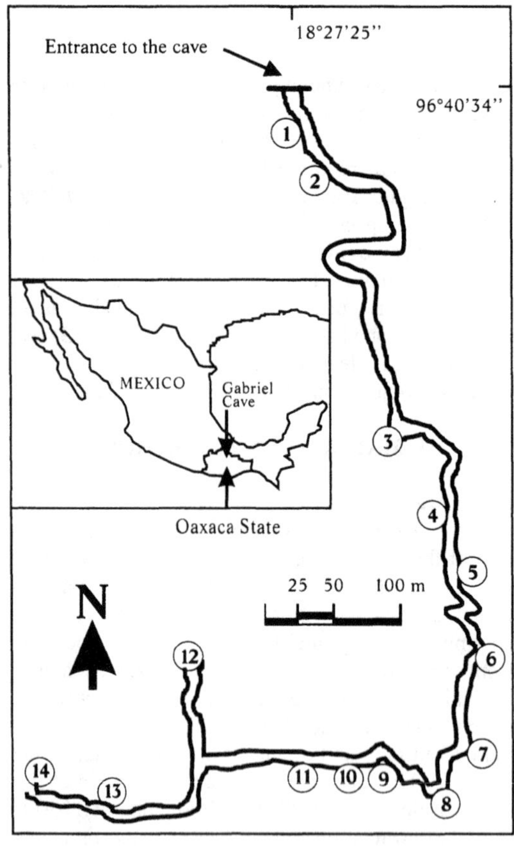

Figure 1. *Study area and sampling sites.*

Sampling methods

Copepods were collected using a plankton net with a 10 μm mesh size, and cylindrical traps with chicken bait, set for 24 h. The organisms were stained with Bengal Rose and fixed with 70% ethanol. In the laboratory, the organisms were identified to genus using mainly the identification keys by Reid (1993), Fiers et al. (1996) and Suárez-Morales et al. (1996). Only the genus, *Diacyclops* Kiefer, 1927, has been determined, while the species is still undescribed.

The oxygen concentration ($\pm$ 0.01 ml/l) and temperature ($\pm$ 0.01°C) were measured in each sampling station with a YSI-5000 oxygen-meter, pH was measured with a digital pH-meter ($\pm$ 0.01) Hanna Instruments-HI991000, and the depth of the pools was determined with a plumb and string marked each 10 cm.

Five cubic centimeters of sediment were taken with a multiple corer at a depth of 8 cm for organic matter determinations. Sediment organic matter content was determined following Luczak (1997) technique.

Data analysis

Multiple regression analysis was used to examine the relationship between copepod abundance and the environmental variables measured inside the cave (depth, temperature, pH, dissolved oxygen concentration, organic matter content in the sediment) using the stepwise regression method (Sokal and Rohlf 1981). The analysis was made using Statgraphics Plus 2 software.

III. RESULTS

Two hundred and twenty five specimens of *Diacyclops* sp. were collected with a 10:1 female to male sex ratio. The distribution pattern within the pools was very heterogeneous. However, the highest abundance was observed in the central portion of the studied area of the cave (stations 6, 7 and 8 with 1.3, 2 and 1.08 ind/cm², respectively). In stations 4 and 5 no specimens were collected, while in the first two pools, 22 and 23 specimens were collected. In the remainder pools, only few organisms were obtained.

The water temperature of the pools was similar in all sampling sites with a mean value of 22°C (Fig. 2a): however, a higher value (24°C) was recorded in station 9 in spite of the proximity to stations 10 and 11. The pH showed an alkaline tendency with values between 7 and 8.3. Only at station 1, the value was 6.3 for the surface layer (Fig. 2b).

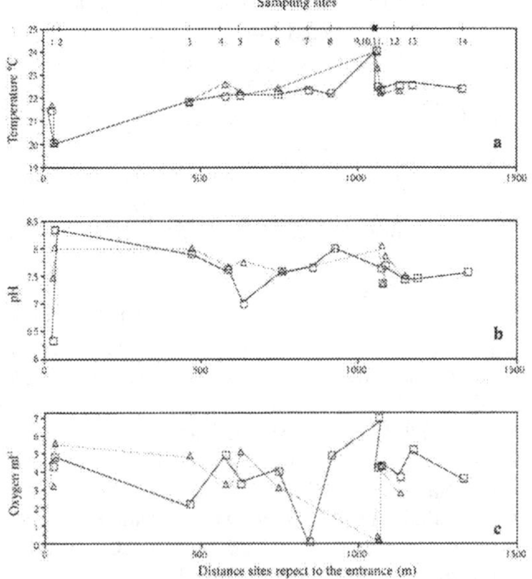

Figure 2. *Abiotic variables within the Gabriel Cave, Oaxaca, Mexico (□ = surface, Δ = bottom).*

The dissolved oxygen concentration was highly variable (Fig. 2c), ranging from 0.1 ml/l (anoxic conditions) to 7 ml/l. The lowest value was detected in station 7, in the bottom of a shallow pool with a high organic matter content (Table 1). In stations 9 and 10, where the temperature was higher, the water had also anoxic conditions in the bottom, with values ranging from 0.1 to 0.2 ml/l.

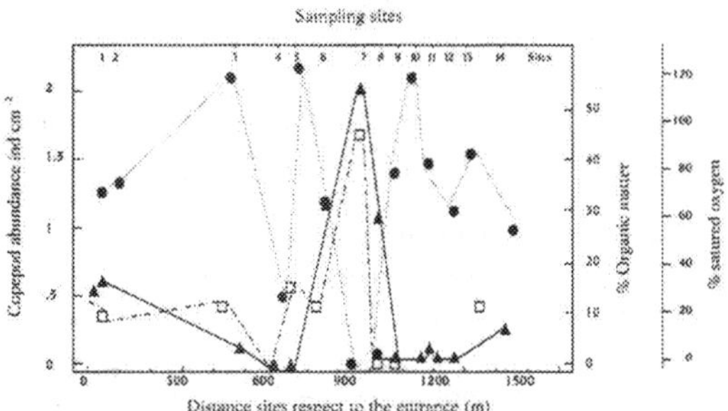

Figure 3. *Pattern of* **Diacyclops** *sp. abundance, organic matter and percent oxygen saturation in Gabriel Cave, Oaxaca, Mexico (▲= copepods; □ = organic matter, ● = percent saturation of oxygen).*

Regarding the multiple regression analysis of copepod abundance in relation to the abiotic variables measured, the results were inconclusive. However, when only the organic matter content and percent of oxygen saturation were considered, 82.6% of the variability in copepod density was explained ($P = 0.0301$). The following model describes the obtained relationship:

No. of organisms = 215,388 – 84,3466*log (PSO) – 3,09845*OM

where PSO = percent of oxygen saturation and OM = organic matter. These results suggest that copepod density is high when organic matter in the pools is abundant and the percent of oxygen saturation is low (Fig. 3).

Table 1. *Copepod abundance and abiotic variables in Gabriel cave, Oaxaca, Mexico.*

Sampling sites	1	2	3	4	5	6	7	8	9	10	11	12	13	14
Distance from the entrance (m)	28	34	465	580	628	748	845	917	1062	1065	1074	1174	1134	1330
No. of organisms ind cm^{-2}	0.611	0.638	0.111	0	0	0.133	2	1.08	0.02	0.05	0.08	0.02	0.02	0.25
Depth (cm)	80	11	60	22	60	15	10	10	32	40	100	40	15	30
Surface Temperature (°C)	21.4	20	21.8	22	22.1	22.1	22.3	22.2	24	22.4	22.3	22.5	22.5	22.3
Botton Temperature (°C)	21.6	20	21.8	22.5	22.2	22.4	*	*	24	23.3	22.2	22.3	*	*
Surface percent satured oxygen	69.9	72.6	115.9	34.7	123	64.9	1.2	6.9	72.5	115.3	78	60	85.3	59.7
Botton percent satured oxygen	51.14	85.6	81.2	54.1	84.2	51.6	*	*	6.3	3.1	71.2	46.2	*	*
Surface Oxygen (ml.l^{-1})	4.3	4.8	2.2	4.9	3.3	4	0.1	4.9	4.2	7	4.3	3.7	5.2	3.6
Botton Oxygen (ml.l^{-1})	3.2	5.6	4.9	3.3	5.1	3.1	*	*	0.4	0.2	4.3	2.8	*	*
Surface pH	6.2	8.3	7.9	7.6	6.9	7.5	7.6	7.9	7.6	7.3	7.6	7.4	7.4	7.5
Botton pH	7.4	8	8	7.6	7.7	7.5	*	*	8	7.3	7.8	7.4	*	*
Organic Matter (%)	11.3	10.5	11.5	*	14.4	11.2	43.9	*	0.001	*	*	*	11	*

** Data are not had for that the depth in occasions was very shallow and they were not considered as bottom and surface and in the organic matter, in some sampling places they were stones*

IV. DISCUSSION

We were in the cave in January 1999, at which time the air temperature was 20°C with high humidity. The climate of the region according to the classification of Köppen (García, 1988) is tropical and humid, with a mean annual temperature of 25.3°C and a mean annual rainfall of 2395 mm. Culver (1982) reports that the temperature in caves is about the mean annual temperature of the region and that in parts of large caves far away from the entrance, the temperature scarcely varies at all. Cave water temperatures show a similar pattern of relative stability. In Gabriel Cave both, air and water, temperatures fit the pattern proposed by Culver (1982).

Three species of *Diacyclops* have been reported from hypogean habitats: *D. chakan*, *D. puuc* and *D. bernardi*, all from cenotes in the Yucatan Peninsula (Fiers et al. 1996). According to Suárez-Morales and Reid (1998), this is the first report of copepods in the karstic zones of the Sierra Madre Oriental. The animals are unpigmented, as an adaptation to cave life, and have not been reported from any nearby epigean body of water. In Gabriel Cave, the organic matter content has its origin, to a large extent, on bat guano that represents the main energy supply for the cave copepod populations. Laboratory observations show that bat feces could provide food of high nutritional value for copepods.

ACKNOWLEDGEMENTS

This research was supported by project ALAC-UAM-Xochimilco No. 34302 and by FOMES Program of the Public Education Ministry No. 983504.

We are very grateful to Mr. Leonidez Guzmán and his family for his kind hospitality, and to Gabriel Muñoz Roa, Sweetia Mendoza Cervantes and Marilú López Mejía for their help during field work.

REFERENCES

Culver DC (1982) Cave life. Evolution and Ecology. Harvard University Press, Cambridge, 189 p

Fiers F, Reid JW, Iliffe TM and Suárez-Morales E (1996) New hypogean cyclopoid copepods (Crustacea) from the Yucatán Peninsula. Smithson Contrib Zool 66:65-102

García E (1988) Modificaciones al sistema de clasificación climática de Köppen. UNAM, México, 217 p

Gillieson D (1996) Caves: processes, development and management. Blackwell Publishers, Oxford, 324 p

Kiefer F (1927) Versuch eines Systems der Cyclopiden. Zool Anz 73:302-308

Mejía-Ortíz LM, Palacios-Vargas JG, Cardona L and Viccon-Pale JA (1997) Microartrópodos de la cueva Gabriel y la cueva del Nacimiento del Río San Antonio, Oaxaca, México. Mundos Subterráneos 8:21-28

Luczak C, Janquin ME and Kupka A (1997) Simple standard procedure for the routine determination of organic matter in marine sediment. Hydrobiologia 345:87-94

Osorio-Tafall B (1941a) Materiales para el estudio del microplancton del Lago de Pátzcuaro (México). An Esc Nal Cien Biol 2:331-383

Osorio-Tafall B (1941b) *Diaptomus cuauhtemoci* nov. sp. de la mesa central de México (Copepoda, Diaptomidae). Ciencia 2:296-298

Osorio-Tafall B (1942a) Un nuevo *Diaptomus* del México Central. Rev Bras Biol 2:147-154

Osorio-Tafall B (1942b) *Diaptomus* (*Microdiaptomus*) cokeri nuevo subgénero y especie de diaptómido de las cuevas de la región de Valles (San Luis Potosí). Ciencia 3:206-210

Osorio-Tafall B (1943) Observaciones sobre la fauna acuática de las cuevas de la región de Valles, San Luis Potosí, México. Rev Soc Mex Hist Nat 4:43-71

Redell JR (1981) A review of the cavernicole fauna of Mexico, Guatemala and Belice. Bull Texas Mem Mus 27:1-327

Reid JW (1993) New records and redescriptions of American species of *Mesocyclops* and of *Diacyclops bernardi* (Petkovski, 1986) (Copepoda: Cyclopoida). Bijdragen tot de Dierkunde 63:173-191

Sokal RR and Rohlf FJ (1981) Biometry. 2nd Edition, WH Freeman and Company, San Francisco

Suárez-Morales E, Reid JW, Iliffe TM and Fiers F (1996) Catálogo de copépodos (Crustacea) continentales de la Península de Yucatán, México. Ecosur and Conabio, México, 296 p

Suárez-Morales E and Reid JW (1998) An updated list of the free-living freshwater copepods (Crustacea) of Mexico. Southwest Nat 43:256-265

DISTRIBUTION AND ECOLOGY OF ISOPODS (CRUSTACEA: PERACARIDA: ISOPODA) OF THE PACIFIC COAST OF MEXICO

Ma. del Carmen Espinosa-Pérez and Michel E. Hendrickx

Unidad Académica Mazatlán, Instituto de Ciencias del Mar y Limnología, UNAM, P.O. Box 811 Mazatlán, 82000 Sinaloa, México

carmene@ola.icmyl.unam.mx

ABSTRACT

The Pacific coast of Mexico extends from 32° 27' N to 14° 32' N, including warm temperate, subtropical and tropical waters. Considering the 200 nautical miles limit, the Mexican Pacific covers approximately 2364200 km² of marine and oceanic environment from estuaries and intertidal to over 2000 m depth. An exhaustive review of published literature and intensive sampling along the coast of the Mexican Pacific and offshore allow us to report a total of 120 species of Isopoda for the area. These species belong to 8 of the 10 currently recognized suborders of Isopoda. Almost half of the species (54) are Flabellifera, while Microcerberidea and Gnathiidea are represented by only one known species each. As expected, there is a clear bathymetric segregation from one group of species to the other. Gnathiidea, Microcerberidea and Oniscidea (10 spp.) are found exclusively in the intertidal, among rocks and on sandy beaches, often associated with algae or decaying material. Anthuridea (10 spp.), Asellota (5 spp.) and Valvifera (20 spp.) occur from the intertidal to 200 m depth, in sandy, muddy or mixed habitats or as commensals. Flabellifera also occur from the intertidal but range to deeper water, with a maximum depth record of 2214 m (*Rocinella belliceps* (Stimpson, 1864)); these are either free-living or parasites and are found in a wide variety of habitats. Epicaridea (21 spp.) are all parasites. Without any doubt, the least studied suborders are Anthuridea (10 known species and at least 2 undescribed species) as well as Gnathiidea and Microcerberidea due to their small size, difficulty in sampling and complexity of available related literature. Contrary to what has been observed with other groups of crustaceans, the diversity of isopods appears to be much higher in Mexican temperate waters, where 59.1% of all known species occur (California Current area), than in the SW tropical coastal area (only 26.7% of the species are registered there). The Gulf of California, on the contrary, features an unusual (relative) richness with up to 73.3% of Mexican species recorded. It should be stressed, however, that most of the collecting effort of the last 30 years has been done in the Gulf of California and along the west coast of Baja California.

I. INTRODUCTION

Isopods are among the most common crustaceans in the tropical and subtropical seas. Marine and brackish water species are part of the more than 9500 species known to date for this order of Crustacea (Brusca pers. comm., April 2000). As a group, marine isopods are one of the most important elements in the marine energy flow for their ability to process decaying organic matter, making it available to other trophic levels. They also represent a factor of economical imbalance; wood-burrowing species can destroy human build structures (e.g., genus *Limnoria* Leach, 1813) and parasitic species affect market quality of fish (Brusca 1981, Markham 1985). The high diversity and success of this group is reflected by their presence in virtually every marine and brackish water habitat, including hard and

E. Escobar-Briones & F. Alvarez Eds.
MODERN APPROACHES TO THE STUDY OF CRUSTACEA
PP. 095-103

soft bottoms, seagrass, algae, mangrove roots and coral reefs (Schultz 1961, Dexter 1972, 1974, 1976, Ribi 1981, Delaney 1984, Kang and Yun 1988, Ellison and Farnsworth 1990, Arrontes and Anadón 1990, Taylor and Moore 1995).

The Mexican Pacific coast, with a total length of about 7150 km (Moreno-Casasola and Castillo 1992) and its adjacent offshore waters, encompasses four currently recognized zoogeographic provinces: the southern portion of the Californian Province (northern border to Magdalena Bay), the Cortez Province (Gulf of California and SW tip of Baja California), the Mexican Province (Corrientes Cape to Tangola Tangola Bay) and the northernmost part of the Panamic Province (Tangola Tangola Bay to southern border) (Fig. 1) (see Brusca and Wallerstein 1979, Hendrickx 1992). The

area extends from 32° 27' N to 14° 32' N and covers a total of 2364200 km^2, from the coastline to the 200 nautical miles offshore limit (Hendrickx 1993), including several oceanic islands. Maximum depth is 6000 m in the Tehuantepec Trench. Considering the entire area, which includes temperate, subtropical and tropical water, isopods find a vast array of habitats suitable for colonization.

Our knowledge of isopods has increased significantly in recent years. This is reflected in the availability of a wide variety of recent researches dealing with large taxonomic groups of isopods or with isopod fauna of selected geographic areas (e.g., Brusca 1980, Wilson 1980, Poore 1984a, 1984b, 1996, Brusca and Iverson 1985, Bruce 1986a, Kensley and Schotte 1989, Brusca et al. 1995, Wetzer and Brusca 1997). In the

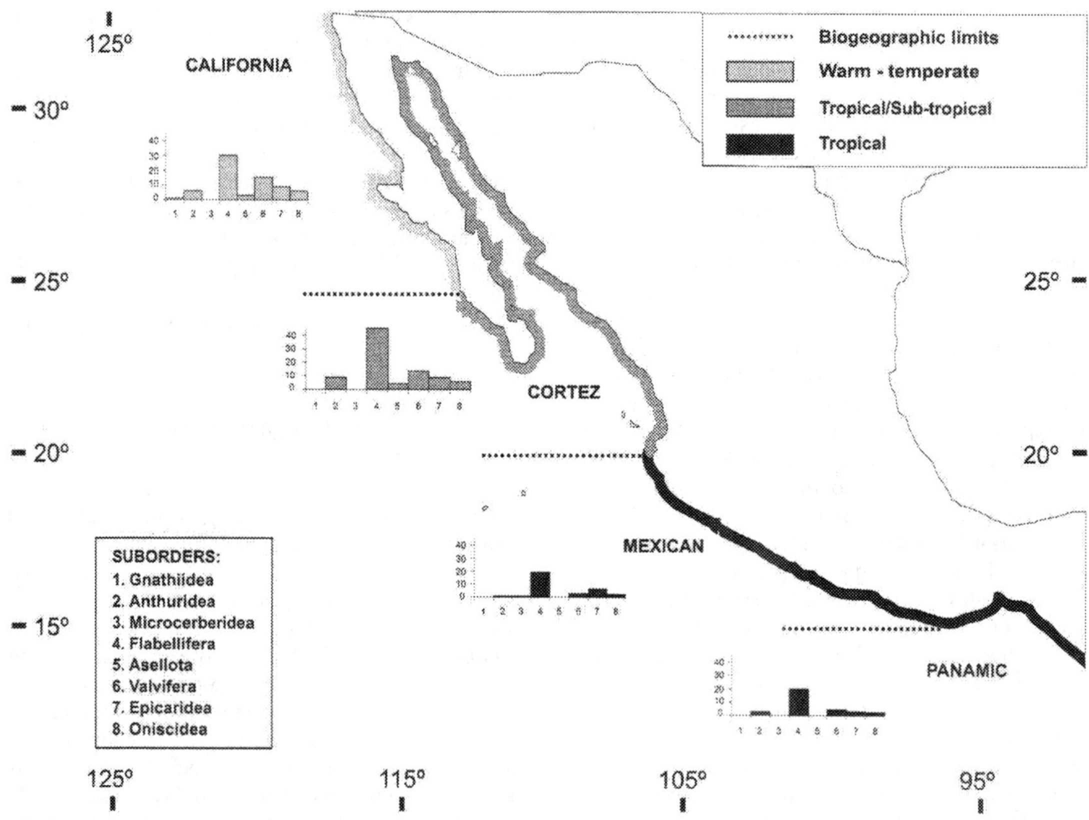

Figure 1. *Number of species of marine and brackish water isopods known from the four biogeographic provinces of the Mexican Pacific (by suborder).*

case of the Mexican Pacific there are few works of synthesis available. Apart from the above-mentioned studies that deal, at least in part, with this area, most papers have focused on the description of new taxa or on autecologic aspects (e.g., Kensley and Kaufman 1978, Carvacho 1983, Carvacho and Hassmann 1984, Wägele 1984, Lombardo 1988, Ruiz and Madrid 1992, Campos et al. 1992, Román-Contreras 1993, 1996, Alvarez and Flores 1997, Hendrickx and Espinosa-Pérez 1998a, 1998b).

Since 1990, an institutional project aimed at the study of the marine and brackish isopods of the Pacific coast of Mexico was initiated. Preliminary results include a review of the distribution of the species occurring in the entire eastern tropical Pacific (Espinosa-Pérez and Hendrickx 2000), new records (Espinosa-Pérez and Hendrickx 1997) and species descriptions (Hendrickx and Espinosa-Pérez 1998a, 1998b, Espinosa-Pérez and Hendrickx 2001). Part of the results that were obtained during this study allow us to present a synthesis of the biodiversity, habitat and geographic distribution by taxonomic group of all isopods known to occur along the Pacific coast of Mexico.

II. MATERIALS AND METHODS

Isopod records included in this study were derived from an exhaustive review of literature dealing mostly with the eastern Pacific and from unpublished data obtained from recent collections along the Pacific coast of Mexico (intertidal to 120 m). A total of 5812 specimens were collected and identified and 1417 specimens were available from three national collections: the Reference Collection of the Laboratorio de Invertebrados Bentónicos, Instituto de Ciencias del Mar y Limnología, UNAM, in Mazatlan; the National Collection of Crustaceans of the Instituto de Biología, UNAM, in Mexico City; and the Crustacean Collection of the Naval Oceanography Department, Secretaría de Marina, Mexico City.

All species records were incorporated into a data base including distribution, depth and substrate data. Biotic substrates include algae, mangroves, sponges and corals; abiotic substrates of mineral origin were divided into gravel, sand and mud. In addition to this, humus (detritus of organic origin) was considered as a separate substrate. In the case of associations with other organisms, three categories were used: commensal, parasite and facultative parasite.

The classification used herein follows Brusca and Wilson (1991). Total number of suborders included in the Isopoda is therefore 10 (Calabozoidea Van Lieshout, 1983, and Phreatoicidea Stebbing, 1893, have no representatives in the Mexican Pacific); some authors, however, believe that the family Gnathiidae Leach, 1814, the only family included in the Gnathiidea Leach, 1814, should form part of the Flabellifera.

III. RESULTS

Species richness

A total of 120 species were recognized for the study area, including nine undescribed species. These species belong to 8 of the 10 currently established suborders of Isopoda (Table 1). Almost half (45%) of the registered species belong to the Flabellifera, and correspond to 8 of the 15 presently recognized families of this suborder. Flabellifera is the largest suborder of Isopoda and contains almost 3000 species (Wetzer and Brusca 1997). They live in a wide variety of habitats, often in shallow water (Brusca and Iverson 1985), and their capture is easier than for other groups due to their relatively large size and abundance (Schultz 1969). Valvifera and Epicaridea represent 20 and 21% of the total number of species, respectively. The high representation of Epicaridea (20 Bopyridae and one species of Dajidae) along the Pacific coast of Mexico is remarkable although the group has not yet been intensively studied in this area. Comparatively, up to 1983, no species of Epicaridea had been reported for the entire Pacific coast of Costa Rica (Brusca and Iverson 1983). On the contrary, Caribbean Epicaridea are much more diverse, with 53 known species (Kensley and Schotte, 1989). Suborders with fewer species are Microcerberidea and Gnathiidae (only one recorded species each). The organisms of these suborders are minute; *Coxicerberus mexicanus* (Pennak 1958), the only Microcerberidea registered in the Mexican Pacific, is only 0.925 mm long (Pennak 1958). If selective sampling with emphasize on small organisms is done, the number of species in these two suborders will certainly increase.

Occurrence by depth and habitat

All Epicaridea are parasites of other crustaceans (Brusca and Iverson 1985, Kensley and Schotte 1989). The rest of the suborders are usually found in several habitats (Fig. 2). *Gnathia steveni* Menzies, 1962, the only species registered for the suborder Gnathiidea, inhabits the intertidal zone of sandy beaches, between algae and sponges (Menzies 1962). The suborder Anthuridae is represented by intertidal and shelf species. One species, *Haliophasma geminatum* Menzies and Barnard, 1959, is registered to about 500 m (Fig. 3). Anthurids of the Pacific coast of Mexico have been registered in sandy substrate, among green and calcareous algae and humus (Fig. 2). The only Microcerberidea, *Coxicerberus mexicanus*, lives in intertidal sandy substrate (Fig. 2). Members of the Asellota are found from the intertidal to at least 200 m (Fig. 3). They do not show a strong affinity for a specific substrate and they have been recorded in six substrates in the area; some are even commensal (e.g., *Maresiella brevicornis* (Carvacho 1983), found in the bivalve *Spondylus calcifer* Carpenter, 1857) (Fig. 2). Records in the literature refer to Asellota as intertidal, shallow water and deep sea organisms (Setubal Pires 1985, Muller 1991, Piertney and Carvalho 1996, Wilson 1997). They are, however, most successful and diverse in the deep sea (Wilson 1989). Little is known of the Mexican Pacific Asellota, particularly because so few deep sea surveys for smaller crustaceans are available for the study area. Two species of Munnidae are reported, including one undescribed species. Most of the Valvifera sampled belong to the family Idoteidae; these are mainly found in the intertidal and to 50 m (at least one record in this bathymetric range exist for 85% of the species) with a maximum depth record for this group of 82 m. As far as habitat is concerned, algae is a favorite hiding place,

Table 1. *Number of species of marine and brackish water isopods known from the Mexican Pacific (by suborder).*

Suborder Isopoda	Number of species and % of total
Gnathiidea Leach, 1814	1 (0.8%)
Anthuridea Leach, 1814	10 (8.3%)
Microcerberidea Lang, 1961	1 (0.8%)
Flabellifera Sars, 1882	54 (45.0%)
Asellota Latreille, 1803	5 (4.2%)
Valvifera Sars, 1882	20 (16.7%)
Epicaridea Latreille, 1831	21 (17.5%)
Oniscidea Latreille, 1803	8 (6.7%)

especially algae of the genus *Sargassum* that are used both as habitat and food supply (Brusca and Wallerstein 1979b). The 20 species of Valvifera recorded from the area occupy 6 different habitats (Fig. 2). The Oniscidea are inhabitants of the intertidal zone and are also associated with sand and humus, the latter used as food and hiding place by species of this group (Kensley and Schotte 1989). The largest group of isopods, the Flabellifera, are distributed from the intertidal zone to a depth of 2214 m (Fig. 4). Flabellifera have been collected in ten different types of substrates (Fig. 2), including on temporary (Aegidae, Corallanidae) or permanent hosts (Cymothoidae). They also include species with commensal relationships (Sphaeromatidae). The 0-10 m depth range contains species of all seven families recorded in the area, while the 11-100 m range includes six families. Aegidae feature the widest bathymetric distribution range, from intertidal to over 2000 m (Figure 4).

Geographic distribution

The information available on the distribution of each species allow us to propose a general pattern of zoogeographic affinities for each group of isopods registered along the Pacific coast of Mexico. Considering the 120 species registered for the area, 71 (59.1%) are found in the Mexican portion of the warm temperate California Province, 88 species (73.3%) occur in the Cortez Province (Gulf of California, s.l.) and 32 species (26.7%) are found in both the Mexican Province and the Mexican portion of the Panamic Province.

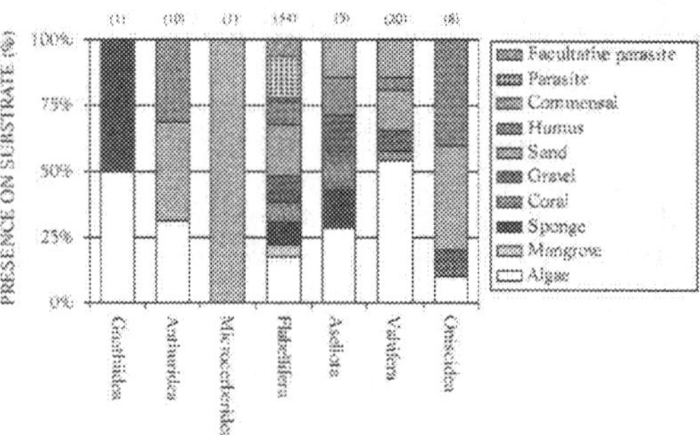

Figure 2. *Species/habitat relationship of marine and brackish water isopods known from the Mexican Pacific (by suborder). Number in parentheses corresponds to number of species in each suborder.*

A detailed analysis of the distribution of the suborders of isopods in the four provinces (Fig. 1) indicates a rather balanced distribution, with an expected dominance of the more abundant groups (e.g., Flabellifera, Valvifera, Epicaridea).

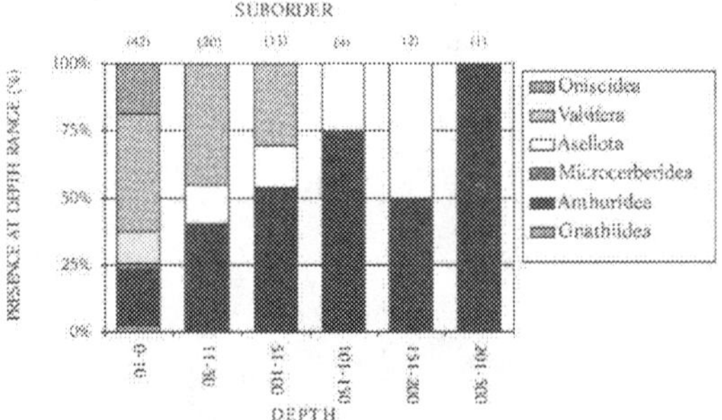

Figure 3. *Species/depth relationship of marine and brackish water isopods known from the Mexican Pacific, by suborder (except Flabellifera). Number in parentheses corresponds to number of species in each depth interval.*

Gnathia steveni, the only Gnathiidea reported for the area, is restricted to the warm-temperate zone (California Province). This actually represents the only record for the genus in the Mexican Pacific (Wetzer et al. 1991). The Anthuridea are slightly better represented in the Cortez Province, despite the fact that they have been considered uncommon in the tropical and subtropical regions (Brusca and Iverson 1985). The species of Flabellifera are well distributed throughout the four biogeographic provinces; their occurrence is higher in the tropical/subtropical Cortez Province. Asellota have only been reported in the warm-temperate and subtropical zones (California and Cortez Provinces), although they are considered cosmopolitan (Cohen and Poore 1994) (Fig. 1). Valvifera, particularly well represented by the Idoteidae, are recorded predominantly in the northern provinces (California and Cortez). Epicaridea and Oniscidea are distributed all along the Mexican Pacific, although the number of species is significantly higher in the warm-temperate and subtropical zones (California and Cortez) (Fig. 1).

Oceanic island species

Oceanic islands are defined here as islands surrounded by water deeper than 200 m, that are located at a distance from the continent that would make non-accidental, sustained recruitment questionable or impossible (see Garth 1946, Vermeij 1978, Hendrickx 1992).

Analysis of the isopods present in oceanic islands of the eastern tropical Pacific and in Mexican Guadalupe Island and Alijos Rocks (both located in the warm temperate California Province) indicates that the Mexican Pacific has no oceanic island endemics. Of the 120 species reported herein, 22 species have been reported in shallow water surrounding at least one island (Table 2). These species belong to six families of Flabellifera, Valvifera and Oniscidea. The highest number of species ocurring in the Mexican Pacific and also present on islands is found in the Galapagos, Ecuador (9 species) and Guadalupe, Mexico (8 species) islands. Only a few species have been recorded on more than one island: *Elthusa menziesi* (Brusca, 1981) and *Ligia exotica* Roux, 1828 occur on two islands; *Rocinela signata* Schiödte and Meinert, 1879 and *Eurydice caudata* Richardson, 1899 have been reported on three islands.

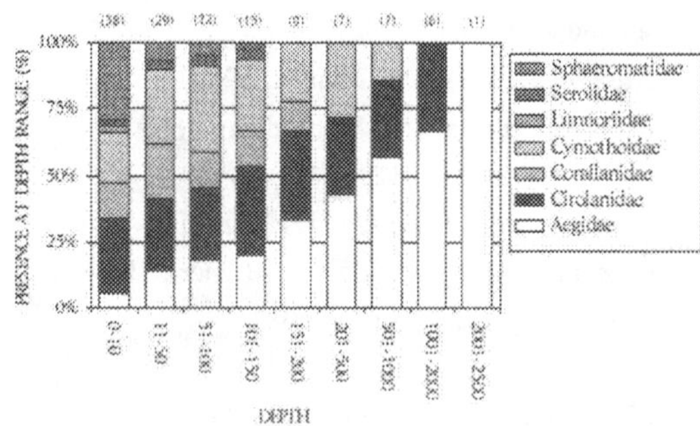

Figure 4. *Species/depth relationship of marine and brackish water isopods known from the Mexican Pacific (suborder Flabellifera). Number in parentheses corresponds to number of species in each interval.*

IV. DISCUSSION

The apparent absence of Asellota and Gnathiidae in the tropical Mexican Pacific may be due to unsuitable sample strategies (see Wetzer and Brusca 1997). On the other hand, the wide distribution of Mexican Flabellifera agrees with previous biogeographic analysis of this group (Brusca 1980, Delaney 1989). Valvifera are mostly found in the California and Cortez Provinces; this agrees with a previous report by Brusca and Wallerstein (1979b), who recognized the group as inhabitants of cold and temperate waters.

	OCEANIC ISLAND								
	GP	RA	IM	RV	CL	CP	CO	MA	GL
FLABELLIFERA Sars, 1882									
Aegidae Dana, 1853									
Rocinela angustata Richardson, 1904	*				*				
Rocinela belliceps (Stimpson, 1864)									
Rocinela hawaiiensis Richardson, 1903	*								
Rocinela signata Schiödte & Meinert, 1879			*	*	*				*
Cirolanidae Dana, 1853									
Anopsilana oaxaca Carvacho & Hassmann, 1984						*			
Cirolana harfordi (Lockington, 1877)	*								
Cirolana parva Hansen, 1890									*
Eurydice caudata Richardson, 1899	*		*	*			*		
Metacirolana costaricensis Brusca & Iverson, 1985									*
Corallanidae Hansen, 1890									
Excorallana houstoni Delaney, 1984									*
Excorallana truncata (Richardson, 1899)									*
Cymothoidae Leach, 1818									
Ceratothoa gaudichaudii (H. Milne-Edwards, 1840)									*
Cymothoa exigua Schiödte & Meinert, 1884									*
Elthusa menziesi (Brusca, 1981)	*	*							
Livoneca bowmani Brusca, 1981			*					*	
Nerocila acuminata Schiödte & Meinert, 1881			*						*
VALVIFERA Sars, 1882									
Holognathidae Thompson, 1904									
Cleantioides occidentalis (Richardson, 1899)									*
Idoteidae Milne-Edwards, 1840									
Colidotea findleyi Brusca & Wallerstein, 1977	*								
Colidotea wallersteini Brusca, 1983	*								
Idotea resecata Stimpson, 1857		*							
ONISCIDEA Latreille, 1803									
Ligiidae Brandt, 1814									
Ligia exotica Roux, 1828	*								
Ligia occidentalis Dana, 1853			*			*			

Table 2. *Presence of isopods species known from the Pacific coast of Mexico on east Pacific oceanic Islands. GP, Guadalupe; RA, Alijos Rocks; IM, Marias; RV, Revillagigedo; CL, Clarion; CP, Clipperton; CO, Coco; MA, Malpelo; GL, Galapagos.*

Five species of isopods have so far been reported from the Tres Marias Islands, in the SE Gulf of California. These islands are considered too close to the continent to be recognized as truly oceanic and might be reachable by continental benthic species with pelagic larvae. They are isolated by water deeper than 200 m from the continental shelf of western Mexico. In the case of isopods, however, the criteria of recruitment through pelagic larvae riding with currents does not apply as they feature direct development. Only in the case of isopods that are obligate or facultative parasites of widely distributed pelagic species (i.e., fishes, pelagic shrimps) or highly mobile benthic species with a wide bathymetric range (*Glyphocrangon spinulosa* Faxon, 1893; *Nematocarcinus agassizzi* Faxon, 1893; *Sclerocrangon procax* Faxon, 1893), and isopods that are rafting on floating algae (i.e., *Synisoma wetzerae* Ormsby, 1991), might one expect a wider dispersal potential through the adult isopods.

As far as our present knowledge goes, the Mexican Pacific isopod fauna shows a clear subtropical affinity with a strong warm-temperate component. This, however, is probably due to a larger sampling effort in northern latitudes (northern and Central Gulf of California and west coast of Baja California) by scientists from the US combined with a much more comprehensive taxonomic study of isopods collected in this area in the last 30 years or so. A more intensive sampling effort and adequate analysis of the collected specimens along the SW coast of Mexico will undoubtedly bring new data on species distribution and basic knowledge of their ecology.

ACKNOWLEDGEMENT

This paper would not have been possible without the help of many colleagues who provided advice and literature during our study of isopods of the Eastern Tropical Pacific. We also thank J.L. Villalobos Hiriart and F. Alvarez Noguera who provided an important series of specimens from the National Collection of Crustaceans, Biology Institute, UNAM, and J.L. Hernández Aguilera who provided specimens from several localities of the Mexican Pacific. Many isopods were obtained during field activities. Collections aboard the R/V "El Puma" were partly supported by CONACyT project CORTES, ICECXNA-021996 (1984086). Shallow water sampling was partly supported by CONABIO project H170. We thank M. Ayon Parente, M. García Guerrero, N. Méndez Ubach and J. Salgado Barragán for their help during field trips and M. Cordero Ruíz for providing information contained in the laboratory data base.

REFERENCES

Alvarez F and Flores M (1997) *Cymothoa exigua* (Isopoda: Cymothoidae) parasitando al pargo *Lutjanus peru* (Pisces: Lutjanidae) en Manzanillo, Colima, México. Rev Biol Trop 45:397-394

Arrontes J and Anadón R (1990) Seasonal variation and population dynamics of isopods inhabiting intertidal macroalgae. Sci Mar 54:231-240

Bruce NL (1986) Cirolanidae (Crustacea: Isopoda) of Australia. Rec Aust Mus 6:1-239

Brusca RC (1980) Common intertidal invertebrates of the Gulf of California. The University of Arizona Press, USA

Brusca RC (1981) A monograph on the Isopoda Cymothoidae (Crustacea) of the Eastern Pacific. Zool J Linn Soc 73:117-199

Brusca RC and Iverson EW (1985) A guide to the marine isopod crustacea of Pacific Costa Rica. Rev Biol Trop 33:1-77

Brusca RC and Wallerstein BR (1979) Zoogeographic patterns of idoteid isopods in the Northeast Pacific, with a review of shallow water zoogeography of the area. Bull Biol Soc Wash 3:67-105

Brusca RC and Wilson GD (1991) A phylogenetic analysis of the Isopoda with some classificatory recommendations. Mem Queen Mus 31:143-204

Brusca RC, Wetzer R and France SC (1995) Cirolanidae (Crustacea: Isopoda: Flabellifera) of the Tropical Eastern Pacific. Proc San Diego Soc Nat Hist 30:1-96

Campos E, de Campos AR and Ramírez J (1992) Remarks on distribution and hosts for symbiotic crustaceans of the Mexican Pacific (Decapoda and Isopoda). Proc Biol Soc Wash 105:753-759

Carvacho A (1983) Asellota del golfo de California, con descripción de dos nuevos géneros y dos nuevas especies (Crustacea, Isopoda). Cah Biol Mar 24:281-295

Carvacho A and Haasmann Y (1984) Isópodos litorales de Oaxaca, Pacífico mexicano. Cah Biol Mar 25:15-32

Cohen BF and Poore GCB (1994) Phylogeny and biogeography of the Gnathiidae (Crustacea: Isopoda)

with descriptions of new genera and species, most from south eastern Australia. Mem Mus Victoria 54:271-397

Delaney MP (1984) Isopods of the genus *Excorallana* Stebbing, 1904 from the Gulf of California, Mexico (Crustacea, Isopoda, Corallanidae). Bull Mar Sci 34:1-20

Delaney MP (1989) Phylogeny and biogeography of the marine isopod family Corallanidae (Crustacea, Isopoda, Flabellifera). Nat Hist Mus Los Angeles Co 409:1-75

Dexter DM (1972) Comparasion of the community structures in a Pacific and Atlantic Panamanian sandy beach. Bull Mar Sci 22: 449-485

Dexter DM (1974) Sandy-beach fauna of the Pacific and Atlantic coast of Costa Rica and Colombia. Rev Biol Trop 22: 51-66

Dexter DM (1976) The sand-beach fauna of Mexico. Southwest Nat 20:479-485

Ellison AM and Farnsworth EJ (1990) The ecology of Belizean mangrove-root fouling communities. I. Epibenthic fauna are barriers to isopod attack of mangrove roots. J Exp Mar Biol Ecol 142:91-104

Espinosa-Pérez MC and Hendrickx ME (1997). New geographic records of two species of Cirolanidae (Crustacea: Isopoda) from the eastern tropical Pacific. An Inst Biol UNAM 68:175-185

Espinosa-Pérez MC and Hendrickx ME (2000) Checklist of the Isopods (Crustacea: Peracarida: Isopoda) from the Eastern Tropical Pacific. Belg J Zool 131:41-54

Espinosa-Pérez MC and Hendrickx ME (2001) A new species of *Exosphaeroma* Stebbing (Crustacea: Isopoda: Sphaeromatidae) from the Pacific coast of Mexico. Proc Biol Soc Wash 114

Garth JS (1946) Distribution Studies of Galapagos Brachyura. Allan Hancock Pac Exp 5:603-638

Hendrickx ME (1992) Distribution and zoogeographic affinities of decapod crustaceans of the Gulf of California, Mexico. Proc San Diego Soc Nat Hist 20:1-11

Hendrickx ME (1993) Crustáceos Decápodos del Pacífico mexicano. In: Salazar-Vallejo SI and González NE (eds) Biodiversidad Marina y Costera de México (pp. 271-318) Com Nal Bio and CIQRO, México

Hendrickx ME and Espinosa-Pérez MC (1998a) A new species of *Cassidinidea* Hansen (Isopoda: Sphaeromatidae) and first record of the genus from the eastern tropical Pacific. Proc Biol Soc Wash 111:295-302

Hendrickx ME and Espinosa-Pérez MC (1998b) A new species of *Excorallana* Stebbing (Crustacea: Isopoda: Corallanidae) from the Pacific coast of Mexico, and additional records for *E. bruscai* Delaney. Proc Biol Soc Wash 111:303-313

Kang YJ and Yun SG (1988) Ecological study on isopod crustaceans in surfgrass beds around Tongbacksum, Haeundae, Pusan. Ocean Res 10:23-31

Kensley B and Kaufman HW (1978) *Cleantioides*, a new genus from Baja California and Panama. Proc Biol Soc Wash 91:658-665

Kensley B and Schotte M (1989) Guide to the marine isopod crustaceans of the Caribbean. Smithsonian Institution Press, Washington, DC

Kensley B & Schotte M (2000) World List of Marine and Freshwater Crustacea Isopoda [on line] (January 2000) Available on Internet: <URL: http://nmnhwww.si.edu/gopher-menus/WorldList ofMarineandFreshwaterCrustaceaIsopoda.html>

Lombardo AC (1988) *Paracerceis richardsoni*, n. sp. di crostaceo isopodo (Sphaeromatidae, Eubranchiatae) delle Coste Pacifiche del Messico. Animalia 15:5-15

Markham JC (1985) A review of the bopyrid isopods infesting caridean shrimps in the northwestern Atlantic Ocean, with special reference to those collected during the Hourglass cruises in the Gulf of Mexico. Mem Hourglass Cruises 7:1-156

Menzies RJ (1962) The marine isopod fauna of Bahia de San Quintin, Baja California, Mexico. Pac Nat 3:338-348

Moreno-Casasola P and Castillo S (1992) Dune ecology on the eastern coast of Mexico. In: Seeliger U (ed) Coastal plant communities of Latin America. Academic Press, New York

Muller HG (1991) Stenetriidae from coral reefs at Reunion Island, southern Indian Ocean. Description of three new species (Crustacea: Isopoda: Asellota). Senckenb Biol 71:303-318.

Pennak WR (1958) A new micro-isopod from a mexican marine beach. Trans Am Microsc Soc 77:298-303

Piertney SB and Carvalho GR (1996) Sex ratio variation in the intertidal isopod, *Jaera albifrons*. J Mar Biol Assoc UK 76:825-828

Poore GCB (1984a) *Colanthura, Califanthura, Cruranthura* and *Cruregens*, related genera of the Paranthuridae (Crustacea: Isopoda). J Nat Hist 18:697-715

Poore GCB (1984b) Redefinition of *Munna* and *Uromunna* (Crustacea: Isopoda: Munnidae), with descriptions of five species from coastal Victoria. Proc R Soc Victoria 96:61-81

Poore GCB (1996) Species differentiation in *Synidotea* (Isopoda: Idoteidae) and recognition of introduced marine species: a reply to Chapman and Carlston. J Crust Biol 16:384-394

Ribi G (1981) Does the wood boring isopod *Sphaeroma terebrans* benefit red mangroves (*Rhizophora mangle*)? Bull Mar Sci 3:925-928

Román-Contreras R (1993) *Probopyrus pacificensis*, a new parasite species (Isopoda: Bopyridae) of *Macrobrachium tenellum* (Smith, 1871) (Decapoda: Palaemonidae) of the Pacific coast of Mexico. Proc Biol Soc Wash 106:689-697

Román-Contreras R (1996) A new species of *Probopyrus* (Isopoda, Bopyridae), parasite of *Macrobrachium americanum* Bate, 1868 (Decapoda, Palaemonidae). Crustaceana 69:204-210

Ruiz A and Madrid J (1992) Estudio de la biología del isópodo parásito *Cymothoa exigua* Schioedte y Meinert, 1884 y su relación con el huachinango *Lutjanus peru* (Pisces: Lutjanidae) Nichols y Murphy, 1922, a partir de capturas comerciales en Michoacán. Ciencias Marinas 18:19-34

Schultz GA (1961) Distribution and establishment of a land isopod in North America. Syst Zool 10:193-196

Schultz GA (1969) How to know the marine isopod crustaceans. W.M.C. Brown Company Publishers, USA

Setubal Pires AM (1985) The ocurrence of *Munna* (Isopoda, Asellota) on the southern Brazilian coast, with a description of two new species. Crustaceana 48:64-73

Taylor AC and Moore PG (1995) The burrows and physiological adaptations to a burrowing lifestyle of *Natatolana borealis* (Isopoda: Cirolanidae). Mar Biol 123:805-814

Vermeij GJ (1978) Biogeography and adaptation. Harvard University Press

Wägele JW (1984) Two new littoral Anthuridae from Baja California and redescription of *Mesanthura occidentalis* (Crustacea, Isopoda). Zool Scr 13:45-57

Wetzer R and Brusca RC (1997) Descriptions of the species of the suborders Anthuridea, Epicaridea, Flabellifera, Gnathiidea, and Valvifera. In: Blake JA and Scott PH (eds) Taxonomic atlas of the benthic fauna of Santa Maria Basin and western Santa Barbara Channel. Vol. 11. The Crustacea. Part 2. The order Isopoda (pp. 9-56) Santa Barbara Museum of Natural History, California, USA

Wilson DFG (1989) A systematic revision of the deep-sea subfamily Lipomerinae of the isopod crustacean family Munnopsidae. Bull Scripps Inst Oceano 27:1-138

Wilson DFG (1997) The suborder Asellota. In: Blake JA and Scott PH (eds) Taxonomic atlas of the benthic fauna of Santa Maria Basin and western Santa Barbara Channel. Vol. 11. The Crustacea. Part 2. The order Isopoda (pp. 59-108) Santa Barbara Museum of Natural History, California, USA

ABUNDANCE AND DISTRIBUTION OF COMMENSAL AMPHIPODS FROM COMMON MARINE SPONGES OF SOUTHEAST FLORIDA

Stacie E. Crowe and James D. Thomas
SEC, Nova Southeastern University Oceanographic Center,
8000 N. Ocean Drive, Dania, Florida 33004 USA

colomastix@hotmail.com

JDT, National Coral Reef Institute, 8000 N. Ocean Drive,
Dania, Florida 33004 USA

ABSTRACT

Marine sponges were examined from southeast Florida and the Florida Keys to determine species composition and distribution of commensal amphipod crustaceans from shallow reef, mangrove, and seagrass habitats. Twenty sponge species were investigated during this study, sixteen of which housed colomastigid and/or leucothoid amphipods. The *Leucothoe spinicarpa* (Abildgaard) complex of species was the most dominant amphipod commensal, representing 63% of the total amphipods collected. Common sponge hosts included *Callyspongia vaginalis*, *Mycale* sp, and *Myriastra kallitetilla*.

I. INTRODUCTION

Commensal relationships are common between sessile plants or animals and motile organisms. Marine environments exhibit a variety of these types of associations. Many marine fish (Tyler and Bohlke 1972), crustaceans (Thomas and Cairns 1984), and polychaete worms (Van Dover et al. 1999) are reported as commensals associated with invertebrate hosts. Details of these associations are little studied in marine ecosystems, yet they are important in determining and evaluating evolutionary changes and maintaining ecosystem balance (Duffy 1992).

Marine sponges are abundant invertebrates that frequently serve as hosts to a variety of commensal organisms including crustaceans, polychaetes, and ophiuroids (Seger and Moran 1996, Uriz et al. 1992, Villamizer and Laughlin 1991). Sponges have several characteristics that can lead to commensal occupation including: an abundance of internal canals that permeate the tissues of many species and supply the commensal with nutrients via currents exchanged with the environment (Duffy 1992), a protective and defensible habitat conducive to eusocial behavior (Spanier et al. 1993), and the protective nature of internal canals as a convenient refuge from predation.

Individual sponges may host entire communities of endocommensal species. Volume of the host as well as size and distribution of internal channels are possible factors that may determine community structure and population size within any given sponge species. In a comparative study between the sponges *Aplysina archeri* and *Aplysina lacunosa*, Villamizer and Laughlin (1991) found more diverse animal communities in *A. lacunosa*, a species with a larger superficial area and volume. The internal cavities of the hosts may also promote eusocial behavior by the commensals as defined by cooperative brood care, reproductive division of labor, and overlap of generations (Seger and Moran 1996).

Amphipod crustaceans are widespread and abundant in many habitats worldwide. Although they are usually free-living, they can be found in sediment, swimming near the substrate, inhabiting a variety of domiciles including empty snail shells, crevices, tubes and burrows produced by other organisms, and coral rubble. Within gammarid amphipods, Colomastigidae and Leucothoidae are prominently mentioned in the literature as sponge commensals (Ortiz 1975,

E. Escobar-Briones & F. Alvarez Eds.
MODERN APPROACHES TO THE STUDY OF CRUSTACEA
PP. 105-110

Biernbaum 1981, Vader 1983, Costello and Myers 1987, Barnard and Karaman 1991, Thomas 1993, LeCroy 1995, Thiel 1999). Both families are cosmopolitan in marine habitats. The Colomastigidae have been studied most extensively in Madagascar (Ledoyer 1979, 1982) and the eastern Gulf of Mexico (LeCroy 1995), while the tropical Indo-Pacific houses the most members of the Leucothoidae. Little information is known concerning host preferences and specific behavior patterns for species of either family.

This study investigates the abundance, distribution, and ecology of amphipods commensal in sponges in southeast Florida and the Florida Keys, an area with abundant and diverse sponge communities.

II. MATERIALS AND METHODS

Sponges were collected between May and September 1999 in southeast Florida and the Florida Keys. Five different habitats were sampled in Broward (2) and Monroe (3) counties. Broward County stations included reef (26° 9.163' N, 26° 5.341' W) and hardbottom (26° 9.774' N, 80° 5.435' W) habitats. Stations in Monroe County included seagrass beds (24° 50.543' N, 80° 49.633' W), a patch reef (24° 50.175' N, 80° 43.740' W), and a mangrove creek (24° 49.567' N, 80° 48.865' W). An initial survey was conducted at each site to determine relative species abundance of sponges. Species consistently sighted at least twice per 3 m² area were considered abundant and ten replicates were collected. Each sponge collected was covered by a plastic bag, cut at its base with a knife, and immediately sealed inside the bag to minimize loss of commensals. Sponge volume was measured by water displacement after draining the canals (error ± 5%). Initial removal of fauna was accomplished by placing sponges in a 4% seawater formalin solution to remove commensals from the canals. The remaining water was passed through a 0.5 mm mesh sieve to insure capture of animals larger than 1 mm. Sponges were then dissected to locate remaining commensals. All amphipods were preserved in a buffered 4% seawater formalin solution for 24 hours, then transferred to 70% ethanol. Mean number of amphipods per sponge is reported followed by ± one standard error. Non-amphipod taxa found within each sponge were not scored. Amphipods were measured from rostrum to telson using a compound microscope with a camera lucida attachment, and identified to species. Host sponges were identified using photographs and spicule preparations.

III. RESULTS

Twenty sponge species were collected from the sampling locations in south Florida (Broward and Monroe counties), sixteen of which housed colomastigid and/or leucothoid amphipods (Fig. 1). Commensals were not reported for four sponge species, *Aplysina fistularis, A. fulva, Agelas sceptrum,* and *Iotrochota birotulata.* Sponges representing several different growth forms were collected including massive (*Myriastra kallitetilla, Spheciospongia vesparium, Ircinia felix, I. campana, Tedania ignis, Amphimedon* sp, *Anthosigmella varians*), branching (*Aplysina fulva, A. fistularis, Agelas sceptrum, Amphimedon compressa, Iotrochota birotulata, Niphates amorpha, N. erecta, Holopsamma helwigi*), and tubular (*Callyspongia vaginalis, N. digitalis*) varieties. Sponges collected at the mangrove creek (*Mycale* sp, *Haliclona* sp, *Tedania ignis,* brown sponge) were all encrusting on the roots of red mangroves (*Rhizopora mangle* L).

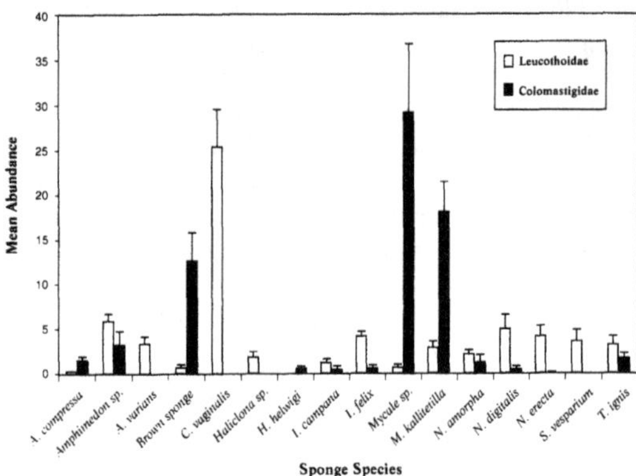

Figure 1. *Mean abundance ± standard error of leucothoid and colomastigid amphipods per sponge from all sites.*

All leucothoid specimens in this study have been identified as one species and are being referred to as a "complex", suggesting the need to clarify the taxonomic ambiguity within this family. Imprecise descriptions within the literature make it very difficult to identify leucothoids to species level. Abildgaard (1789) originally described *Leucothoe spinicarpa* from the Skagerrak Sea off Denmark. Since that time, several authors have labeled amphipods as *L. spinicarpa* which do not show consistency with the type description.

There are probably several species of commensal or cryptic amphipods identified by various authors under the name *L. spinicarpa* in the southeast Florida region (Thomas 1993).

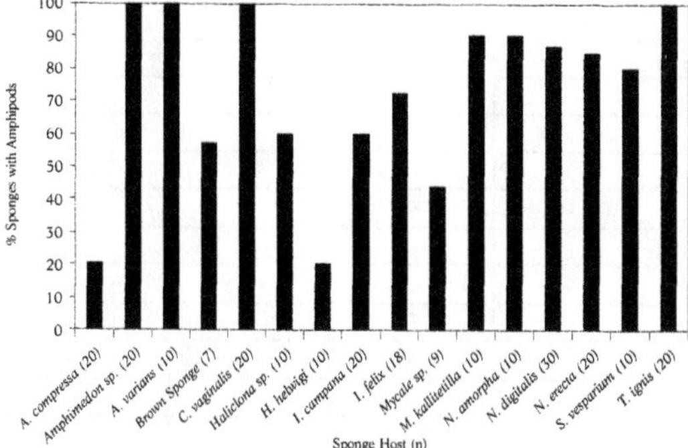

Figure 2. *Percent of individual sponges with* **L. spinicarpa** *amphipods; n=number of individual sponges examined.*

While the scope of this paper is not taxonomic in nature, the persistent problem of *Leucothoe spinicarpa* identifications must be addressed before detailed ecological and behavioral studies are initiated. For the purposes of this project, the designation of *L. spinicarpa* is used with complete recognition of the shortcomings of this approach. For example, one of the more common species in this study is characterized by multiple long setae on the anterior edge of article 2 of gnathopod 2. Long setae on the basis of gnathopod 2 are not indicative of *L. spinicarpa* according to the original type description. Also, two distinct color morphs were found in this study, a completely white morph (displayed by the majority of *L. spinicarpa*) and one occurrence of a deep purple morph found in *Spheciospongia vesparium*.

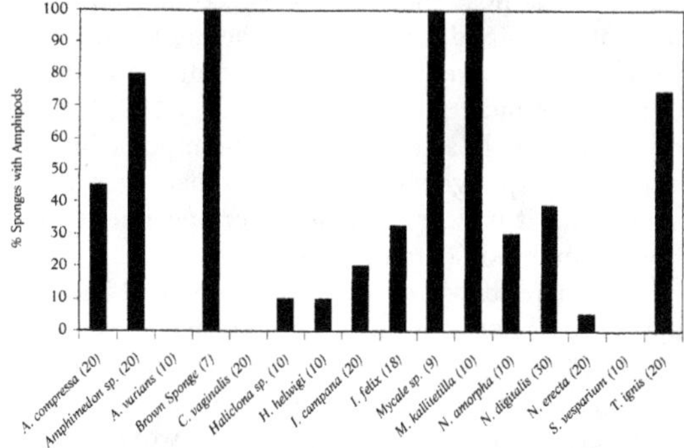

Figure 3. *Percent of individual sponges with* **Colomastix** *amphipods; n = number of individual sponges examined.*

From 244 sponges sampled, 1930 amphipods of six species were collected (Table 1). Specimens identified as the *Leucothoe spinicarpa* (Abildgaard) complex of species were the most abundant commensal amphipods, comprising 63% (n = 1215) of the total. The mean number of *L. spinicarpa* per sponge at all sites varied from less than one (*Holopsamma helwigi, Amphimedon compressa*) to 25.4 ± 4.15 in *Callyspongia vaginalis* (Fig. 1). The maximum number of *L. spinicarpa* found in one solitary sponge was 53 in *C. vaginalis* from the shallow reef area in Broward County. Leucothoids were present in all sponge species containing amphipod commensals. Overall, 73% of the individual host sponges harbored *L. spinicarpa* (Fig. 2).

The mean number of colomastigids per sponge at all sites varied from less than one (*Niphates digitalis, Holopsamma helwigi, N. erecta, Ircinia felix, Haliclona* sp) to 29.2 ± 7.52 in *Mycale* sp from the mangrove creek in Monroe County (Fig. 1). A total of 715 colomastigid amphipods were collected including five species: *Colomastix bousfieldi* LeCroy, *C. falcirama* LeCroy, *C. halichondriae* Bousfield, *C. irciniae* LeCroy, and *C. janiceae* Heard and Perlmutter (Table 1). The maximum number of *Colomastix* individuals found in one solitary sponge was 76 *C. falcirama* in *Mycale* sp. *Myriastra kallitetilla* and *Mycale* sp were the most common hosts of *Colomastix* individuals (Table 1). *Colomastix falcirama* was the most abundant *Colomastix* species, comprising 52% (n = 375) of the population, 97% of which were collected from the mangrove creek. *Colomastix* amphipods inhabited 41% of individual sponge hosts (Fig. 3). Colomastigids were not found in three sponge species, *Anthosigmella varians*, *Callyspongia vaginalis*, and *Spheciospongia vesparium*.

IV. DISCUSSION

Factors affecting the occupation of host sponges by amphipod species are virtually unknown. Host characteristics such as size, presence or absence of secondary metabolites that may deter predators and structural complexity may affect the occupation of host species by symbiotic associates. Other parameters may include host species, location of the host, host densities, water temperature, salinity, and depth.

Several studies have investigated sponge size in relation to abundance of associated fauna (Westinga and Hoetjes 1981, Costello and Myers 1987). Esti-

Host Sponge	Habitat	Colomastix bousfieldi	Colomastix falcirama	Colomastix halichondriae	Colomastix irciniae	Colomastix janiceae	Leucothoe spinicarpa
Amphimedon compressa	[1]HB (n=10)					1	1
	[2]BR (n=10)					27	4
Amphimedon sp.	HB (n=10)			28			70
	BR (n=10)			35			52
Anthosigmella varians	[3]SB (n=10)						34
Brown Sponge	[4]MC (n=7)		89				5
Callyspongia vaginalis	HB (n=10)						210
	BR (n=10)						319
Haliclona sp.	MC (n=10)		2				19
Holopsamma helwigi	HB (n=10)					1	4
Ircinia campana	HB (n=10)				2		7
	SB (n=10)				4		15
Ircinia felix	HB (n=10)						6
	SB (n=8)				11		77
Mycale sp.	MC (n=9)		262				6
Myriastra kallitetilla	SB (n=10)					186	29
Niphates amorpha	HB (n=10)	12					21
Niphates digitalis	HB (n=10)					8	35
	BR (n=10)		3				63
	[5]PR (n=10)		7				48
Niphates erecta	BR (n=10)			1			52
	HB (n=10)						31
Spheciospongia vesparium	SB (n=10)						38 (*9)
Tedania ignis	SB (n=10)					25	45
	MC (n=10)		11				24

Table 1. *Abundance and distribution of* **Colomastix** *and* **Leucothoe** *species within sponge hosts ([1]Hardbottom Area, [2] Broward Reef, [3] Seagrass Beds, [4] Mangrove Creek, [5] Patch Reef,*indicates purple morph).*

mates include volume displacement (LeCroy 1995), and dry weight (Biernbaum 1981, Thiel 1999). These methods measure sponge tissue, not the internal spaces where the commensals are commonly located. Therefore, while volume displacement gives a baseline idea of sponge size in relation to other specimens, it may not provide an accurate correlation when compared to abundance of fauna. A better way to assess the amount of space available for inhabitants would be to measure channel (canal) diameter. Duffy (1992) found that *Synalpheus* species occurred more commonly in a host with a greater diversity of canal widths, such as *Spheciospongia vesparium*. A study by Villamizer and Laughlin (1991) also suggests that a sponge that offers a system of channels and meanders, which provide a more complex domicile, may support a greater number of individuals and diversity of taxa.

Colomastigid and leucothoid amphipods appear to be quite common in the study areas. Out of 20 potential host sponge species, 16 housed commensal amphipods. Sponges of the genus *Aplysina* previously documented as containing amphipods (Villamizer and Laughlin 1991, LeCroy 1995), were not found to host commensal amphipods in this study. The sponge specimens examined were quite small, and it is possible that larger specimens may house amphipod commensals.

The *Leucothoe spinicarpa* "complex" does not appear to be host specific. They inhabited all potential hosts indicating that the availability of a host may be the most important factor in determining distribution and occurrence. Further taxonomic work on the leucothoids in this study is needed in order to determine specific host occupancy patterns.

Occupancy levels of colomastigids within individual hosts were relatively low at 41%. It appears that a factor other than host species may be important in determining occupancy. For example, *Colomastix janiceae* and *C. falcirama* showed a very generalized distribution among hosts. One exception to this was *C. irciniae*, that preferred hosts of the genus *Ircinia*, although did not prefer one species of host to another. This is consistent with patterns of *C. irciniae* reported from the Gulf of Mexico (LeCroy 1995).

Further analysis of volume and/or channel diameter of host species, population structure within the host individuals and presence or absence of secondary metabolites is needed to evaluate which factors may affect the patterns and distributions exhibited in this study.

ACKNOWLEDGEMENTS

We wish to thank the Secretaria de Educacion Publica (SEP) for financial support for Stacie E.Crowe to attend the 2000 Summer Meeting of the Crustacean Society held in Puerto Vallarta, Mexico.

REFERENCES

Abildgaard PC (1789) In: Müller OF (ed) Zoologica Danica seu Animalium Daniae et Norvegiae …Havniae 3:1-71

Barnard JL and Karaman GS (1991) The Families and Genera of Marine Gammaridian Amphipoda (Except Marine Gammaroids). Rec Aust Mus, Suppl 13, 417 p

Biernbaum CK (1981) Seasonal changes in the amphipod fauna of *Microciona prolifera* (Ellis and Solander) (Porifera: Demospongiae) and associated sponges in a shallow salt-marsh creek. Estuaries 4:85-96

Bousfield EL (1973) Shallow-water Gammaridean Amphipoda of New England. Cornell University Press, Ithaca, New York, 313 p

Costello MJ and Myers AA (1987) Amphipod fauna of the sponges *Halichondria panicea* and *Hymeniacidon perleve* in Lough Hyne, Ireland. Mar Ecol Prog Ser 41:115-121

Duffy JE (1992) Host use patterns and demography in a guild of tropical sponge-dwelling shrimps. Mar Ecol 90:127-138

Heard RW and Perlmutter DG (1977) Description of *Colomastix janiceae* n. sp., a commensal amphipod (Gammaridea: Colomastigidae) from the Florida Keys, U.S.A. Proc Biol Soc Wash 90:30-42

LeCroy SE (1995) Memoirs of the Hourglass Cruises: Amphipod Crustacea III Family Colomastigidae. Fla Dept Env Pro V IX, part II, 139 p

Ledoyer M (1979) Les gammariens de la pente externe du grand recif de Tulear (Madagascar), (Crustacea: Amphipoda). Memorie del Museo Civico di Storia Naturale di Verona 2:1-149

Ledoyer M (1982) Faune de Madagascar, 59(1), Crustaces amphipodes gammariens, familles des Acanthonotozomatidae a' Gammaridae. Centre National de la Recherche Scientifique, 598 p

Ortiz LM (1975) Algunos datos ecológicos de *Leucothoe spinicarpa* Abildgaard (Amphipoda, Gammaridea) en aguas Cubanas. Ciencias, Series 8, Investigaciones Marinas 16:1-12

Seger J and Moran NA (1996) Snapping social swimmers. Nature 381:473-474

Spanier E, Cobb JS and James M (1993) Why are there no reports of eusocial marine crustaceans? Oikos 67:573-576

Thiel M (1999) Host-use and population demographics of the ascidian-dwelling amphipod *Leucothoe spinicarpa* indication for extended parental care and advanced social behaviour. J Nat Hist 33:193-206

Thomas JD (1993) Identification manual for marine amphipoda (Gammaridea): I. Common coral reef and rocky bottom amphipods of south Florida. Dept. of Environmental Protection. Tallahassee, Florida, 83 p

Thomas JD and Cairns KD (1984) Discovery of a majid host for the commensal amphipod *Stenothoe symbiotica* Shoemaker, 1956. Bull Mar Sci 34:484-485

Tyler JC and Bohlke JE (1972) Records of sponge-dwelling fishes, primarily of the Caribbean. Bull Mar Sci 22:601-642

Uriz M, Rosell D and Maldonado M (1992) Parasitism, commensalism or mutualism? The case of Scyphozoa (Coronatae) and horny sponges. Mar Ecol Prog Ser 81:247-255

Vader W (1983) Associations between amphipods (Crustacea: Amphipoda) and sea anemones (Anthozoa, Actiniaria). Mem Aust Mus18:141-153

Van Dover CL, Trask J, Gross J and Knowlton A (1999) Reproductive biology of free-living and commensal polynoid polychaetes at the Lucky Strike hydrothermal vent field (Mid-Atlantic Ridge). Mar Ecol Prog Ser 181:201-214

Villamizer E and Laughlin RA (1991) Fauna associated with the sponges *Aplysina archeri* and *Aplysina lacunosa* in a coral reef of the Archipielago de Los Roques, National Park, Venezuela. In: Reitner J and Keupp H (eds) Fossil and Recent Sponges, pp 522-542. Springer-Verlag, Berlin

Westinga E and Hoetjes P (1981) The intrasponge fauna of *Spheciospongia vesparium* (Porifera, Demospongiae) at Curacao and Bonaire. Mar Biol 62:139-150

POPULATION STRUCTURE OF THE MACROBENTHIC AMPHIPOD *HYALELLA AZTECA* SAUSSURE (CRUSTACEA: PERACARIDA) ON THE LITTORAL ZONE OF SIX CRATER LAKES

Javier Alcocer, Elva Escobar-Briones, Laura Peralta and Fernando Álvarez

JA, LP, *Proyecto de Conservación y Mejoramiento del Ambiente, UIICSE, FES Iztacala, UNAM, Av. de los Barrios s/n, Los Reyes Iztacala, Tlalnepantla 54090, Estado de México, México*

jalcocer@servidor.unam.mx

EEB, *Unidad Académica en Sistemas Costeros y Oceanográficos, Instituto de Ciencias del Mar y Limnología, Universidad Nacional Autónoma de México, México 04510, D.F., México*

FA, *Colección Nacional de Crustáceos, Instituto de Biología, Universidad Nacional Autónoma de México, México 04510, D.F., México*

ABSTRACT

The population structure of the amphipod *Hyalella azteca* in the littoral zone of six crater lakes in Puebla (i.e., Alchichica, Atexcac, Quechulac, La Preciosa, Aljojuca and Tecuitlapa), Central Mexico, was studied throughout a yearly cycle. *Hyalella azteca* was the only species found residing in the littoral area of the lakes. Mean annual density ranged from 1042 ± 3544 ind/m^2 up to 13496 ± 20740 ind/m^2. Seasonal (dry and rainy seasons) changes did not significantly affect density. In terms of abundance, juveniles dominated (55-64%) over females (13-20%), males (10-17%), and ovigerous females (8-14%). Ovigerous females held from 1 up to 38 eggs. Sex ratio (male: female) ranged from 1:1 (1.0) to 1:3.25 (0.31). *Hyalella azteca* developed along the ample range of environmental conditions present in the littoral zone of the crater lakes of Puebla. No statistically significant correlations between abiotic factors and *H. azteca* abundance in these lakes were found.

I. INTRODUCTION

Hyalella azteca (Saussure) was originally described from a cistern in Veracruz and from Chapultepec Lake in Mexico City (Saussure 1858). It is of strict epigean facies –although occasionally found in caves– and commonly occurs on surface fresh as well as brackish habitats. *Hyalella azteca* is a widely distributed species in Mexico as well as in North and South America, and the Caribbean area (Holsinger 1984). A literature search could result in hundreds of papers dealing with "*Hyalella azteca*", most of them dealing with toxicological tests and less extensively with genetics. However, its population biology and ecology is poorly known. To help filling this gap, this study examines the population structure of *H. azteca* inhabiting the littoral area of a cluster of six crater-lakes located in the southeastern portion of the Mexican Plateau.

II. MATERIALS AND METHODS

A quarterly sampling program devoted to study the benthic macrofaunal structure was carried out throughout a year with a total of five samples obtained in the littoral zone of the six lakes. At each sampling site, the vegetation coverage and type, exposure to wave action and sediment grain size were characterized. Previous results (Alcocer et al. 1993, Alcocer 1995) have recognized that plant type and cover and salinity are the most important environmental variables explaining the differences between sampling sites in the littoral zone of the six lakes.

Kluwer Academic/Plenum Publishers

E. Escobar-Briones & F. Alvarez Eds.
MODERN APPROACHES TO THE STUDY OF CRUSTACEA
PP. 111-115

Table 1. *Environmental variables of the littoral areas of ALChicha, ATExcac, LA Preciosa, QUEchulac, ALJjojuca and TECuitlapa crater lakes. Values represent mean ± one standard deviation. DO = dissolved oxygen, TEMP = temperature, SAL = salinity, OM = organic matter, CO_3 = carbonates, ST = sediment texture, EM = emergent macrophytes, SM = submerged macrophytes and BA = benthic algae.*

Variable	ALC	ATE	LAP	QUE	ALJ	TEC
pH	8.9-9.1	8.2-8.6	8.8-9.5	8.7-9.1	8.9-9.3	9.7-9.9
DO (mg L^{-1})	6.5-12.3	5.4-8.4	5.5-12.8	4.1-7.7	3.7-9.1	6.5-12.1
DO (% Sat)	84-196	78-122	80-197	57-111	55-135	96-178
TEMP (°C)	18.3-24.9	18.7-21.7	17.4-24.5	15.5-19.5	19.4-23.4	26.3-21.9
SAL (g L^{-1})	6.0-7.4	6.0	1.0	0.1	0.1	1.0
OM (%)	3-8	0-7.2	6-18	2-4	1-2	7-8
CO_3 (%)	2-29	4-5	5-24	0.5-1	0.5-1	3-6
ST (ø)	0.2-2.3	0.9-1.1	0.1-2.6	0.2-0.3	0.2-0.3	2-2.2
EM (%)	0	50-60	0-50	80-100	0	0
SM (%)	25-100	70-75	0-75	0	20-25	90-100
BA (%)	0-25	20-25	25-75	10-25	50-60	0

Sediment samples were collected (three replicates) in the littoral zone with an Ekman grab (0.00225 m² sampling area). The sediment samples were sieved both in the field and subsequently in the laboratory through a 590 μm mesh. The amphipods were sorted from the rest of the benthic macrofauna under a dissecting microscope, and identified using the identification keys published by Bousfield (1973) and Barnard and Barnard (1983).

Density (ind/m²), sex ratio (males: females), and reproductive state (ovigerous females, presence and number of eggs, proportion of juveniles) of all samples were obtained. Data were analyzed by means of classification analyses. A cluster analysis using complete linkage was carried out on amphipod densities and interpreted with Euclidean distances as the measure of the degree of affinity among lakes. Data were logarithmically transformed according to Kunz (1988) and Mirza and Gray (1981). Sampling sites were grouped based on amphipod density, using the statistical software Statistica (release 6.0, 1997). Seasonal differences among amphipod densities in the six lakes were graphically analyzed.

Study area

The crater lakes Alchichica (ALC), Atexcac (ATE), Quechulac (QUE), La Preciosa (LAP), Aljojuca (ALJ) and Tecuitlapa (TEC) are located in the Cuenca Oriental (18° 56' 51"-19° 43' 25" N and 97° 07' 10"- 98° 03' 04" W, at a mean altitude of 2,312 m) (Fig. 1). The climate is temperate dry with a dry summer and small temperature fluctuations in the northern Llanos de San Juan where ALC, ATE, QUE and LAP are situated. The climate is subhumid temperate with summer rains in the southern portion of the Llanos de San Andrés, where ALJ and TEC are found. García (1988) has recognized two dominant seasons in the region: a cold dry season from November to April, and a warm rainy season from May to October.

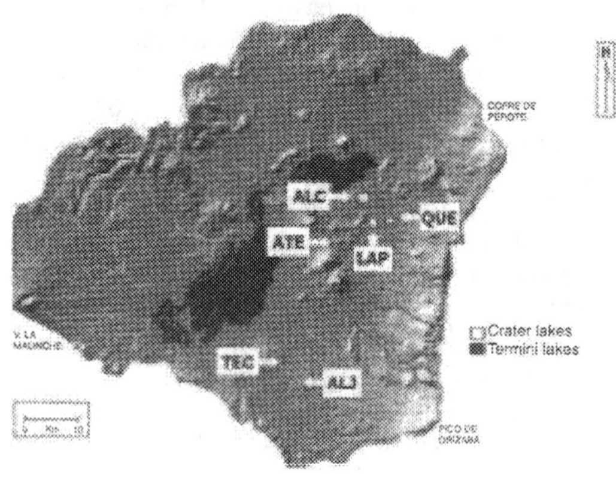

Figure 1. *Geographic location of the crater lakes of Puebla, Mexico.*

The littoral zones of all six crater lakes (Table 1) have mean water temperatures above 18°C, an alkaline pH > 8, and a high dissolved oxygen concentration (> 72%) through daylight time (8:30-17:30 h). ALC and ATE are saline lakes (7.4 and 6 g/l, respectively), while QUE and ALJ (0.1 g/l) and LAP and TEC (1.0 g/l) are freshwater lakes. The sediment, although covered by a thin, superficial mud layer, is composed of sand and gravel, with low to medium concentrations of calcium carbonate (> 0.1%) and medium to high concentrations of organic matter (> 1.5%). Rooted submerged and emergent macrophytes as well as benthic algae (diatoms and filamentous chlorophytes and cyanobacteria) characterize the aquatic vegetation in the littoral bottoms. The submerged macrophyte *Ruppia maritima* L. commonly covers the sampling sites of ALC, while *Potamogeton pectinatus* L. is patchily distributed in ALJ, LAP and QUE. The emergent *Cyperus laevigatus* L. borders the littoral zone of ALC, *Phragmites australis* (Cav.) Trin. ex Steud. and *Cyperus laevigatus* L. characterize ATE, *Typha domingensis* Pers. Emer is typical in ALJ, *Elocharis montevidensis* Kunth and *Juncus andicola* Hook were recorded in TEC, *Juncus andicola* Hook in LAP, and *Scirpus californicus* (Mey.) Steud in QUE and LAP (Ramírez-García and Vázquez-Gutiérrez 1989). A benthic macrofauna rich in species (> 70 species) inhabits the littoral area of these lakes (Alcocer et al. 1993, Alcocer 1995, Alcocer et al 1998); *H. azteca* constitutes from 16.5% (ALJ) to 63.8% (QUE) of the total benthic macroinvertebrate abundance.

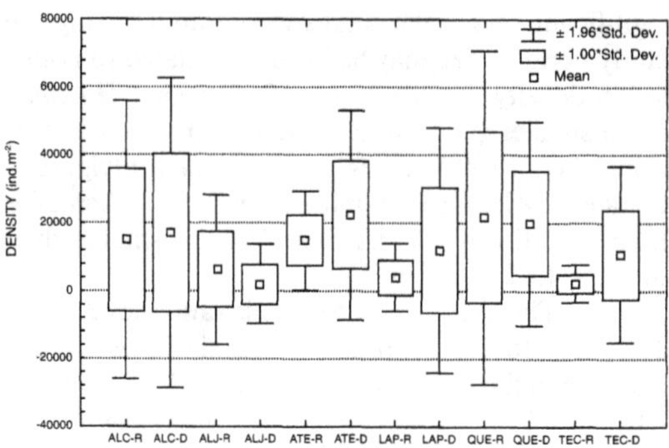

Figure 2. *Box and whisker plots of **H. azteca** density (ind.m⁻²) in the rainy (R) and dry (D) seasons (ALC = Alchichica, ALJ = Aljojuca, ATE = Atexcac, LAP = La Preciosa, QUE = Quechulac, TEC = Tecuitlapa).*

III. RESULTS AND DISCUSSION

Hyalella azteca is the second most abundant species (26.3%) after the tubificid worm *Limnodrilus hoffmeisteri* Claparède (Alcocer et al. 1993, Alcocer 1995), occurring over a wide range of environmental conditions. Regarding water quality, the species occurs in highly alkaline waters (pH = 8.2 in ATE up to 9.9 in TEC), and from fresh (0.1 g/l in QUE and ALJ) to saline water (7.4 g/l in ALC). Timms et al. (1986) considered *H. azteca* as a typical freshwater organism, which can tolerate saline conditions (5.5-22.5 g/l).

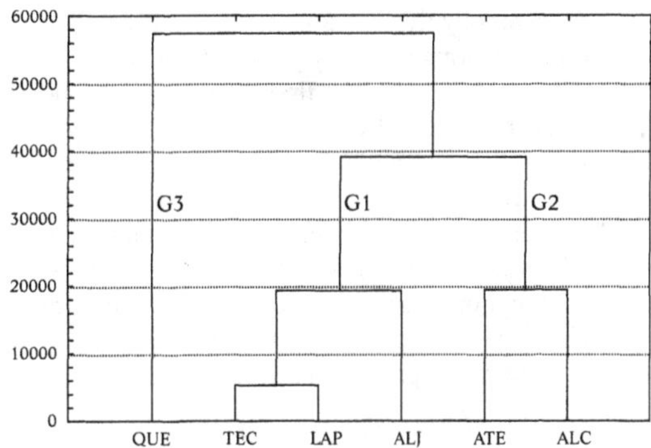

Figure 3. *Similarity dendrogram (euclidean distances, complete linkage) of the six crater lakes.*

Regarding the sediment characteristics, *H. azteca* inhabited very fine (clay and silt) as well as coarse textures (gravel), with organic matter contents varying from 0% to 18%. The amphipods were found in a wide range of vegetation types and coverage, from exposed sediment (lacking vegetation) to bottoms totally covered by submerged or emergent aquatic macrophytes, or swimming around filamentous benthic algae. Fewer organisms were observed in non-vegetated areas; this is probably associated to a higher predation rate as found by Richardson et al. (1998).

Among the benthic macrofaunal components, *Hyalella azteca* (Saussure) is the only amphipod found to inhabit the littoral zone of the six crater lakes. Mean amphipod density ranged from 1042 ± 3544 ind/m² in ALJ to 13496 ± 20740 ind/m² in ALC. The values obtained for TEC (1445 ± 3538 ind/m²), LAP (1833 ± 4518 ind/m²), ATE (4475 ± 5983 ind/m²) and QUE (5271 ± 8211 ind/m²) were intermediate within these two extremes. The densities reached by *H. azteca* in

ALC are similar to those reported for *Diporeia hoyi* (11400 ind/m²) by Nalepa et al. (1988 in Fitzgerald and Gardner 1993) in Lake Michigan. However, Alcocer (1988), Pickard and Benke (1996) and Mathias (1971) reported much lower densities for *H. azteca* from the Chapultepec Lake in Mexico-type locality–(9-104 ind/m²), a small wetland pond in southeastern United States (904 ind/m²), and the Marion Lake in British Columbia (1952 ind/m²), respectively. Pennak (1978) considered densities above 10000 ind/m² as "unbelievably abundant".

Amphipod density remained constant throughout the annual cycle ($P < 0.05$) with no significant differences among the five samplings (Fig. 2). The lakes were grouped along a salinity gradient, with three clusters recognized (Fig. 3). The first cluster (G1) included the freshwater lakes TEC, LAP and ALJ (salinity < 3 g/l) with the lowest amphipod densities. The second cluster (G2) included the two saline lakes, ALC and ATE (salinity > 3 g/l) with mid to high amphipod densities. Finally, QUE (G3), a freshwater lake with intermediate amphipod densities, forms a separate group by itself. A narrow littoral zone that is densely covered by emerging macrophytes characterizes QUE.

Table 2. *Population structure (%) of* **H. azteca** *in the crater lakes of Puebla. (JUV = juveniles, ♂ = males, ♀ = females, ♀♀ = ovigerous females).*

Crater Lake	JUV	♂	♀	♀♀
Alchichica	55	16	17	12
Aljojuca	64	14	14	8
Atexcac	59	17	13	11
La Preciosa	57	17	15	12
Quechulac	61	13	16	10
Tecuitlapa	56	10	20	14

The population structure of *H. azteca*, showed general pattern in which juveniles are numerically dominant (55-64%) (Table 2). In contrast, females (13-20%), males (10-17%), and ovigerous females (8-14%) are poorly represented. The lakes ATE and LAP showed slight changes from this general pattern, but with no significant differences, in which males (17% in both lakes) were proportionally more abundant than females (13% and 15%, respectively). This fact suggests reproduction was continuous throughout the year similar to the findings of Pickard and Benke (1996).

The number of eggs carried by the ovigerous females ranged from 1 to 38. The mean number of eggs throughout the study period showed no significant changes in any of the six lakes: 4.5 ± 2.8 in ALC, 6.1 ± 3.6 in ATE, 6.4 ± 3.4 in QUE, 6.4 ± 3.5 in LAP, 7.7 ± 5.4 in ALJ, and 8 ± 5.6 in TEC. Fecundity was positively correlated to female size (Fig. 4). These numbers are considered low, since Pennak (1978) mentioned that *H. azteca* averages about 18 eggs per brood.

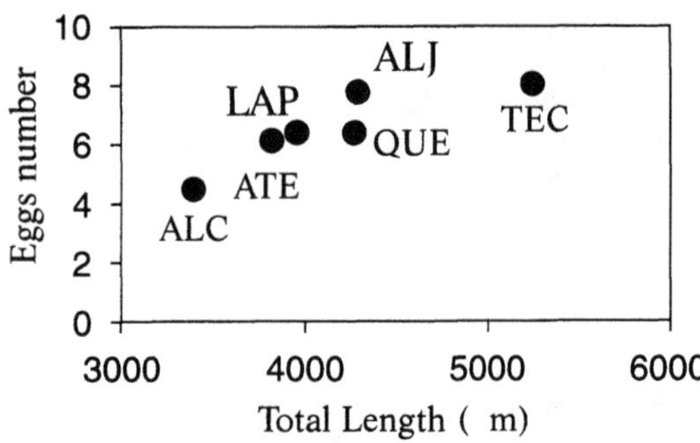

Figure 4. *Total length (µm) vs egg number of H. azteca ovigerous females inhabiting the crater lakes of Puebla (ALC = Alchichica, ALJ = Aljojuca, ATE = Atexcac, LAP = La Preciosa, QUE = Quechulac, TEC = Tecuitlapa).*

Although mean ovigerous females total length varied among lakes: 3399.2 ± 504.4 µm in ALC, 3823 ± 699.9 µm in ATE, 3954.1 ± 573.8 µm in LAP, 4262.9 ± 555.9 µm in QUE, 4280.4 ± 684.3 µm in ALJ, and 5238 ± 820.9 µm in TEC; no significant seasonal changes were found during the study period. Size may be inversely related to population density. Kubitz et al. (1996) found *H azteca* to be smaller (i.e., 16-20%) under crowded conditions (i.e., 14000 versus 7000 ind/m²). Ovigerous females in ALC (13496 ind/m²) were 11.1% smaller than in ATE (4475 ind/m²) and 20.1% smaller than in QUE (5,271 ind/m²).

The overall male to female ratio ranged from 1:1 (1.0) to 1:3.25 (0.31) in all six lakes. Ratios changed in the following pattern ATE (1:1.33, 0.75), ALJ (1:1.51, 0.66), LAP (1:1.6, 0.63) and ALC (1:1.75, 0.57) showed ratios lower than 1:2 (0.5), while QUE (1:2.02, 0.5), and TEC (1: 3.25, 0.31) exceeded the 1:2 ratio. Extreme cases were recorded in TEC with 7.17 females per male (0.14), and in ALJ with two males per female (2.0).

Live *H. azteca* specimens in the crater lakes of Puebla are light brown to greenish. This coloration could be resulting from its diet mainly composed of benthic microphytes (Escobar Briones et al. 1998), in spite of this, *H. azteca* has been considered primarily an omnivore, scavenger, herbivore and detritus feeder (Cole and Watkins 1977).

Many previous studies have failed to find a consistent relationship between abiotic factors and species abundance of particular taxa (Williams 1996). We also did not find any statistically significant ($P > 0.05$) correlations between any of the considered abiotic factors and *H. azteca* abundance in the crater lakes. However, the cluster analysis suggests that biotic interactions, especially with vegetation type, may play a major role in explaining *H. azteca* size and abundance.

ACKNOWLEDGEMENTS

CONACyT project 25340-T partially supported this study. A. Lugo, M. R. Sánchez, A. Salas, L. A. Oseguera and M. J. Montoya assisted in the field and laboratory.

REFERENCES

Alcocer J (1988) Caracterización hidrobiológica de los lagos de Chapultec, México. Tesis de Maestría en Ciencias del Mar, UACPyP, CCH, UNAM, 88 p

Alcocer J, Lugo A, Estrada S, Ubeda M and Escobar E (1993) La macrofauna bentónica de los axalapazcos mexicanos. Actas del VI Congreso Español de Limnología 33:409-415

Alcocer J (1995) Análisis holístico de la comunidad de macroinvertebrados bentónicos litorales de seis lagos-cráter con un gradiente de salinidad. Tesis Doctoral, Facultad de Ciencias, UNAM, México, 106 p

Alcocer J, Escobar E, Lugo A and Peralta L (1998) Littoral benthos of the saline crater lakes of the basin of Oriental, Mexico. Int J Salt Lake Res 7:87-108

Barnard JL and Barnard CM (1983) Freshwater amphipods of the world. Mayfield Associates. Virginia, 830 p

Bousfield EL (1973) Shallow-water Gammaridean amphipods of New England. Cornell University Press, London, 1389 p

Cole GA and Watkins RL (1977) *Hyalella montezuma*, a new species (Crustacea: Amphipoda) from Montezuma Well, Arizona. Hydrobiologia 52:175-184

Escobar-Briones E, Alcocer J, Cienfuegos E and Morales P (1998) Carbon stable isotopes of pelagic and littoral communities in Alchichica crater-lake, Mexico. Int J Salt Lake Res 7:345-355

Fitzgerald SA and Gardner WS (1993) An algal carbon budget for pelagic-benthic coupling in Lake Michigan. Limnol Oceanogr 38:547-560

Holsinger JR (1982) Amphipoda (pp 209-214) In: Hurlbert SH and Villalobos-Figueroa A (eds) Aquatic Biota of Mexico, Central America and the West Indies. San Diego State University, San Diego, 529 p

Kubitz JA, Besser JM and Giesy JP (1996) A two-step experimental design for a sediment bioassay using growth of the amphipod *Hyalella azteca* for the test end point. Environ Toxicol Chem 15:1783-1792

Mathias JA (1971) Energy flow and secondary production of the amphipod *Hyalella azteca* and *Crangonyx richmondensis occidentalis* in Marion Lake, British Columbia. J Fish Res Bd Can 28:711-726

Pickard DP and Benke AC (1996) Production dynamics of *Hyalella azteca* (Amphipoda) among different habitats in a small wetland in the southeastern USA. J N Am Benthol Soc 15:537-550

Richardson WB, Zigler SJ and Dewey MR (1998) Bioenergetic relations in submerged aquatic vegetation: An experimental test of prey use by juvenile bluegills. Ecol Freshwat Fish 7:1-12

Saussure H (1858) Memoire sur divers crustaces noveaux de Antilles et du Mexique. Mem Soc Phys Hist Nat 14 P 2:417-496

Williams DD (1996) Environmental constraints in temporary fresh waters and their consequences for the insect fauna. J N Am Benthol Soc 15:634-650

COMPOSITION AND ABUNDANCE OF SHRIMP SPECIES (PENAEIDEA AND CARIDEA) IN FORTALEZA BAY, UBATUBA, SÃO PAULO, BRAZIL

Adilson Fransozo, Rogério C. Costa, Fernando L. M. Mantelatto, Marcelo A. A. Pinheiro and Sandro Santos

NEBECC,Group of Studies on Crustacean Biology, Ecology and Culture
AF & RCC, Depto. de Zoologia, IBB, UNESP, s/n, 18618-000, Botucatu, SP, Brasil

fransozo@ibb.unesp.br

FLMM, Depto. de Biologia – FFCLRP - Universidade de São Paulo (USP) -
Av. Bandeirantes, 3900 - 14040-901, Riberão Preto, SP, Brasil
MAAP, Depto de Biologia Aplicada – FCAV – UNESP – 14870-000, Jaboticabal, SP, Brasil
SS, Depto de Biologia – CCNE – Universidade Federal de Santa Maria, 97119-900,
Santa Maria, RS, Brasil

ABSTRACT

The abundance and spatio-temporal distribution of the caridean and penaeid fauna from Fortaleza Bay, Ubatuba, São Paulo, Brazil, were analyzed. Seven transects were sampled over a one year period from November 1988 to October 1989. A total of 17047 shrimps were captured, representing 13 species belonging to 5 families. The interaction of temperature and type of sediment was fundamental in determining the presence and abundance of shrimp species in the area.

I. INTRODUCTION

Since human populations are growing along coastlines and their resultant anthropogenic impact thus increasing, gathering accurate information on benthic communities is urgently needed for proper management and conservation of coastal ecosystems (Alongi 1989).

Along the Brazilian coast, the shrimp fauna is represented by 61 species of Dendrobranchiata (D'Incao 1995), 38 species of Alpheoidea (Holthuis 1993) and 48 species of Palaemonoidea (Ramos-Porto 1986). In São Paulo State, dendrobranchiates, palaemonoids and alpheoids comprise 19, 14, and 23 species, respectively (Christoffersen 1982, Holthuis 1993, D'Incao 1995, Costa et al. 2000).

In spite of their economic importance, only a few studies have been conducted so far on the shrimp fauna of São Paulo, namely those on their taxonomy (Christoffersen 1982, D'Incao 1995, Costa et al. 2000), abundance (Pires 1992, Nakagaki et al. 1995) and population biology (Rodrigues et al. 1993, Chacur and Negreiros-Fransozo 1998, Nakagaki and Negreiros-Fransozo 1998, Costa and Fransozo 1999).

The aim of the present paper is to characterize the caridean and penaeid fauna from Fortaleza Bay, Ubatuba (SP) Brazil with emphasis on their abundance and spatio-temporal distribution as a function of the most relevant environmental factors, i.e. temperature, salinity, depth, organic matter contents and sediment grain-size composition.

II. MATERIALS AND METHODS

Fortaleza Bay is located in the northern coast of the State of São Paulo (23° 29'30" to 23° 32'30" S and 45° 06'30" to 45° 10'30" W). According to Castro-Filho et al. (1987) this region is affected by three water masses with different distribution patterns in summer and winter. The Coastal Water (CW) mass is characterized by a high temperature and low salinity, the Tropical Water (TW) mass by a high temperature and high salinity, and the South Atlantic Central Water (SACW) mass by a low temperature and salinity following an annual cycle.

E. Escobar-Briones & F. Alvarez Eds.
MODERN APPROACHES TO THE STUDY OF CRUSTACEA
PP. 117-123

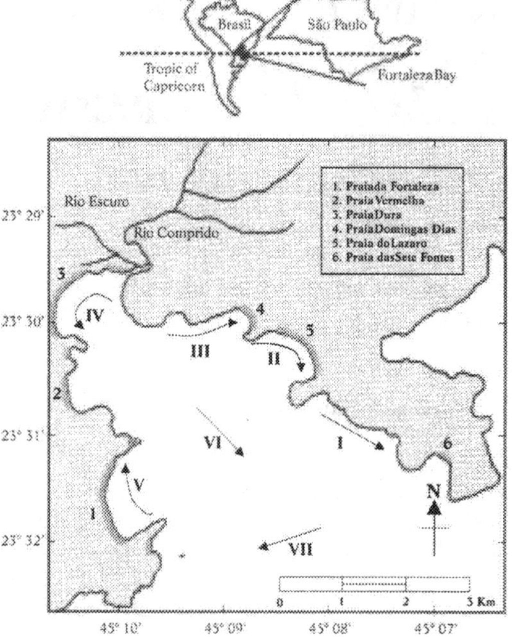

Figure 1. *Map of the Fortaleza Bay.*

Shrimps were sampled in Fortaleza Bay, using a 7.5 m long double-rig net with a 10 mm mesh size cod end. Samplings were performed monthly over a year, from November 1988 to October 1989. During the monthly samplings, 7 trawls (transects) of 1 km each (Fig. 1) were performed.

Average depth at Fortaleza Bay is 9 m (range 4.4 - 13.3 m). The mean annual values for temperature and salinity were 23.5°C (range 21 - 28.1°C) and 34.4‰ (range 32.4 - 35.6‰). The mean value of organic matter content for the whole sampled year was 4.07% (range 2.3 - 6.6%) and the sediment grain-size composition, expressed in Phi values (ϕ) (Suguio 1973), showed a predominance of fine and very fine sand fractions for most transects (Table 1). A detailed description of sampling methods and the analysis of environmental factors at Fortaleza Bay, during the same period, are available elsewhere (Negreiros-Fransozo et al. 1991).

The relationship between species abundance and the variation of studied abiotic factors was assessed by means of examining Pearson's correlation coeffficients at the 5% significance level (Zar 1996).

The species constancy index (C) was calculated according to Dajoz (1983): C=px100/P, where "p" is the number of samples in which a given species was recorded, and "P" the total number of samples analyzed. Species were classified in three different categories: constant (C > 50%), accessory (25% < C < 50%), and accidental (C < 25%). For diversity analyses, the Shannon-Wiener (H') and equitability (J') indices were calculated as indicated by Krebs (1989).

Table 1. *Average values of abiotic factors obtained in each transect during the period from November 1988 to October 1989. Sediment grain composition is expressed as Phi (ϕ) values.*

Transect	Depth (m)	Phi ϕ	Salinity (‰)	Organic Matter (%)
I	11.2 ± 0.9	2.9	34.8 ± 0.8	4.4 ± 2.5
II	7.0 ± 0.9	2.5	34.3 ± 1.3	6.7 ± 2.4
III	8.5 ± 0.9	3.8	34.4 ± 1.1	2.3 ± 1.3
IV	4.4 ± 0.6	3.3	33.3 ± 1.5	1.8 ± 1.3
V	7.1 ± 0.8	2.7	34.4 ± 1.1	3.5 ± 1.4
VI	11.1 ± 1.2	3.4	34.4 ± 1.1	5.1 ± 1.8
VII	13.3 ± 1.6	3.5	34.9 ± 1.7	4.6 ± 3.6

III. RESULTS

A total of 17047 specimens was obtained comprising 13 species belonging to 5 families: Penaeidae, Solenoceridae, Sicyoniidae, Palaemonidae, and Hippolytidae. The family Penaeidae showed the highest species richness (6 species) and abundance (16610 individuals).

Xiphopenaeus kroyeri (Heller, 1862) was a constant species (78%), followed by the accessory species *Artemesia longinaris* Bate, 1888 (18.1%). The species *Farfantepenaeus brasiliensis* (Latreille, 1817), *Farfantepenaeus paulensis* (Pérez Farfante, 1967), *Litopenaeus schmitti* (Burkenroad, 1936), *Rimapenaeus constrictus* (Stimpson, 1874), *Pleoticus muelleri* (Bate, 1888), *Sicyonia dorsalis* Kingsley, 1878, *Sicyonia typica* (Boeck, 1864), *Sicyonia laevigata* Stimpson, 1871, *Nematopalaemon schmitti* (Holthuis, 1950), *Palaemon pandaliformis* (Stimpson, 1871) and *Exhippolysmata oplophoroides* (Holthuis, 1948) were considered as accidental contributing with only 3.9% of the total abundance (Table 2).

With the exception of *X. kroyeri*, recorded throughout the year, the occurrence of shrimp species varied throughout the sampling period (Table 3). *Farfantepenaeus brasilensis* and *F. paulensis* were present in samples from

Table 2. *Species composition and number of shrimps each transect in Fortaleza bay. (C=constancy; Co=constant; Ac=accessory and Ad=acccidental).*

Families/Species	Transects							Total	C
	I	II	III	IV	V	VI	VII		
PENAEIDAE									
Xiphopenaeus kroyeri	2263	434	2224	1160	394	2378	4445	13298	Co
Farfantepenaeus brasiliensis	9	61	4	0	0	1	5	80	Ad
Farfantepenaeus paulensis	7	47	1	0	1	1	0	57	Ad
Litopenaeus schmitti	12	6	7	16	4	4	3	52	Ad
Rimapenaeus constrictus	1	3	0	0	1	3	27	35	Ad
Artemesia longinaris	1151	5	276	172	0	435	1049	3088	Ac
Subtotal	**3443**	**556**	**2512**	**1348**	**400**	**2822**	**5529**	**16610**	
SOLENOCERIDAE									
Pleoticus muelleri	9	21	1	0	5	10	105	151	Ad
Subtotal	**9**	**21**	**1**	**0**	**5**	**10**	**105**	**151**	
SICYONIIDAE									
Sicyonia dorsalis	6	8	2	0	0	9	26	51	Ad
Sicyonia typica	5	5	0	0	0	4	2	16	Ad
Sicyonia laevigata	0	0	0	0	1	0	0	1	Ad
Subtotal	**11**	**13**	**2**	**0**	**1**	**13**	**28**	**68**	
PALAEOMONIDAE									
Nematopalaemon schmitti	0	0	4	0	0	0	1	5	Ad
Palaemon pandaliformes	0	1	0	0	0	0	0	1	Ad
Subtotal	**0**	**1**	**4**	**0**	**0**	**0**	**1**	**6**	
HIPPOLYTIDAE									
Exhippolysmata oplophoroides	52	3	60	0	1	8	88	212	Ad
Subtotal	**52**	**3**	**60**	**0**	**1**	**8**	**88**	**212**	
TOTAL	**3515**	**594**	**2579**	**1348**	**407**	**2853**	**5771**	**17047**	

March to May (autumn), *L. schmitti* and *E. oplophoroides* during late autumn and winter and, finally, *A. longinaris* recorded during summer and winter months.

Diversity and equitability indices showed great variation throughout the study period being higher from November to January and from May to August, when lower mean temperatures were recorded (Fig. 2).

The occurrence of only three species, i.e. *X. kroyeri*, *A. longinaris* and *E. oplophoroides*, was shown to be correlated to environmental factors, chiefly to depth (Table 4). *Xiphopenaeus kroyeri* was captured in areas where Phi values for grain-size composition varied between 3 and 4. In the case of *A. longinaris*, abundance was negatively correlated with temperature and positively correlated with salinity.

Highest species richness and diversity were found in transects I and II. At those sites, sediments are mainly composed of very fine sand associated to other larger grains, and a high organic content in the sediment. High richness values were also obtained at transects VI and VII, but differences among the remaining transects were not significant. At those latter sites, relatively lower abundance values reflect lower diversity with a strong prevalence of *X. kroyeri* and *A. longinaris* compared to sites I and II. Lowest diversity and equitability indices were obtained at transect V (Fig. 2).

Table 3. *Number of shrimps collected from November 1988 to October 1989 in Fortaleza Bay.*

Families/Species	N	D	J	F	M	A	M	J	J	A	S	O	TO
PENAEIDAE													
Xiphopenaeus kroyeri	946	383	628	383	968	1468	1063	848	1364	3066	1180	1001	13298
Farfantepenaeus brasiliensis	1	3	0	0	30	32	14	0	0	0	0	0	80
Farfantepenaeus paulensis	0	0	0	0	28	11	17	1	0	0	0	0	57
Litopenaeus schmitti	1	1	0	0	0	0	11	16	16	4	3	0	52
Rimapenaeus constrictus	26	1	0	0	0	1	2	3	0	0	1	1	35
Artemesia longinaris	145	810	863	108	1	2	15	199	684	261	0	0	3088
Subtotal	**1119**	**1198**	**1491**	**491**	**1027**	**1514**	**1122**	**1067**	**2064**	**3331**	**1184**	**1002**	
SOLENOCERIDAE													
Pleoticus muelleri	92	0	24	0	0	0	12	0	8	0	0	15	151
Subtotal	**92**	**0**	**24**	**0**	**0**	**0**	**12**	**0**	**8**	**0**	**0**	**15**	
SICYONIIDAE													
Sicyonia dorsalis	9	25	2	3	2	0	1	1	1	5	0	2	51
Sicyonia typica	0	3	9	0	0	0	0	1	3	0	0	0	16
Sicyonia laevigata	0	0	0	0	0	0	0	0	0	0	1	0	1
Subtotal	**9**	**28**	**11**	**3**	**2**	**0**	**1**	**2**	**4**	**5**	**1**	**2**	
PALAEOMONIDAE													
Nematopalaemon schmitti	2	0	0	0	0	0	0	1	0	1	1	0	5
Palaemon pandaliformes	0	0	0	0	0	0	0	0	0	0	0	1	1
Subtotal	**2**	**0**	**0**	**0**	**0**	**0**	**0**	**1**	**0**	**1**	**1**	**1**	
HIPPOLYTIDAE													
Exhippolysmata oplophoroides	18	3	1	0	11	4	34	43	1	61	1	34	212
Subtotal	**18**	**3**	**1**	**0**	**11**	**4**	**34**	**43**	**1**	**61**	**1**	**34**	
TOTAL	**1240**	**1229**	**1527**	**494**	**1040**	**1518**	**1169**	**1113**	**2077**	**3398**	**1187**	**1055**	**17047**

(The "Months" span heads columns N through TO.)

IV. DISCUSSION

From the 61 species of Dendrobranchiata already recorded along the Brazilian coast, 10 species were captured at Fortaleza Bay. Considering the limited area covered in the present study, it may be concluded that the shrimp fauna is well represented at the study area.

Within the dendrobranchiates, *X. kroyeri* is the dominant species. It represents the second most important fishery resource along the coast of the State of São Paulo, and its trophic relationships may be essential in maintaining the stability of benthic communities in the studied region (Pires 1992). The species *Litopenaeus schmitti*, *Farfantepenaeus brasiliensis* and *F. paulensis* were less abundant, probably due to the lack of large estuarine areas within the northern coast of the State of São Paulo (Costa and Fransozo 1999), where certain penaeid species often constitute large populations (Stoner 1988).

The diversity of the caridean group was comparatively lower. Near Ubatuba Bay, where overall abiotic characterisctics are similar, six other caridean species were additionally found (Costa et al. 2000). The relative low richness at Fortaleza Bay may be associated to certain ecological features of caridean shrimps.

120

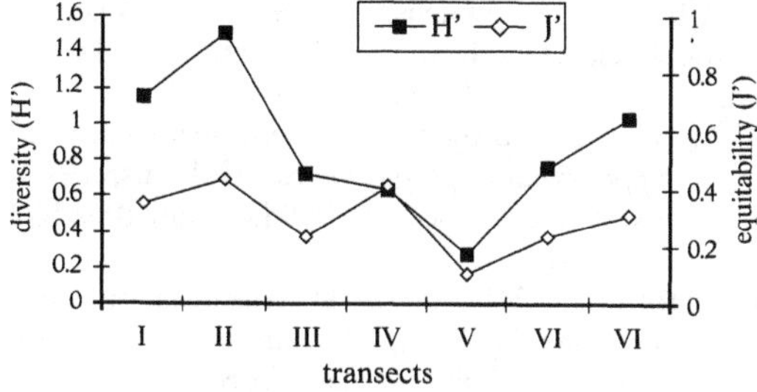

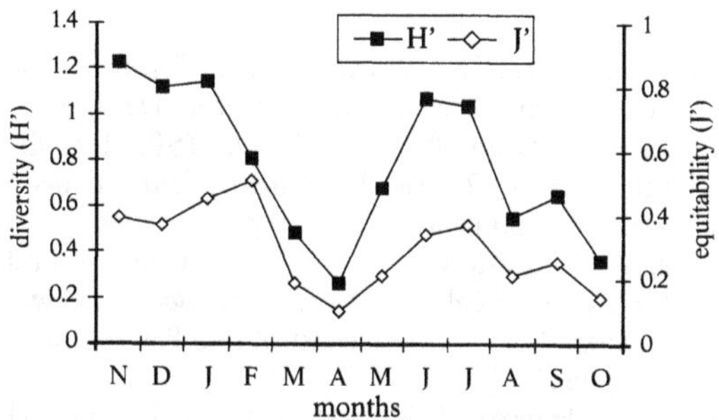

Figure 2. *Spatial and temporal variation of the Shannon-Wiener diversity Index (H') and equitability (J') during the study period (November, 1988 to October, 1989) at Fortaleza Bay.*

Table 4. *Pearson's linear correlation coefficients between the abundance of shrimps and abiotic factors at Fortaleza Bay. (* indicates significant correlations, P < 0.05)*

Species / Abiotic factors	*Xiphopenaeus kroyeri*	*Artemesia longinaris*	*Pleoticus muelleri*	*Exhippolysmata oplophoroides*
Temperature	- 0,16	- 0,3*	- 0,16	- 0,16
Salinity	0,14	0,32*	0,09	0,07
Deep	0,42*	0,32*	0,19	0,23*
Organic matter	0	0,4	0,1	- 0,03
Phi	0,34*	0,11	0,6	0.18

Many caridean species occur on microhabitats associated to the rocky shallow subtidal. They often use algal canopy, shallow burrows and hydroid substrates as commensals during their life cycle or just as shelter to avoid predation.

The shrimp community at Fortaleza Bay is largely dependent on the migration events of *A. longinaris*, the second most abundant species, since the presence of the seabob *X. kroyeri* is constant through-

out the year. High diversity and equitability indices were found from November, 1988 to January, 1989 and from June to July, 1989, when large numbers of *A. longinaris*, *P. muelleri*, *E. oplophoroides* and *L. schmitti* were captured. The lowest indices were found in April, 1989 and from August to October, 1989, due to the strong prevalence of *X. kroyeri* at Fortaleza Bay.

Pires (1992) studied the benthic macrofauna off the coast of Ubatuba and showed that permanent benthic communities rely on trophic relationships in which *X. kroyeri* is a key-role species. The author also found that the abundance of *X. kroyeri* is related to the hydrologic dynamics taking place in the study region. During fall and winter, bottom temperature of the CW water mass ranges from 22 to 25°C, providing favourable conditions for the establishment of *X. kroyeri* populations. In contrast, average bottom temperature falls to values lower than 20°C during the summer period due to the influence of the SACW water mass. The present study corroborates the data obtained by Nakagaki et al. (1995) and Costa (1997), who observed seasonal changes in the abundance of several shrimp species at Ubatuba Bay, such as *A. longinaris* and *P. muelleri*, which entered the bay during late fall and remained in the bay throughout the winter. The presence of these species was also noted during summer, when water temperature decreases due to the emergence of the SACW water mass.

The highest number of species registered during the fall is mainly related to the presence of *F. paulensis*, *F. brasiliensis* and *L. schmitti*. These species occupy the bay during short periods, co-occurring in this area only during May. They need to migrate to sheltered areas for the completion of their life cycle.

Regarding spatial distribution, the highest diversity and species richness were observed at transect II, where coarse sand contains high percentage of organic matter. In transects II and V, fewer *X. kroyeri* and *A. longinaris* were captured compared to the other transects, probably because sediments at those sites are mainly composed by coarser sediments known to be unfavourable for the establishment of those species at Ubatuba Bay (Castro 1997, Costa 1997).

According to Boschi (1963, 1989), *P. muelleri* and *A. longinaris* do not require low salinities to complete their life cycles but prefer lower temperatures from 15 to 21°C. Such results would explain the seasonal occurrence of those species and their presence in deeper areas.

Most of the recorded species are present in Fortaleza Bay because very fine sand is the predominant sediment fraction at this area (phi > 4). *Rimapenaeus constrictus, S. dorsalis, P. muelleri* and *X. kroyeri* are known to be associated to muddy sediments (Sanchez and Soto 1987, Boschi 1989, Dall et al. 1990, Nakagaki et al. 1995, Castro 1997, Costa 1997).

Based on the present findings and previous research (Fransozo et al. 1992, 1998, Negreiros-Fransozo et al. 1997, Costa et al. 2000), it is suggested that environmental conditions at Fortaleza Bay are favourable for the establishment and development of a diverse shrimp guild. Continuing research on both inshore and offshore areas in this subtropical region will provide a more accurate characterization of the shrimp diversity and contribute to a better understanding of their life cycles.

ACKNOWLEDGEMENTS

We are grateful to the "Fundação de Amparo à Pesquisa do Estado de São Paulo (FAPESP), Conselho Nacional de Desenvolvimento Científico e Técnologico (CNPq) and Fundação para o Desenvolvimento da UNESP (FUNDUNESP)" for financial support. We are thankful to Dr. Maria Lúcia Negreiros-Fransozo for her constructive comments on early drafts of the manuscript. All experiments conducted in this study comply with current applicable state and federal laws.

REFERENCES

Alongi DM (1989) Ecology of tropical soft-bottom benthos: a review with emphasis on emerging concepts. Rev Biol Trop 37:85-100

Boschi EE (1963). Los camarones comerciales de la familia Penaeidae de la costa Atlántica de America del Sur. Bol Inst Biol Mar 3:1-39

Boschi EE (1989) Biología pesquera del langostino del litoral patagónico de Argentina (*Pleoticus muellleri*). Contrib Inst Nac Invest Desarro Pesq Mar del Plata, Argentina, 646:5-71

Castro-Filho BM, Miranda LB and Myao SY (1987) Condições hidrográficas na plataforma continental ao largo de Ubatuba: variações sazonais e em média escala. Bolm Inst Oceanogr 35:135-151

Castro RC (1997) Padrões distribucionais do camarão *Xiphopenaeus kroyeri* (Heller, 1862) (Crustacea: Decapoda: Penaeidae) na enseada de Ubatuba, Ubatuba, SP. Unpublished M.Sc. Thesis, Instituto de Biociências - UNESP – Botucatu (SP), 143 p

Chacur MM and Negreiros-Fransozo ML (1999) Aspectos biológicos do camarão-espinho *Exhippolysmata oplophoroides* (Holthuis, 1948) (Crustacea, Caridea, Hippolytidae). Rev Bras Biol 59:173-181

Christoffersen ML (1982) Distribution of warm water alpheoid shrimp (Crustacea: Caridea) on the continental shelf of eastern South America, between 23° and 35° Lat. S. Bolm Inst Oceanogr 31:93-112

Costa RC (1997) Composição e padrões distribucionais dos camarões Penaeoidea (Crustacea: Decapoda), na Enseada de Ubatuba, Ubatuba (SP). Unpublished M.Sc. Thesis, Instituto de Biociências - UNESP – Botucatu (SP), 129 p

Costa RC and Fransozo A (1999) A nursery ground for two tropical pink-shrimp *Penaeus* species: Ubatuba Bay, northern coast of São Paulo, Brazil. Nauplius 7:73-81

Costa RC, Fransozo A, Mantelatto FLM and Castro RH (2000). Occurrences of shrimps (Natantia: Penaeidea and Caridea) in Ubatuba Bay, Ubatuba, São Paulo, Brazil. Proc Biol Soc Wash 113:776-781

Dajoz R (1983) Ecologia Geral. Editora Vozes, EDUSP, São Paulo, 472 p

Dall W, Hill BJ, Rothlisberg PC and Staples DJ (1990) In: Blaxter JHS and Southward AJ (ed) Advances in Marine Biology. San Diego. Academic Press, Vol 27, 489 p

D'Incao F (1995) Taxonomia, padrões distribucionais e ecológicos dos Dendrobranchiata (Crustacea: Decapoda) do litoral brasileiro. Unpublished Ph.D. Thesis, Universidade Federal do Paraná, Curitiba, 365 p

Fransozo A, Negreiros-Fransozo ML, Mantelatto FLM, Pinheiro MAA and Santos S (1992) Composição e distribuição dos Brachyura (Crustacea; Decapoda) do sublitoral não consolidado na Enseada da Fortaleza, Ubatuba (SP). Rev Bras Biol 52:667-675

Fransozo A, Mantelatto FLM, Bertini G, Fernandes-Góes LC and Martinelli JM (1998) Distribution and assemblages of anomuran crustaceans in Ubatuba Bay, north coast of São Paulo State, Brazil. Acta Biol Venez 18:17-25

Holthuis LB (1993) The recent genera of the caridean and stenopodidean shrimps (Crustacea, Decapoda), with an appendix on the Order Amphionidacea. Leiden, Nationaal Natuurhistorisch Museum, 328 p

Krebs CJ (1989) Ecological Methodology. New York, Harper and Row, 645 p

Nakagaki JM and Negreiros-Fransozo (1998) Population biology of *Xiphopenaeus kroyeri* (Heller, 1862) (Decapoda: Penaeidae) from Ubatuba bay, São Paulo, Brazil. J Shell Res 17:931-935

Nakagaki JM, Negreiros-Fransozo ML and Fransozo A (1995) Composição e abundância de camarões marinhos (Crustacea: Decapoda: Penaeidae) na Enseada de Ubatuba, Ubatuba, Brasil. Arq Biol Tecnol 38:583-591

Negreiros-Fransozo ML, Fransozo A, Mantelatto FLM, Pinheiro MAA and Santos S (1991) Caracterização física e química da enseada da Fortaleza, Ubatuba, SP. Rev Bras Geogr 21:114-120

Negreiros-Fransozo ML, Fransozo A, Mantelatto FLM, Pinheiro MAA and Santos S (1997) Anomuran species (Crustacea, Decapoda) and their ecological distribution at Fortaleza bay sublitoral Ubatuba, São Paulo, Brazil. Iheringia, Sér Zool 83:187-197

Pires AMS (1992) Structure and dynamics of benthic megafauna on the continental shelf offshore of Ubatuba, southeastern, Brazil. Mar Ecol Prog Ser 86:63-76

Ramos-Porto M (1986) Crustáceos decápodos marinhos do Brasil: família Palaemonidae. Unpublished M.Sc. Thesis, Universidade Federal de Pernambuco (UFPE), Recife, 347 p

Rodrigues ES, Pita JB, Graça-Lopes R, Coelho JA and Puzzi A (1993) Aspectos biológicos e pesqueiros do camarão sete-barbas (*Xiphopenaeus kroyeri*) capturados pela pesca artesanal no litoral do Estado de São Paulo. Bolm Inst Pesca 19:67-81

Sánchez AJ and Soto LA (1987) Camarones de la Superfamilia Penaeoidea (Rafinesque, 1815) distribuidos en la Plataforma continental de suroeste del Golfo del México. An Inst Cienc Mar Limnol, Univ Nac Autón Mex 14:157-180

Stoner AW (1988) A nursery ground for four tropical *Penaeus* species: Laguna Joyuda, Puerto Rico. Mar Ecol Prog Ser 42:133-141

Zar JH (1996) Biostatistical analysis. Prentice Hall, Upper Saddle River, 662 p

POPULATION STUDY OF *MACROBRACHIUM TENELLUM* (SMITH 1871) IN COYUCA DE BENÍTEZ LAGOON, GUERRERO, MEXICO

Giséle Signoret and David Brailovsky

GS, Deparmento El Hombre y su Ambiente, Universidad Autónoma Metropolitana-Xochimilco, Calzada del Hueso 1100, Colonia Villa Quietud, México, D.F., CP 04960, México

gsigno@cueyatl.uam.mx

DB, Facultad Ciencias, Universidad Nacional Autónoma de México, Av. Universidad 3000, México, D.F., México

ABSTRACT

The population of *Macrobrachium tenellum* (Smith 1871) in Coyuca de Benítez lagoon, Guerrero, Mexico, was studied throughout one year. The following population parameters were recorded for 569 organisms: a sex ratio of 2:1 females/males, an average total length of 7.52 cm (8.02 cm for males, 7.01 cm for females) and an average total weight of 4.28 g (5.95 g for males, 4.02 g for females). The species reproduces throughout the year with an average fecundity of 984 eggs for small females (3.80 cm long) and of 900 to 1200 eggs for bigger specimens (6.70 to 8.00 cm long).

I. INTRODUCTION

Freshwater prawns represent one of the most important aquatic resources of Mexico (Signoret et al. 2000). Four of the 17 native species found in fresh and brackish water bodies, are commercially exploited: *Macrobrachium acanthurus* (Wiegmann 1836), *M. americanum* (Bate 1868), *M. carcinus* (Linnaeus 1758) and *M. tenellum* (Smith 1871). This last species is distributed from Baja California to the State of Chiapas, in the southwest Mexico (Holthuis, 1952, Rodríguez de la Cruz 1965). Throughout their distribution range, these species support important fisheries that represent an important source of income for local fisher-

man. In spite of this, no studies have been done to provide appropriate management strategies for this resource. As prawn culture practices for the species are only extensive (e.g., juveniles are collected from their natural habitats and are fed to increase their length and weight up to commercial values), natural populations are affected by the environmental changes and mainly depend on the natural stocks (Gutiérrez 1992).

This is the case of *M. tenellum,* which supports a yearly-long exploitation (Granados 1984), which results in the reduction of its natural populations. Ecological and physiological studies are necessary to establish better exploitation and culture practices for this species (Mallasen and Controni 1998).

Freshwater prawns have been the object of various studies (Villalobos and Nates 1990). Despite the economic importance of *M. tenellum* as a commercially exploited species, only a few papers have dealt with this prawn directly. The fecundity of *M. tenellum* was described by Negrete (1977) and Cabrera et al. (1979). Guzmán et al. (1982) studied some of its reproductive aspects and Román (1991) carried out ecological studies. With regard to physiological aspects, Signoret and Soto (1995, 1997) studied the salinity tolerance and osmotic behavior of *M. tenellum* and other palaemonid prawns.

In order to extend the knowledge on *Macrobrachium tenellum* in Coyuca de Benítez lagoon,

Guerrero, Mexico (16° 54', 16° 58' N; 99° 58', 100° 08' W) (Fig. 1), its population structure and reproductive information (fecundity) throughout one year is presented in the following pages.

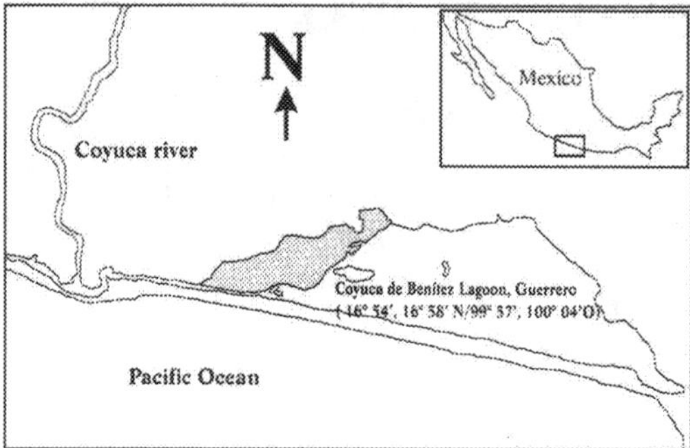

Figure 1. *Map of the study area.*

II. MATERIALS AND METHODS

Regular monthly collections of prawns were made from December 1995 to November 1996 (with the exception of October 1996), using a dip-net "atarraya", with a 2 m diameter and 1 cm mesh size. The sampling locality was approximately 3 km to the east from the river mouth. Direct observations were carried out in the lagoon and temperature and salinity were measured using conventional methods. Water depth ranged from 0 to 3.5 m, temperature from 20 to 32°C, salinity from 0 to 5 ppt, and average pH was 8.1. Organisms were collected from muddy areas near mangroves, *Typha latifolia* and *Eichhornia* sp.

In total, 569 *M. tenellum* were collected (152 males, 292 females and 125 ovigerous females), of which 58.17 % carried the parasitic isopod *Probopyrus pacificensis* (Román 1993). Individuals were separated, sexed and fixed in 10 % formaline solution for their analysis in the laboratory. Length and weight data were recorded for both sexes. Specimens were measured with a precision caliper to the nearest 0.1 mm. Total body length (TL cm) was measured from the tip of the rostrum to the tip of the telson. An electronic scale was used to obtain wet weight (WW g) to the nearest 0.1 g. Specimens were grouped according to sex and TL in 13 classes (each 1 cm long). Fecundity was determined from 125 ovigerous females. The egg mass from each ovigerous female was gently removed and dried in the

laboratory oven at 60°C. The number of eggs for each ovigerous female was estimated by the volumetric method with the aid of a dissecting binocular microscope. The eggs were measured in a Carl Zeiss microscope with the aid of a micrometer scale.

III. RESULTS

Population characteristics

The size of the prawns varied from 1.60 to 13.90 cm, the mean total length (TL) was of 7.00±0.50 cm. The TL for the males varied from 3.50 cm (April) to 13.90 cm (June), the mean was of 8.70±2.30 cm; the females presented a TL from 1.60 cm (February) to 10.70 cm (June), with a mean of 6.15±3.50 cm. The TL for ovigerous females was between 3.80 cm (January) and 10.60 cm (December), with a mean of 7.20±1.80 cm. The size distribution by sex shows that almost 66% occupied the three middle classes (Fig. 2), and that males have a larger body size (3.50 to 13.90 cm) than females (1.60 to 10.70 cm). The size ranges of both sexes in this study fall within those reported in the literature (Guzmán 1987).

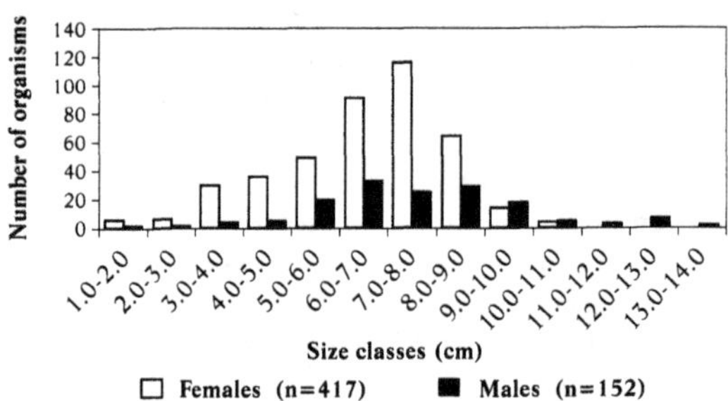

Figure 2. *Size distribution by sex of* **Macrobrachium tenellum**.

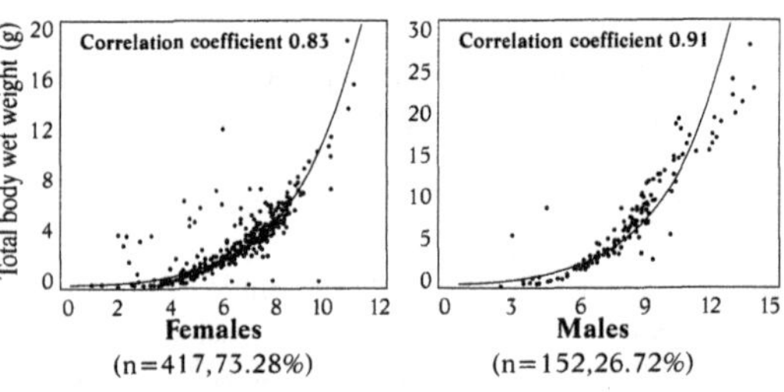

Figure 3. *Relationship between total length (TL cm) and total body wet weight (W g) in* **Macrobrachium tenellum** *females and males.*

The weight (W) varied from 0.20 to 27.40 g , the average was of 4.00±0.80 g. The W for the males varied from 0.20 g (September) to 27.40 g (July), with a mean of 13.80±4.20 g, while the females presented a W from 0.20 g (November) to 10.90 g (March), with a mean of 5.55±3.70 g. W for the ovigerous females was between 0.70 and 13.70 g (December), with a mean of 8.75±1.90 g. Most individuals have a W between 2.60 and 5.00 g, with only few specimens weighing between 25.70 and 27.40 g.

The relationship between TL and W can be explained with an exponential model y= exp(a+bx). The correlation coefficient was 0.87 for the whole population, 0.91 for the males and 0.83 for the females (Fig. 3).

Population composition

Significant differences were recorded in the population structure of both female and male prawns throughout the year (Fig. 4). The numerical predominance of females over males increased towards the end of summer. Although the minimum and maximum values of TL differ between the sexes, males reached a larger size than females in winter (13.90 *vs.* 8.40 cm), spring (12.30 *vs.* 9.90 cm) and summer (12.30 vs. 8.40 cm), but not in autumn (5.50 vs. 9.52 cm) (Table 1).

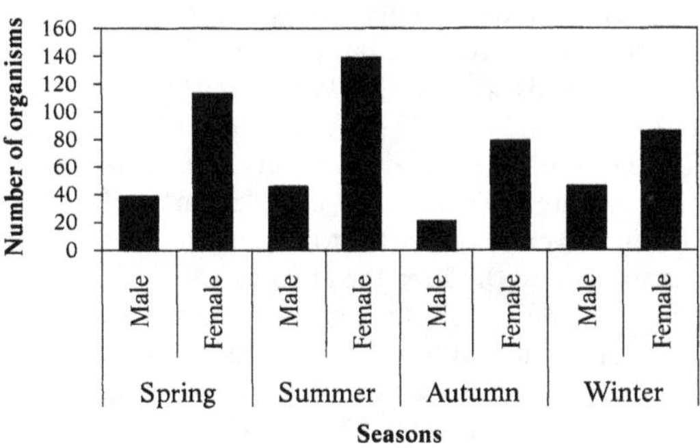

Figure 4. *Population structure of **Macrobrachium tenellum** throughout the year.*

Sex ratio

The sex ratio of *M. tenellum* population revealed a higher number of females (n=417, 63.29%) over males (n=152, 26.71%). This is in agreement with Román (1991) who found a 53.46% of female of composition in contrast to 46.54% of male specimens of *M. tenellum* in the same study area. The average sex ratio (females: males) for the whole sample was 2:1, with the lowest (1:3) and highest (3:1) ratios observed in December and July respectively (Fig. 5).

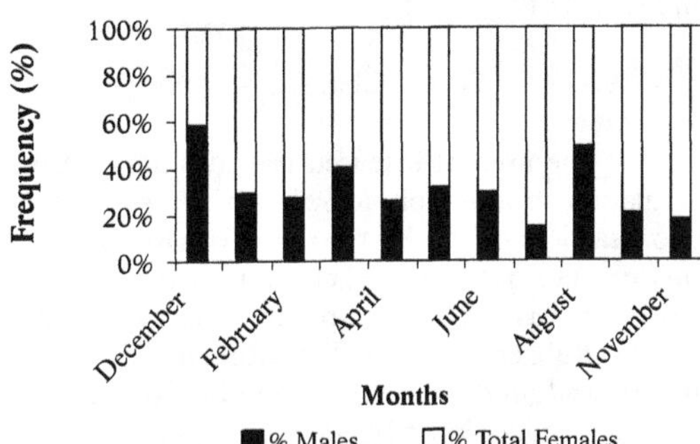

Figure 5. *Sex ratio of **Macrobrachium tenellum** throughout the year.*

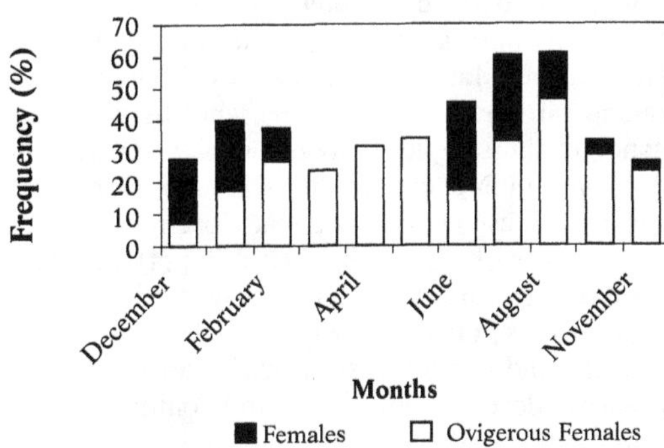

Figure 6. *Occurrence of ovigerous females throughout the year.*

Ovigerous females

Ovigerous females were present in every sample indicating continuous reproductive activity for this population (Fig. 6). The total length of the ovigerous females varied from 3.80 cm (January) to 10.60 cm (December). Negrete (1977) reported the smallest *M. tenellum* ovigerous female with a total length of 3.80 cm in Tres Palos lagoon, Mexico.

Fecundity

Reproduction occurs at the beginning of spring, a time that is favorable in terms of temperature and availability of food according to Guzmán et al. (1982). Eggs were numerous and small (average 500 μm), and there is a positive correlation between fecundity and female body size, as in *Atya scabra* (Galvao and Bueno 2000). The average fecundity is represented by 984 eggs per females with a TL=3.80 cm and 900 to 1200 eggs for specimens with a TL=6.70 to 8.00 cm respectively.

IV. DISCUSSION

The present research provides data on the population structure of *Macrobrachium tenellum* in a coastal lagoon of Mexico. In *Macrobrachium tenellum* most references support the equal presence of two sexes as has been found for other palaemonid prawns (Ruiz et al. 1996). Therefore, in the present surveys we found that females dominated quantitatively the population all-year round (sex ratio: 2:1 females/males).

As already stated by Román (1979) and Guzmán et al. (1982), *Macrobrachium tenellum* follows a well defined seasonal mobility pattern. The species is more abundant on the mud mangroves areas during March and July, as consistent with our catches at the end of summer, whereas the abundance of the samples increase in July (n=81) and decrease in November (n=41) and December (n=35). These changes show a close relation with the presence of a low salinity level (0 ppt) and moderate temperatures (20 °C) in the lagoon during July. As in various prawns the sex ratio in *Macrobrachium tenellum* did not differ statistically during the first phases of development (Souza and Fontoura 1996). Important variations might occur in the mature animals that live on the mud due to the specific conditions for each sex, mainly related to the reproductive behavior. In relation with these, important sex an size segregation exist in adult females that are present all year round in different habitats, peaking in July, whereas the abundance of males is higher in December. This change in the sex ratio is due to the economical importance of *Macrobrachium tenellum* in the study area, since mainly males are captured.

Size ranges of both sexes in the present study fall within the ranges given in the literature for similar prawn species. Thus the greater number of the females observed in the current study is considered a prominent feature of this species. In particular, *M. tenellum* shows a large size difference between the sexes; males reach a bigger size than females and the first pereiopods are longer in males than in females. Males attaining larger size

than females are also commonly observed in palaemonid prawns. This condition has been attributed to the amount of energy expended in reproduction, which is larger in females than in males, causing a slower growth of the females, particularly during breeding periods.

A continuous reproductive pattern is quite common in caridean shrimps that inhabit tropical regions (Galvao and Bueno 2000), while the subtropical species usually show periodic reproductive cycles. The studied species is considered within the first group. Females spawn in April and the number of females breeding first eggs and the presence of hatchings were small in our catches (Odinez and Rabelo 1996). The fecundity was determined from 125 ovigerous females with eggs in the early embryonic state, since loss of eggs during the incubation period is common in caridean shrimps (Balasundaram and Pandian 1982).

V. CONCLUSIONS

The field study of the *Macrobrachium tenellum* population (n=569, TL=7.52 cm, W=4.2g) revealed a greater number of females (n=417, 73.28%) than males (n=152, 26.72%). The sex ratio (females:males) was 2:1 for the whole sample. Males reach a larger body size (LT=8.02 cm) and body wet weight (W=5.95 g) than females (LT=8.0 cm, W=4.02 g). The relationship between TL and W can be explained by an exponential model (y= exp(a+bx)), and the correlation coefficients obtained were 0.87 for the total population, 0.91 for the males and 0.83 for the females.

The average fecundity is represented by 984 eggs for females with a TL=3.80 cm and 900 to 1200 eggs for specimens with a TL= 6.70 to 8.00 cm respectively. Additionally, 58.17% of the total population showed the parasitic isopod *Probopyrus pacificensis*.

ACKNOWLEDGEMENTS

The authors are grateful to the Universidad Autónoma Metropolitana where this research was carried out, as well as to Dra. Andrea Raz-Guzmán for reviewing an earlier English version, and to anonymous reviewers for providing valuable comments on earlier drafts of this manuscript.

REFERENCES

Balasundaram C and Pandian TJ (1982) Egg loss during incubation in *Macrobrachium nobilli* (Henderson & Mathai). J Exp Mar Biol Ecol 59:289-299

Cabrera JJ, Chávez A and Martínez CA (1979) Fecundidad y cultivo de *Macrobrachium tenellum* (Smith) en el laboratorio. An Inst Biol, Univ Nal Autón México 50, Ser Zool 1:127-150

Galvao R and Bueno S (2000) Population structure and reproductive biology of the Camacuto shrimp, *Atya scabra* (Decapoda, Caridea, Atyidae), from Sao Sebastiao, Brazil. (pp 291-299) In: JC von Vaupel Klein and Schram FR (eds) Crustacean Issues 12: The Biodiversity Crisis and Crustacea. AA Balkema, Rotterdam

Granados BA (1984) Biología, ecología y pesquerías de los langostinos de México. Universidad y Ciencia I:5-27

Gutiérrez LG (1992) Biología y cultivo de los langostinos nativos (*Macrobrachium* spp) de importancia comercial en México. Tesis Profesional, Escuela Nacional de Ciencias Biológicas, Inst Politécn Nal México 63 p

Guzmán AM, Rojas JL and González LD (1982) Ciclo anual de maduración y reproducción del "chacal" *Macrobrachium tenellum* y su relación con los factores ambientales en las lagunas costeras de Mitla y Tres Palos, Guerrero, México (Decapoda: Palaemonidae). An Inst Cienc Mar y Limnol UNAM 9:67-80

Guzmán AM (1987) Biología, ecología y pesca de *Macrobrachium tenellum*. Tesis Doctoral, Instituto de Ciencias del Mar y Limnología, UNAM México 306 p

Holthuis LB (1952) A general revision of the Palaemonidae (Crustacea Decapoda Natantia) of the Americas. II. The subfamily Palaemoninae. Allan Hancock Foundation Publishers Occasional Paper 22:1-396

Mallasen M and Controni W (1998) Comparison of artificial and natural, new and reused, brackish water for the larviculture of the freshwater prawn *Macrobrachium rosenbergii* in a recirculating system. J World Aquac Soc 29:345-350

Negrete AR (1977) Fecundidad en el langostino *Macrobrachium tenellum* (Smith, 1871) (Decapoda: Palaemonidae) en la Laguna de Tres Palos, Guerrero, México. Tesis Profesional, Fac Ciencias, Univ Nal Autón México 70 p

Odinez C and Rabelo YH (1996) Variation in egg size of the fresh-water prawn *Macrobrachium amazonicum* (Decapoda, Palaemonidae) J Crust Biol 16:684-688

Rodríguez de la Cruz I (1965) Contribución al conocimiento de los palemónidos de México. II Palaemónidos del Atlántico y vertiente oriental de México con descripción de dos especies nuevas. An Inst Nal Inv Biol Pesq, SIC, México 1:71-112

Román CR (1979) Contribución al conocimiento de la biología y ecología de *Macrobrachium tenellum* (Smith) (Crustacea, Decapoda, Palaemonidae). An Inst Cienc Mar y Limnol UNAM 6:137-160

Román CR (1991) Ecología de *Macrobrachium tenellum* (Decapoda: Palaemonidae) en la Laguna de Coyuca, Guerrero, Pacífico de México. An Inst Cienc Mar y Limnol UNAM 18:109-121

Román CR (1993) *Probopyrus pacificensis*, a new parasite species (Isopoda: Bopyridae) of *Macrobrachium tenellum* (Smith, 1871) (Decapoda: Palaemonidae) of the Pacific coast of México. Proc Biol Soc Wash 106:689-697

Ruiz MA, Peña JC and López YS (1996) Morfometría, época reproductiva y talla comercial de *Macrobrachium americanum* (Crustacea: Palaemonidae) en Guanacaste, Costa Rica. Rev Biol Trop 44:127-132

Signoret G and Soto E (1995) Tolerancia de *Macrobrachium tenellum* (Smith) a diferentes salinidades. Univ Cienc y Tecnol 4:20-24

Signoret G and Soto E (1997) Comportamiento osmorregulador de *Macrobrachium tenellum* y *Macrobrachium acanthurus* (Decapoda Palaemonidae) en diferentes salinidades. Rev Biol Trop 45:1085-1091

Signoret G, Ortega L and Brailovsky D (2000) Partially abbreviated larval development in an undescribed freshwater palaemonid prawn of the genus *Macrobrachium* from Chiapas, Mexico. Crustaceana 73:273-282

Souza GD and Fontoura NF (1996) Reproduçao, longevidade e razão sexual de *Macrobrachium potiuna* (Müller, 1880) (Crustacea, Decapoda, Palaemonidae) no arroio Sapucaia, Municipio de Gravataí, Rio Grande do Sul. Nauplius 4:49-60

Villalobos JL and Nates JC (1990) Dos especies nuevas de camarones de agua dulce del género *Macrobrachium* Bate (Crustacea, Decapoda, Palaemonidae) de la vertiente occidental de México. An Inst Biol Univ Nal Autón México Ser Zool 61:1-11

GASTROPOD SHELLS OR GASTROPOD TUBES? SHELTER CHOICE IN THE HERMIT CRAB *CALCINUS VERRILLI*

Lisa J. Rodrigues, David W. Dunham and Kathryn A. Coates

Department of Zoology, University of Toronto, 25 Harbord Street, Toronto, Ontario, Canada, M5S 3G5
Bermuda Biological Station for Research, 17 Biological Lane, St. George's, Bermuda GE 01

rodrigul@sas.upenn.edu

ABSTRACT

Calcinus verrilli, a hermit crab endemic to Bermuda, is unusual in that it inhabits both gastropod tubes (*Dendropoma* spp. and *Vermicularia* spp.) and gastropod shells (*Cerithium litteratum*). In the laboratory, when given a choice between shells and tubes, neither males nor females showed a significant preference for either shelter. In a Y-maze test of attraction to chemical extracts from tube- versus shell-forming gastropods, males showed no discrimination, but females collected in tubes preferred the tube-associated extract. Therefore, previous experience may be important in determination of adult preference in females.

This is contribution #1563 of the Bermuda Biological Station for Research and contribution # 38 of the Bermuda Biodiversity Project.

I. INTRODUCTION

The majority of hermit crab species utilize gastropod shells, which provide protection from predation, physical stress and damage (Reese 1969). The size and shape of gastropod shells influence both the growth (Markham 1968) and reproductive stress (Hazlett and Baron 1989) of the inhabitant. A hermit crab's fitness is also affected by the size and condition of the shell (Bertness 1981a), as crabs in poor shells tend to be more vulnerable to predation (Vance 1972). Therefore, shells are essential to the survival of the hermit crab and also directly affect its fitness.

Early experiments sought to determine the most important shell dimensions being used by hermit crabs to select a particular shell over another. Weight (Reese 1963) and volume (Bertness 1980) were suggested as the most important dimensions to the hermit crab. Preferences have also been shown for other shell variables, including the presence (Conover 1976) or absence (Grant and Ulmer 1974) of hydroids on the shell, depending on the hermit crab species being tested. Past experience, i.e. whether a conspecific or the individual presently choosing has previously occupied the shell type, has been shown to be important for at least some species (Grant 1963, Abrams 1978).

Preference is not static, it can change and be dependent on environmental conditions. Higher predation rates result in hermit crabs choosing heavier shells with thicker apertures (Bertness 1980, 1981b). Hermit crabs that live in areas with high rates of water flow select significantly heavier shells than those conspecifics from still water sites (Hahn 1998). In areas where the risk of desiccation is severe (high intertidal habitats), individuals prefer shells that have high spires, which retain more water than low spired-shells, and thereby minimize thermal stress (Bertness 1981b).

In the past, the focus of research has been on the use of gastropod shells by hermit crabs, but not all hermit crabs utilize gastropod shells. There are many species that use alternative shelters, rather than traditional gastropod shells (see Gherardi 1996a, for a review). *Calcinus verrilli* is a hermit crab that inhabits alternative shelters. It is unusual among the alternative shelter dwellers because it utilizes both gastropod shells and gastropod tubes in the field (Markham 1977, Rodrigues et al. 2000), while most alternative shelter dwellers do not use gastropod shells. Therefore, it is an ideal species to use in studies of preference for traditional shelters (*Cerithium litteratum* shells) versus alternative shelters (*Vermicularia knorii* and *Vermicularia spirata* tubes).

We investigated the preference of *Calcinus verrilli* for either *C. litteratum* shells or *Vermicularia* spp. tubes and asked the following questions: (1) What are the shelter preferences of *C. verrilli* in laboratory experiments? (2) How does the use of shelter types determined in the field (Rodrigues et al. 2000) compare to the choices made in the laboratory?

II. MATERIALS AND METHODS

This work was carried out in Bermuda from May to September 1999 and April to May 2000 at the Bermuda Biological Station for Research (BBSR). Depending on the location, *Calcinus verrilli* were collected by wading in tide pools, snorkeling, or SCUBA diving. Animals found in tubes were collected with a chisel and hammer, those housed in shells were collected by hand. Individuals were isolated from each other in partitioned plastic containers while in the field and the laboratory, to ensure that shell exchanges did not occur.

All crabs were removed from their shelters in a similar manner: the tip of the shell or the base of the tube was broken with Vice Grip brand pliers, and monofilament fishing line was inserted to gently urge the animal out. Shield length (Markham 1968), measured to the nearest 0.01mm, and sex were determined with a light microscope and ocular micrometer .

Shelter choice experiment

Prior to experimentation, naked crabs (without shelters) were isolated for 24 hours in plastic cylindrical containers (diameter = 10 cm, height = 8 cm) with seawater, sand substrate, and compressed air. Only healthy, intact animals were used in the experiments.

The same plastic isolation containers were covered with a fiberglass mesh lid and 26 were submerged in a large glass aquarium supplied with a constant flow-through of unfiltered seawater from Ferry Reach near the BBSR, and maintained on the local (natural) light/dark cycle. Gastropod tubes, *Vermicularia knorii* and *Vermicularia spirata*, and gastropod shells, *Cerithium litteratum*, were obtained by collecting individuals and sacrificing the live gastropod with injections of sodium hydroxide so as not to damage the shells. These were washed and rinsed daily for three to four weeks prior to being used in the experiment.

The sizes of the tubes and shells that were offered to each hermit crab were determined from regression analyses of shelter size versus hermit crab shield length (Rodrigues et al. 2000). The empty tube and shell were placed next to each other on the substrate at the centre of the container that already held a naked hermit crab. The preference for shells versus tubes was tested on a total of 95 individuals consisting of: 30 males collected in shells, 5 males collected in tubes, 30 females collected in shells, and 30 females collected in tubes. Once an individual inhabited a shelter, the individual's shelter occupancy was recorded every hour for the next four to six hours, followed by recordings made twice per day (at 10 am and 2 pm) for the next two days. Each crab was tested only once and then returned to its collection site.

Chemical choice experiment

A Y-maze of clear Plexiglas with two arms at right angles to each other (15 x 6 x 10 cm, each) and a base (19.5 x 6 x 10 cm) was used for the chemical choice experiment. The snail chemical extracts of *Cerithium litteratum* and *Vermicularia* spp. were prepared using a method similar to that of Orihuela et al. (1992). Gastropods were placed on ice for approximately fifteen minutes. Removal of the soft parts (whole animal excluding the foot and operculum) of the gastropods from their respective shells was conducted using Vice Grip brand pliers and forceps.

The tissue was placed into separate glass vials and frozen. The next day, the frozen tissue was ground, and weighed with an electronic balance to the nearest 0.1 g. Unfiltered seawater was added in a ratio of 1 ml for each 0.1 g of snail tissue. The extract was removed and kept on ice throughout the day of the experiment. Fresh samples of both extracts were prepared each day for the duration of the experiment.

A naked crab that had been isolated for 24 h was placed into an open-topped mesh cage (5 x 6 x 10 cm) at the centre of the Y-maze. Then, 0.5 ml of each gastropod extract was simultaneously introduced to the far end of a different arm using a pipette. Thirty seconds after extract introduction, the mesh cage was lifted and the following data recorded: (1) reached end of left arm and extract type, (2) reached end of right arm and extract type, (3) remained in base, (4) remained in the middle of an arm or no movement after 5 minutes. The chemical choice experiment was performed on a total of 93 crabs consisting of: 30 males collected in shells, 3 males collected in tubes, 30 females collected in shells, and 30 females collected in tubes.

Following each trial, the Y-maze was washed thoroughly with fresh seawater and the substrate was redistributed. *Cerithium* and *Vermicularia* extract presentation arms were alternated between trials to eliminate any possible directional bias. Each animal was given the shelter it was originally collected in and returned to its collection site.

Statistical analysis

X^2 tests were calculated by hand; a table from Sokal and Rohlf (1987) was used to determine P values. Student's t-tests (for normally distributed data) or Mann-Whitney rank tests (for non-normal data) were calculated using Sigma Stat software version 2.0 by Jandel. Significant P values were those ≤ 0.05.

III. RESULTS

Field collections

The distribution of males and females among shelters showed significant differences (Fig. 1). Significantly more males than females inhabited *C. litteratum* shells ($X^2 = 34.68$, DF $= 1$, $P \leq 0.001$), while more females than males were collected in both *Vermicularia* spp. tubes ($X^2 = 147.43$, DF $= 1$, $P \leq 0.001$) and *Dendropoma* spp. tubes ($X^2 = 42.19$, DF $= P \leq 0.001$).

Shelter choice experiment

X^2 tests conducted on choices made by males from shells, females from shells, males from tubes, and females from tubes found no significant differences in shelter type selected (Table la).

From observations made during the trials, the first choice that was made was not reversed later. In the next four to six hours, and over the following two days, no individual was observed to have changed shelters.

Chemical choice experiment

As in the shelter choice experiment, there was no significant difference found among the choices made by males in shells, females in shells, or males in tubes (Table 1b). However, females in tubes selected the *Vermicularia* spp. extract significantly more than the *C. litteratum* extract ($X^2 = 6.5$, DF $= 1$, $P < 0.05$).

IV. DISCUSSION

Males and females showed no significant discrimination between shells and tubes in the shelter choice experiment. Observations of the trials suggested that hermit crabs inhabited the first shelter that they approached. In many cases, one or the other of the shelters had not been obviously disturbed, so it may be that the individual never investigated both shelter types. None of the study animals ever changed from the shelter they initially selected, which suggests that they may not have made a true choice, as other hermit crabs are reported to "try on" several available shells before making a final decision (Hazlett 1966).

Therefore, the experiment using chemical stimuli was conducted. In that experiment, we could be certain that each individual was exposed to both stimuli. The set-up was based on published studies of gastropod predation sites, where hermit crabs locate

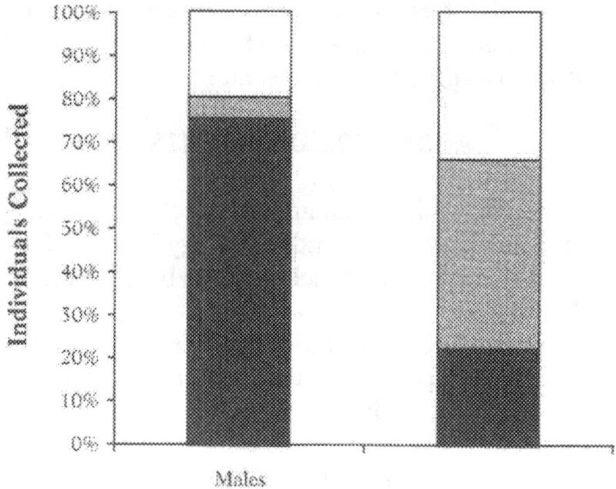

Figure 1. *Distribution of **Calcinus verrilli** males and females among three different shelter types in the field, **Dendropoma** spp. tubes (open bars), **Vermicularia** spp. tubes (grey bars), and **Cerithium litteratum** shells (solid bars).*

Table 1. *The choices made by **Calcinus verrilli** in (a) the Shelter Choice Experiment between **Cerithium litteratum** shells and **Vermicularia** spp. tubes and (b) the Chemical Choice Experiment between **C. litteratum** shell extract and **Vermicularia** spp. tube extract.*

(a)

	Sex and shelter collected in			
Choice	Males in shells	Females in shells	Males in tubes	Females in tubes
Shells	19	18	3	18
Tubes	11	12	2	12

(b)

	Sex and shelter collected in			
Choice	Males in shells	Females in shells	Males in tubes	Females in tubes
Shells	16	20	1	8
Tubes	14	10	2	22

new gastropod shells from molecules released from gastropod flesh during predation events (McLean 1974, Rittschof 1980). They have been shown to only attend sites containing gastropod flesh and are not attracted to other flesh sources.

When both extracts were presented together in the chemical choice experiment, males did not show a preference between *C. litteratum* and *Vermicularia* spp. extracts. However, for females, choice was linked to the shelter type they were collected in. Females collected in tubes showed a significant preference for the *Vermicularia* spp. presentation arm, while females collected in shells showed a slightly significant preference for the *C. litteratum* presentation arm. This result is in agreement with the conclusions made by Grant (1963) and Abrams (1978) that past experience is an important determinant of shelter preference for some species.

Comparisons with other alternative shelter dwellers

Of the few studies that have investigated shelter preferences in hermit crabs that utilize alternative shelters, only one (Busato et al. 1998) concerned a species (*Calcinus tubularis*) that naturally uses both gastropod shells and tubes. All other studies of shelter preference in alternative shelter-dwellers (Gherardi 1996a, b, Imafuku and Ando 1999) observed species that do not utilize gastropod shells in the field, although shells were offered as a choice in the laboratory.

Busato et al. (1998) concluded that *C. tubularis*'s use of alternative shelters is an example of "making the best of a bad situation," as the hermit crabs preferred mobile shells over sessile tubes, despite using both in the field. However, the authors did not consider the shelter type each individual was collected in when they recorded choice, and so it is not known whether previous experience may have affected laboratory choice. We can not conclude that *Calcinus verrilli* females are "making the best of a bad situation," because females selected the *Vermicularia* spp. extract in the chemical choice study.

Our results for females are similar to those found by Gherardi (1996b), who observed that adults of *Discorsopagurus schmitti* prefer tubes to shells in the laboratory , and are rarely, if ever, found in shells in the field. However, *D. schmitti* were found to prefer loose tubes to sessile tubes, although loose tubes are not normally found in nature. Gherardi (1996b) also found that postlarval stages preferred to settle in shells rather than tubes. Previous experience is also important for that species, but preference may change with developmental stage.

Although preference of post-megalopa stages has not been tested in *C. verrilli*, it would make an interesting avenue for future research. The preference of *C. verrilli* postlarval stages for either shells or tubes, rather than adult preferences, may be more important for determining the observed sexual differences in shelter occupancy that have been observed in the field. Intersexual competition seems to also be an important determinant of shelter occupancy for *Calcinus verrilli* (Rodrigues et al. in prep).

ACKNOWLEDGEMENTS

We thank M. Hammond for help with collecting and members of the Benthic Ecology Research Program at BBSR for use of their research boat, especially S.R. Smith and S. dePutron.

Funding was provided by a Bermuda Government Scholarship and the Munson Foundation through the BBSR Graduate Intern Program.

REFERENCES

Abrams PA (1978) Shell selection and utilization in a terrestrial hermit crab, *Coenobita compressus* (H. Milne Edwards). Oecologia 34:239-253

Bertness MD (1980) Shell preference and utilization patterns in littoral hermit crabs of the Bay of Panama. J Exp Mar Biol Ecol 48:1-16

Bertness MD (1981a) The influence of shell-type on hermit crab growth rate and clutch size (Decapoda, Anomura). Crustaceana 40:197-205

Bertness MD (1981b) Predation, physical stress, and the organization of a tropical rocky intertidal hermit crab community. Ecology 62:411-425

Busato P, Benvenuto C and Gherardi F (1998) Competitive dynamics of a Mediterranean hermit crab assemblage: the role of interference and exploitative competition for shells. J Nat Hist 32:1447-1451

Conover MR (1976) The influence of some symbionts on the shell-selection behaviour of the hermit crabs, *Pagurus pollicarpus* and *P. longicarpus*. Anim Behav 24:191-194

Gherardi F (1996a) Non-conventional hermit crabs: pros and cons of a sessile, tube-dwelling life in *Discorsopagurus schmitti* (Stevens). J Exp Mar Biol Ecol 202:119-136

Gherardi F (196b) Gastropod shells or polychaete tubes? The hermit crab *Discorsopagurus schmitti*'s housing dilemma. Ecosci 3:154-164

Grant WC Jr (1963) Notes on the ecology and behavior of the hermit crab *Pagurus acadianus*. Ecology 44:767-771

Grant WC Jr and Ulmer KM (1974) Shell selection and aggressive behavior in two sympatric species of hermit crabs. Biol Bull 146:32-43

Hahn DR (1998) Hermit crab shell use patterns: response to previous shell experience and to water flow. J Exp Mar Biol Ecol 228:35-51

Hazlett BA (1966) Social behavior of the Paguridae and Diogenidae of Curaçao. Stud Fauna Curaçao 88:1-143

Hazlett BA and Baron LC (1989) Influence of shells on mating behaviour in the hermit crab *Calcinus tibicen*. Behav Ecol Sociobio 124:369-376

Imafuku M and Ando T (1999) Behaviour and morphology of pagurid hermit crabs (Decapoda, Anomura) that live in tusk shells (Mollusca, Scaphopoda). Crustaceana 72:129-144

Markham JC (1968) Notes on growth-patterns and shell-utilization of the hermit crab *Pagurus bernhardus* L. Ophelia 5:189-205

Markham JC (1977) Preliminary note on the ecology of *Calcinus verrilli*, an endemic Bermuda hermit crab occupying attached vermetid shells. J Zool London 181:131-136

McLean RB (1974) Direct shell acquisitions by hermit crabs from gastropods. Experientia 30:206-208

Orihuela B, Diaz H, Forward RB and Rittschof D (1992) Orientation of the hermit crab *Clibanarius vittatus* (Bosc) to visual cues: effects of mollusc chemical cues. J Exp Mar Biol Ecol 164:193-208

Reese ES (1963) The behavioral mechanisms underlying shell selection by hermit crabs. Behaviour 21:78-126

Reese ES (1969) Behavioral adaptations of intertidal hermit crabs. Am Zool 9:343-355

Rittschof D (1980) Chemical attraction of hermit crabs and other attendants to gastropod predation sites. J Chem Ecol 6:103-118

Rodrigues LJ, Dunham DW and Coates KA (2000) Shelter preferences in the endemic Bermudian hermit crab, *Calcinus verrilli* (Rathbun, 1901) (Decapoda, Anomura). Crustaceana 73:737-750

Rodrigues LJ, Dunham DW and Coates KA (in prep) Intraspecific shell competition in the endemic Bermudian hermit crab, *Calcinus verrilli*.

Sokal RR and Rohlf FJ (1987) Introduction to Biostatistics 2nd Ed. WH Freeman and Co, New York

Vance RR (1972) The role of shell adequacy in behavior interactions involving hermit crabs. Ecology 53:1075-1083

HERMIT CRAB FAUNA FROM THE INFRALITTORAL ZONE OF ANCHIETA ISLAND (UBATUBA, BRAZIL)

Fernando L. M. Mantelatto and Renata B. Garcia

Departamento de Biologia, Faculdade de Filosofia, Ciências e Letras de Ribeirão Preto (FFCLRP), Universidade de São Paulo (USP), Av. Bandeirantes 3900, Ribeirão Preto (SP), Brasil, CEP. 14040-901

flmantel@usp.br

ABSTRACT

The checklist and biological aspects of hermit crab communities living on the infralittoral rocky bottom (Anchieta Island, Ubatuba) were studied through monthly samplings from January to December 1998. The specimens were captured by three persons using scuba diving methods. Data on seasonal distribution and abundance of each species are presented. Nine species were recorded: *Calcinus tibicen, Dardanus insignis, D. venosus, Paguristes calliopsis, P. erythrops, P. tortugae, Petrochirus diogenes, Pagurus brevidactylus,* and *P. criniticornis*. The species recorded on Anchieta Island correspond to 42.9% of the total hermit crab species catalogued for the State of São Paulo. The taxocoenosis was controlled by *P. tortugae* and *P. brevidactylus* during all seasons. In the present study *D. venosus* was recorded for the first time in the State of São Paulo. The large amount of resources utilized by the hermit crabs on Anchieta Island reveal a locality with high shell availability, constituting an important site for crabs to find, use and exchange shells.

I. INTRODUCTION

Reliable regional checklists of marine species have multiple uses, especially for biodiversity studies on zoogeographic regions or provinces in specific habitats, which serve as points of departure for studying several aspects of the community ecology (Hendrickx and Harvey 1999). In this sense, the literature on decapods from the Brazilian coast is often limited to restricted areas, which are important to understand the ecology of the benthic community. Among them, during the last 12 years a long-term effort has been made to identify and characterize the biology (i.e., larval phases, growth, reproduction, population dynamics, relationships with abiotic factors) of decapod crustaceans commonly and uncommonly occurring in the Ubatuba region, from the intertidal zone to 20 m of depth. This region is located along the northern coast of the State of São Paulo, and is considered an important area for crustacean investigations because it comprises a zone of faunal transition and presents a mixture of faunas of both tropical and Patagonian origins (Sumida and Pires-Vanin 1997). For this reason, an impressive number of studies are available on the intertidal zone and continental shelf, focusing on the crab, shrimp and hermit crab species composition from a variety of habitats in the Ubatuba area.

The Anomura represent a highly significant group among marine crustaceans, including approximately 800 species of hermit crabs, which have undergone considerable revision (Ingle 1993), and are now classified in 6 families with 86 genera. Hermit crabs represent a very important group within the intertidal community and also a significant taxon in the benthic sublittoral habitat, playing an important role in the marine trophic chain (Fransozo et al. 1998). Of the 46 hermit crab species recorded from Brazilian waters, 20 were recorded on the northern coast of the State of São Paulo, 13 of them belonging to the family Diogenidae and 7 to the family Paguridae (Melo 1999). Among the available reports, the following represent

E. Escobar-Briones & F. Alvarez Eds.
MODERN APPROACHES TO THE STUDY OF CRUSTACEA
PP. 137-143

particularly important contributions to the knowledge of the hermit fauna of Ubatuba: Hebling et al. (1994) reported the crabs sampled in the soft bottom of Anchieta Island region; Negreiros-Fransozo et al. (1997) established the species composition and distribution of crabs along the non-consolidated sublittoral region of Fortaleza Bay; Mantelatto and Souza-Carey (1998) reported the species inhabiting the bryozoan colonies of *Schyzoporella unicornis* (Johnston, 1847), and recently Fransozo et al. (1998) provided an anomuran check-list and described ecological aspects of this fauna in terms of their composition and distribution along the non-consolidated sublittoral region of Ubatuba Bay.

Despite this promising situation, and a widespread interest in the faunal composition and ecology of the crustacean species inhabiting different areas of Ubatuba, the island community was poorly known until recently, mainly due to the distance from the coast and the sampling difficulties. Islands are typically species-poor systems because of their reduced area in comparison to mainland areas, and this poverty is accentuated by increasing isolation and decreasing island relief and altitude (Whittaker 1998). However, island communities possess characteristics, such as population growth that differ from populations in the continental coastal zone. Thus, the island biotope offers good perspectives for future studies in Brazil, considering the high number of islands found along the coast. The knowledge of the communities may contribute to the elucidation of evolution, speciation and distribution problems with respect to marine species (MacArthur 1972). On this basis, we studied the composition and some biological aspects of the hermit crab communities living on the infralittoral rocky bottom in Anchieta Island, Ubatuba, as a contribution to the study of the biodiversity of Anomura from the São Paulo coast.

II. MATERIALS AND METHODS

Study area

Anchieta Island (23° 33'S, 45° 05'W) was recently declared an ecological reserve of the State of São Paulo. The island has a total area of about 10 km², which makes it the largest island off the coast of the State of São Paulo north of São Sebastião Island. This island is separated from the coast by a 300 m long and 35 m deep canal. There are 6 small beaches, and almost the entire shore area is rocky with an irregular surface and areas with large boulders. This collection area is important in view of the significant anthropogenic activity, which has led to the expansion of the tourist center. The physical and chemical features of the area have been described by Oliveira (1983) and Medeiros (1992). However, scientific information about the crustacean fauna from this area is sparse.

Sampling and analysis

Hermit crabs were obtained monthly from January to December 1998, in the infralittoral rocky shores of South, East and Small Beaches and Wind Bay of Anchieta Island, whose surface is irregular, with many large boulders. Every month, at least one of these sites was sampled depending on ocean and weather conditions. Specimens were captured during the daytime by three persons using scuba over a period of 30 min over the same area of about 850 m².

Animals were frozen and transported to the laboratory where they were carefully removed from their shells and measured for cephalothoracic shield length (CSL = from the tip of the rostrum to the V-shaped groove at the posterior edge). Sex was determined from the gonopore position. Measurements were made with a caliper rule (0.1 mm) or by drawing with the aid of a camera lucida. Shell species were identified according to Rios (1994) and by Dr. Osmar Domaneschi.

The constancy index (C) for each species was calculated according to Dajoz (1983): C = p X 100/ P, where "p" is the number of samples belonging to a given species, and "P" is the total number of samples analyzed. Species were then classified into three different constancy categories: constant (C $\geq$ 50 %), accessory (25 % < C < 50 %) and accidental (C < 25 %). Diversity was calculated using the Shannon-Weaver index (Shannon and Weaver, 1963): H' = $\sum_i^s P_i . \log_2 P_i$, where "s" is the number of species and "P_i" is the proportion of ith species. The equitability index (J') was calculated as indicated by García Raso and Fernandez Muñoz (1987): J' = H' / $\log_2$ s.

III. RESULTS

A total of 4,133 specimens were collected (Table 1), comprising nine hermit crab species and two families. Diogenids outnumbered pagurids. The anomuran taxocoenosis was dominated by a few species (constant species). Four of those had a relative abundance higher than 3%, representing together 95.5%

Table 1. *Number of individuals collected from January to December 1999 in Anchieta Island (CN = constancy: Co = constant; Ac = accessory; Ad = accidental).*

SPECIES	MONTHS													
	Jan	Feb	Mar	Apr	May	Jun	Jul	Aug	Sep	Oct	Nov	Dec	CN	Total
C. tibicen	16	02	13	01	05	04	18	09	14	17	07	18	Co	124
D. insignis	-	-	-	01	-	-	-	18	-	02	-	-	Ad	21
D. venosus	-	-	01	-	-	-	01	02	-	02	-	-	Ac	06
P. calliopsis	-	-	-	-	-	-	-	-	-	40	42	34	Ad	116
P. erythrops	-	05	02	05	02	-	-	97	-	111	02	32	Co	256
P. tortugae	41	144	178	207	161	173	36	406	187	334	281	281	Co	2429
P. diogenes	-	-	-	-	-	-	-	01	-	-	-	-	Ad	01
P. brevidactylus	08	83	81	123	79	86	61	91	67	120	121	215	Co	1135
P. criniticornis	-	01	01	05	03	01	01	-	01	-	07	25	Co	45
TOTAL	65	235	276	342	250	264	117	624	269	626	460	605		4133

of the total sample. The family Diogenidae was represented by seven species, contributing with 71.5% of all individuals: *Calcinus tibicen* (Herbst, 1791), *Dardanus insignis* (Saussure, 1858), *D. venosus* (H. Milne Edwards, 1848), *Paguristes calliopsis* Forest and Saint Laurent, 1967, *P. erythrops* Holthuis, 1959, *P. tortugae* Schmitt, 1933, and *Petrochirus diogenes* (Linnaeus, 1758). Pagurids were represented by *Pagurus brevidactylus* (Stimpson, 1859) and *P. criniticornis* (Dana, 1852). These species represented 28.5% of the total sample.

The species showed different distribution patterns on Anchieta Island. The most abundant species were *P. tortugae* and *P. brevidactylus*, with an outstanding occurrence on rocky areas covered by algae. The third ranked species was *P. erythrops*, almost always found on soft bottoms where the sediments are mainly composed of very coarse sand or coarse sand and medium sand, followed by *C. tibicen*, which was especially abundant in rock crevices. *Paguristes calliopsis* occurs in the same area as the previous more abundant species, while *D. insignis* and *D. venosus* were found with *P. erythrops*. Almost all the specimens of *P. criniticornis* were found on sediments composed mainly of very coarse sand.

All constant species, except *P. erythrops*, showed a continuous occurrence throughout the sampling period. Number of species and their respective frequency varied seasonally, both increasing during the warmest months and decreasing during the coldest ones (Fig. 1).

The diversity index ranged from 1.05 to 1.85, depending more on equitability than on richness (Table 1, Fig. 1). During the study period, both indexes showed a wide variation, decreasing from January to June and increasing from July to December.

The hermit crab species were found occupying, in different percentages, 29 species of gastropod shells, belonging to 23 genera (Table 2). *Cerythium atratum*, *Pisania auritula* and *Morula nodulosa* were the shells most commonly occupied by hermit crabs at the study site, accounting for more than 74.9% of the collected shells.

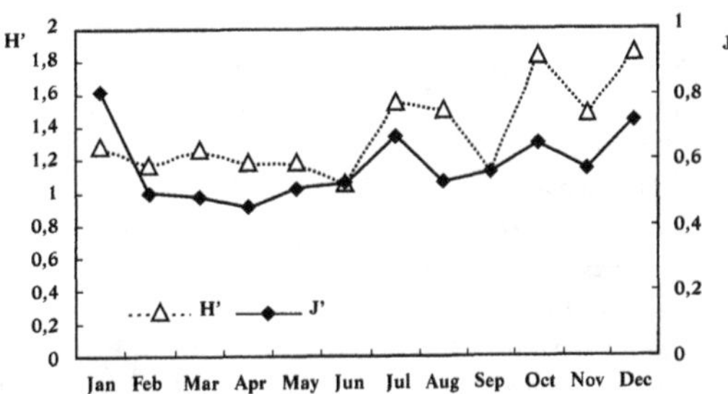

Figure 1. *Monthly oscillation of diversity (H') and equitability (J') of hermit crab species during the study period (January to December, 1998) in Anchieta Island.*

IV. DISCUSSION

The sampling methodology used in this study, which provided large amounts of material from irregular surfaces, should be used in other areas to make comparisons possible. The efficiency of this sampling method reveals that the diversity of hermit crab species on Anchieta Island is considerably high. More than 40% of the 21 species recorded for the São Paulo coast, have been captured in this area. Among them, *D. venosus* was recorded for the first time in the State of São Paulo (Mantelatto et al. 2001). This may be regarded

Table 2. *Total number of occupied shells by the hermit crab species collected in the studied area (Ct:* **Calcinus tibicen***; Di:* **Dardanus insignis***; Dv:* **Dardanus venosus***; Pc:* **Paguristes calliopsis***; Pe:* **Paguristes erythrops***; Pt:* **Paguristes tortugae***; Pd:* **Petrochirus diogenes***; Pb:* **Pagurus brevidactylus***; Pcr:* **Pagurus criniticornis***).*

Shells	Ct	Di	Dv	Pc	Pe	Pt	Pd	Pb	Pcr	Total
Adelomelon beckii	-	-	-	-	-	-	1	-	-	1
Anachis lyrata	-	-	-	1	-	-	-	17	-	18
Astraea sp.	1	-	-	-	-	-	-	1	-	2
Astraea latispina	4	-	1	1	3	25	-	-	-	34
Astraea olfersii	23	9	2	4	5	40	-	-	-	83
Astraea phoebia	3	-	-	4	1	18	-	2	-	28
Buccinanops gradatum	-	-	-	-	3	1	-	-	-	4
Calliostoma bullisi	-	-	-	1	-	-	-	-	-	1
Cerithium atratum	17	-	-	88	6	676	-	334	37	1158
Chicoreus tenuivaricosus	2	1	-	-	35	13	-	2	-	53
Coralliophila aberrans	-	-	-	-	-	5	-	2	-	7
Cymatium parthenopeum	3	-	-	-	11	22	-	1	-	37
Cypraea acicularis	-	-	-	-	1	-	-	-	-	1
Favartia cellulosa	-	-	-	-	-	1	-	1	-	2
Fusinus brasiliensis	1	2	-	2	20	9	-	-	-	34
Leucozonia nassa	6	-	-	-	12	166	-	15	-	199
Modulus modulus	1	-	-	1	-	1	-	-	-	3
Morula nodulosa	-	-	-	7	-	308	-	645	5	965
Muricopsis necocheanus	-	-	-	-	-	1	-	-	-	1
Oliva reticularis	-	-	-	-	2	-	-	-	-	2
Olivancillaria urceus	-	5	-	-	8	-	-	-	-	13
Phalium granulatum	-	-	1	-	-	1	-	-	-	2
Pisania auritula	43	-	-	-	48	862	-	17	1	971
Pisania pusio	2	-	-	-	5	59	-	3	-	69
Polinices hepaticus	-	-	-	-	3	2	-	-	-	5
Polinices lacteus	-	-	-	-	1	4	-	-	1	6
Stramonita haemastoma	9	4	-	1	57	68	-	11	-	150
Strombus pugilis	1	-	2	-	31	-	-	-	-	34
Tegula viridula	8	-	-	6	4	131	-	8	1	158
no identification	-	-	-	-	-	16	-	76	-	92
Total	124	21	6	116	256	2429	1	1135	45	4133

as a very high diversity, considering the relatively small area of the island compared to the range of the Brazilian coast, with a total of 46 species recorded (Melo 1999).

The structure and dynamics of the hermit crab community on Anchieta Island basically depend on the seasonal abundance of the dominant species, *P. tortugae* and *P. brevidactylus*, with a moderate influence of *P. erythrops* and *C. tibicen*. This pattern of dominance of a few species in the community was also recorded during previous studies on brachyurans and anomurans from a non-consolidated substratum in the Ubatuba region (Fransozo et al. 1992, Negreiros-Fransozo et al. 1997, Fransozo et al. 1998). Recently Hebling et al. (1994), working on a non-consolidated area on Anchieta Island, recorded a hermit crab community composed of 11 species, including four (*P. criniticornis, P. erythrops, P. tortugae*, and *D. insignis*) of those captured in the present study. These authors reported relatively low numbers of individuals for most of the species, except for *D. insignis*. This fact confirms the preference of this species for rocky areas close to sand substrates. The presence of different types of substrates may also allow for the coexistence of various species by means of their differential use of space. While some species are able to use the substratum as shelter, others might use it as a feeding ground or even as a food medium from which organic particles can be obtained, thus reducing competitive interactions among species (Abele 1976).

In general, the anomuran taxocoenosis in the different areas of Ubatuba Bay is qualitatively similar, with small differences in species composition. This similarity is probably related to the fact that all areas are exposed to the influence of the same prevailing water masses, which directly affect the dynamics of environmental factors (Castro-Filho et al. 1987). Number of individuals and species diversity decrease during the summer months, a fact probably correlated with raising temperatures during this period, and tend to increase during the winter months. In the Ubatuba region, the interaction of two main water masses, the Coastal Water (CW) and the South Atlantic Central Water (SACW), results in a mixing zone with temporal and spatial effects dependent on penetration intensity of SACW (Pires 1992). The SACW has a strong influence on the seabed temperature, especially on the inner shelf during summer. These conditions contribute to the development of the local community providing nutrients and favoring recruitment, allowing each species to occupy a favorable habitat for growth and reproduction, a fact that probably occurs on Anchieta Island. Evidence for such a scenario is the high number of young individuals of all species captured.

Once established in a locality, hermit crab populations can adapt to local conditions and coexist with other species inhabiting the area. Specifically for hermit crabs, these relationships involve an important aspect, which is the sharing of shells by various species (Gherardi and Nardone 1997). The resource partition can directly affect the distribution of individuals, possibly reducing the abundance of the population (Miller 1967).

In the natural environment, almost all hermit crab populations occupy a wide variety of shell types and sizes as a function of local availability (Bertness 1980, Reddy and Biseswar 1993), which is considered a limiting factor for hermit crab survival (Kellogg 1976). On the basis of the large quantities of resources used by hermit crabs on Anchieta Island (Table 2), we can infer that shells are readily available in this locality, representing an important site for crabs to find, use, and exchange shells. Ohmori et al. (1995) postulated that shell type availability in nature is determined by the relative abun-

dance of living gastropods and their mortality rates. The abundance of living gastropods reflects directly on the shell stock since the most available shells are generally the most occupied in the field (Mantelatto and Garcia 2000).

Hermit crabs are mainly deposit feeders (Schembri 1982), and their abundance is strongly associated with food (Bertness 1981) and shell availability (Raimondi and Lively, 1986). We can infer that these factors may be responsible for the seasonal abundance pattern of hermit crabs on Anchieta Island, as recorded for other anomurans studied in the non-consolidate substrate of Ubatuba Bay (Fransozo et al. 1998).

The seasonal availability of shells for shelter renewal in adults and the abundance of suitable shells for new settlers may also affect hermit crab distribution (McLean 1983). Hermit crabs may obtain new housing by exchanging shells among themselves, seeking empty shells, or even by migrating to specific areas where gastropod predation is intense (Rittschoff 1980).

Detailed information on the population structure and reproductive biology of the species reported herein will provide important data to elucidate questions about the dynamics of the anomuran community on this island. In addition, comparative data for biodiversity studies are provided, which can also be used to detect the potential impact of anthropogenic activities in the area.

ACKNOWLEDGEMENTS

The authors are grateful to FAPESP (Grants # 98/07454-5 and 00/02554-3) for financial support during the field work and to CNPq (Grants # 520646/98-3 and 450728/00-5) for financial support to attend the Crustacean Society 2000 Summer Meeting, held in Puerto Vallarta, Mexico. We thank Secretaria do Meio Ambiente do Estado de São Paulo, IBAMA and Parque Estadual da Ilha Anchieta for permission (Proc. # 42358/98) and facilities during the field work. Special thanks are due to Dr. Nilton José Hebling (Zoology Department, UNESP – Rio Claro, Brazil) and Dr. Osmar Domaneschi (Zoology Department, IB – USP) for assistance with hermit crab species and shell identification, respectively, and to Jussara Moretto Martinelli and NEBECC-RP members for their help during field and laboratory work. All experiments conducted in this study comply with current applicable state and federal laws.

REFERENCES

Abele LG (1976) Comparative species composition and relative abundance of decapod crustaceans in marine habitats of Panama. Mar Biol 38:263-270

Bertness MD (1980) Shell preference and utilization patterns in littoral hermit crabs of the Bay of Panama. J Exp Mar Biol Ecol 48:1-16

Bertness MD (1981) Competitive dynamics of a tropical hermit crab assemblage. Ecology 62:751-761

Castro Filho BM, Miranda LB and Myao SY (1987) Condições hidrográficas na plataforma continental ao largo de Ubatuba: variações sazonais e em média escala. Bolm Inst oceanogr 35:135-151

Dajoz R (1983) Ecologia Geral. Ed. Vozes, EDUSP, São Paulo, 472 p

Fransozo A, Negreiros-Fransozo ML, Mantelatto FLM, Pinheiro MAA and Santos S (1992) Composição e distribuição dos Brachyura (Crustacea, Decapoda) do sublitoral não consolidado na Enseada da Fortaleza, Ubatuba (SP). Rev Brasil Biol 52:667-675

Fransozo A, Mantelatto FLM, Bertini G, Fernandez-Góes L and Martinelli JM (1998) Distribution and assemblages of anomuran crustaceans in Ubatuba Bay, north coast of São Paulo State, Brazil. Acta Biol Venez 18:17-25

Garcia Raso JE and Fernandez Muñoz R (1987) Estudio de una comunidad de Crustáceos Decápodos de fondos "coralígenos" del alga calcárea *Mesophyllum lichenoides* del sur de España. Invest Pesq 51 (Supl. 1):301-322

Gherardi F and Nardone F (1997) The question of coexistence in hermit crabs: population ecology of a tropical intertidal assemblage. Crustaceana 70:608-629

Hendrickx ME and Harvey AW (1999) Checklist of anomuran crabs (Crustacea: Decapoda) from the eastern tropical Pacific. Belg J Zool 129:363-389

Hebling NJ, Mantelatto FLM, Negreiros-Fransozo ML and Fransozo A (1994) Levantamento e distribuição de braquiúros e anomuros (Crustacea, Decapoda) dos sedimentos sublitorais da região da Ilha Anchieta, Ubatuba (SP). Bolm Inst Pesca 21:1-9

Ingle R (1993) Hermit Crabs of the Northeastern Atlantic Ocean and the Mediterranean Sea. An Illustrated Key. Ed. Chapman and Hall, 495 p

Kellogg CW (1976) Gastropod shells: a potentially limiting resource for hermit crabs. J Exp Mar Biol Ecol 22:101-111

MacArthur RH (1972) Geographical ecology: patterns in distribution of species. Harper and Row, New York, 269 p

Mantelatto FLM and Souza-Carey MM (1998) Caranguejos anomuros (Crustacea, Decapoda) associados à *Schizoporella unicornis* (Bryozoa, Gymnolaemata) em Ubatuba (SP), Brasil. Anais do "IV Simpósio de Ecosistemas Brasileiros", ACIESP, 2:200-206

Mantelatto FLM and Garcia RB (2000) Shell utilization pattern of the hermit crab *Calcinus tibicen* (Diogenidae) from southern Brazil. J Crust Biol 20:460-467

Mantelatto FLM, Garcia RB, Martinelli JM and Hebling NJ (2001) On a record of *Dardanus venosus* (H. Milne Edwards, 1848) (Crustacea, Anomura) from the São Paulo State, Brazil. Revta Bras Zool 18:71-73

Mclean R (1983) Gastropod shells: A dynamic resource that helps shape benthic community structure. J Exp Mar Biol Ecol 69:151-174

Medeiros LRA (1992) Meiofauna de praia arenosa da Ilha Anchieta, São Paulo: I. Fatores físicos. Bolm Inst oceanogr 40:27-38

Melo GAS (1999) Manual de identificação dos Crustacea Decapoda do litoral brasileiro: Anomura, Thalassinidea, Palinuridea e Astacidea. Plêiade Editora, São Paulo, 551 p

Miller RS (1967) Pattern and process in competition. Adv Ecol Res 4:1-74

Negreiros-Fransozo ML, Fransozo A, Mantelatto FLM, Pinheiro MAA and Santos S (1997) Anomuran species (Crustacea, Decapoda) and their ecological distribution at Fortaleza Bay sublitoral, Ubatuba, São Paulo, Brazil. Iheringia, Sér Zool 83:187-194

Ohmori H, Wada S, Goshima S and Nakao S (1995) Effects of body size and shell availability on the shell utilization pattern of the hermit crab *Pagurus filholi* (Anomura: Paguridae). Crust Res 24:85-92

Oliveira MMA (1983) Moluscos de praias da Ilha Anchieta (Ubatuba, São Paulo). Arq Biol Tecnol 26:383-390

Pires AMS (1992) Structure and dynamics of benthic megafauna on the continental shelf offshore of Ubatuba, southeastern Brazil. Mar Ecol Progr Ser 86:63-76

Raimondi PT and Lively CM (1986) Positive abundance and negative distribution effects of a gastropod on an intertidal hermit crab. Oecologia 69:213-216

Reddy T and Biseswar R (1993) Patterns of shell utilization in two sympatric species of hermit crabs from the Natal coast (Decapoda, Anomura, Diogenidae). Crustaceana 65:13-24

Rios EC (1994) Seashells of Brazil. Rio Grande do Sul. Fundação cidade do Rio Grande, Instituto Acqua, Museu Oceanográfico de Rio Grande, Universidade de Rio Grande. 2ª Ed. 368p. + 113pl.

Rittschoff D (1980) Chemical attraction of hermit crabs and other attendants to simulated gastropod predation sities. J Chem Ecol 6:103-118

Schembri PJ (1982) Feeding behaviour of fifteen species of hermit crabs (Crustacea: Decapoda: Anomura) from the Otago region, southeastern New Zealand. J Nat Hist 16:859-878

Shannon CE and Weaver W (1963) The mathematical theory of communication. University of Illinois Press, Urbana 177 p

Sumida PYG and Pires-Vanin AMS (1997) Benthic associations of the shelf break and upper slope off Ubatuba-SP, southeastern Brazil. Estuar Coast Shelf Sci 44:779-784

Whittaker RJ (1998) Island Biogeography: Ecology, Evolution and Conservation. Oxford University Press, New York 285 p

BREEDING SEASON OF THE HERMIT CRAB *PETROCHIRUS DIOGENES* (ANOMURA: DIOGENIDAE) IN THE NORTH COAST OF SÃO PAULO STATE, BRAZIL

Giovana Bertini and Adilson Fransozo

*NEBECC, Group of Studies on Crustacean Biology, Ecology and Culture
Departamento de Zoologia, Instituto de Biociências – IBB - Universidade Estadual Paulista, UNESP,
18618-000 Botucatu (São Paulo), Brasil*

fransozo@ibb.unesp.br

ABSTRACT

Petrochirus diogenes, one of the largest hermit crabs in the western Atlantic, was studied in the north coast of the State of São Paulo, Brazil, to determine its breeding season through the analysis of gonad development at a macroscopic level. The analysis of 999 crabs collected from 1995 to 1999, based on four categories of gonad development, showed that males present gonads in the advanced stage during all seasons, and although females present developed ovaries all year as well, the peak incidence is in the summer. The peak of recruitment occurs in the winter.

I. INTRODUCTION

One of the goals of studies on the reproductive biology of benthic invertebrates is to test generalizations related to latitudinal variation in reproductive patterns and recruitment (Bauer and Vega 1992). The most common model for invertebrates from tropical areas is the continuous reproduction one, which becomes more restricted with increasing latitude (Sastry 1983, Grahame and Branch 1985).

The reproductive activity of hermit crabs has been analyzed based mainly on the frequency of appearance of ovigerous females during the year (Reese 1968, Ameyaw-Akumfi 1975, Asakura and Kikuchi 1984, Lancaster 1988, Imazu and Asakura 1994,

Manjón-Cabeza and Garcia-Raso 1995), and secondarily by gonad analysis (Varadarajan and Subramonian 1982, Lancaster 1990).

Although ecologically important, the reproductive biology of hermit crabs has been scarcely studied, particularly the southwestern Atlantic species. Only *Calcinus tibicen* has been studied by Fransozo and Mantelatto (1998), and Mantelatto and Garcia (1999) in the Ubatuba region.

Petrochirus diogenes (L.) can be considered the largest hermit crab that occupies gastropod shells in the western Atlantic (Gasparini and Floeter 1999). The first study related to this species addressed its larval development (Provenzano 1968). More recently, Bertini and Fransozo (1999a, b, c, 2000) studied the same species concerning its relative growth, spatial and seasonal distribution, population dynamics and patterns of shell utilization.

The aim of the present study is to determine the breeding season of *P. diogenes* based on the analysis of gonad development (macroscopic morphology) and to provide a better understanding of the reproductive biology of the species.

II. MATERIALS AND METHODS

Hermit crabs were collected in the Ubatuba region (between 23° 20' S to 23° 35' S and 44° 50' W to 45° 14' W) from 1995 to 1999 using a shrimp fishing boat

---

Kluwer Academic/Plenum Publishers

E. Escobar-Briones & F. Alvarez Eds.
MODERN APPROACHES TO THE STUDY OF CRUSTACEA
PP. 145-150

equipped with double-rig trawling nets. Hermit crabs were grouped according to seasons of the year: summer (January to March), fall (April to June), winter (July to September) and spring (October to December). After removal from the shell, the sex of each specimen was registered (based on gonopore position and number of pleopods) and the cephalothoracic shield length (SL = from the tip of the rostrum to the mid point of the posterior margin of the shield) was measured in millimeters with a caliper. Afterwards, the soft abdomen was dissected to verify gonad development stages for both sexes. Gonad observation was done under a stereomicroscope. Fifteen size classes of 2.5 mm wide were used for the analysis in a range from 2.5 to 40 mm of SL.

Table 1. *Description of developmental stages for male and female gonad of **Petrochirus diogenes** (see Fig. 1).*

	SEX	
Stages	**Males**	**Females**
Immature	Gonad thin and translucent	Ovaries pale and with no obvious coloration
Rudimentary	Gonad initial stages with two small white ducts	Ovaries cream to light yellow opaque
Developing	Initial of filament winding of coloration white pale	Ovaries are reddish orange large and conspicuous
Advanced	Strong winding of the testis with dominant visible of the coloration white pale and vas deferens is white	Ovaries are bright red. The dominant visible internal organ

III. RESULTS

A total of 999 hermit crabs were captured, of which 608 were males and 391 females. Twenty-four ovigerous females were collected in the following months: January (2), February (4), March (6), April (6), September (2), October (2), November (1) and December (1). Four gonad development stages were characterized for each sex: immature, rudimentary, developing and advanced (Table 1, Fig. 1). Physiologically the onset of sexual maturity occurred near 8.0 mm SL. The majority of ovigerous females were found presenting gonads in the ad-

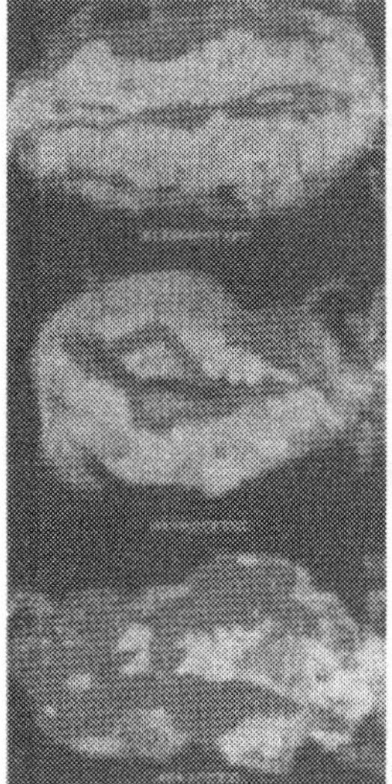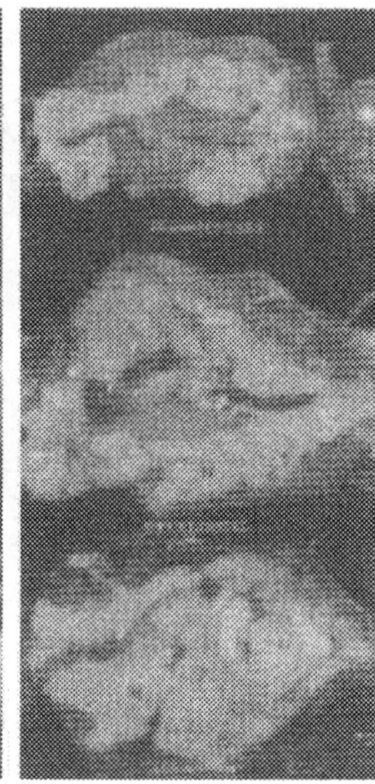

Figure 1. *P. diogenes. Dissected abdomen showing gonad development stages for each sex. (G = gonad; H = hepatopancreas). Animals in the figures are not of the same size.*

vanced stage of development. The recruitment peak (48%) occurred during winter. The abundance of each gonad development stage throughout the year is presented in Table 2.

In Table 3 the size of the hermit crabs in each interest group (juvenile males, adult males, juvenile females, non ovigerous adult females and ovigerous females) can be observed. The frequency distribution of hermit crabs in size classes related to gonad development stages shows that juveniles are present in the three first classes (Fig. 2). Males presented gonads in the advanced stage during all seasons. Although females presenting developed ovaries appear all year, the peak incidence is in the summer. The reproductive period of *P. diogenes* based on gonad analysis is described in Figure 3.

IV. DISCUSSION

Based on a macroscopical gonad analysis of *P. diogenes*, it was observed that both sexes reached physiological sexual maturity around 8.0 mm of SL. This result suggests a lower limit for sexual maturity compared to that proposed by Bertini and Fransozo (1999c) of 10 mm of SL, based on the smallest ovigerous females obtained.

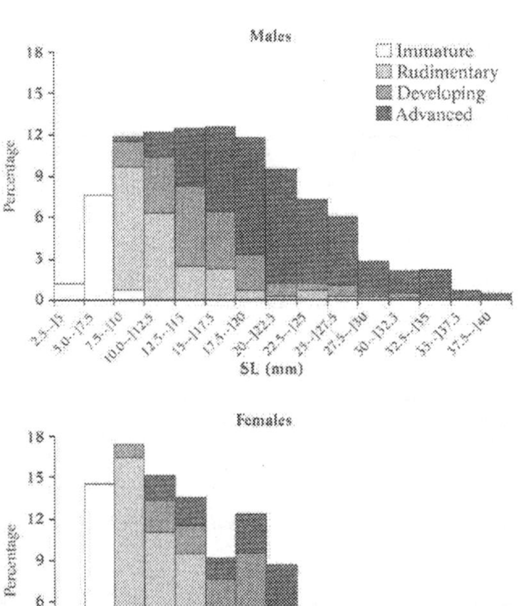

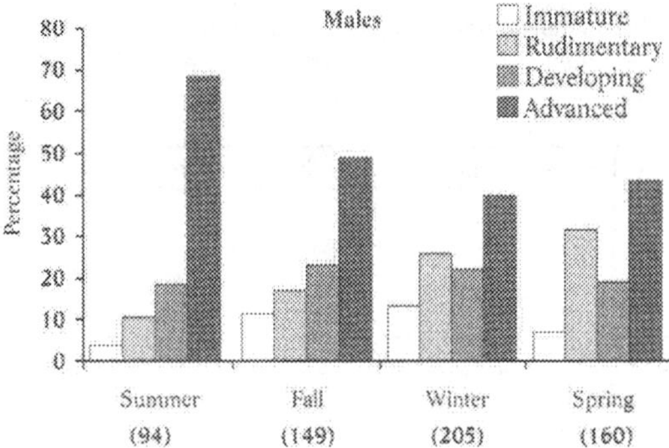

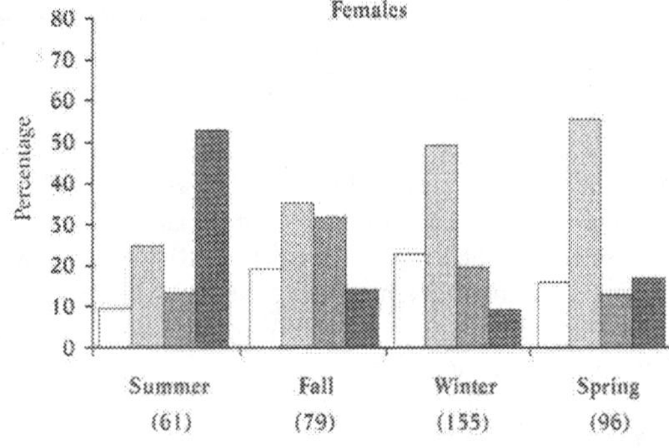

Figure 2. *P. diogenes. Size frequency distribution based on gonad development stages*

Figure 3. *P. diogenes. Frequency in each gonad development stage for each sex and season.*

For *P. diogenes*, it was observed that hepatic development in relation to gonad development is not an antagonistic process as it has been shown for brachyuran crabs. Varadarajan and Subramoniam (1982) studying the reproductive activity of *Clibanarius clibanarius*, found that the hepatic index was almost always higher and relatively more stable compared to the gonad index.

The majority (75%) of the ovigerous females of *P. diogenes* presented their gonads in an advanced stage of development, suggesting that reproduction takes place immediately after the hatching of the larvae. Such behavior has also been observed for *C. clibanarius* by Varadarajan and Subramonian (1982) and for *Pagurus bernhardus* by Lancaster (1990).

In the present study, it was found that the greatest incidence of females in the advanced go-nad development stage occurred during summer, which coincided with the highest abundance of ovigerous crabs. The absence of ovigerous females during May and August can be explained by the low number of hermit crabs with advanced gonad development, probably because of the short period of embryonic development of *P. diogenes*. The same was also observed for carideans by Bauer (1989).

Based on the gonad analysis and the presence of ovigerous females, it can be inferred that *P. diogenes* has a continuous reproduction pattern with a peak of activity during summer. According to Negreiros-Fransozo and Fransozo (1992), hermit crabs from subtropical regions can reproduce throughout the year with higher intensity during summer months. This trend may be correlated to favorable temperature and high food availability during that period, ensuring adequate conditions for larval development.

Table 2. *Petrochirus diogenes* *abundance throughout the study period (1995 to 1999) in relation to gonad developmental stage (IM = immature; RU = rudimentary; DE = developing; AD = advanced; Number in parenthesis means ovigerous females).*

Sex	Stages	Jan	Feb	Mar	Apr	May	Jun	Jul	Aug	Sep	Oct	Nov	Dec	Total
							Months							
Females	IM	3		3	2	7	6	6	24	5	8	4	3	71
Females	RU	8	4	3	11 (1)	7	10	21	42	13	23	11	19	172
Females	DE	5 (1)	2 (1)	1 (1)	5 (1)	13	7	19	10	1	5 (1)	4	3	75
Females	AD	8 (1)	12 (3)	12 (5)	6 (4)	2	2	4	7	3 (2)	3 (1)	3 (1)	11 (1)	73
Subtotal		24	18	19	24	29	25	50	83	22	39	22	36	391
Males	IM	1	1	1	4	8	5	7	15	5	4	4	3	58
Males	RU	7	3		4	12	9	19	31	2	17	10	23	137
Males	DE	6	8	3	16	11	7	15	21	9	10	6	14	126
Males	AD	23	30	11	24	19	30	50	23	8	13	19	37	287
Subtotal		37	42	15	48	50	51	91	90	24	44	39	77	608
Total		61	60	34	73	79	76	141	173	46	83	60	113	999

However, variations in the breeding season can be related to several other factors, such as: availability of food for adults (Goodbody 1965), larval ecology (Reese 1968), shell supply (Bertness 1981), synchrony among sexual maturity, mating, gonad development and incubation period (Sastry 1983) and sea water temperature (Lancaster 1990).

Continuous reproduction with breeding peaks is a well-known pattern present in many other tropical marine crustaceans (e.g., Goodbody 1965, Reese 1968, Nyblade 1987, Tunberg et al. 1994). The lack of ovigerous females in certain periods (winter) was also observed in the same region for *Calcinus tibicen* (Fransozo and Mantelatto, 1998). Other species of hermit crabs reproduce mainly during the winter, such as *Pagurus geminus* and *Pagurus lanuginosus* (Imazu and Asakura 1994), and *Discorsopagurus schmitti* (Gherardi and Cassidy 1995). However, the reproductive cycle of those hermit crabs was based only on the presence of ovigerous females, lacking data on gonad development.

Lancaster (1990), who studied the reproductive cycle of *Pagurus bernhardus* in England, based on the analysis of gonads and frequency of ovigerous females, evidenced a seasonal reproductive cycle with higher breeding activity during the winter. The author also showed that females caught in summer, and bearing rudimentary gonads, can develop into fully mature individuals if maintained in temperature and photoperiod conditions typical of winter months. These factors might operate in the opposite way for *P. diogenes* causing a decrease of breeding activity under low temperature and photoperiod, assumed to be adverse environmental conditions.

Table III. *Petrochirus diogenes* *size (mm) based on cephalothoracic shield length.*

Sex	N	Mean ± sd	Minimum	Maximum
Juvenile males	58	6.3 ± 1.0	4.0	7.8
Adult males	550	17.9 ± 6.7	8.0	40.0
Juvenile females	71	6.3 ± 1.0	3.6	8.0
Non ovigerous adult females	296	14.4 ± 4.5	8.2	32.1
Ovigerous females	24	18.3 ± 5.4	9.6	29.3

Giese (1959) mentioned that a prolonged reproductive period indicates that individuals produce several successive broods during the year or breed asynchronously, that is some are in the earlier stages of maturation, some are getting ready to spawn, some are spawning, and still others are already spent. Such reproductive pattern can be applied to *P. diogenes*, since individuals in all the different gonad developmental stages are found all year-round.

According to Provenzano (1968), the larval development of *P. diogenes* lasts 37 to 50 days, at 25°C, with the cephalothoracic length of the megalopae varying from 0.98 to 1.34 mm. Considering other studies on juvenile development, e.g., *Clibanarius sclopetarius*, *Eriphia gonagra* and *Petrolisthes armatus* studied by Brossi-Garcia (1987), Fransozo and Negreiros-Fransozo (1987) and Brossi-Garcia and Moreira, (1996), respectively, it can be inferred that molt increments for such decapods in the first stages is around 12 to 15%. Taking into account that the intermolt period is shorter in the first juvenile stages and becomes longer in subsequent stages, it can be inferred that it takes approximately four months for *P. diogenes* to attain the size (SL = 3.6 mm) of the smallest captured specimen, and can go through seven molts before reaching this size.

Petrochirus diogenes presented differential recruitment rates during all seasons, but the highest peak occurred during the winter. Different reproductive seasons can result in distinct recruitment periods, reducing the inter and intraspecific competition for shells.

ACKNOWLEDGEMENTS

We are grateful to the FAPESP (Fundação de Amparo à Pesquisa do Estado de São Paulo) and CNPq (Conselho Nacional de Desenvolvimento Científico e Tecnológico). We are also thankful to the NEBECC coworkers for their help during fieldwork and to Dr. Maria Lucia Negreiros Fransozo who made helpful comments on earlier drafts of the manuscript.

REFERENCES

Ameyaw-Akumfi C (1975) The breeding biology of two sympatric species of tropical intertidal hermit crabs, *Clibanarius chapini* and *C. senegalensis*. Mar Biol 29:15-28

Asakura A and Kikuchi T (1984) Population ecology of the sand dwelling hermit crab, *Diogenes nitidimanus* Terao 2. Migration and life history. Publ Amakusa Mar Biol Lab Kyushu Univ 7:109-123

Bauer RT (1989) Continuous reproduction and episodic recruitment in nine shrimp species inhabiting a tropical seagrass meadow. J Exp Mar Biol Ecol 127:175-187

Bauer RT and Vega LWR (1992) Pattern of reproduction and recruitment in two sicyoniid shrimp species (Decapoda: Penaeoidea) from a tropical seagrass habitat. J Exp Mar Biol Ecol 161:223-240

Bertini G and Fransozo A (1999a) Relative growth of *Petrochirus diogenes* (Crustacea, Anomura, Diogenidae) in the Ubatuba region, São Paulo, Brazil. Rev Brasil Biol 59:617-625

Bertini G and Fransozo A (1999b) Spatial and temporal distribution of *Petrochirus diogenes* (Anomura, Diogenidae) in the Ubatuba Bay, São Paulo, Brazil. Iheringia, Sér Zool 86:145-150

Bertini G and Fransozo A (1999c) Population dynamics of *Petrochirus diogenes* (Crustacea, Anomura, Diogenidae) in the Ubatuba region, São Paulo, Brazil. (pp 331-342) In: Klein JCVV and Schram FR (eds) The Biodiversity Crisis and Crustacea, Crustacean Issues 12, AA Balkema, Rotterdam, Brookfield

Bertini G and Fransozo A (2000) Patterns of shell utilization in *Petrochirus diogenes* (Linnaeus, 1758) (Decapoda, Anomura, Diogenidae) in the Ubatuba region, São Paulo, Brazil. J Crust Biol 20:470-475

Bertness MD (1981) Pattern and plasticity in tropical hermit crab growth and reproduction. Am Nat 117:754-773

Brossi-Garcia AL (1987) Juvenile development of *Clibanarius sclopetarius* (Herbst, 1796) (Crustacea, Paguridae, Diogenidae) in the laboratory. J Crust Biol 7:338-357

Brossi-Garcia AL and Moreira RG (1996) Estudos biométricos e morfológicos dos primeiros estágios juvenis de *Petrolisthes armatus* (Gibbes, 1850) (Decapoda, Porcellanidae) em laboratório. Rev Brasil Biol 56:231-243

Fransozo A and Negreiros-Fransozo ML (1987) Morfologia dos primeiros estágios juvenis de *Eriphia gonagra* (Fabricius, 1781) e *Eurypanopeus abbreviatus* (Stimpson, 1860) (Crustacea, Decapoda, Xanthidae), obtidos em laboratório. Pap Avulsos Zool 36:257-277

Fransozo A and Mantelatto FLM (1998) Population structure and reproductive period of the tropical hermit crab *Calcinus tibicen* (Decapoda, Diogenidae) in the Ubatuba region, São Paulo, Brazil. J Crust Biol 18:446-452

Gasparini JL and Floeter SR (1999) The giant hermit crab *Petrochirus diogenes* (Linnaeus, 1758) (Crustacea: Anomura) maximum size. Nauplius 7:185-186

Gherardi F and Cassidy PM (1995) Life history patterns of *Discorsopagurus schmitti*, a hermit crab inhabiting polychaete tubes. Biol Bull 188:68-77

Giese AC (1959) Comparative physiology: annual reproductive cycles of marine invertebrates. A Rev Physiol 21:547-576

Goodbody H (1965) Continuous breeding in populations of two tropical crustaceans, *Mysidium columbiae* (Zimmer) and *Emerita portoricensis* Schmidt. Ecology 46:195-197

Grahame J and Branch GM (1985) Reproductive patterns of marine invertebrates. Oceanogr Mar Biol Annu Rev 23:373-398

Imazu M and Asakura A (1994) Distribution, reproduction and shell utilization patterns in three species of intertidal hermit crabs on a rocky shore on the Pacific coast of Japan. J Exp Mar Biol Ecol 184:41-65

Lancaster I (1988) *Pagurus bernhardus* (L.) - An introduction to the natural history of hermit crabs. Field Studies 7:189-238

Lancaster I (1990) Reproduction and life history strategy of the hermit crab *Pagurus bernhardus*. J Mar Biol Ass UK 70:129-142

Manjón-Cabeza ME and García Raso JE (1995) Study of a population of *Calcinus tubularis* (Crustacea, Diogenidae) from a shallow *Posidonia oceanica* meadow. Cah Biol Mar 36:277-284

Mantelatto FLM and Garcia RG (1999) Reproductive potential of the hermit crab *Calcinus tibicen* (Anomura) from Ubatuba, São Paulo, Brazil. J Crust Biol 19:268-275

Negreiros-Fransozo ML and Fransozo A (1992) Estrutura populacional e relação com a concha em *Paguristes tortugae* Schimtt, 1933 (Decapoda, Diogenidae), no litoral norte do estado de São Paulo, Brasil. Naturalia 17:31-42

Nyblabe CF (1987) Phylum or Subphylum Crustacea, class Malacostraca, order Decapoda, Anomura. (pp 441-450) In: Strathmann MF (ed) Reproduction and development of marine invertebrates of the northen Pacific coast. University of Washington Press, Seattle, Washington

Provenzano AJ (1968) The complete larval development of the West Indian hermit crab *Petrochirus diogenes* (L.) (Decapoda, Diogenidae) reared in the laboratory. Bull Mar Sci 18:143-181

Reese ES (1968) Annual breeding seasons of three sympatric species of tropical intertidal hermit crabs, with a discussion of factors controlling breeding. J Exp Mar Biol Ecol 2:308-318

Sastry AN (1983) Ecological aspects of reproduction. (pp 79-270) In: Waterman TH (ed) Biology of Crustacea. VIII Environmental adaptations. Academic Press

Tunberg BG, Nelson EG and Smith G (1994) Population ecology of *Pagurus maclaughlinae* García-Gomes (Decapoda: Anomura: Paguridae) in the Indian River Lagoon, Florida. J Crust Biol 14:686-699

Varadarajan S and Subramonian T (1982) Reproduction of the continuously breeding tropical hermit crab *Clibanarius clibanarius*. Mar Ecol Prog Ser 8:197-201

DISTRIBUTION AND ABUNDANCE OF THE FAMILY CALAPPIDAE OFF THE COAST OF JALISCO AND COLIMA, MEXICO

Judith Arciniega Flores and Victor Landa Jaime

Centro de Ecología Costera, Universidad de Guadalajara
Gómez Farías No. 82, San Patricio-Melaque, 48980 Jalisco, México

judith@nautilus.melaque.udg.mx

ABSTRACT

Samples of the family Calappidae were collected from soft bottoms on the continental shelf off the coast of Jalisco and Colima, Mexico. The examined material (2300 specimens) comes from five cruises, conducted during 1995 and 1996. The samples were obtained with shrimp trawls, at seven collection stations along the coast and four depths: 20, 40, 60 and 80 m. Five genera and six species were identified: *Calappa convexa, C. saussurei, Cycloes bairdii, Hepatus kossmanni, Osachila lata* and *Platymera gaudichaudi*. The most abundant species was *C. bairdii* (1269 specimens) followed by *H. kossmanni* (551) and *P. gaudichaudi* (387). The highest abundance (892 specimens) was registered during the June-July 1996 cruise, capturing a higher number of specimens at 60 m and the lowest number (112) was collected during the November-December 1996 cruise at 20 m depth. Despite the significant variation in abundance, as well as in the number of species found from one cruise to the other, members of the family Calappidae are widely distributed from Cuitzmala, Jalisco to Cuyutlan, Colima.

I. INTRODUCTION

Thirteen species in the family Calappidae have been recorded from the eastern tropical Pacific (Hendrickx 1995). Members of this family (e.g., *Hepatus kossmanni, Platymera gaudichaudi* and *Calappa convexa*) are some of the most common crabs of the crustacean communities that inhabit the continental shelf in the Mexican Pacific (Hendrickx 1993, 1997). Although these species are common, they are not economically important, essentially because of their reduced size and the lack of a local market demand. However, some species, like *C. convexa*, are used by local artisans in some areas of the eastern tropical Pacific, while other species could represent a potential food source in the future (Hendrickx 1995). From a biogeographical and an ecological perspective, there is a lack of knowledge about this species, in contrast with systematic studies, which have been more common in the past years. Thus, there is a need for ecological studies based on systematic sampling programs for a great number of species that inhabit the tropical Pacific region. Such studies will not only help to complement carcinological inventories, but will also provide valuable ecological information necessary for the conservation, management and possible exploitation of these species.

II. MATERIALS AND METHODS

Specimens of the family Calappidae were collected during five cruises which took place in 1995 and 1996 (May-June 1995, November-December 1995, March 1996, June-July 1996 and November-December 1996) along the coasts of Jalisco and Colima, Mexico, on the ship BIP-V of the Coastal Ecology Center, of the University of Guadalajara. The night trawls took place on soft bottoms, at seven sampling stations and four different depths (20 m, 40 m, 60 m and 80 m) (Fig. 1). Semi-portuguese shrimp nets were used, with an estimated mouth opening of 6.9 m, a headline

E. Escobar-Briones & F. Alvarez Eds.
MODERN APPROACHES TO THE STUDY OF CRUSTACEA
PP. 151-156

height of 1.15 m and a stretched mesh size at the cod-end of 38 mm. Each trawl lasted 30 minutes at an average speed of two knots. The samples were frozen and processed in the laboratory. The total number of individuals per species was recorded and identified according to Rathbun (1937), Rodríguez de la Cruz (1987) and Hendrickx (1997). An analysis of variance (ANOVA) was made to test for differences in the capture among sampling stations, depths, and time of the year. A Spearman correlation analysis (Zar 1996) was conducted to determine if abiotic variables as temperature, salinity and dissolved oxygen concentration, were associated to the number of individuals captured per species.

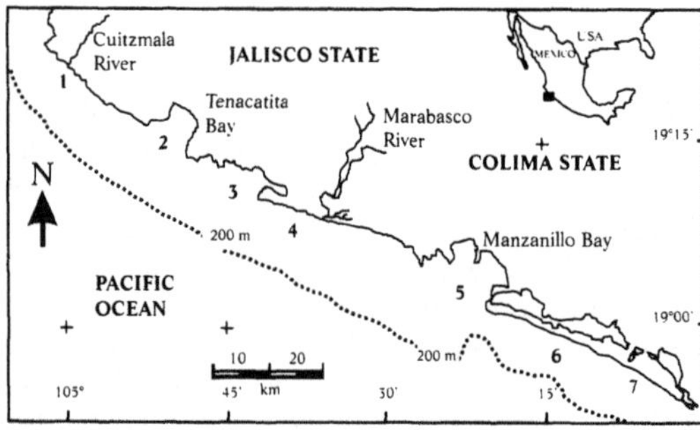

Figure 1. *Study area and sampling stations on the continental shelves off Jalisco and Colima coast during 1995-1996.*

III. RESULTS

Six of the 10 species recorded in the Mexican Pacific region were found along the Jalisco and Colima coastline: *Calappa convexa, C. saussurei, Cycloes bairdii, Hepatus kossmanni, Osachila lata,* and *Platymera gaudichaudi.* Overall, the most abundant species was *C. bairdii* (1269 organisms). The June-July 1996 cruise produced the most abundant sample of calappids with 892 individuals.

The spatial and temporal distribution of calappids varied from one cruise to the other (Fig. 2). The most abundant samples were obtained during the May-June 1995 and the June-July 1996 cruises, at a depth of 60 m. Stations 2 and 5 contributed with the largest samples, as well as station 4 in the June-July 1996 cruise. Nevertheless, the ANOVA results indicate that only during the May-June 1995 cruise, the number of calappids was significantly different among the different depths ($F_{3,15}$= 4.151, P = 0.025) and sites ($F_{6,15}$= 3.021, P = 0.0385).

The species presented noticeable variations in their distributions, according to the sampling station and depth. Significant differences in bathymetric distribution were obtained for *C. bairdii* ($F_{3,43}$= 5.745, P = 0.0021) and *P. gaudichaudi* ($F_{3,15}$= 3.443, P = 0.0439) which were more common at 60 m and 80 m, respectively. *Cycloes bairdii* was significantly more abundant during the June-July 1996 cruise along station 5; *Hepatus kossmanni* was more numerous during the May-June 1995 cruise in station 1; and *P. gaudichaudi* during the March 1996 cruise in station 2 (Fig. 3). These three species were present at all depths. *Cycloes bairdii* and *H. kossmanni* were more abundant at 60 m while *P. gaudichaudi* at 80 m (Fig. 4).

Platymera gaudichaudi abundance showed a significant negative correlation with temperature during the May-June 1995 cruise (P = 0.005) and the November-December 1995 cruise (P = 0.031). During the June-July 1996 cruise, positive correlations were found with temperature (P = 0.004) and dissolved oxygen (P = 0.006). During November-December 1996 cruise *C. bairdii* abundance was significantly correlated to all parameters: temperature (P = 0.026), salinity (P = 0.013), and dissolved oxygen (P = 0.014). *Hepatus kossmanni* abundance was correlated with temperature in the May-June 1995 cruise (P = 0.041). The remaining species did not show significant correlation with any of the abiotic variables.

The temperature recorded at depths between 20 and 80 m fluctuated considerably from one cruise to the other (14-28.5°C). During the March 1996 cruise the lowest temperatures of the sampling period were recorded, and two high abundance areas were found. One at a temperature between 14-18°C with a 1 ml/l dissolved oxygen concentration in which the two dominant species were *O. lata* and *P. gaudichaudi,* and the other at temperatures between 19 and 22°C in which the most numerous species were *C. bairdii* and *H. kossmanni* (Fig. 5). During the June-July 1996 cruise the highest abundance values appeared between 23 and 28°C with dissolved oxygen concentrations which varied between 3-6 ml/l, and again the dominant species were *C. bairdii* and *H. kossmanni.* The November-December 1996 cruise had similar results to the June-July 1996 cruise (Fig. 5).

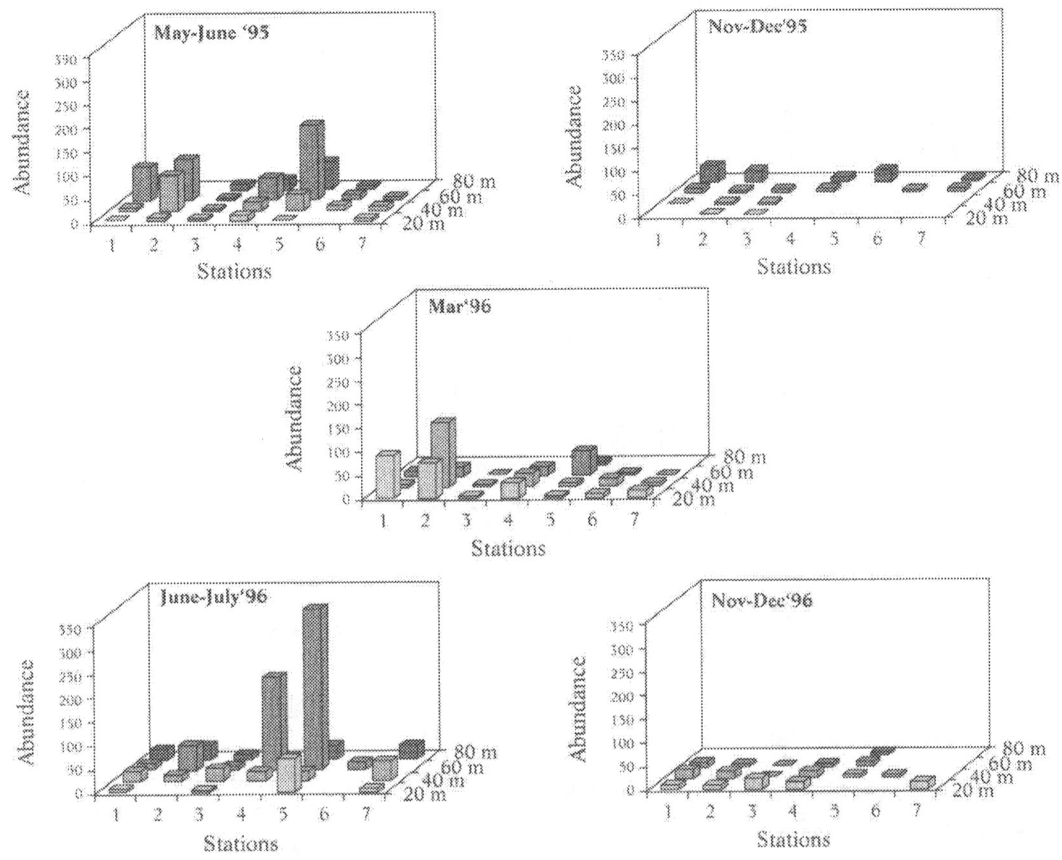

Figure 2. *Total abundance of calappids from the five cruises off Jalisco and Colima during 1995-1996.*

IV. DISCUSSION

Cycloes bairdii was the most abundant and frequently collected species throughout the study. In contrast, in the Gulf of Tehuantepec *H. kossmanni* was the most abundant calappid followed by *P. gaudichaudi* while only one individual of *C. bairdii* was collected (Hendrickx and Vázquez-Cureño 1998). Comparatively, in the Jalisco and Colima continental shelf, *H. kossmanni* was the second most abundant species. Hendrickx (1997) has reported both species as very abundant in the shrimp by-catch along the Mexican Pacific. In southern Sinaloa, *C. convexa* is very abundant and common in traps and gill-nets used in the fishery of the spiny lobster *Panulirus* (Ayón-Parente 1997). *Calappa convexa* is a very frequent species captured commonly with penaeid shrimps in the eastern tropical Pacific (Hendrickx 1995); however, in the coast of Jalisco and Colima it resulted very scarce, only 40 individuals were collected during the 1995 and 1996 cruises.

The specimens of *H. kossmanni* and *P. gaudichaudi* captured in this study represented the first records in the waters off Jalisco and Colima. Although both species have considerable distribution areas (*H. kossmanni* from the Gulf of California, Mexico to La Libertad, Ecuador and *P. gaudichaudi* from the Gulf of California, Mexico to Chile), there were no previous records for either species in this zone.

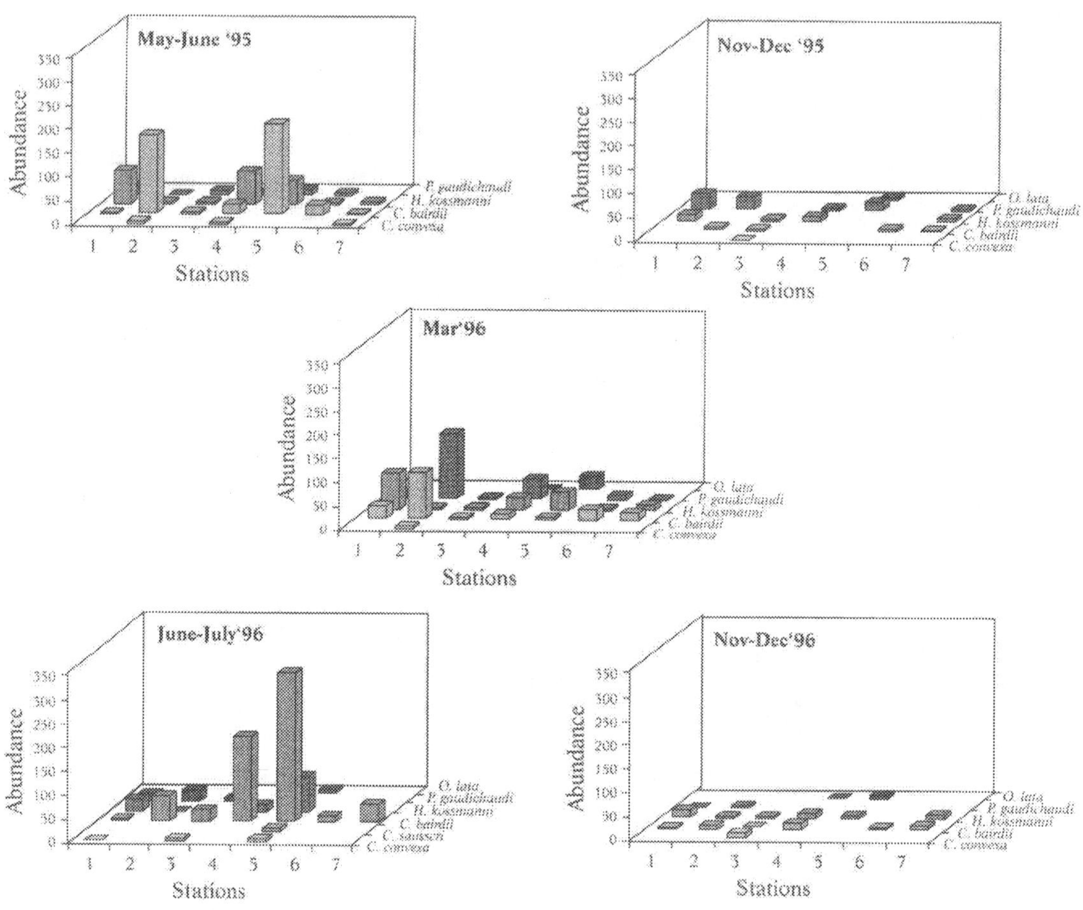

Figure 3. *Abundance of calappids by species and sampling stations off Jalisco and Colima during 1995-1996.*

All organisms used in this study were collected between 20 and 80 m, a range that represents the known depths at which these species have been found elsewhere (Hendrickx 1997). Spatial variation of abundance of calappids was affected by depth, with the highest number of organisms captured at 60 m. *Platymera gaudichaudi* was most abundant at 80 m, agreeing with the available records for the Gulf of Tehuantepec (Hendrickx and Vázquez-Cureño 1998); *H. kossmanni*, however, was most abundant at 60 m, contrasting with the Gulf of Tehuantepec records, where the highest abundances were recorded between 23 and 28 m.

The temperature at which calappid species were collected in this study ranged from 14 to 28.5°C. Hendrickx (1997) reported *C. convexa* to occur at 13.5°C, *C. bairdii* at 13.5-21°C and *H. kossmanni* between 13.5-25.5°C. However, our results show that *C. convexa* occurs at 20-28.5°C and *C. bairdii* and *H. kossmanni* at 14-28°C.

The depth and temperature records obtained show that these species can adapt locally to a wide range of conditions, allowing them to exploit the continental shelf of most of the tropical and subtropical eastern Pacific.

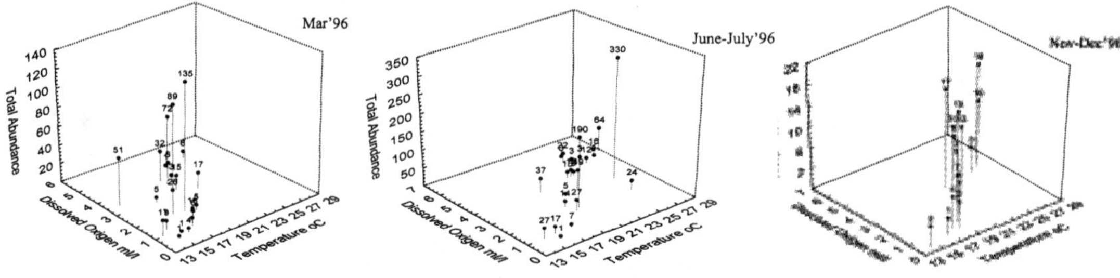

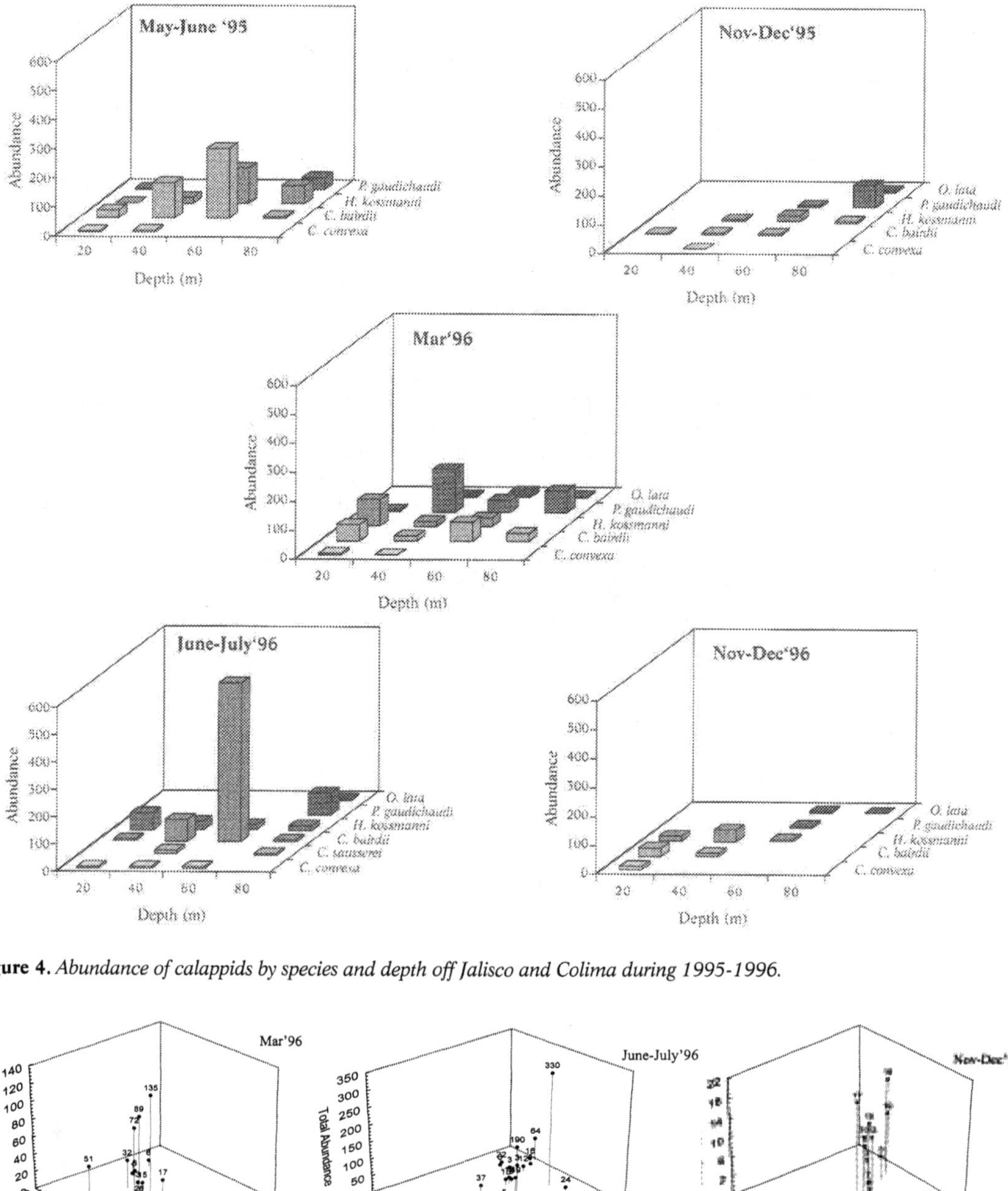

Figure 4. *Abundance of calappids by species and depth off Jalisco and Colima during 1995-1996.*

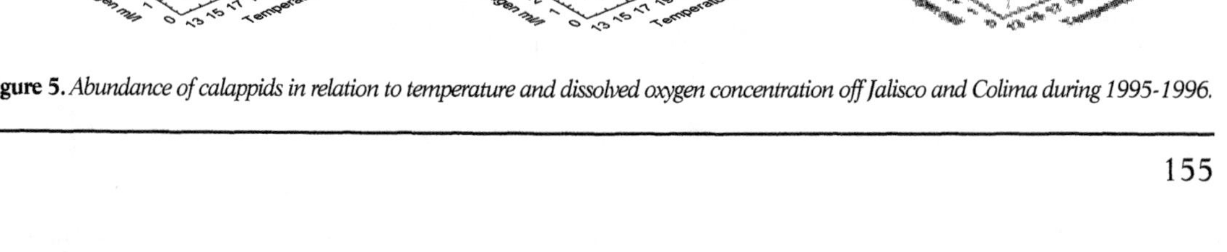

Figure 5. *Abundance of calappids in relation to temperature and dissolved oxygen concentration off Jalisco and Colima during 1995-1996.*

ACKNOWLEDGEMENTS

We thank our colleagues at the Coastal Ecology Center and the students who participated in the separation of the samples. We also thank the crew of the BIP-V for the effort made to obtain the material examined and especially to Siân Gower and José Mariscal for their suggestions and review of the manuscript. This study was made under permit SEMARNAP-150995-214-03 for fisheries foment and research.

REFERENCES

Ayón-Parente M (1997) Crustáceos decápodos asociados a la captura de la langosta *Panulirus* spp (White, 1847) en el sur de Sinaloa. Tesis de Licenciatura. Univ Autón Sinaloa Fac Cienc Mar 92 p

Hendrickx ME (1993) Crustáceos decápodos del Pacífico mexicano. In: Salazar-Vallejo SI and González NE (eds) Biodiversidad marina y costera de México (pp 271-318) CONABIO and CIQRO, México

Hendrickx ME (1995) Cangrejos. In: Fisher W, Krupp F, Schneider W, Sommer C, Carpenter KE and Niem VH (eds) Guía FAO para la identificación de especies para los fines de la pesca, Pacífico centro-oriental. Vol 1 (pp 565-636). Plantas e Invertebrados, FAO Roma

Hendrickx ME (1997) Los cangrejos braquiuros (Crustacea: Brachyura: Dromiidae, hasta Leucosiidae) del Pacífico Mexicano. CONABIO and ICMyL, UNAM, México 178 p

Hendrickx ME and Vázquez-Cureño LA (1998) Composition and zoogeographical affinities of the stomatopod and decapod crustacean fauna collected during the Ceemex P4 Cruise in the Gulf of Tehuantepec, Mexico. Bull Inst Royal des Sciences naturelles de Belgique, Biologie 68:135-144

Rathbun MJ (1937) The oxystomatous and allied crabs of America. US Nat Mus Bull 166:1-278

Rodríguez de la Cruz MC (1987) Crustáceos decápodos del Golfo de California. Secretaría de Pesca, México 306 p

Zar JH (1996) Biostatistical Analysis. Third Ed. Prentice-Hall, Inc., Englewood Cliffs, New Jersey 662 p

SEXUAL MATURITY OF *EURYTIUM LIMOSUM* (SAY, 1818) FROM A SUBTROPICAL MANGROVE IN BRAZIL

Fernanda Jordão Guimarães and Maria Lucia Negreiros-Fransozo

NEBECC, Group of Studies on Crustacean Biology, Ecology and Culture
Depto. de Zoologia, IBB, UNESP C.P. 510
Botucatu, SP, Brazil

fguimaraes@laser.com.br

ABSTRACT

The size at onset of maturity of *Eurytium limosum* from a subtropical mangrove in Brazil was investigated. In this species, sexual maturity for males can be indicated by the allometric growth and gonopod length. For females, the morphological sexual maturity can only be externally verified through the relative quantity of setae along the abdominal margins and pleopods. Internally, gonad development was also examined. The size at which half of the population is physiologically mature was 11.6 mm of CW for females and 12.3 mm of CW for males. The values for the morphological and physiological maturity are very similar, indicating that the development of the secondary sexual characters is synchronized with the achievement of the physiological maturity for *E. limosum*.

I. INTRODUCTION

Estimates of size at the onset of sexual maturity, mainly in the case of edible crab species, are among the most important information concerning a given population (Pinheiro and Fransozo 1998). Several reproductive and growth patterns of crustaceans are adaptive, and have been shaped by evolutionary pressures to maximize the survival of offspring in the next generation (Hartnoll and Gould 1988). The onset of sexual maturity, defined as a measure of size, can vary considerably (Fonteles-Filho 1989, Hines 1989).

Among brachyuran crabs the assessment of sexual maturity has been accomplished by associating the developmental stage of the gonads with the size of individuals (Santos and Negreiros-Fransozo 1996, Mantelatto and Fransozo 1994, 1999, Pinheiro and Fransozo 1998, Reigada and Negreiros-Fransozo 1999). Morphometric data have also been extensively used in Crustacea for studies on relative growth (Hartnoll 1974, 1982), specifically to detect changes of allometric levels on relative growth analyses, which are known to be related to some ontogenetic events, namely sexual maturity, when crabs undergo the puberty molt.

In spite of the abundance of papers on the reproductive biology of brachyuran crabs, very little is known in the case of xanthids. Among the available data, the papers by Finney and Abele (1981) on *Trapezia ferruginea*, Vannini and Gherardi (1988) on *Eriphia smithi* and Góes and Fransozo (1997) on *Eriphia gonagra* provide useful information on this issue.

The xanthid *Eurytium limosum* (Say, 1818) is a mud crab commonly found in estuarine or mangrove areas. The species is particularly abundant inside burrows near mangrove roots along the northern coast of the State of São Paulo, Brazil. According to Melo (1996), *E. limosum* is found in the Western Atlantic, from Florida (USA) to Santa Catarina (Brazil). The present paper, provides estimates of the size at the onset of sexual maturity for a subtropical population of *E. limosum*.

Kluwer Academic/Plenum Publishers

E. Escobar-Briones & F. Alvarez Eds.
MODERN APPROACHES TO THE STUDY OF CRUSTACEA
PP. 157-161

II. MATERIALS AND METHODS

Monthly collections, from January to December 1999, were carried out during ebbing tides at the mangrove area bordering the Guaratuba River (23° 45' S, 45° 53' W), located in Bertioga, State of São Paulo, Brazil.

Each specimen was sexed and the following measures taken: carapace width (CW), chelar propodus length (PL), chelar propodus height (PH), abdomen width (AW) and, in the case of males, gonopod length (GL). All crabs were dissected in order to verify the gonad development by macroscopic analysis (Costa and Negreiros- Fransozo 1998, Pinheiro and Fransozo 1999) and to determine the size at which half of the population is physiologically mature ($CW_{50\%}$) in both sexes (Vazoller 1982).

Table 1. *Relationships between morphometric parameters and carapace width in* **E. limosum** *using allometric equation.*

Variable	Sex	N	$Y=a.b^X$	r^2	Allometry level
PL	TM	246	$PL=0.5804CW^{1.0946}$	0.9885	+
	TF	280	$PL=0.6470CW^{1.0292}$	0.9632	=
PH	TM	246	$PH=0.2313CW^{1.1482}$	0.9578	+
	TF	274	$PH=0.2917CW^{1.0371}$	0.9010	=
	JM	57	$AW=0.1669CW^{1.001}$	0.7821	=
AW	AM	183	$AW=0.2861CW^{0.7818}$	0.9095	-
	TF	282	$AW=0.1319CW^{1.2407}$	0.9495	+
GL	JM	57	$GL=0.0617CW^{1.7051}$	0.8125	+
	AM	166	$GL=0.4015CW^{0.9051}$	0.9496	-

M= males; F= females; J = juvenile; A = adult; T = total; r^2 = determination coefficient,PL= cheliped propodus length, PH= cheliped propodus height, AW= Abdomen width, GL= 2nd gonopod length and CW= carapace with.

The allometric function, $Y = aX^b$ (Huxley 1950), was used to estimate the allometric growth of all dimensions in relation to carapace size. Determination coefficients and Student's t-test for departures from isometry (Sokal and Rohlf 1979) were calculated for all relationships. Analyses of covariance (ANCOVA) were performed to compare the allometry level between adults and juveniles (Zar 1996). The abdomen and pleopods of adult and juvenile females were observed under a stereomicroscope and their morphological differences described.

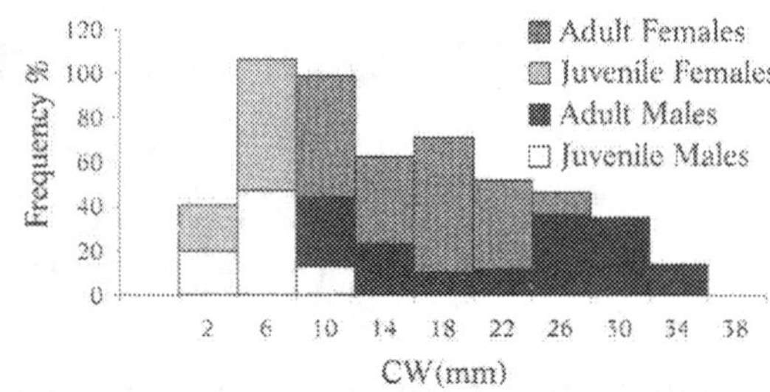

Figure 1. *Size frequency distribution of sampled individuals of E. **limosum** in Guaratuba Mangrove, Bertioga, Brazil.*

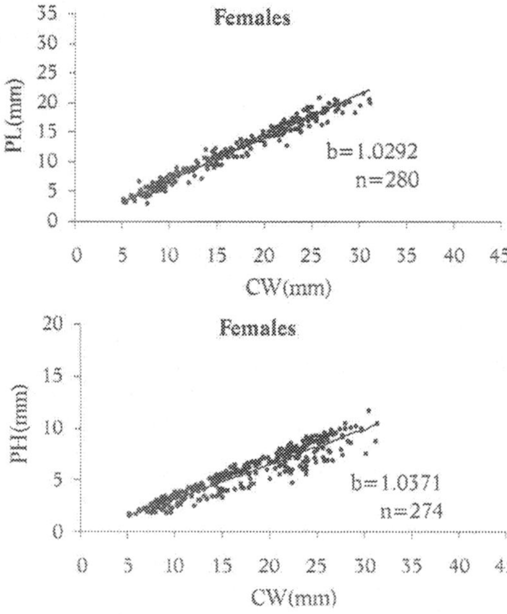

Figure 2. *Eurytium limosum morphometric relationship between chelar propodus length (PL) and chelar propodus height (PH) with carapace width for females (b = slope, n = number of crabs).*

III. RESULTS

A total of 522 crabs was captured, including 240 males and 282 females. Size ranged from 6.3 to 40.1 and 5.2 to 31.2 mm CW, for males and females, respectively (Fig. 1). The smallest ovigerous female obtained measured 12.5 mm CW.

The allometric function describing cheliped dimensions (PL and PH) shows an isometric growth for females and a positive allometric growth for males (Figs. 2, 3). However, they did not show any difference in the relative growth rate during the maturation phase for both sexes. The relationships obtained from the regressions are presented in Table 1.

For males, the abdomen allometric growth differs considerably between juveniles (b = 1.001) and adult crabs (b = 0.7818) (Fig. 4). With reference to the relative growth of gonopods (Fig. 5), a positive allometric growth was found for juveniles (b = 1.7051) and a negative allometric growth (b = 0.9051) for adults. In both relationships, the size at the onset of sexual maturity is estimated to vary from 12 to 13 mm CW.

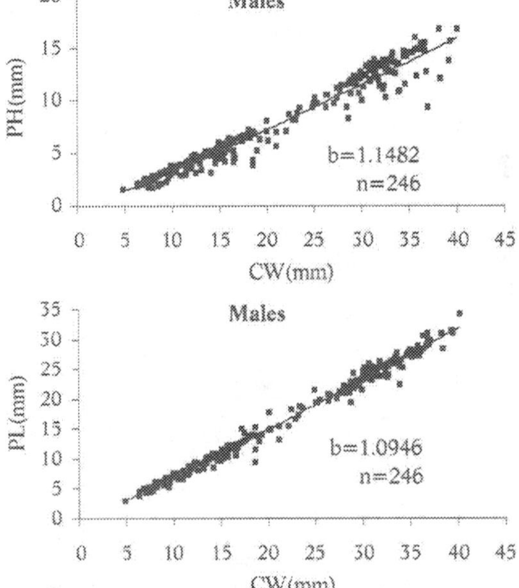

Figure 3. **Eurytium limosum** *morphometric relationship between chelar propodus length (PL) and chelar propodus height (PH) with carapace width for males (b = slope, n = number of crabs).*

For juvenile and adult females, there is no difference in the abdomen relative growth (Fig. 4). However, mature females bearing developing gonads present more setose abdominal margins and pleopods (Fig. 6), compared to immature females (Fig. 7). An analysis of such morphological features indicates that sexual maturity in females may be achieved between 11 and 12 mm CW.

Regarding the physiological maturity, the size at which 50% of the population is already mature is 11.6 mm of CW for females and 12.3 mm of CW for males (Fig. 8).

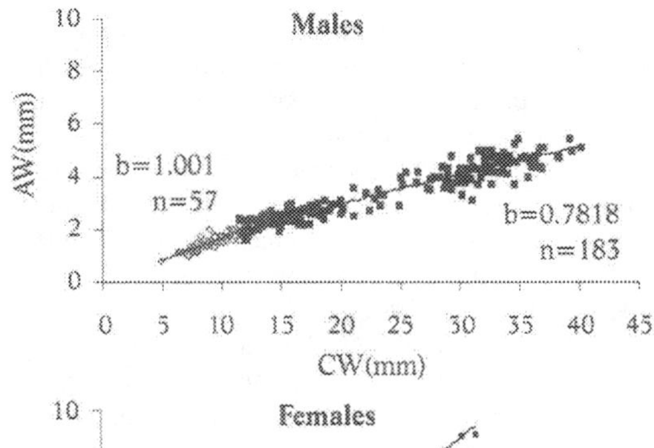

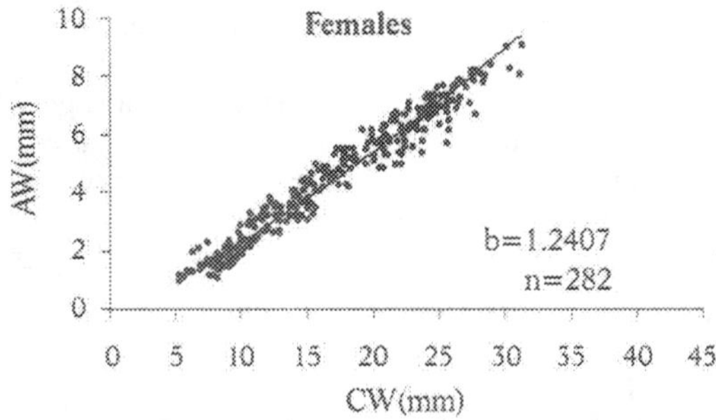

Figure 4. **Eurytium limosum** *morphometric relationships between abdomen and carapace width for both sexes (b = slope, n = number of crabs).*

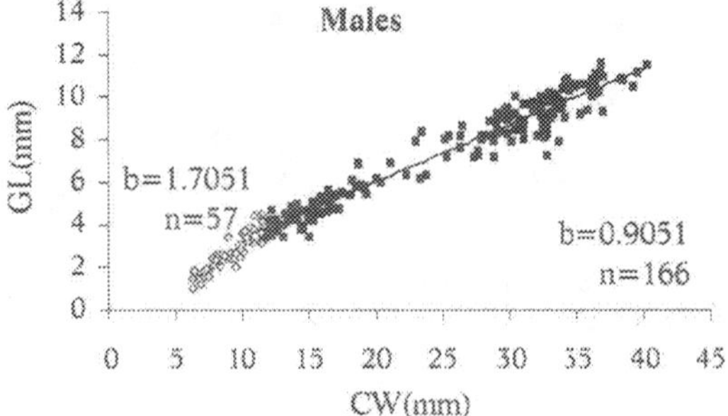

Figure 5. **Eurytium limosum** *male morphometric relationship between gonopod length and carapace width for males (b = slope, n = number of crabs).*

IV. DISCUSSION

Besides the larger size attained by males, sexual dimorphism in *E. limosum* is restricted to the abdominal morphology. Among brachyuran crabs, males usually mature at a larger size than females. This fact has been attributed to mating behavior since males protect females during pre- and post-copulatory events. It is usually assumed that larger males may more efficiently perform guarding.

For the allometry of both chelar propodus height and length, isometric growth was found for females and positive allometric growth for males. This pattern is commonly found among brachyurans. Positive allometry in males has been attributed to the utilization of chelipeds during agonistic behavior, territorial combats and courtship, when chelipeds are displayed to females (Hartnoll 1974, Vannini and Gherardi 1988, Góes and Fransozo 1997).

The relative growth of the abdomen commonly presents remarkable differences during the ontogeny of brachyurans. However, this transition separating growth phases occurs only for males in *E. limosum*. The function of the male abdomen is to cover and protect the gonopods, for what its relative growth pattern may be remarkably similar to the allometric growth of the first pleopods (Hartnoll 1974).

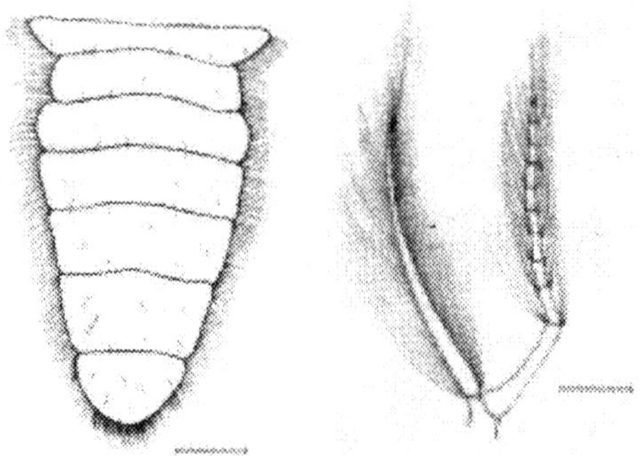

Figure 6. *Abdomen and second pleopod of mature females of* **E. limosum** *(scale bars = 1 mm).*

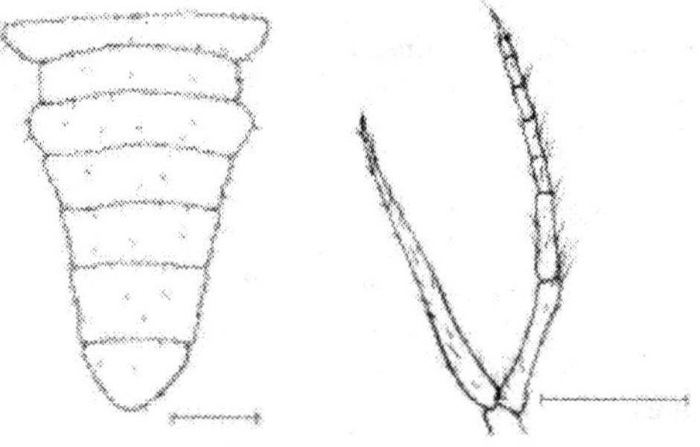

Figure 7. *Abdomen and second pleopod of immature females of* **E. limosum** *(scale bars = 1 mm).*

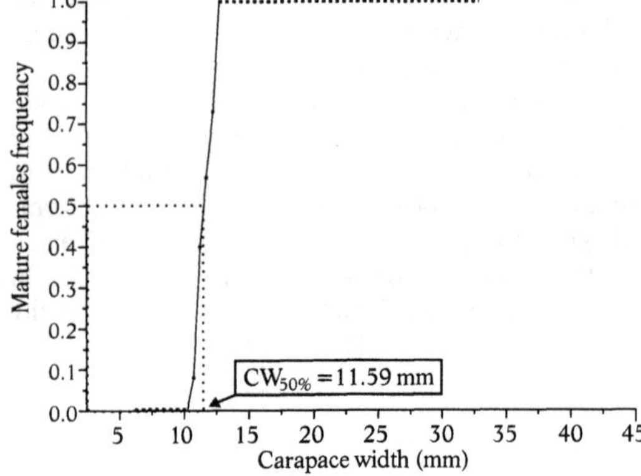

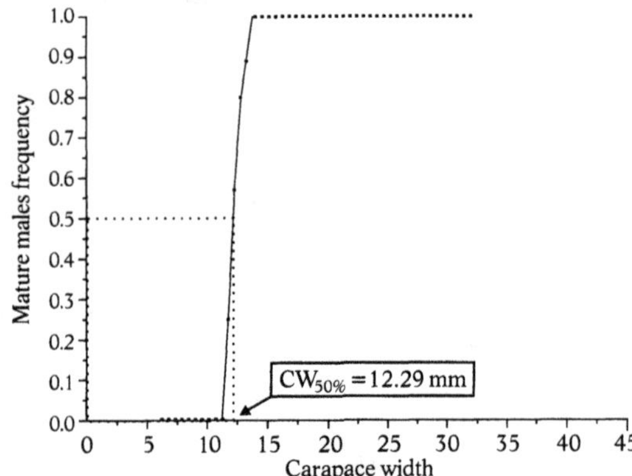

Figure 8. *Sexual maturity of* **E. limosum** *females and males based on percentage of gonad maturation.*

The relationship between GL and CW showed a positive allometric growth during the juvenile period and a negative allometric growth during the adult development. This pattern indicates that gonopods grow in higher rate during the juvenile phase than the adult phase. After reaching maturity, the growth of gonopods slows down, for what larger males are still capable to mate with females within a relatively wide size range (Hartnoll 1982, Finney and Abele 1981).

In the relationships CW vs. AW and CW vs. GL, differences in the allometric level between juvenile and adult crabs provide estimates of size at the onset of sexual maturity for males around 12 to 13 mm CW. Although no apparent differences were found for the allometric growth of the abdomen in females, setal changes in the abdomen seem to be co-occurent to gonad development, thus providing a good indication of size at sexual maturity for *E. limosum*. Similarly, the pleopods of mature females become more setose, what is probably related to the function of setae in egg adherence. The examination of these alternate features indicates that females reach maturity at a size between 11 and 12 mm CW. The size of the smallest ovigerous female obtained is very similar to the estimated size, indicating that these morphologic characteristics are good indicators of the onset of sexual maturity.

The morphometric analyses for males also provide accurate estimates of size at the onset maturity. For *E. limosum*, development of secondary sexual characters are synchronous with physiological maturity.

ACKNOWLEDGEMENTS

We are grateful to CAPES, FAPESP (#94/4878-8; #98/03134-6) and SESC for their financial support to conduct field work. To Dr. Gustavo Augusto Schmitt de Melo, for the identification of the crabs, to Augusto A. V. Flores for his helpful coments on the manuscript and to the NEBECC members for helping in field and laboratory studies.

REFERENCES

Costa TM and Negreiros-Fransozo ML (1998) The reproductive cycle of *Callinectes danae* Smith, 1869 (Decapoda, Portunidae) in Ubatuba region, Brazil. Crustaceana 71:615-627

Finney WC and Abele LG (1981) Allometric variation and sexual maturity in the obligate coral commensal *T. ferruginea* Latreille (Decapoda, Xanthidae). Crustaceana 41:114-130

Fonteles-Filho AA (1989) Recursos Pesqueiros. Biologia e dinâmica populacional. Imprensa Oficial do Ceará, 296 p

Góes JM and Fransozo A (1997) Relative growth of *Eriphia gonagra* (Fabricius, 1781) (Crustacea, Decapoda, Xanthidae) in Ubatuba, State of São Paulo, Brazil. Nauplius 5:85-98.

Hartnoll RG (1974) Variation in growth pattern between some secondary sexual characters in crabs (Decapoda Brachyura). Crustaceana 27: 130-136.

Hartnoll RG (1982) Growth. In: Bliss DE (ed) The Biology of Crustacea, Embriology, Morphology and Genetics. Vol. 2 (pp. 111-196) Academic Press, New York

Hartnoll RG and Gould P (1988) Brachyuran live history strategies and the optimization of egg production. Symp Zool Soc Lond 59:1-9

Hines AH (1989) Geographic variation in size of maturity in brachyuran crabs. Bull Mar Sci 45:356-364

Huxley JS (1950) Relative growth and form transformation. Proc R Soc Lon 137:465-469

Mantelatto FLM and Fransozo A (1994) Crescimento relativo e dimorfismo sexual de *Hepatus pudibundus* (Herbst, 1785) (Decapoda, Brachyura) no litoral paulista. Papéis Avulsos Zool 39:33-48

Mantelatto FLM and Fransozo A (1999) Relative growth of the crab *Sesarma rectum* Randall 1840 (Decapoda, Brachyura, Grapsidae) from Bertioga, São Paulo, Brazil. Pak J Mar Biol (Mar Res) 5:11-21

Melo GAS (1996) Manual de identificação dos Brachyura (Caranguejos e Siris) do litoral brasileiro. Plêiade, São Paulo 603 p

Pinheiro MAA and Fransozo A (1998) Sexual maturity of the speckled swimming crab *Arenaeus cribrarius* (Lamarck, 1818) (Crustacea, Brachyura, Portunidae) in Ubatuba Coast, State of São Paulo, Brazil. Crustaceana 71:434-452

Pinheiro MAA and Fransozo A (1999) reproductive behavior of the swimming crab *Arenaeus cribrarius* (Lamarck, 1818) (Crustacea, Brachyura, Portunidae) in capivity. Bull Mar Sci 64:243-253

Reigada ALD and Negreiros-Fransozo ML (1999) Maturidade sexual em *Hepatus pudipundus* (Decapoda, Brachyura, Calappidae). Iheringia, Sér Zool 86:159-164

Santos S and Negreiros-Fransozo ML (1996) Maturidade fisiológica em *Portunus spinimanus* Latreille, 1819 (Crustacea, Brachyura, Portunidae). Papéis Avulsos Zool 39:365-377

Sokal R and Rohlf FJ (1979) Biometria: Principios y métodos estadisticos en la investigación biológica. H. Blume Ediciones, Madrid 832 p

Vannini V and Gherardi F (1988) Studies on the pebble crab, *Eriphia smithi* Macleay, 1838 (Xanthoidea, Menippidae): patterns of relative growth and populations structure. Trop Zool 1:203-206

Vazoller AEAM (1982) Manual the métodos para estudos biológicos de populações de peixes. Reprodução e Crescimento. CNPq -Programa Nacional de Zoologia, Brasil 108 p

Zar JH (1996) Bioestatistical analysis. Prentice-Hall, Upper Saddey River 662 p

PANOPEID CRABS (CRUSTACEA: BRACHYURA: PANOPEIDAE) ASSOCIATED WITH PROP ROOTS OF *RHIZOPHORA MANGLE* L. IN A TROPICAL COASTAL LAGOON OF THE SE GULF OF CALIFORNIA, MEXICO

José Salgado-Barragán and Michel E. Hendrickx

Unidad Mazatlán, Instituto de Ciencias del Mar y Limnología,Universidad Nacional Autónoma de México P.O. Box 811, Mazatlán, 82000 Sinaloa, México

jsb@ola.icmyl.unam.mx

ABSTRACT

Xanthid crab communities associated with aerial roots of the mangrove *Rhizophora mangle* L. remain poorly known. Aspects of their ecology were studied in a survey from 1993 to 1995 in the upper portion of the Urías lagoon, Mazatlan, an anti-estuary type coastal lagoon in NW Mexico. The species collected were *Panopeus mirafloresensis* (81.3%), *Eurypanopeus canalensis* (15%), *P. chilensis* (1.3%), *P. purpureus* (1.1%), *Hexapanopeus beebei* (< 1%) and *P.* aff. *gatunensis* (< 1%). Salinity varied from 9 to 45‰ and temperature ranged from 19 to 32 °C. Krüskal-Wallis analysis showed no statistical differences in salinity and temperature records among the sampling stations. *Panopeus mirafloresensis*, while very common in various substrata in coastal lagoons of NW Mexico, was not an exclusive species of these ecosystems. On mangrove roots in the Estero de Urías coastal lagoon, the occurrence of this species is related to the presence of the bivalve *Mytella strigata*, a dominant epizooid. *Panopeus mirafloresensis* is, however, less abundant regardless of the bivalve next to a shrimp farm located at the head of the system. Ovigerous females were found in all samples. The sex ratio of *P. mirafloresensis* was F>M (49%), M>F (37%) and F=M (14%). *Eurypanopeus canalensis* was more abundant in the upper lagoon where a 1:1 sex ratio was encountered.

I. INTRODUCTION

According to Lacerda et al. (1993), mangrove forests along the Pacific coast of Mexico are comprised of only five species: *Rhizophora mangle* L., *R. harrisonii* Leechm , *Avicennia germinans* (L.), *Laguncularia racemosa* Gaertn and *Conocarpus erectus* L. As in other *Rhizophora* species, *R. mangle* is characterized by its arching prop roots. These roots are partly or entirely submerged into channel water during high tide; some well-developed roots make contact with the bottom of the channels while younger roots barely reach the surface of the water. Once they make contact with water, they provide a substrate for settlement of epiphytic species and the establishment of complex epibiotic communities (Lacerda et al. 1993, Bingham and Young 1995). These communities have been studied in several tropical areas (e.g., Sutherland 1980, Rodriguez and Stoner 1990, Orihuela et al. 1991, Ellison and Farnsworth 1992, Reyes and Campos 1992, Diaz et al. 1992), but results are often incomplete due to the difficulty to properly identify their components.

Aquatic macrofauna associated with *Rhizophora* aerial roots is comprised of primarily Porifera, Cnidaria, Bryozoa, Polychaeta, Mollusca, Crustacea and Urochordata (Hubbard-Zamudio 1983, Hendrickx 1984, Díaz et al. 1992, Lacerda et al. 1993, Chapman 1998). Biology and ecology of many species representing these higher groups have never been stud-

Kluwer Academic/Plenum Publishers

E. Escobar-Briones & F. Alvarez Eds.
MODERN APPROACHES TO THE STUDY OF CRUSTACEA
PP. 163-169

ied; in many cases, taxonomic placement of the species is still uncertain. Inhabitants of aerial root communities include several species of brachyuran crabs of the family Xanthidae. These crabs do not normally venture on to mangrove trunks and branches, as does *Aratus pisonii* (H. Milne Edwards, 1837), a species of Grapsidae that permanently lives and feeds on *Rhizophora* (Conde and Díaz 1989, Díaz and Conde 1989). Xanthids settle or migrate once the community has reached certain complexity, thus providing shelter and food for the crabs. The brachyuran crabs of the Pacific coast of Mexico include 74 species of the family Xanthidae (s.l., or Xanthoidea sensu Guinot 1978) (Hendrickx 1993, Salgado-Barragán and Hendrickx 1996); of these, 14 species belonging to the genera *Panopeus*, *Eurypanopeus* and *Hexapanopeus* have been assigned to the family Panopeidae, sensu Guinot (1978). Several of these species have previously been reported associated with mangrove forests or in adjacent coastal lagoons.

The purpose of this paper is to report the distribution patterns of panopeid crabs associated with aerial roots of the red mangrove, *Rhizophora mangle*, in a tropical coastal lagoon located in the SE Gulf of California, Mexico.

II. MATERIALS AND METHODS

The Estero de Urías coastal lagoon is a shallow, saline, vertically mixed body of water of approximately 18 km², located in the coastal plain of the Pacific coast of Mexico, SE Gulf of California. The system has an average mixed tidal amplitude of 1.5 m and a positive gradient of salinity towards the interior during most of the year (negative estuary, as defined by McLusky 1971). It is partly bordered by mangrove forests, best developed in the upper portion. The mangrove forest and its associated waterways are for the most part tidally dominated with fresh water input limited to drainage occurring during the rainy season. During this short period, the water input exceeds the evaporation and the system behaves like a positive estuary (Villalba 1986). The four most common species of trees known to the mangrove forests of western Mexico are present in the system: *Rhizophora mangle*, *Avicennia germinans*, *Conocarpus erectus* and *Laguncularia racemosa* (Agraz-Hernández 1999). Beyond the mangrove forest a semi-arid vegetation has developed inland. Tidal flats and flooding areas are found at the head of the system. Part of these areas, however, has recently been used for extensive shrimp farming activities. Shrimp farms use water from one of the tidal channels at the head of the system and discharge most effluents to another adjacent channel (Fig. 1). In 1994, culture ponds in this area occupied approximately 150 ha. Discharges have increased steadily in the last few years due to increase of the pond areas which presently extends over ca. 250 ha.

Six sampling stations were located in the upper part of the lagoon where mangrove forests still exist; sampling sites were selected according to an increasing distance from the head of the system (Fig. 1). At each station subsurface salinity and water temperature were measured every two weeks. Mangrove prop roots of *R. mangle* were collected every three months. At each site, mangrove roots were randomly selected and two hanging roots (sampling unit) were cut above the level of the epibiont assemblages. Ground roots were not selected in order to avoid the presence of crabs living on channel bottoms that may visit the prop roots temporarily. Collected roots were marked with a code number and kept in separate plastic bags. Macrofauna was later removed from the root and separated by species. Crabs were preserved in 70% ethanol, identified to species, counted and sexed. Abbreviations in this paper are: M, male; F, female; sex ratio is expressed as M:F.

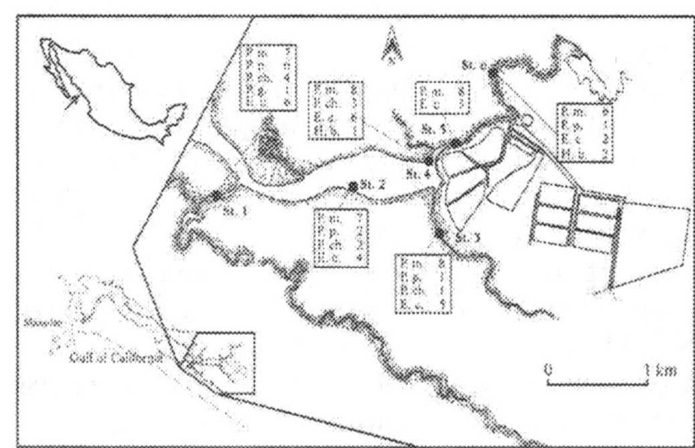

Figure 1. *Study site and sampling stations in Estero de Urías coastal lagoon, Sinaloa, Mexico. Occurrence of species during the study is indicated for each station. Number of samples is 8, except for station 6, which was sampled 7 times (P.m. =* **Panopeus mirafloresensis;** *P.p. =* **Panopeus purpureus;** *P.ch. =* **Panopeus chilensis;** *P.g. =* **Panopeus aff gatunensis;** *E.c. =* **Eurypanopeus canalensis;** *H.b. =* **Hexapanopeus beebei**).

Table 1. *Number of organisms of each species of Panopeidae collected on prop roots (sampling units) of* **Rhizophora mangle**. *Percentage per month and for the entire survey is indicated in parenthesis.*

	May 93	Jul 93	Nov 93	Feb 94	May 94	Aug 94	Nov 94	Feb 95	Total
Panopeus mirafloresensis	294	143	88	121	46	22	49	149	912
	(93.9)	(87.7)	(85.4)	(70.8)	(80.7)	(29.7)	(70.0)	(90.3)	(81.7)
Panopeus purpureus	1	3	1	2	2	-	2	1	12
	(0.3)	(1.8)	(1.0)	(1.2)	(3.5)		(2.9)	(0.6)	(1.1)
Panopeus chilensis	6	1	2	-	1	2	-	3	15
	(1.9)	(0.6)	(1.9)		(1.8)	(2.7)		(1.8)	(1.3)
Panopeus aff. gatunensis	1	-	-	-	-	-	-	-	1
	(0.3)								(0.1)
Eurypanopeus canalensis	10	15	11	48	8	50	19	12	173
	(3.2)	(9.2)	(10.7)	(28.1)	(14.0)	(67.6)	(27.1)	(7.3)	(15.5)
Hexapanopeus beebei.	1	1	1	-	-	-	-	-	3
	(0.3)	(0.6)	(1.0)						(0.3)
Total	313	163	103	171	57	74	70	165	1116

III. RESULTS

Environmental data

Throughout the year, salinity varied considerably. Typical mesohaline brackish water (9‰) was sampled during the rainy season and hyperhaline conditions (45‰) were found during the warmer, dry season. Water temperature varied from 19 °C in winter to 32 °C in summer. There was, however, no statistical difference (Kruskal-Wallis analysis) among stations for salinity and temperature measured each month; both parameters featured a synchronized variation throughout the entire survey period.

Distribution and abundance of Panopeidae

A total of six species of Panopeidae were collected from the prop roots. All samples considered, the relative proportion of each species (number of specimens collected on two hanging roots compared to all collected Panopeidae) indicated dominance by *Panopeus mirafloresensis* Abele and Kim, 1989 (81.3% of examined specimens), followed by *Eurypanopeus canalensis* Abele and Kim, 1989 (15%), *Panopeus chilensis* H. Milne Edwards and Lucas, 1844 (1.3%), *P. purpureus* Lockington, 1877 (1.1%), *Hexapanopeus beebei* Garth, 1961 (<1%), and *Panopeus* aff. *gatunensis* Abele and Kim, 1989 (<1%) (Table 1).

Panopeus mirafloresensis was by far the most abundant species and 913 specimens were collected, all stations included. It was outnumbered by *E. canalensis* (36.7% vs. 60.0%) only once, in August 1993. Otherwise, it represented from 69.7 to 93.9% of all panopeids (Table 1). *Panopeus mirafloresensis* also was a frequent species, present in 44 out of 47 samples (Fig. 1), although a sharp decrease in abundance per root was observed in stations 5 and 6, close to the head of the lagoon system (Fig. 2). *Eurypanopeus canalensis* is also associated with *R. mangle* prop roots and with hard substrata in the Estero de Urías lagoon. With a total of 173 specimens collected, *E. canalensis* was the second species both by abundance (15% of all panopeids collected) and by occurrence (present in 32 out of 48 samples). It was consistently collected throughout the study (Table 1) and although it was not as abundant as *P. mirafloresensis*, it clearly outnumbered every other species of Panopeidae collected on the roots (Fig. 1). *Panopeus purpureus* and *P. chilensis* were scarce (12 and 15 specimens) (Table 1). Most specimens of these two species (14 out of 29) were collected in stations 1 and 2, closer to the mouth of the system. Although total numbers of specimens were small, both species were collected throughout the study. Only three male specimens of *Hexapanopeus beebei* were collected (Table 1) during the first year of the study at stations 4 (1 specimen) and 6 (2 specimens). A male specimen, very similar to *Panopeus gatunensis*, but with small differences in the gonopod structure, was also captured. Salinity conditions at the station where *P.* aff. *gatunensis* was collected were hypersaline (40 ‰ in May 1993).

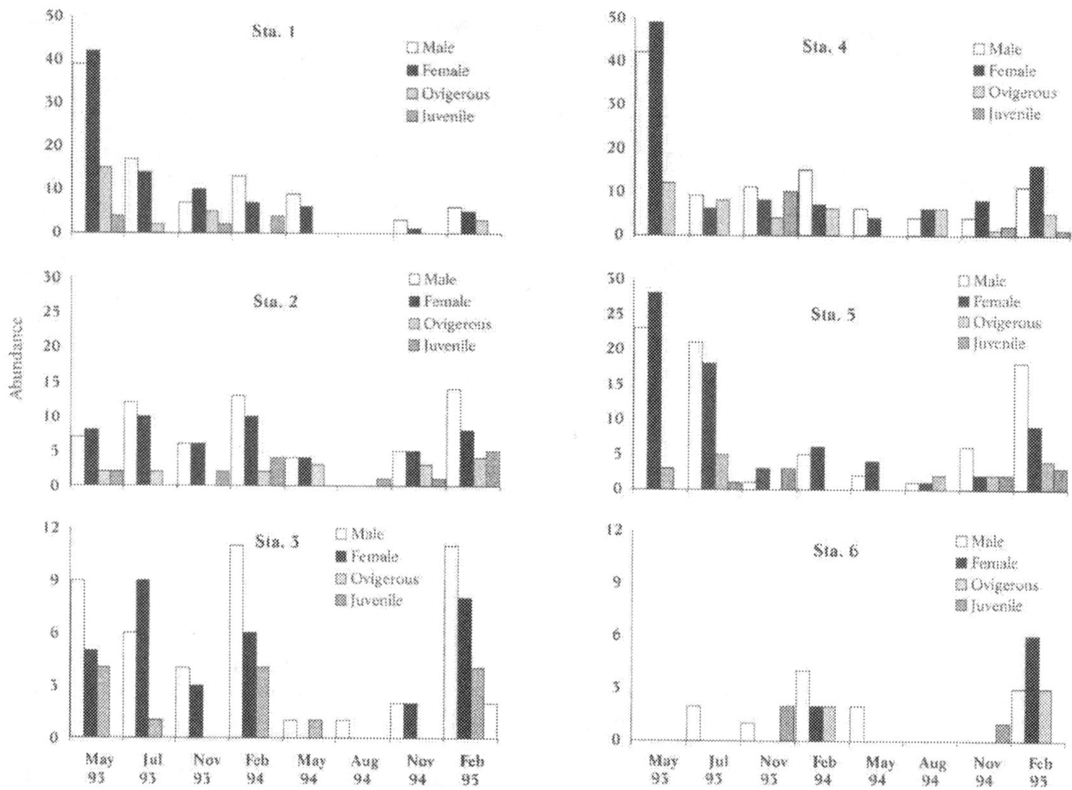

Figure 2. *Abundance (number of organisms per sampling unit) of males, females, ovigerous females and juveniles of* **Panopeus mirafloresensis** *collected on* **Rhizophora mangle** *roots in six sampling locations in Estero de Urías coastal lagoon, Sinaloa, Mexico (scale variable).*

Associated species

The prop roots that were sampled support a diverse sessile flora and fauna. In the case of *P. purpureus, P. chilensis, P. gatunensis* and *H. beebei*, numbers of collected specimens and frequency of occurrence on roots were both too small to detect a specific association with sessile organisms. In the case of the two dominant species, neither *P. mirafloresensis* nor *E. canalensis* show a clear correlation with the presence of the sessile biota. Many roots serve as support for populations of the striate mangrove mussel *Mytella strigata* (Hanley, 1843), the usually dominant mangrove mussel in the region. Although these two species were not detected on roots devoid of mussels, there was no significant statistical relationship between *Mytella* abundance and the presence of crabs.

Reproduction period and sex ratio

Of the 913 specimens of *P. mirafloresensis* collected, 391 were males, 352 non-ovigerous females, 118 ovigerous females and the rest (52 specimens) were juveniles. Ovigerous females were present in at least two of the six stations visited during each sampling trip; high occurrences of ovigerous females were registered in February, May and July. Ovigerous females and/or juveniles were observed in 34 of the 48 samples, thus indicating that the species breed all year round (Fig. 3). Sex ratio of *P. mirafloresensis* was variable. In 14% of the cases the ratio was 1:1; females outnumbered males in 49% of the samples and males outnumbered females in 37%. There was no clear difference in sex ratio among stations or seasons. Of the 173 specimens of *E. canalensis* collected, 67 were males, 81 non oviger-

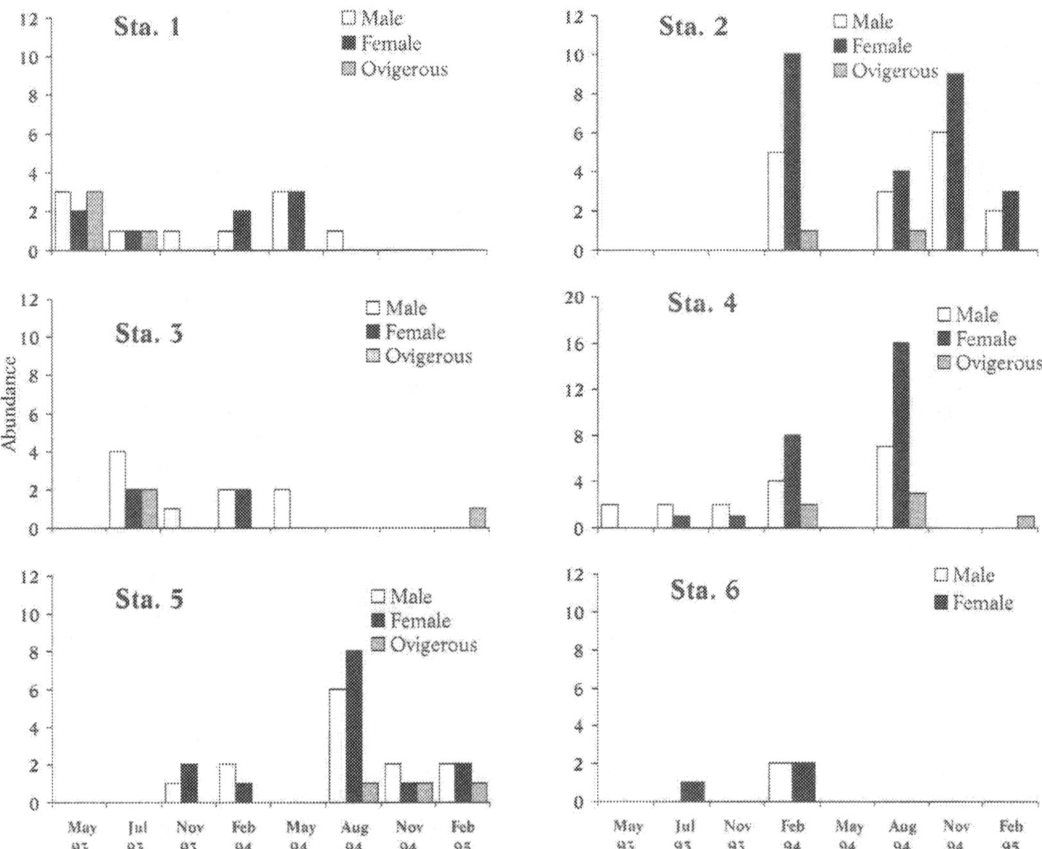

Figure 3. *Abundance (number of organisms per sampling unit) of males, females and ovigerous females of **Eurypanopeus canalensis** collected on **Rhizophora mangle** roots in six sampling locations at Estero de Urías coastal lagoon, Sinaloa, Mexico.*

ous females, 21 ovigerous females, and only 4 juveniles. The monthly sum of females (non ovigerous and ovigerous) was always higher than the corresponding number of males, except in May 1994. Ovigerous females of *E. canalensis* were not collected at station 6 and they were less frequent than *P. mirafloresensis* in stations 1-5 (Fig. 3). Females of *E. canalensis* were found during every sampling month, except in November 1993 and May 1994, thus indicating that the species tends to reproduce year-round. Sex ratio of *E. canalensis* was variable. In five cases the ratio was 1:1; females outnumbered males in 10% of the samples and males outnumbered females in 3% of the samples. There was no clear difference in sex ratio among stations or seasons.

IV. DISCUSSION

Panopeid species recorded during this study are known to use a wide variety of hard substrates as habitats. Although they are commonly recognized as mud crabs, they also occur among pebbles and oysters, below dead wood, on pier walls and along rocky shores throughout their distributional range (Rathbun 1930, Crane 1947, Abele and Kim 1989). Prop roots represent a relatively safe and permanent habitat for many sessile and mobile species; at the same time, stress imposed by peculiar environmental conditions (e.g., strong tidal currents, turbidity, and wide variations of water temperature and salinity) favors well-adapted species. Where mangrove mussels have settled and established, they offer a compact, dominant habitat that protects small and medium-size mobile species from predators.

Our salinity and temperature data indicate that the two dominant species of panopeids found on prop roots in the system, *P. mirafloresensis* and *E. canalensis*, are euryhaline and eurythermic. Presence of large reproducing populations of both species 14 degrees of latitude north of their type locality indicated that they probably occur in every mangrove forest ecosystem within their geographic range. Interconnecting lagoons and estuaries as well as coastal currents, would favor dispersion of both species.

Gatun Locks, the type locality of *P. gatunensis*, are located east of Gatun Lake, which provides a barrier between the Atlantic and the Pacific for passage of marine (stenohaline) organisms (Abele and Kim 1989). The shape of the carapace, chelipeds and walking legs as well as the shape of the first gonopod of the Urías material are very similar to what has been described for the only known specimen of *gatunensis*. A variability of characters similar to the one reported in *P. mirafloresensis* by Abele and Kim (1989), and also observed in Urías material of *mirafloresensis*, could possibly occur with specimens of *gatunensis*. Should our tentative identification be confirmed, *P. gatunensis* would represent an amphiamerican species. However, it would be difficult to explain why it has not been found on the Pacific coast of Panama.

The upper portion of Urías coastal lagoon is under the constant influence of effluents from the shrimp farms (Páez-Osuna et al. 1997). Discharges from the farm are particularly large and provoke a short, stronger impact when ponds are emptied during shrimp harvest (Hendrickx et al. 1996, Páez-Osuna et al. 1997). Increase of organic matter input into the adjacent waters affect their quality. According to Gowen et al. (1990) this may lead to hypernutrification, increases the biochemical oxygen demand, changes the bacterial populations and ultimately disturbs the wildlife. The presence of reproducing populations of the two dominating panopeids in the area where discharges occur (stations 3-5) indicates that potential disturbance due to excess of organic matter (particulated and/or dissolved) does not drastically affect the reproduction process of the panopeid populations, although there is no evidence related to hatching success and larval survival rate.

ACKNOWLEDGEMENTS

The authors wish to thank Benito Mejía, Arturo Núñez Pastén, Lucía Tron Mayén, Marcela Ruiz and Marco A. Meda for their assistance in sampling and sorting the specimens, Clara Ramírez J. helped in the search of literature and Germán Ramírez and Carlos Suárez provided technical computing support. We also would like to thank the late Ray Manning for the loan of specimens of *P. mirafloresensis* and *E. canalensis*. This work was partly financed by the Consejo Nacional de Ciencia y Tecnología, Mexico (CONACyT, grant 0625-N9110).

REFERENCES

Abele LG and Kim W (1989) The decapod crustaceans of the Panama Canal. Smithson Contrib Zool 482:1-50

Agraz-Hernández CM (1999) Reforestación experimental de manglares en ecosistemas lagunares estuarinos de la costa noroccidental de México. PhD Thesis, Universidad Autónoma de Nuevo León, Mexico 111 p

Bingham BL and Young CM (1995) Stochastic events and dynamics of a mangrove root epifaunal Community. Mar Ecol 16:145-163

Chapman MG (1998) Relationship between spatial patterns of benthic assemblages in a mangrove forest using different levels of taxonomic resolution. Mar Ecol Prog Ser 162:71-78

Conde JE and Diaz H (1989) The mangrove tree crab *Aratus pisonii* in a tropical estuarine coastal lagoon. Estuar Coast Shelf Sci 28:639650

Crane J (1947) Eastern Pacific expeditions of the New York Zoological Society. XXXVIII. Intertidal brachygnathous crabs from the west coast of tropical America with special reference to ecology. Zoologica 32:69-95

Díaz H and Conde JE (1989) Population dynamics and life history of the mangrove crab *Aratus pisonii* (Brachyura, Grapsidae) in a marine environment. Bull Mar Sci 45:148-163

Díaz H, Conde JE and Orihuela B (1992) Estimating the species number and cover of a mangrove-root community: A comparison of methods. Aust J Mar Freshwater Res 43:707-714

Ellison AM and Farnsworth EJ (1992) The ecology of Belizean mangrove-root fouling communities: patterns of distribution and abundance and effects on root growth. Hydrobiologia 247:87-98

Gowen RJ, Bradbury NB and Brown JR (1990) The use of simple models in assessing two of the interactions between fish farming and the marine environment. In: De Pauw N, Jaspers, E Ackefors andWilkins N (eds) Aquaculture – A biotechnology in progress. European Aquaculture Society, Bredene, Belgium

Guinot D (1978) Principes d´une classification évolutive des crustacés décapodes brachyoures. Bull Biol France et Belgique 112:211-292

Hendrickx ME (1984) Studies of the coastal marine fauna of southern Sinaloa, Mexico. II. The decapod crustaceans of Estero El Verde. An Inst Biol, Univ Nac Autón México 11: 23-48

Hendrickx ME (1993) Crustáceos decápodos del Pacífico mexicano. In: Salazar Vallejo SI and González NE (eds) Biodiversidad Marina y Costera de México (pp 271-318) CONABIO and CIQRO, Mexico

Hendrickx ME (1995) Checklist of brachyuran crabs (Decapoda: Brachyura) from the eastern tropical Pacific. Bull Inst Roy Sci Nat Belgique 65:125-150

Hendrickx ME, Salgado-Barragán J and Meda-Martínez MA (1996) Abundance and diversity of macrofauna (fishes and decapod crustaceans) associated to culture ponds of *Penaeus vannamei*, western Mexico. Aquaculture 143:61-73

Hubbard-Zamudio W (1983) Estudio de los crustáceos decápodos y moluscos en el Estero de Urías, puerto de Mazatlán, Sinaloa, en relación con la presencia del mangle. Profesional Thesis. Universidad Autónoma de Guadalajara, México 83 p

Lacerda LD, Conde JE, Alarcón C, Alvarez-León R, Bacon PR, D´Croz L, Kjerfve B, Polaina J and Vannucci M (1993) Mangrove ecosystems of Latin America and the Caribbean: a summary. In: Lacerda LD (ed) Conservation and sustainable utilization of mangrove forests in Latin America and Africa regions. ITTO/ISME Project PD114/90(F) International Society for Mangrove Ecosystems and International Tropical Timber Organization

McLusky D S (1971) Ecology of Estuaries. HEB Ltd, London 144 p

Orihuela B, Díaz H and Conde JE (1991) Mass mortality in a mangrove roots fouling community in a hypersaline tropical lagoon. Biotropica 23: 592-601

Páez-Osuna F, Guerrero-Galván SR, Ruiz-Fernández AC and Espinoza-Angulo R (1997) Fluxes and mass balances of nutrients in a semi-intensive shrimp farm in north western Mexico. Mar Poll Bull 34:290-297

Rathbun MJ (1930) The cancroid crabs of America of the families Euryalidae, Portunidae, Atelecyclidae, Cancridae and Xanthidae. US Natl Mus Bull, 152

Reyes R and Campos NH (1992) Macroinvertebrados colonizadores de raíces de *Rhizophora mangle* en la Bahía de Chengue, Caribe colombiano. An Inst Invest Mar Punta Betin 21:101-116

Rodriguez C and Stoner AW (1990) The epiphyte community of mangrove roots in a tropical estuary: distribution and biomass. Aquat Bot 36:117-126

Salgado-Barragán J and Hendrickx ME (1996) Decapod crustaceans from the Pacific coast of Mexico, including new records, taxonomic remarks and ecological data. Rev Biol Trop 44/45:680-683

Sutherland JP (1980) Dynamics of the epibenthic community on roots of the mangrove *Rhizophora mangle*, at Bahia de Buche, Venezuela. Mar Biol 58:75-84

Villalba LA (1986) Descripción general del estero de Urías, Mazatlán, Sinaloa. Rev Ciencias del Mar 8:32-37

DAILY RHYTHM OF FEEDING ACTIVITY OF THE FRESHWATER CRAB *DILOCARCINUS PAGEI PAGEI* IN THE RIO PILCOMAYO NATIONAL PARK, FORMOSA, ARGENTINA

Verónica Williner and Pablo A. Collins

Instituto Nacional de Limnología, José Macía 1933, CP3016, Santo Tomé, Santa Fe, Argentina

inali@ceride.goc.ar

I. INTRODUCTION

Dilocarcinus pagei pagei (Simpson) is one the several Argentinean freshwater crabs; its geographical distribution extends from northeastern Argentina to Buenos Aires (Morrone and Lopretto 1995, Magalhães and Türkay 1996). This crab is an important component of the littoral fauna in the lagoons of the Río Pilcomayo National Park, in the province of Formosa; however, there is little information in the literature about its biology and ecology (Lopretto 1981, 1995, 1998, Magalhães and Türkay 1996).

All ecosystems exhibit a complex periodical variation. Variations in activity rhythms over a time period also differ among species (Margalef 1986). Invertebrates, in particular crustaceans, have circadian rhythms determined by endogenous and exogenous components. These factors may produce changes in movement, color, secretory activity, feeding, trophic spectrum, and shelter selection (Lewis et al. 1966, Hopkin and Nott 1980, Wassemberg and Hill 1987, Collins 1995, 1997). Certainly, trophic activity in animal populations can occur at any time during the day, but its intensity varies according to food availability, intra and interspecific competition and the risk of predation. As a consequence, organisms have adjusted their daily activities to these factors. Further, it is often thought that evolution has shaped these patterns in an optimal manner (Stenseth 1983).

The aim of this study was to examine the diet composition and its possible daily variation in the freshwater crab *D. pagei pagei* in the "Laguna Blanca" lake, in the Río Pilcomayo National Park, Argentina. The obtained information may be an essential start point to study other aspects of the ecology and conservation of freshwater crabs.

II. MATERIALS AND METHODS

Study area

The study site was located in the "Laguna Blanca" lake at the Río Pilcomayo National Park (25° 30' S, 58° 30' W), province of Formosa, northern Argentina (Fig. 1). The park covers an area of about 51889 ha and the lake over 700 ha. The protected area is located in a region classified as "hyperseasonal Chaco" with a humid - semihumid weather regime.

The lake border is covered with floating and emergent macrophytes: *Eichhornia crassipes, E. azurea, Pontederia cordata, P. lanceolata, Nymphoides indica, Pistia stratiotes,* and *Azolla* spp. There is also a tall grass area where the dominant species are: *Thalia geniculata, T. multiflora, Canna glauca, Cyperus giganteus,* and *Juncus* spp (Reca and Pujalte 1986).

Sampling

Samples of *D. pagei pagei* were taken from under the vegetation at the border of the lake in two different sites with identical characteristics, at the same depth (0.5 m), and with the presence of *E. crassipes, N. indica, P. stratiotes,* and *Azolla* spp. The samples were taken every four hours during three days on May, 1999. The net used to sample had a 0.2 m² mouth area and a 1mm mesh. The same sampling effort was used

E. Escobar-Briones & F. Alvarez Eds.
MODERN APPROACHES TO THE STUDY OF CRUSTACEA
PP. 171-178

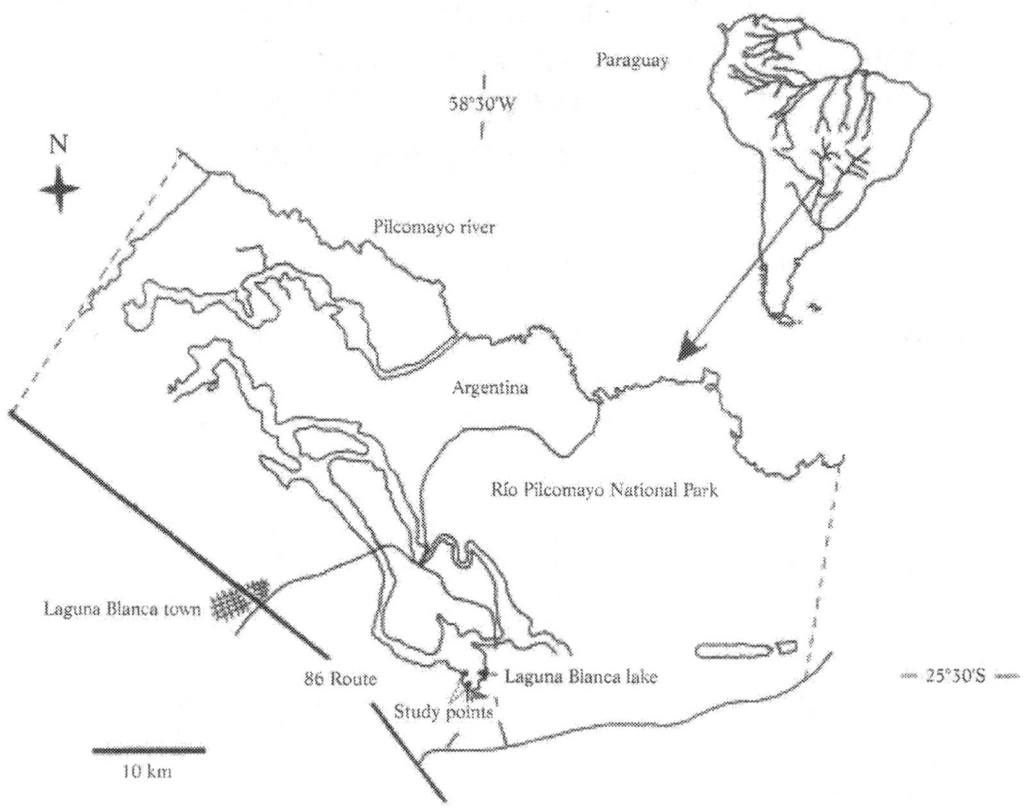

Figure 1. *Sampling sites in the "Laguna Blanca" lake, in the Río Pilcomayo National Park, Formosa, Argentina.*

for all the samples. Crabs were immediately anaesthetized by cooling, and later preserved in 70% alcohol.

The dissolved oxygen concentration of the sampling sites was monitored with a portable oxygenmeter. Temperature was measured with a conventional mercury thermometer. Conductivity and pH were determined using a Beckman conductimeter and a Helige pH-meter, respectively. Transparency was measured with a Secchi disc. During the sampling, these parameters were registered every 4 hours.

Individual analysis

Crabs were separated into females, males or juveniles according to the abdomen morphology (Lopretto 1976). Both, caparace length and width, were obtained from every crab. After opening the alimentary tract of each crab, its fullness was estimated; in this case a subjective scale ranking from 1 (empty) to 3 (full) was used. Later, foreguts were opened and the contents were placed in Khan-tubes with 70 % alcohol and stained with eritroxine. Aliquots were put on slides and examined under the compound microscope. Prey items were identified as precisely as possible.

Numerical analysis

Trophic diversity was estimated for every sample (Figs. 2, 3) using the Simpson index (Legendre and Legendre 1979). Food items were analyzed with the Index of Relative Importance (IRI), IRI = (Cv + Cn) x F (Pinkas et al. 1971); where Cv is the volumetric content of prey, Cn the number of prey, and F the frequency of occurrence of each prey item. The volume of the items was calculated by approximation to a geometric shape or transformed from values obtained from the literature (Edmondson and Winberg 1971, Dumont et al. 1975, Ruttner-Kolisko 1977). Caparace width (CW) and IRI values were tested for significance using analysis of variance (ANOVA) (Zar 1996).

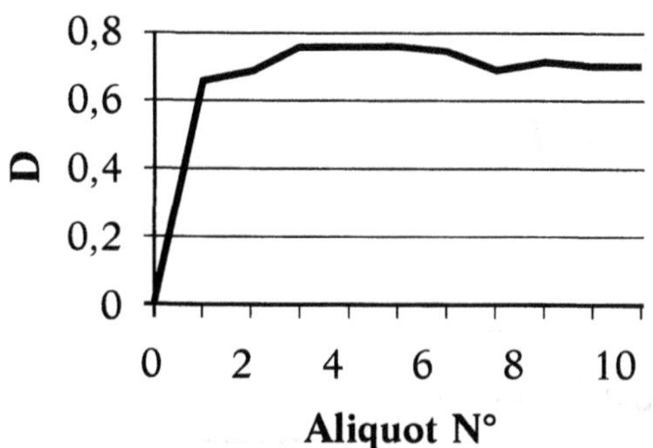

Figure 2. *Mean of cumulative trophic diversity for aliquots obtained from **Dilocarcinus pagei pagei** stomach contents, calculated with the Simpson index.*

III. RESULTS

Among the water quality parameters recorded, the ones that showed the greatest variation were temperature, oxygen, and transparency (Table 1). Marked changes in the temporal abundance of crabs during daytime were also observed (Table 1). The highest number of crabs was recorded during the hottest part of the day (15 h) (7.2 ± 2.54 crabs/m²) and the minimum (0.8 ± 0.92 crabs/m²) was at daybreak (7 h).

Differences in the size distribution were not significant during the day (ANOVA, $P > 0.05$). The minimum size recorded was 4.4 mm CW and the maximum 39.4 mm CW (Fig. 4); adults were present in all samples, whereas juveniles were only registered at 3 h and 19 h. Males were more abundant than females, which represented between 0% and 17.6 % of the total abundance.

Crabs grouped according to the fullness index showed that the percentage of empty stomachs declined from midday to sunset and night (Fig. 5). The minimum number of stomachs examined varied between 4 and 8. More animals were captured and examined in the night samples (Fig. 3). Although the trophic activity was evident during the whole 24 h, the differences between day and night were significant (ANOVA, $P < 0.05$).

Diet composition of *D. pagei pagei* consisted of macrophyte remains, algae (filamentous, diatoms, and other unicellular groups), fungi, amoebas, ciliates, rotifers, oligochaetes and microcrustaceans (Table 2). In spite of the damage, most organisms found in the stomach contents were identified to genus or species. Sediment and sand particles were also found among the stomach contents. The high occurrence items in stomach contents were algae (*Spirogira* sp, *Mougeotia* sp, *Navicula* sp, *Bacillaria* sp, *Amphora* sp, *Oedogonium* sp, *Ulotrix* sp), plant material, and sand (Table 2).

Macrophyte remains were found in all stomachs, becoming the item with the highest IRI value.

Hour	T °C	Oxygen mg/l	Conductivity Ohms/cm	pH	Secchi disc cm	Densities Crabs/m2	Male /fem ratio
7 h	15.0±0.50	7.5±1.00	320.0±10.00	7.5 ± 0.12	54.0±5.29	0.8±0.92	-
11 h	17.5±0.50	7.2±0.91	323.3±5.77	7.6 ± 0.00	48.3±4.16	5.0±2.40	5
15 h	21.2±1.04	8.0±0.35	326.7±20.82	7.5 ± 0.12	46.0±3.61	7.2±2.54	13
19 h	18.7±0.5	8.8±1.48	311.7±2.89	7.4 ± 0.00	41.5±2.53	4.2±2.16	-
23 h	17.2±0.29	7.6±0.96	320.0±10.00	7.5 ± 0.12	-	2.6±1.96	-
3 h	15.8±0.29	7.2±0.70	320.0±10.00	7.4 ± 0.00	-	4.2±2.30	4

Table 1. *Mean values and standard deviation of water quality parameters and **Dilocarcinus pagei pagei** abundance, during the sampling in "Laguna Blanca" lake, Pilcomayo National Park, Formosa, Argentina.*

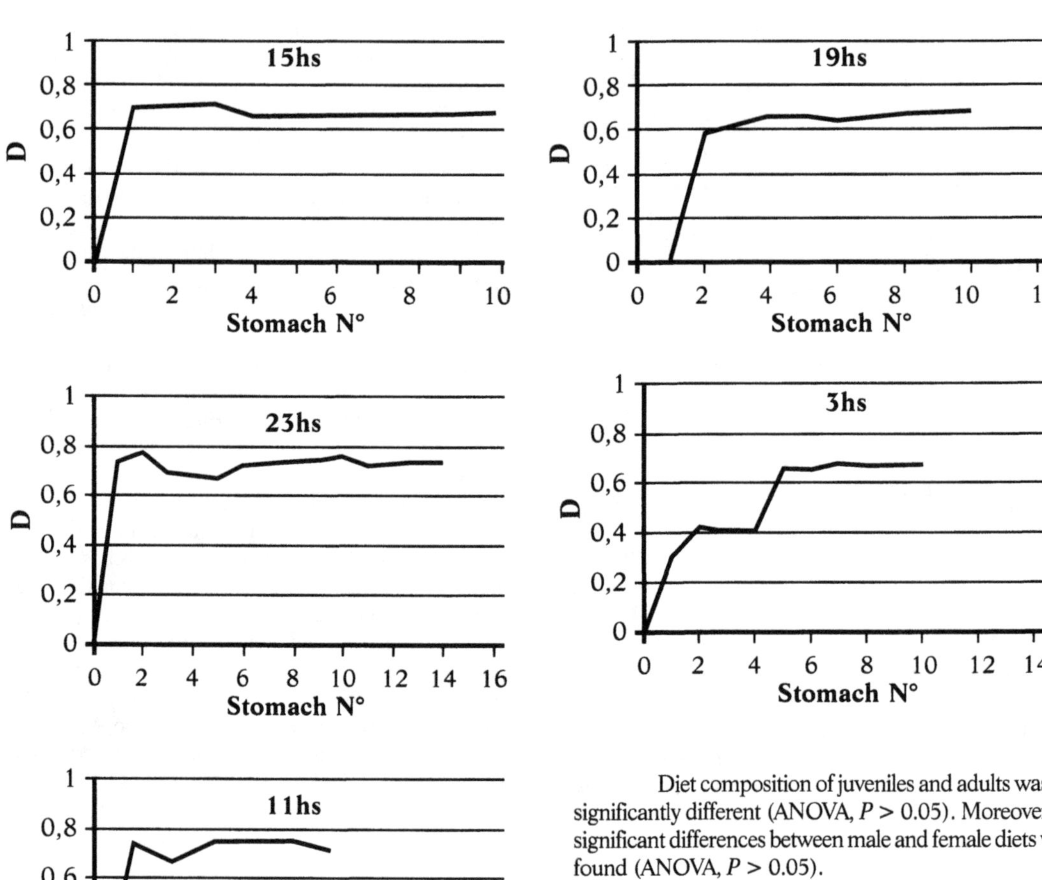

Figure 3. *Mean of cumulative trophic diversity for stomachs of* **Dilocarcinus pagei pagei** *calculated with the Simpson index, during a 24 h period.*

These were followed, in order of importance, by filamentous algae and diatoms (Fig. 6). This analysis showed that the diet composition at 11 h was significantly different (ANOVA, $P < 0.05$) from those during the night (19 h, 23 h, 3 h).

Diet composition of juveniles and adults was not significantly different (ANOVA, $P > 0.05$). Moreover, no significant differences between male and female diets were found (ANOVA, $P > 0.05$).

IV. DISCUSSION

From the analyzed data, it is possible to conclude that *D. pagei pagei* is an omnivorous crab, wich incorporates an important percentage of plant material into its diet. According to Barnes et al. (1993), *Dilocarcinus pagei pagei* may be classified as a grazer or browser, since the most important food item found was macrophyte remains. Similar diets have been found for other freshwater macro-crustaceans (e.g., *Euryrhynchus amazoniensis, Macrobrachium nattereri,* Kensley and Walker 1982).

Consumption of the available food components varied throughout the day. Food items of animal origin were more commonly found in the night samples; although there appearance was volumetrically and numerically low. The finding of oligochaetes, amoebas,

Table 2. *List of food items recorded in the stomach contents of* **Dilocarcinus pagei pagei** *from "Laguna Blanca" lake Río Pilcomayo National Park, Formosa, Argentina.*

Items	Frequency	Items	Frequency
ALGAE	***	Ulothricophyceae	***
Cyanophyceae	*	*Oedogonium* sp.	***
Chrococcus sp.	*	*Ulotrix* sp.	***
Euclorophyceae	**	*Bulbochaeta* sp.	*
Coelastrum sp.	*	*Uronema* sp.	*
Ankistrodesmus sp.	**	Xanthophyceae	*
Chlorella sp.	*	*Tribonema* sp.	*
Rizoclonium sp.	**	Charophyceae	*
Platydorina sp.	*	*Microthamnion* sp.	*
Volvox sp.	*	Fungi	*
Cladophora sp.	*	Hiphomycetes	*
Euglenophyceae	*	*Candelobrum* sp.	*
Trachelomonas sp.	*	VEGETAL REMAINS	***
Zygophyceae	***	PROTOZOA	**
Zignema sp.	**	Rizopoda	**
Spirogira sp.	***	*Diflugia* sp.	**
Mougeotia sp.	***	*Centropyxis* sp.	**
Bacillariophyceae	***	*Arcella* sp.	**
Fragilaria sp.	*	Ciliophora	**
Synedra sp.	**	*Vorticella* sp.	**
Navicula sp.	***	*Urocentrum* sp.	**
Asterionella sp.	*	ROTIFERA	**
Bacillaria sp.	***	*Keratella* sp.	**
Cymbella sp.	**	OLIGOCHAETA	*
Surirella sp.	*	CRUSTACEA	*
Rhoicosphenia sp.	*	Ostracoda	*
Nitschia sp.	**	Cladocera	*
Melosira sp.	**	*Macrothrix* sp.	*
Amphora sp.	***	Copepoda Cyclopoida	*
Meridium sp.	**	Copepoda Calanoida	*
Cocconeis sp.	*	INSECTA (larvae)	*
Gomphonema sp.	*	Diptera	*
Aulacoseira sp.	*	*Chironomus* sp.	*
Desmidiaceae	**	Sand	***
Desmidium sp.	**		

** low frequency; ** medium frequency; *** high frequency*

and rotifers could be related to the ingestion of sand and sediment particles. These organisms are members of the benthic and pleuston fauna in freshwater systems (Paporello 1976, Ezcurra de Drago 1980, Poi de Neiff and Bruquetas 1983, Paporello de Amsler 1987, Bruquetas and Neiff 1991, Poi de Neiff and Neiff 1991, Poi de Neiff and Carignan 1997). Although the crabs were supposed to be in the benthos, all the samples were taken from the superficial layer of the submerged vegetation. The other animals found in the stomach contents may have been ingested together with plant

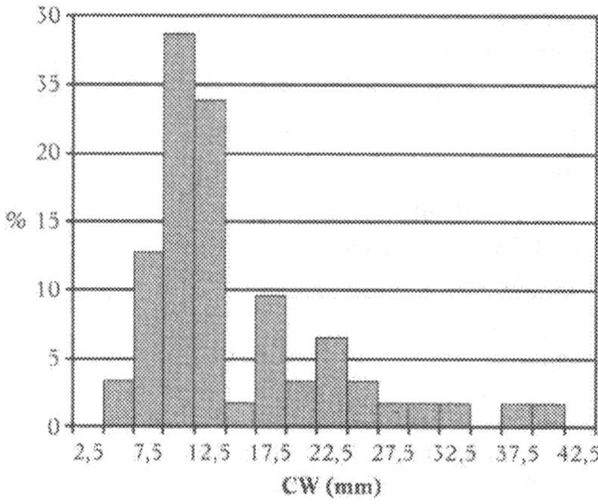

Figure 4. *Size (CW) frequency distribution, expressed as percentage, of **Dilocarcinus pagei pagei**.*

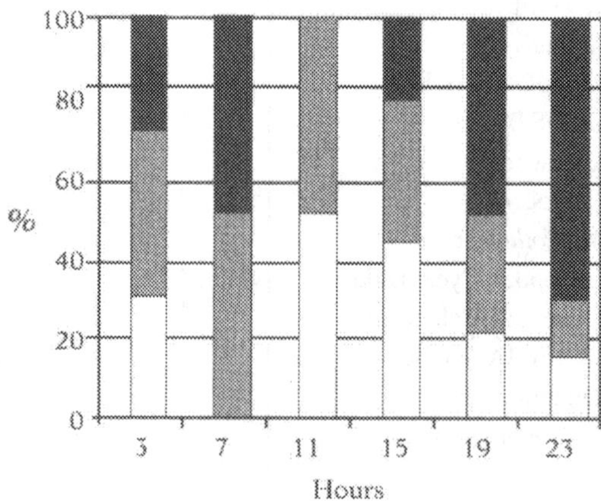

Figure 5. *Stomach filling index of **Dilocarcinus pagei pagei**. Open bars represent empty stomachs, gray bars correspond to partially full stomachs, and solid bars represent full stomachs.*

material. Ostracods (*Herpetocypris* sp), for example, are members of the littoral community associated to the vegetation, while ciliates (*Vorticella* sp) are epiphytic organisms (Foggetta 1995, Moguilevsky and Whatley 1995). Cladocerans (Family Macrothricidae) are both, benthic and associated to plants (Paggi 1980, Paggi and José de Paggi 1990). These findings may suggest that *D. pagei pagei* moves vertically through the water column to feed.

The presence of sand grains in the digestive tract could increase the maceration efficiency, as earlier studies have shown for freshwater shrimps (Jayachandran and Joseph 1989). Moreover, sediment particles may provide some of the required mineral and organic elements (Margalef 1986, Nyström et al. 1996). On the other hand, most of the filamentous algae and diatoms found in the stomach tracts are epilithic or epiphytic (Vélez and Maidana 1995).

The analysis on the variation of the stomach fullness indicates the existence of a feeding periodicity. Although the highest IRI values of plant material were registered at 11 h, feeding activity was at the minimum level. This result could be reflecting an earlier ingestion and the difficulty to digest plant material. In the tropical freshwater crayfish, *Cherax quadricarinatus*, the foregut evacuation process occurs in about 4 h, but this period differed according to food availability (Loyajavellana et al. 1995). No such data on evacuation times exists for *D. pagei pagei*, or for any other freshwater omnivorous crab. Information on marine decapods suggests a variable digestion time depending on the type of food (Barker and Gibson 1977, Hopkin and Nott 1980, Al-Mohanna and Nott 1987, Hoyt et al. 2000).

During daytime, animal items were almost absent from the stomach contents. Differences between day and night may be due to the differential availability of resources and to endogenous factors. Another intervening factor may be the predator avoidance response of many animal species during the day (Sogard and Able 1994, McNeil et al. 1995).

Growth and maintenance of *D. pagei pagei* are possible with a herbivorous diet, as it occurs with the herbivorous crab *Grapsus albolineatus* (Kennish 1996). In this case, energy storage and reproductive performance may require a nutrient supply that is found in food of animal origin. Other herbivorous crabs have similar strategies, *Gecarcinus lateralis* and *Cardisoma guanhumi* complement their plant-based diets with animal matter (Wolcott and Wolcott 1984, 1987).

Finally, symbiotic organisms housed in specialized gut compartments (Wainwright and Mann 1982, Barnes et al. 1993, Donachie et al. 1995) may be present in the alimentary canals of *D. pagei pagei* aiding in the digestion of plant matter.

REFERENCES

Al-Mohanna SY and Nott JA (1987) R-cells and the digestive cycle in *Penaeus semisulcatus* (Crustacea: Decapoda). Mar Biol 95:129-137

Barker PL and Gibson R (1977) Observations on the feeding mechanism, structure of the gut, and digestive physiology of the European lobster *Homarus gammarus* (L.) (Decapoda: Nephropidae). J Exp Mar Biol Ecol 26:297-324

Barnes RSK, Calow P, Olive PJW and Golding DW (1993) The Invertebrates, A new synthesis. Blackwell Scientific Publications, Oxford

Bruquetas de Zozaya IY and Neiff JJ (1991) Decomposition and colonization by invertebrates of *Typha latifolia* L. litter in Chaco cattail swamp (Argentina). Aquatic Botany 40:185-193

Collins P (1995) Variaciones diarias de la actividad trófica en una población de *Palaemonetes argentinus* (Crustacea, Decapoda). Rev Asoc Cienc Nat Litoral 26:57-66

Collins P (1997) Ritmo diario de alimentación en el camarón *Macrobrachium borellii* (Decapoda, Palaemonidae). Iheringia Sér Zool 82:19-24

Donachie SP, Saborowski R, Peters G and Buchholz F (1995) Bacterial digestive enzyme activity in the stomach and hepatopancreas of *Meganyctiphanes norvegica* (M Sars, 1987). J Exp Mar Biol Ecol 188:151-165

Dumont HJ, van de Velde I and Dumont S (1975) The dry weight estimate of biomass in a selection of Cladocera, Copepoda and Rotifera from the plankton, periphyton and benthos of continental waters. Oecologia 19:79-97

Edmonson WF and Winberg GG (1971) A manual on methods for the assessment of secondary productivity in fresh waters. Blackwell Scientific Publications, Oxford

Ezcurra de Drago I (1980) Campaña limnológica "Keratella" en el río Paraná medio. Complejo bentónico del río y ambientes leníticos asociados. Ecología 4:89-91

Fogetta M (1995) Ciliophora. In: Lopretto E and Tell G (eds) Ecosistemas de aguas continentales. Metodología para su estudio. T 2. (pp.557-582) Sur, La Plata

Hopkin SP and Nott JA (1980) Studies on the digestive cycle of the shore crab *Carcinus maenas* (L.) with special reference to the B-cells in the hepatopancreas. J Mar Biol Assoc UK 60: 891-907

Hoyt M, Fleeger JW, Siebeling R and Feller RJ (2000) Serological estimation of prey-protein gut-residence time and quantification of meal size for grass shrimp consuming meiofaunal copepods. J Exp Mar Biol Ecol 248:105-119

Jayachandra KV and Joseph NY (1989) Food and feeding habits of the slender river prawn, *Macrobrachium idella* (Hilgendorf, 1898) (Decapoda, Palaemonidae). Mahasagar 22: 121-129

Kennish R (1996) Diet composition influences the fitness of the herbivorous crab *Grapsus albolineatus*. Oecologia 105:22-29

Kensley B and Walker I (1982) Palaemonid shrimps from the Amazon basin, Brazil (Crustacea: Decapoda: Natantia). Smithson Contrib Zool 362:1-28

Legendre L and Legendre P (1979) Ecologie Numerique. Masson, Paris

Lewis JB, Ward J and McIver A (1966) The breeding cycle, growth and food of the fresh water shrimp *Macrobrachium carcinus* (Linnaeus). Crustaceana 10:48-52

Lopretto E (1976) Morfología comparada de los pleópodos sexuales masculinos en los Trichodactylidae de la Argentina (Decapoda, Brachyura). Limnobios 1:67-94

Lopretto E (1981) Discusión sobre las presuntas subespecies de *Dilocarcinus* (*D.*) *pagei* (Crustacea, Brachyura, Trichodactylidae). Redescripción y referencia a su polifenismo. Physis B39(97):21-31

Lopretto E (1995) Crustacea Eumalacostraca. In: Lopretto E and Tell G (ed.) Ecosistemas de aguas continentales. Metodología para su estudio. T 3. (pp. 1001-1021) Sur, La Plata

Lopretto E (1998). Eucarida. In: Morrone J and Coscarón S (eds) Biodiversidad de Artrópodos Argentinos. Una perspectiva biotaxonómica. (pp. 536-544) Sur, La Plata

Loyajavellana GN, Fielder DR and Thorne MJ (1995) Foregut evacuation, return of appetite and gastric fluid secretion in the tropical freshwater crayfish, *Cherax quadricarinatus*. Aquaculture 134:295-306

McNeil R, Diaz O, Liñero I and Rodríguez R (1995) Day- and night- time prey availability for waterbirds in a tropical lagoon. Can J Zool 73:869-878

Magalhães C and Türkay M (1996) Taxonomy of the neotropical freshwater crab family Trichodactylidae I. The generic system with description of some new genera. Senckenberg Biol 75:63-95

Margalef R (1986) Ecología. Omega, Barcelona

Moguilevsky A and Whatley R (1995) Crustacea Ostracoda. In: Lopretto E and Tell G (eds) Ecosistemas de aguas continentales. Metodología para su estudio. T 3. (pp. 973-999) Sur, La Plata

Morrone JJ and Lopretto E (1995) Parsimony analysis of endemicity of freshwater Decapoda (Crustacea: Malacostraca) from southern South America. Neotropica 41:3-8

Nyström P, Brömark C and Graméli W (1996) Patterns in benthic food webs: a role for omnivorous crayfish. Freshwater Aquatic 36:631-646

Paggi JC (1980) Campaña limnológica "Keratella I" en el tramo Paraná medio (Argentina): zooplancton de ambientes leníticos. Ecología 4:77-88

Paggi JC and Jose de Paggi S (1990) Zooplankton of the lotic and lentic environments of the middle Parana River. Acta Limnologica Brasiliensia 3:685-719

Paporello G (1976). Estudio preliminar del pleuston en el madrejón "El Negro" (Provincia de Santa Fe). Rev Asoc Cienc Nat Litoral 7:185-192

Paporello de Amsler G (1987) Fauna asociada a las raices de *Eichornia crassipes* en una laguna del valle aluvial del río Paraná ("Los Matadores", Santa Fé, Argentina). Rev Asoc Cienc Nat Litoral 18:93-103

Pinkas L, Olipham MS and Iversor ILK (1971) Food habits of albacore, bluefin tuna and bonito in California waters. Fish Bull 152:1-105

Poi de Neiff A and Bruquetas I (1983) Fauna fitófila de *Eichornia crassipes* en ambientes leníticos afectados por las crecidas del río Parana. Ecosur 10:127-137

Poi de Neiff A and Carignan R (1997) Macroinvertebrates on *Eichornia crassipes* roots in two lakes of the Paraná river floodplain. Hydrobiologia 345:185-196

Poi de Neiff A and Neiff JJ (1991) Dry weight loss and colonization by invertebrates of *Eichornia crassipes* litter under aerobic conditions. J Trop Ecol 30:175-182

Reca A and Pujalte JC (1986). Criterios para el relevamiento de unidades ecológicas en Parques Nacionales. APN Serie del Cincuentenario 9:1-25

Ruttner-Kolisko A (1977) Suggestions for biomass calculation of plankton rotifers. Archiv für Hydrobiologie 8:71-76

Sogard SM and Able KW (1994) Diel variation in immigration of fishes and decapods crustaceans to artificial seagrass habitat. Estuaries 17:622-630

Stenseth NC (1983) A coevolutionary theory for communities and food web configurations. Oikos 41:487-495

Vélez CG and Maidana NI (1995) Algae. In: Lopretto E and Tell G (eds) Ecosistemas de aguas continentales. Metodología para su estudio. T 2 (pp. 379-442) Sur, La Plata

Wainwright PF and Mann KH (1982) Effect of antimicrobial substance on the ability of the mysid shrimp *Mysis stenolepsis* to digest cellulose. Mar Ecol Prog Ser 7:309-313

Wasemberg TJ and Hill BJ (1987) Natural diet of the tiger prawns *Penaeus esculentus* and *P. semisulcatus*. Aust J Mar Fresh Res 38:169-182

Wolcott DL and Wolcott TG (1984) Food quality and cannibalism in the red land crab *Gecarcinus lateralis*. Physiol Zool 57:318-324

Wolcott DL and Wolcott TG (1987) Nitrogen limitation in the herbivorous land crab *Cardisoma guanhumi*. Physiol Zool 60:262-268

Zar JH (1996) Bioestatistical Analysis. Prentice Hall, New York

POPULATION STRUCTURE OF THE FRESHWATER CRAB *TRICHODACTYLUS FLUVIATILIS* LATREILLE, 1828 (DECAPODA, TRICHODACTYLIDAE) IN UBATUBA, NORTHERN COAST OF SAO PAULO STATE, BRAZIL

Daniela Trigueirinho Alarcon, Maria Helena de Arruda Leme and Valter José Cobo

*University of Taubaté, Department of Biology, Laboratory of Zoology,
Pça. Marcelino Monteiro, 63, Taubaté, São Paulo, Brazil, 12.030-010
NEBECC – Group of Studies on Crustacean Biology, Ecology and Culture*

vjcobo@infocad.com.br

ABSTRACT

The available information on the biology of freshwater crabs, in contrast with marine crabs, remains very scarce making it an important and interesting target for new studies. The objective of the present study was to determine basic parameters of the population structure of *Trichodactylus fluviatilis* in Ubatuba, northern coast of the State of São Paulo, Brazil. Monthly nocturnal collections were carried out by hand, from March 1998 to February 1999. A total of 306 individuals were captured, 138 males and 168 females. Ovigerous females were not caught during the sampling period, which is probably the result of their cryptic behavior during egg incubation. Female mean size (23.53 ± 4.8 mm CW) was significantly larger than males (22.32 ± 4.8 mm CW). There was no significant deviation from a 1:1 sex ratio during the studied period and the size distribution of the sampled population presented a normal distribution.

I. INTRODUCTION

Brachyuran crabs are found in marine, estuarine, freshwater and terrestrial habitats. Most brachyurans living in freshwater belong to the superfamily Potamoidea, which comprises 11 families (Bowman and Abele 1982, Hartnoll 1988).

Compared to marine crabs, freshwater and terrestrial species show the following evolutionary trends: habitat specialization, limited geographic range, high rates of endemism, low fecundity with small numbers of large eggs, brood protection, and abbreviated larval development (Hartnoll 1988).

Trichodactylus fluviatilis Latreille, 1828 is a freshwater crab that belongs to the family Trichodactylidae. It presents a wide geographic distribution in Brazil, and some degree of polimorphism according to Mello (1967). It is found in mountain streams or in logs in wet forests.

Information on the population biology of freshwater crabs is very scarce (Gherardi et al. 1987, Gherardi et al. 1988, Gherardi and Micheli 1989, Gherardi and Vannini 1989), with very few studies focusing on the New World species (Mello 1967). Regarding *T. fluviatilis*, the available studies treat only its geographic distribution and taxonomic status (Mello 1967, Magalhães and Türkay 1986, Magalhães 1987).

The aim of the present study was to describe the population structure, sex ratio, and recruitment of *T. fluviatilis* from Brazil based on the analysis of size frequency distributions.

E. Escobar-Briones & F. Alvarez Eds.
MODERN APPROACHES TO THE STUDY OF CRUSTACEA
PP. 179-182

II. MATERIALS AND METHODS

According to Mello (1967), *T. fluviatilis* is active and feeds at night. Based on this habit, monthly collections of crabs were made at night from March 1998 to February 1999, in Ubatuba county (23° 27' 22" S, 45° 03' 00" W), on the northern coast of the State of São Paulo, Brazil.

Crabs were collected by hand during a two-hour period by two people. In the laboratory their sex was determined according to their abdominal morphology and the presence of eggs or young in the females was recorded. For all crabs carapace width (CW) was measured to the nearest 0.1 mm using a Vernier caliper.

The statistical analysis consisted of applying the Shapiro-Wilks' test for normality of the size distribution, Students t-test to compare male and female sizes, and the binomial distribution to test for differences in the sex ratio (Zar 1996).

III. RESULTS

A total of 306 individuals was obtained over a twelve-month period, which included 138 males and 168 females. Crabs fell into 11 size classes with an interval of 3 mm CW, the size range was 5.0 to 35.7 mm CW. The obtained population sample presented a normal distribution (Shapiro-Wilks' W = 0.98, P > 0.05), which was unimodal and slightly asymmetric with a deviation to the left (Fig. 1).

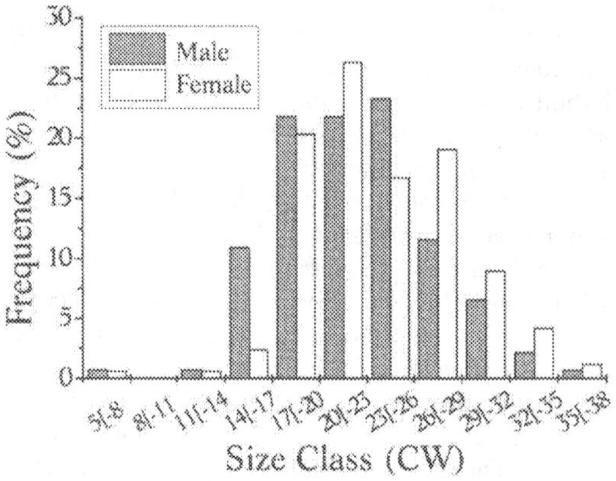

Figure 1. *Frequency distribution by size and sex of* **Trichodactylus fluviatilis** *from Ubatuba, Brazil.*

Mean carapace width of males was 22.32 ± 4.82 mm and that of females was 23.57 ± 4.80 mm, which resulted significantly larger than males (t = -2.59; P < 0.05). Adult crabs were dominant in the samples, while young crabs were sporadically recorded throughout the study period and no conspicuous recruitment cohort was verified. No ovigerous females were registered, however females carrying young on the abdomen were collected in August (n = 1) and September (n = 2) (Fig. 2).

The sex ratio for the total sample was 1:1.2, and was not significantly different from a 1: 1 ratio (c^2 = 1.28, P > 0.05), although more males were collected in the 14 - 17 mm CW class (Table 1). In general, approximately the same number of males and females were collected every month, except for January 1999, when females were significantly more numerous than males (Table 2).

Table 1. *Sex ratio by month of* **Trichodactylus fluviatilis** *from Ubatuba, Brazil, over a twelve-month period (* = P < 0.05).*

Month	Male		Female		Total	Sex ratio	χ^2
	N	%	N	%		M/F	
Mar	11	61.1	7	38.9	18	1:0.6	ns
Apr	21	63.6	12	36.4	33	1:0.6	ns
May	10	41.7	14	58.3	24	1:1.4	ns
Jun	16	61.5	10	38.5	26	1:0.6	ns
Jul	16	66.7	8	33.3	24	1:0.5	ns
Aug	11	52.4	10	47.6	21	1:0.9	ns
Sep	9	33.3	18	66.7	27	1:2.0	ns
Oct	10	38.5	16	61.5	26	1:1.6	ns
Nov	10	52.7	9	47.3	19	1:0.9	ns
Dec	11	29.7	26	70.3	37	1:2.4	ns
Jan	7	20.6	27	79.4	34	1:3.9	s*
Feb	6	35.3	11	64.7	17	1:1.8	ns
Total	**138**	**45.1**	**168**	**54.9**	**30**	**1:1.2**	**ns**

IV. DISCUSSION

In this study, females were larger than males, in contrast to what has been observed for other freshwater crabs, like *Potamon fluviatile* (Gherardi *et al.* 1987) and *Potamon potamios palestinensis* (Gherardi and Micheli 1989). Populations with larger females have already been observed for *Dilocarcinus pagei* in southeastern Brazil (Taddei 1999).

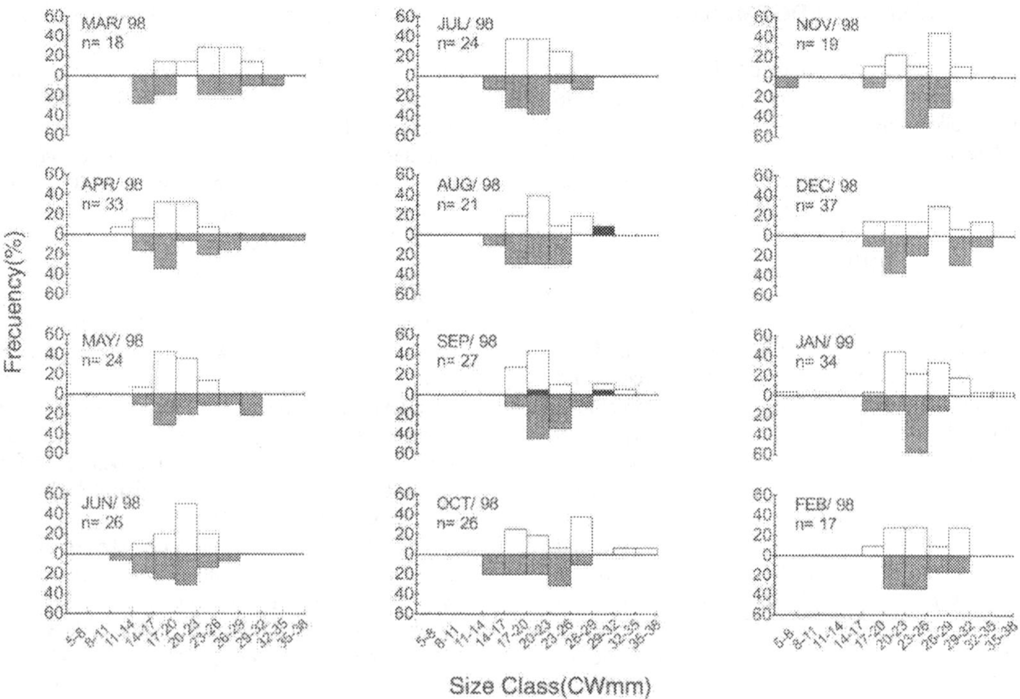

Figure 2. *Monthly frequency distributions of males, females, and females carrying eggs, by size of* **Trichodactylus fluviatilis** *from Ubatuba, Brazil (open bars = females, gray bars = males, solid bars = females carrying young).*

Table 2. *Ratio of males to females by size class of* **Trichodactylus fluviatilis** *from Ubatuba, Brazil (* = P < 0.05).*

Size Class	Male		Female		Total	Sex ratio	χ^2
	N	%	N	%		M/F	
5[-8	1	50.0	1	50.0	2	1:1.0	ns
8[-11	0	0	0	0	0	0	ns
11[-14	1	50.0	1	50.0	2	1:1.0	ns
14[-17	15	78.9	4	21.1	19	1:0.3	s*
17[-20	30	46.9	34	53.1	64	1:1.2	ns
20[-23	30	40.5	44	59.5	74	1:1.5	ns
23[-26	32	53.3	28	46.7	60	1:0.9	ns
26[-29	16	33.3	32	66.7	48	1:2.0	s*
29[-32	9	37.5	15	62.5	24	1:1.7	ns
32[-35	3	30.0	7	70.0	10	1:2.3	ns
35[-38	1	33.3	2	66.7	3	1:2.0	ns
Total	**138**	**45.1**	**168**	**54.9**	**306**	**1:1.2**	**ns**

The single mode found in the size distribution of this population is characteristic of tropical and subtropical populations with continuous recruitment throughout the year (Warner 1967, Díaz and Conde 1989). The cryptic behavior of mature females may explain the absence of ovigerous females and the low number of females carrying young in the samples, a similar behavior has been recorded for the terrestrial crab *Geosesarma notophorum* (Ng and Tan 1995).

The higher number of males observed in the 14 - 17 mm CW class, coincides with the size at which the puberty molt occurs (Cobo, in prep) suggesting that males grow faster than females during the immature phase (Wenner 1972, Trott 1998). The higher number of females in larger size classes may reflect a differential mortality rate, a sampling error, or a differential microhabitat selection with respect to males (Wilson and Pianka 1963, Wenner 1972, Trott 1998).

ACKNOWLEDGEMENTS

The authors are indebted to Prof. Itamar Alves Martins to for his scientific contribution and his help with field collections. We are also grateful to colleagues in the Zoology laboratory for help with the field collections, and to the University of Taubaté to the financial support to develop of this research.

REFERENCES

Bowman TE and Abele LG (1982) Classification of the recent Crustacea. In: Bliss DE (ed) The Biology of Crustacea. Systematics, the fossil record, and biogeography (pp 1-27) Academy Press, New York

Díaz H and Conde JE (1989) Population dynamics and life history of the mangrove crab *Aratus pisonii* (Brachyura, Grapsidae) in a marine environment. Bull Mar Sci 45:148-163

Gherardi F and Micheli F (1989) Relative growth and population structure of the freshwater crab, *Potamon potamios palestinensis*, in the Dead Sea area (Israel). Israel J Zool 36:133-145

Gherardi F and Vannini M (1989) Spatial behavior of the freshwater crab, *Potamon fluviatile*: a radio-telemetric study. Biol Behav 14:28-45

Gherardi F, Guidi S and Vannini M (1987) Behavioral ecology of the freshwater crab, *Potamon fluviatile*: Preliminary observations. Inv Pesq 51:389-402

Gherardi F, Micheli F, Monaci F and Tarducci L (1988) Note sulla biologia ed ecologia del granchio di fiume, *Potamon fluviatile*. Bull Mus St Nat Lunigiana 6:169-174

Hartnoll RG (1988) Evolution, systematics, and geographical distribution. In: Burggren WW and McMahon BR (eds) Biology of the land crabs (pp 6-53) Cambridge University Press, Cambridge

Magalhães C (1987) Notes on some Pseudothelphusidae crabs from Venezuela, Ecuador and Mexico found in the collection of the Museu de Zoologia da Universidade de São Paulo, São Paulo, Brazil. Revta Bras Zool 4:55-58

Magalhães C and Türkay M (1986) *Brasiliothelphusa*, a new Brazilian freshwater crab genus (Crustacea: Decapoda: Pseudothelphusidae). Sencken Biol 66:371-376

Mello GAS (1967) Diferenciação Geográfica e dimorfismo sexual de *Trichodactylus* (*Trichodactylus*) *fluviatilis* Latreille, 1825 (Crustacea, Brachyura). Pap Avulsos Zool S Paulo 3:13-44

Ng PKL and Tan CGS (1995) *Geosesarma notophorum* sp. nov. (Decapoda, Brachyura, Grapsidae, Sesarminae), a terrestrial crab from Sumatra, with novel brooding behavior. Crustaceana 68:390-395

Taddei FG (1999) Biologia populacional e crescimento do caranguejo de água doce *Dilocarcinus pagei* Stimpson, 1861 (Crustacea, Brachyura, Trichodactylidae) da represa municipal de São José do Rio Preto, SP. Dissertação (mestrado) – Instituto de Biociências, Universidade Estadual Paulista, Botucatu, 107 p

Trott TJ (1998) On the sex-ratio of the painted ghost crab *Ocypode gaudichaudii* H. Milne Edwards & Lucas, 1843 (Brachyura, Ocypodidae). Crustaceana 71:47-56

Warner GF (1967) The life history of mangrove tree crab, *Aratus pisonii*. J Zool London 153:321-335

Wenner AM (1972) Sex ratio as a function of size in marine Crustacea. Am Nat 106:321-350.

Wilson MF and Pianka ER (1963) Sexual selection, sex-ratio and mating system. Am Nat 97:405-407

Zar JH (1996) Biostatistical Analysis. 3rd Ed Prentice-Hall, New Jersey

SIZE VARIATION IN THE GRAPSID CRAB *ARATUS PISONII* (H. MILNE-EDWARDS, 1837) AMONG POPULATIONS OF DIFFERENT SUBTROPICAL MANGROVES

Maria Lucia Negreiros-Fransozo

NEBECC, Group of Studies on Crustacean Biology, Ecology and Culture,
Depto. de Zoologia, IBB, UNESP C.P. 510
Botucatu, SP 18618-000, Brazil

mlnf@ibb.unesp.br

ABSTRACT

Size variation in the adult population of the grapsid crab *Aratus pisonii* (H. Milne-Edwards, 1837) was studied by comparing median values obtained from six different populations from nearby mangrove areas. The mangrove structure at those sites was also examined and related to size differences among populations. Disregarding sex, crab size is apparently related to the mangrove structure. Smallest and largest crab sizes were recorded at the sites where tree density and tree diameter were minimum and maximum, respectively. Estimates of size at the onset of maturity, considering the size of the smallest ovigerous female, followed the same trend. These results support the hypothesis that average crab size and size at the onset of maturity can be associated to mangrove productivity.

I. INTRODUCTION

Mangrove forests are complex, intertidal, soft-substrate habitats that occur circum-tropically (Wilson 1989) and constitute one of the major ecosystems of the biosphere (Por 1984). There are probably two extreme mangrove types: soft bottom estuarine and hard bottom euhaline-metahaline mangroves (Por and Dor 1975). The soft mangrove is best developed in brackish estuaries along the equatorial, wet tropical and subtropical coasts (Por 1984).

Despite of their structural variability, mangroves are areas of high primary production holding complex food chains, whether estuarine or coastal, in which mineral nutrients are recycled into organic material (Robertson 1991). Food resources for the rich mangrove fauna, largely constituted by secondary consumers, are almost entirely provided by the mangrove trees and resulting production of particulate matter, since other autotrophic food sources are scarce (Rodríguez 1987). In this sense, the role of the microflora and meiofauna in the decompositon of leafs is very important (Reice et al. 1984).

Along the Brazilian coast, seven species of mangrove trees can be found belonging to the genera *Rhizophora* (3), *Laguncularia* (1), *Conocarpus* (1) and *Avicennia* (2) as listed by Lacerda et al. (1993).

Grapsid crabs, in particular *Sesarma* and *Aratus*, consume leaves and algae by direct grazing while ocypodid deposit feeders assist in the breakdown of particulate organic material both by exposing new surfaces of the swamp floor to microbial attack and by producing faeces rich in carbon and nitrogen (Jones 1984).

Crabs can be extraordinarily abundant in many mangrove forests. Species from six crab families, most of them belonging to the families Grapsidae and Ocypodidae, are known to be adapted to the mangrove ecosystem, their associated tidal flats or bordering areas (Jones 1984, Melo 1996). In the subtropical Brazilian mangroves, the most common genera are *Aratus*,

E. Escobar-Briones & F. Alvarez Eds.
MODERN APPROACHES TO THE STUDY OF CRUSTACEA
PP. 183-188

Sesarma, Armases and *Goniopsis* within the Grapsidae, and *Uca* and *Ucides* in the Ocypodidae.

The grapsid *Aratus pisonii* (H. Milne Edwards, 1837) is one of the most common crustaceans living in the mangroves of the New World, and perhaps the only true marine arboreal crab in this region (Lacerda et al. 1993). Throughout its adult life it inhabits mangrove swamps, actively climbing the trees close to the seaward margins (Warner 1967). This crab can be found on the emergent roots and branches of the mangrove trees (Díaz and Conde 1988), being often regarded as a herbivore specialized on mangrove leaves (Beever et al.1979). Yet, fresh wood pulp, algae and debris of fishes, among other organisms commonly found attached to mangrove roots, may be also part of the diet of *A. pisonii*, for what it has been considered as an opportunistic omnivore (Díaz and Conde 1988, Lacerda et al. 1993). Conde and Díaz (1989) and Conde et al. (1995) have examined several populations of *A. pisonii* in western Venezuela reporting a close relationship between habitat structure and crab size in metahaline mangrove forests.

The aim of this study was to investigate the size variation and the size at onset of maturity of the grapsid crab *A. pisonii* in six nearby estuarine (oligohaline) mangrove areas (Itapanhaú, Itaguaré, Guaratuba, Comprido, Indaiá and Ubatumirim rivers) and its association with mangrove vegetation structure.

II. MATERIALS AND METHODS

The study area is located in southeastern Brazil (23° 20' S, 44° 53' W and 33° 40' S, 46° 09' W). Six sites were chosen according to size and density of the mangrove vegetation and the presence of *A. pisonii*. Sites at Itapanhaú, Itaguaré and Guaratuba rivers are located in Bertioga, while sampling sites at Comprido, Indaiá and Ubatumirim are located in Ubatuba, State of São Paulo (Fig. 1). The position of each locality was obtained using a Geo Positioning System (GPS).

Crabs were collected monthly over a 1-year period. At each site, samples were obtained using a catch per unit effort of two collectors over a 30-min ebb tide period. Maximum carapace width (CW), regarded herein as crab size, was measured to the nearest 0.1 mm and sex was determined according to abdominal morphology. The oviparous condition was also checked and, for each site, crabs larger than the smallest egg-carrying female were considered adults. Percentage of crabs larger than 22.2 mm CW, the size of the largest crab found at the site with lowest median CW, was used as a parameter to compare the percentage of large crabs at each site. The size at onset of maturity was based on the size of the smallest ovigerous female obtained in each site.

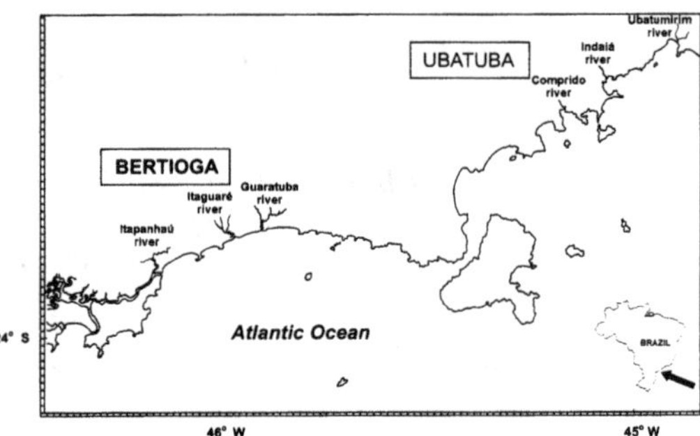

Figure 1. *Localization of each studied site.*

Sampling for estimating the density of the mangrove vegetation followed the procedure used by Schaeffer-Novelli and Cintrón (1990). At each collecting site, an area of 0.1 ha was delimited, and within it a 24 by 48 m rectangle (1,152 m²) was positioned perpendicular to the river margin and subdivided into 12 contiguous parallel bands of six 4 x 4 m (16 m²) quadrats. Quadrat size was determined following the criteria of Green (1979), taking into account the mean size of trees recorded in a pilot study. The number of quadrats effectively sampled was determined based on density estimates obtained in previous studies (Zar 1996, Höisaeter and Matthiesen 1979, Conde and Díaz 1985). Within each quadrat, mangrove trees were also identified. Additionally, the height of at least the ten highest trees in each sampling site was measured with a clinometer and the diameter of all trees in the quadrats was recorded with a caliper, as suggested by Citrón and Schaeffer-Novelli (1984).

The grain size of sediments was determined based on three samples collected at each site in each of the four seasons of the year. The American granulometric scale was followed (Wentworth 1922).

Statistical comparisons among the mangrove features and size of crabs were performed using analysis of variance or Kruskal-Wallis test in cases of non-normal data (Sokal and Rohlf 1995). Spearman correlation analyses between median size of crabs and tree density were also accomplished (Sokal and Rohlf

1995). All statistical analyses were carried out at the 5% significance level.

III. RESULTS

Laguncularia racemosa was predominant in all mangroves, except at Itapanhaú where *Rhizophora mangle* was the most abundant. The mangroves at Itaguaré and Guaratuba present a similar proportion of *R. mangle*. In the case of *Avicennia shaueriana*, the highest density and the largest trees of this species were observed at Guaratuba. The relative density of tree species for each site is presented in Figure 2.

Regarding tree height, the lowest trees were found at Itapanhaú and Itaguaré, differing significantly from the remaining mangrove sites. The highest trees were found at Guaratuba, but no statistical difference was found between those and the ones sampled at the sites in Ubatuba. The lowest density of trees was found at Guaratuba. At Itapanhaú, where the lowest trees were recorded together with Itaguaré, tree density was intermediate and median diameter the smallest. The comparisons of mangrove height, density and diameter among sites are presented in Table 1.

Table 1. *Mangrove features at each study site. All results are median values. Statistical results are the outcome of Kruskal-Wallis tests (values in the columns sharing at least one letter did not differ statistically among sites; P > 0.05).*

SITES	TREES HEIGHT (m)	DENSITY (trees/ha)	TREES DIAMETER (mm)
ITAPANHAÚ	7.01 c	4,320 bc	70.8 ab
ITAGUARÉ	7.45 b	3,600 bc	74.3 a
GUARATUBA	11.06 a	2,160 c	54.2 abc
UBATUMIRIM	10.60 a	7,200 b	47.3 bc
COMPRIDO	9.63 b	3,600 bc	64.0 ab
INDAIÁ	10.0 ab	10,800 a	45.5 c

Fine and very fine sand prevailed at most sites, while silt-clay was the most representative fraction at Itapanhaú mangrove (Fig. 3).

Mean size (CW) estimated for adults of *A. pisonii* were: 16.4 ± 2.2 mm at Itapanhaú (n = 687), 17.9 ± 2.4 mm at Itaguaré (n = 522), 18.8 ± 2.5 mm at Guaratuba (n = 597), 19.1 ± 2.6 mm at Ubatumirim (n = 601), 19.4 ± 2.2 mm at Comprido (n = 386) and 21.0 ± 2.4 mm at Indaiá (n = 407). Modal size increased from 16.6 mm at Itapanhaú to 21.3 mm in Indaiá (Fig. 4).

The smallest specimens were found at Itapanhaú mangrove, where the modal class was 16.64 mm of CW. Only 0.15% of the specimens collected in this locality were larger than 22.2 mm CW. The largest specimens were found at Indaiá, where the modal class was 21.3 mm CW, being 15% of the total specimens obtained larger than 22.2 mm CW. Concerning the ovigerous females, the smallest one was obtained at Itapanhaú, while the largest one was collected at Indaiá. The highest percentage of ovigerous females was also found at Indaiá. The descriptive data on the *A. pisonii* populations from the different sampling sites are presented in Table 2.

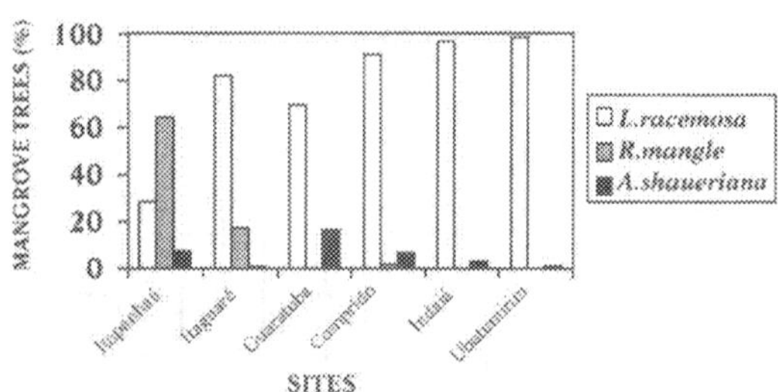

Figure 2. *Relative density of mangrove trees at each studied site.*

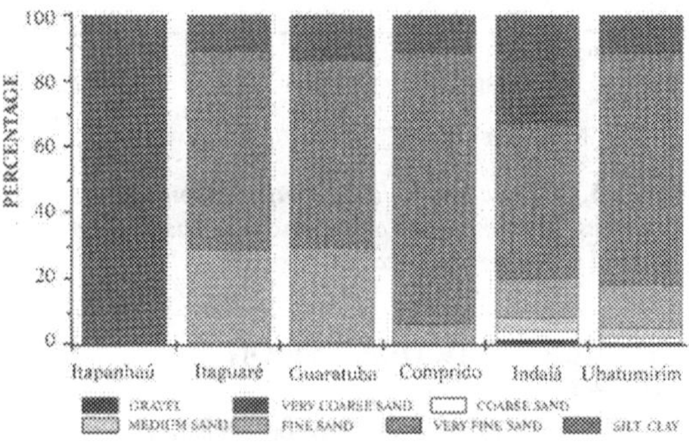

Figure 3. *Texture composition of the sediment in each mangrove area sampled.*

Disregarding sex and ovigerous condition, crabs from Itapanhaú were the smallest ($P < 0.05$) while those collected at Indaiá were the largest ($P < 0.05$) (Figs. 5, 6, 7). Spearman's correlation coefficient (0.81) indicates that size of crabs increases with tree density ($P < 0.05$).

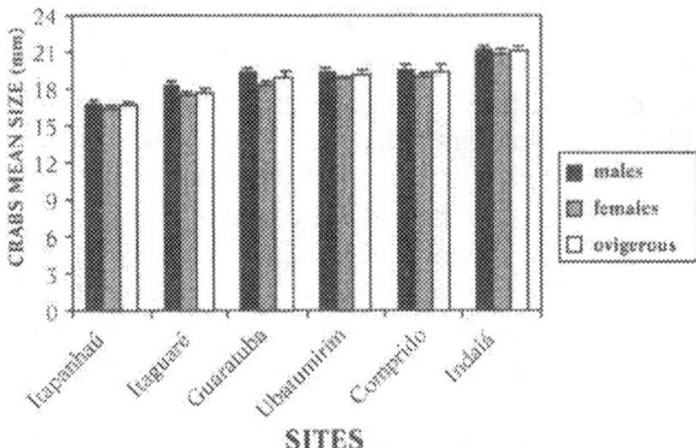

Figure 4. *Mean size of **Aratus pisonii** in each studied site, error bars indicate the 95% confidence interval.*

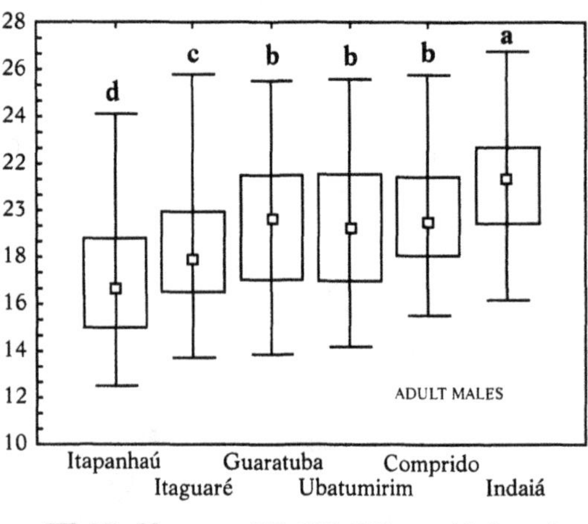

Figure 5. *Size of adult males of **Aratus pisonii** at the sampling sites. Median values above box plots sharing the same letter did not differ statistically (P > 0.05).*

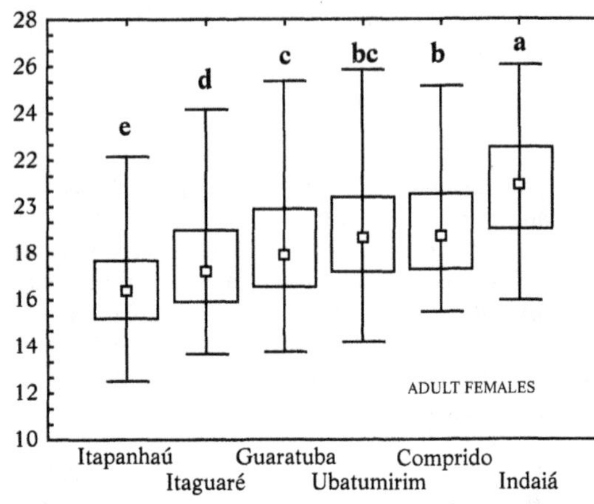

Figure 6. *Size of adult females of **Aratus pisonii** at the sampling sites. Median values above box plots sharing the same letter did not differ statistically (P > 0.05).*

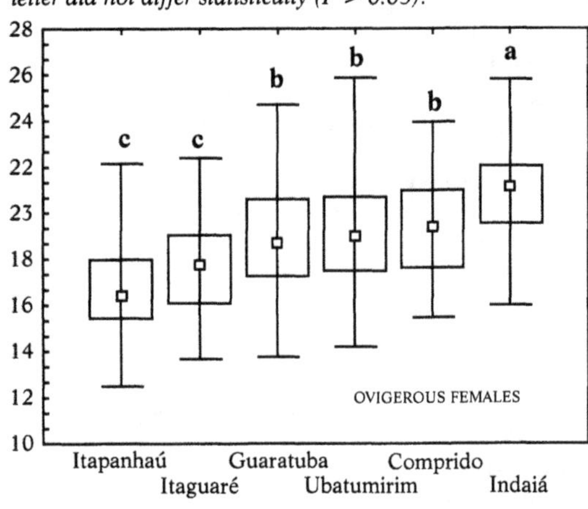

Figure 7. *Size of ovigerous females of **Aratus pisonii** at the sampling sites. Median values above the box plots sharing the same letter did not differ statistically (P > 0.05).*

IV. DISCUSSION

All studied sites may be considered as soft-bottom mangroves as described by Por and Dor (1975), but there are some structural differences among the sampled sites, mainly concerning the relative density of mangrove species. *Rhizophora mangle* is most abundant in the site where silt and clay are predominant.

Since leaves and other vegetal items are the principal food source of *A. pisonii* (Warner 1967, Díaz and Conde 1988, Lacerda et al. 1993), it can be presumed that the nutritional requirements for growth and reproduction should be largely obtained from the man-grove trees. Growth and reproductive potential are highly dependent on the quality of the ingested food. Therefore, for a given foraging effort the amount of energy available would be greatly different under different nutritional conditions.

Reduced size of *A. pisonii* at Itapanhaú mangrove might be related to the reduced size of the mangrove trees and the great proportion of *R. mangle* found at this site. The largest specimens of *A. pisonii* were found at

Table 2. ***Aratus pisonii***, *descriptive statistics of the populations examined.*

SITES	Males (N)	Females (N)	Largest crab (mm)	Ovigerous crabs (N)	Smallest ovigerous crab (mm)	Number (%) of crabs ≥ 22.2 mm
ITAPANHAÚ	314	373	22.2	128	12.5	1 (0.15)
ITAGUARÉ	238	284	24.2	81	13.7	12 (2.30)
GUARATUBA	294	303	25.4	90	13.8	18 (3.02)
UBATUMIRIM	258	343	25.9	150	14.2	32 (5.32)
COMPRIDO	207	179	25.2	56	15.5	16 (4.15)
INDAIÁ	188	219	26.1	102	16.2	61 (15.00)

Indaiá mangrove, where vegetation density is highest, presumably providing a greater amount of litter.

The evident size variation presented in this study indicates that crabs settling in better conditions of food supply may grow larger. Phenotypic plasticity is regarded as a species-specific feature which in the case of size variation may reflect changes in habitat productivity as observed for metahaline mangroves (Conde and Díaz 1989).

In this study, it was shown that size variation of *A. pisonii* according to the mangrove forest structure may also take place in estuarine (oligohaline) populations. Additionally, the reproductive potential of the studied populations can also be dependent on the available resources as the abundance of ovigerous females and the size at onset of maturity showed variation among mangroves.

ACKNOWLEDGEMENTS

The author is indebted to FAPESP (Fundação de Amparo à Pesquisa no Estado de São Paulo) for financial support (contracts number 94/4878-8; 95/8520-3; 98/03134-6 and 00/03170-4). Thanks to Augusto A. V. Flores for his helpful comments on the manuscript.

REFERENCES

Beever JW III, Simberloff SD and King LL (1979) Herbivory and predation by the mangrove tree crab *Aratus pisonii*. Oecologia 43:317-328

Cintrón G and Schaeffer-Novelli Y (1984) Methods for studying mangrove structure. In: Snedaker SC and Snedaker JG (eds) The mangrove ecosystem: research methods (pp. 91-113) UNESCO, Monographs on Oceanographic Methodology 8, 251 p

Conde JE and Díaz H (1985) Diseño de muestreo aleatorio estratificado aplicado al estudio de poblaciones del género *Uca* (Brachyura: Ocypodidae). Inv Pesq 49:567-579

Conde JE and Díaz H (1989) Productividad del habitat e historias de vida del cangrejo de mangle *Aratus pisonii* (H. Milne Edwards) (Brachyura, Grapsidae). Bol Inst Oceanogr Venezuela Univ Oriente 28:113-120

Conde JE, Alarcón C, Flores S and Díaz H (1995) Nitrogen and tannins in mangrove leaves might explain interpopulation variations in the crab *Aratus pisonii*. Avance de investigación, Acta Cient Venez 46:245-246

Díaz H and Conde J E (1988) On the food sources for the mangrove tree crab *Aratus pisonii* (Brachyura, Grapsidae). Biotropica 20:348-350

Green RH (1979) Sampling design and statistical methods for environmental biologists. Wiley & Sons, New York

Höisaeter T and Matthiesen A (1979) Report on some statistical aspects of marine biological sampling based on a UNESCO sponsored training course in sampling design for marine biologists. Univ of San Carlos Publ., Cebú, Phillipines, 118 p

Jones DA (1984) Crabs of the mangal ecosystem. In: Por FD and Dor I (eds) Hydrobiology of the mangal (pp. 89-109) Dr. W. Junk Publishers, Boston

Lacerda D, Conde JE, Alarcón C, Alvarez-Léon R, Bacon PR, D'Acroz L, Kjerfve B, Polanía J and Vannucci M (1993) Ecosistemas de manglar de América Latina y el Caribe: Sinopsis (pp. 1-38) Informe técnico del proyecto conservación y aprovechamiento sostenible de bosque de manglar en las regiones America Latina y Africa y Japón

Melo GAS (1996) Manual de Identificação dos Brachyura (Caranguejos e Siris) do Litoral Brasileiro. Plêiade ed, São Paulo, 604 p

Por FD (1984) The ecosystem of the mangrove: general considerations. In: Por FD and Dor I (eds) Hydrobiology of the mangrove (pp 1-14) The ecosystem of the mangrove forests. Dr W Junk Publishers, Boston, 260 p

Por FD and Dor I (1975) The hard bottom mangroves of Sinai, Red Sea. Rapp Comm int Mer Medit 23:145-147

Reice SR, Spira Y and Por FD (1984) Decomposition in the mangrove of Sinai: the effect of spatial heterogeneity. In: Por FD and Dor I (eds) Hydrobiology of the mangrove (pp 193-199) The ecosystem of the mangrove forests. Dr W Junk Publishers, Boston, 260 p

Robertson AI (1991) Plant-animal interactions and the structure and function of mangrove forest ecosystems. Aust J Ecol 16:433-443

Rodríguez G (1987) Structure and production in neotropical mangroves. Tree 2:264-267

Schaeffer-Novelli Y and Cintón G (1990) Status of mangrove research in Latin America and the Caribbean. Bolm Inst Oceanogr 38:93-97

Sokal RR and Rohlf FJ (1995) Biometry. 3rd Edition (Third Printing), WH Freeman and Co, New York

Warner GF (1967) The life history of the mangrove tree crab *Aratus pisonii*. J Zool Lond 153:321-335

Wentworth CH (1922) A scale of grade and class terms for clastic sediments. J Geol 30:377-392

Wilson KA (1989). Ecology of mangrove crabs: predation, physical factors and refuges. Bull Mar Sci 44:263-273

Zar JH (1996) Biostatistical Analysis. Prentice-Hall, USA

MORPHOMETRIC STUDY OF THE GHOST CRAB *OCYPODE QUADRATA* (FABRICIUS, 1887) (BRACHYURA, OCYPODIDAE) FROM UBATUBA, SÃO PAULO, BRAZIL

Adilson Fransozo, Maria Lucia Negreiros-Fransozo and Giovana Bertini

*NEBECC, Group of Studies on Crustacean Biology, Ecology and Culture
Departamento de Zoologia, Instituto de Biociências C.P. 510 – Universidade Estadual Paulista, UNESP,
18618-000 Botucatu (São Paulo), Brasil*

fransozo@ibb.unesp.br

ABSTRACT

A morphometric study of the ghost crab *Ocypode quadrata* from Ubatuba, São Paulo, Brazil, focusing on the relationships between carapace width and carapace length, abdomen width, gonopod length and propodus length and height of the major cheliped is presented. An allometric positive growth was found for most of the analyzed relationships; however, important changes in the gonopod and abdomen were observed at the size at which sexual maturity is attained. Morphological (external) sexual maturity for both sexes is achieved at a smaller size than that at which gonad maturity is reached. Concerning heterochely, 53.3% of the analyzed specimens exhibited a longer propodus length in the right cheliped.

I. INTRODUCTION

The value of morphometric studies is well recognized in studies on the Crustacea, and relative growth data have been extensively utilized to evaluate developmental stages of the Crustacea (Haley 1973). In several families, secondary sexual characters (chela, abdomen, and pleopods) were seen to have differential growth rates after the attainment of maturity (Hartnoll 1978). In *Ocypode quadrata* the cheliped is used during courtship to maximize mating, besides being used to feed, as protection from predators and for competing over territory. Thus the growth of such appendage can evidence some change during the puberty molt (Hartnoll 1974, Mantelatto and Fransozo 1994, Tushida and Fujikura 2000).

Ghost crabs are commonly found in tropical and subtropical sand beaches around the world. *Ocypode quadrata* (Fabricius, 1787) represents the genus in the Western Atlantic, occurring from Rhode Island (42° N and 70° W), USA to Rio Grande do Sul (30° S and 50° W), Brazil (Melo 1996; Alberto and Fontoura 1999).

Although a number of studies have been conducted on the relative growth of brachyurans in Brazil (Pinheiro and Fransozo 1993, Mantelatto and Fransozo 1994, Negreiros-Fransozo et al. 1994, Hiyodo and Fransozo 1995, Góes and Fransozo 1997, Cobo and Fransozo 1998), none of them have dealt with ocypodid crabs. Previous studies on the morphometry of ocypodid crabs include those by Sandon (1937), Crane (1941), and Haley (1969, 1973). However, other aspects of the biology of *O. quadrata* have been investigated in Brazil, such as its reproductive cycle (Negreiros-Fransozo et al. 2002), in which the functional sexual maturity is achieved around 23.0 mm of carapace width for females and 20.0 mm of carapace width for males. The goal of this study is to analyze the relative growth of *O. quadrata*, focusing on the morphometric relationships between carapace width with the dimensions of other appendages. Through the analyses of these

E. Escobar-Briones & F. Alvarez Eds.
MODERN APPROACHES TO THE STUDY OF CRUSTACEA
PP. 189-195

parameters it is possible to characterize the allometry levels and to make inferences about the sizes at which males and females reach sexual maturity, thus evidencing their puberty molt. The frequency of occurrence of a major right cheliped in each sex and their growth are also investigated.

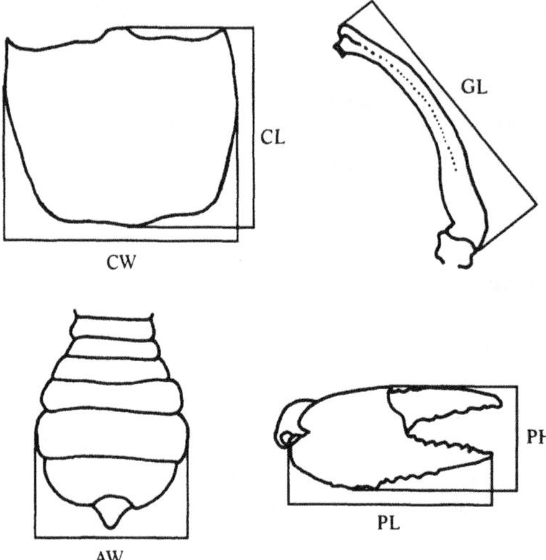

Figure 1. *Ocypode quadrata. Indication of each measurement taken (CL= carapace length; CW= carapace width; PL= cheliped propodus length; PH= cheliped propodus height; AW= abdomen width and GL= gonopod length).*

II. MATERIAL AND METHODS

The region of Ubatuba is characterized by a tropical - subtropical climate. The mean temperature of the surface water is around 18° C in the winter and 29° C in the summer (Negreiros-Fransozo et al. 1999). Nevertheless, the deeper water layer near the coast presents lower temperatures during summer due to the entrance of the South Atlantic Central Current waters (Pires 1992).

Crabs were collected by hand at night during low tide. Collections were carried out monthly during 1998 in the "Vermelha do Sul" beach (23° 27' 42" S and 45° 02' 48" W). Crabs were individually kept in plastic bags, labeled and frozen. In the laboratory, crabs were counted and numbered. The secondary sexual characters present in the abdominal appendages were examined under the stereomicroscope.

Body dimensions were measured with a caliper (0.01 mm). Measurements included: carapace width (CW), carapace length (CL), abdomen width (AW), gonopod length (GL), length (PL) and height (PH) of the propodi of both chelipeds (Fig. 1). A stereomicroscope provided with camera lucida was used to measure the smallest crabs and gonopods. The manus area was calculated obtaining the product of PL . PH and the carapace area from CL . CW.

Male and female size was compared with the Mann-Whitney test (a = 0.05) (Sokal and Rohlf 1995). The size at sexual maturity was obtained using of Somerton's Mature I and Mature II software (Somerton 1980a, b). Analysis of covariance (ANCOVA) was performed to compare slopes and intercepts of the obtained regressions within each allometric relationship (Zar 1996). Departures from isometry (H_o: b=1) were tested using a Student's t-test on the obtained slope values (a = 0.05).

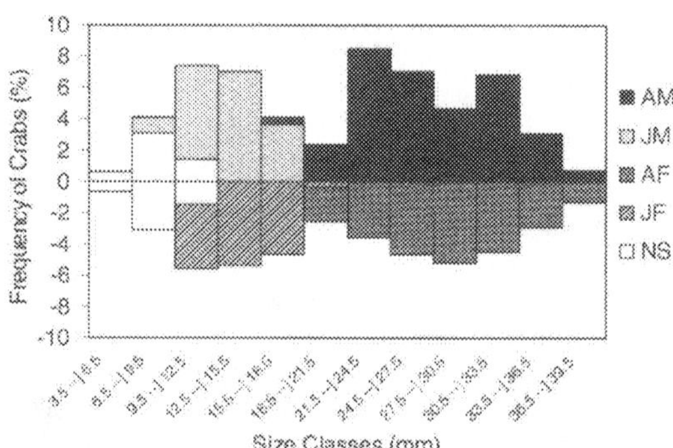

Figure 2. *Size frequency distribution of the sampled population in Vermelha beach, Ubatuba, SP, Brazil. (NS= non-sexable specimens; JF =juvenile females; AF = adult females; JM = juvenile males; AM = adult females).*

III. RESULTS

A total of 543 specimens was captured and their size frequency distribution obtained (Fig. 2). The size of *O. quadrata* ranged from 5.8 to 12 mm for non-sexable specimens (N= 56); from 8.5 to 37.5 mm for males (N=278), and from 9.5 to 39.2 for females (N=209). The smallest and the largest ovigerous females found measured 24.5 and 39.0 mm of CW, respectively. The median size of adult males (23.7 mm) and females (23.1mm) did not differ statistically ($P > 0.05$).

Table 1. *Regression analyses of morphometric data of* **Ocypode quadrata**. *Carapace width (CW) was used as the independent variable.*

Variable	Demographic Category	N	Power Equation $Y = a.X^b$	Linearized equation $LogY = loga + blogX$	r^2	Allometry level *
CL	JM	95	$CL = 0.837.CW^{0.984}$	$LogCL = 0.984 \, LogCW - 0.083$	0.97	=
	AM	180	$CL = 0.623.CW^{1.086}$	$LogCL = 1.086 \, LogCW - 0.206$	0.99	+
	JF	74	$CL = 0.833.CW^{0.977}$	$LogCL = 0.977 \, LogCW - 0.079$	0.96	=
	AF	134	$CL = 0.647.CW^{1.066}$	$LogCL = 1.066 \, LogCW - 0.189$	0.98	+
PL	JM	95	$PL = 0.507.CW^{1.137}$	$LogPL = 1.137 \, LogCW - 0.295$	0.94	+
	AM	182	$PL = 0.264.CW^{1.351}$	$LogPL = 1.351 LogCW - 0.578$	0.97	+
	JF	73	$PL = 0.743.CW^{0.990}$	$LogPL = 0.99 \, LogCW - 0.129$	0.89	=
	AF	121	$PL = 0.429.CW^{1.176}$	$LogPL = 1.176 \, LogCW - 0.368$	0.96	+
PH	JM	95	$PH = 0.223.CW^{1.157}$	$LogPH = 1.157 \, LogCW - 0.652$	0.94	+
	AM	182	$PH = 0.122.CW^{1.365}$	$LogPH = 1.365 \, LogCW - 0.914$	0.97	+
	JF	73	$PH = 0.240.CW^{1.125}$	$LogPH = 1.125 \, LogCW - 0.620$	0.96	+
	AF	121	$PH = 0.187.CW^{1.207}$	$LogPH = 1.207 \, LogCW - 0.728$	0.98	+
AW	JM	91	$AW = 0.324.CW^{0.876}$	$LogAW = 0.88 \, LogCW - 0.484$	0.81	−
	AM	181	$AW = 0.173.CW^{1.096}$	$LogAW = 1.096 \, LogCW - 0.762$	0.93	+
	JF	76	$AW = 0.154.CW^{1.204}$	$LogAW = 1.204 \, LogCW - 0.813$	0.86	+
	AF	128	$AW = 0.050.CW^{1.673}$	$LogAW = 1.673 \, LogCW - 1.301$	0.92	+
GL	JM	62	$GL = 0.003.CW^{2.642}$	$LogGL = 2.642 \, LogCW - 2.523$	0.91	+
	AM	126	$GL = 0.217.CW^{1.122}$	$LogGL = 1.122 \, LogCW - 0.664$	0.94	+

*t test; $\mu = 0.05$

(CL = carapace length; PL = Major cheliped propodus length; PH = Major cheliped propodus height; AW = abdomen width; GL = gonopod length; CW = carapace width; JM = juvenile male; AM = adult male; JF = juvenile female; AF = adult female; N = number of specimens; = isometry; + positive allometry; - negative allometry).

The regression parameters (slope and intercept) of all allometric relationships, for both sexes in juveniles and adults, were statistically different ($P < 0.05$), thus evidencing that each demographic category has a distinct growth pattern. The relative growth equations obtained for males and females are shown in Table I. All analyzed relationships for adults of both sexes presented positive allometry (Figs. 3, 4, 5).

The relationships that best indicate the change in the allometric coefficient between juveniles and adults are CW vs. AW for females and CW vs. GL for males. However, the CW vs. PL relationship, despite its lower allometric coefficient, can also be used to classify crabs into juveniles or adults.

According to Somerton's technique, the size at which 50% of the females reached maturity is 19.2 mm of CW based on the CW vs. AW relationship, while males reached maturity at 18.3 mm of CW, based on the CW vs. GL relationship. With reference to the cheliped - propodus growth, 53.3% of the analyzed specimens had a larger right propodus (Table II). The relative growth of the two cheliped types (right or left major manus), using manus area vs. carapace area are represented in figure 6. Although both sexes present a positive allometric growth for manus area, males have a higher allometric coefficient. Within the same sex, the allometric coefficient varies depending on which side the major cheliped is on.

IV. DISCUSSION

For brachyurans in general, when two carapace measurements are correlated (e.g., CL and CW), changes during the ontogeny that tend to be isometric are hard to notice. Such fact has been frequently observed, for example for *Portunus spinimanus* (Santos et al. 1995) and *Sesarma rectum* (Mantelatto and Fransozo 1999). For *O. quadrata*, such pattern has only been observed for juveniles of the same sex, because after the puberty molt CW has larger growth increments than CL.

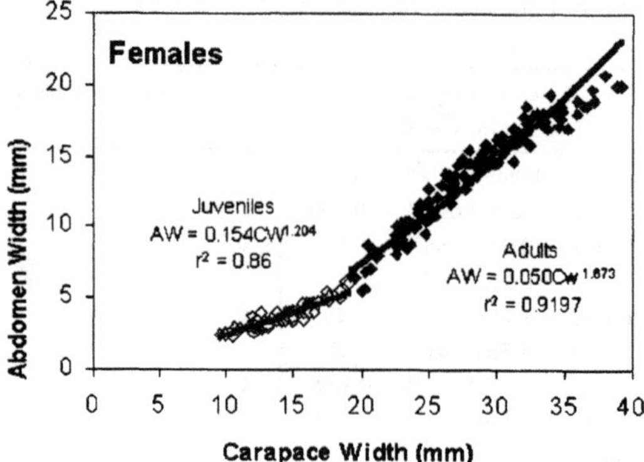

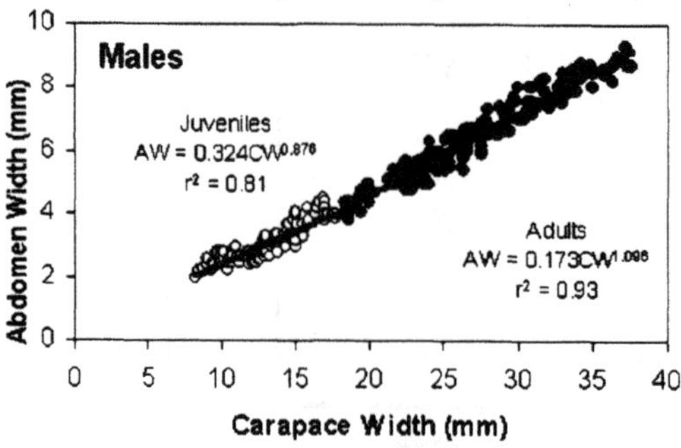

Figure 3. *Scatter diagram and adjusted curve for the relationship CW vs. AW in* **Ocypode quadrata**, *upper graphic for females and lower graphic for males (□○=juveniles;* ■ ● *= adults).*

Relationships that indicate the importance of chelar dimensions (i. e., chelar propodus lenght or height) to characterize sexual dimorphism or maturation in crabs have been previously reported (Hartnoll 1974, Vannini and Gherardi 1988). These studies are especially relevant for males, because the chelae that are used in intra or interspecific combats can achieve disproportionately large sizes. Another advantage for males would be during reproduction, when females are held and manipulated with the chelipeds during copulation (Warner 1970).

The PL vs. CW relationship in males showed a clear increase in propodus length soon after the puberty molt, which may occur at approximately 20 mm of CW, due to an increase in the level of allometric growth. The juvenile phase also exhibits positive allometry, but it is less pronounced. According to our data, the analyzed population of *O. quadrata* from

Ubatuba acquires external sexual maturity at a smaller size than that at which they achieve gonad maturity (20 mm for males, 23 mm for females, Negreiros-Fransozo et al. 2002).

The growth of AW usually characterizes a distinct sexual dimorphism in representatives of the infraorder Brachyura. As far as *O. quadrata* is concerned, the relationship AW vs. CW follows the same pattern found for the majority of crab species. Based on the regression AW vs. CW, it can be concluded that the puberty molt for females of this species occurs at 19.2 mm of CW, which corresponds to the size at which 50% reach sexual maturity. Haley (1969) found that the puberty molt of the same species occurred at 24 mm for males and 26 mm of CW for females in a Texas population.

Table 2. *Proportions of major cheliped propodus in each sex.*

SEX	N	Major Cheliped	%
Males	52	Right	28.0
	126	Left	23.2
Females	96	Right	17.7
	113	Left	20.8
ns	41	Right	7.6
	15	Left	2.7
Total	289	Right	53.3
	254	Left	46.8

The differences observed concerning the puberty molt could be explained by the maximum size range attained by the Texas population, which is 35% larger than the Ubatuba population. Further, the ghost crabs studied by Haley (1969) could have had more available energy than the crabs from Ubatuba, and this could have delayed the puberty molt.

Ocypode quadrata shows positive allometric growth for most of the analyzed relationships. However, remarkable changes during the ontogeny were observed in the gonopod and abdomen, indicating that their development is closely connected to reproductive aspects. In many brachyurans, male chelipeds become larger than female ones immediately after maturity, as secondary sexual characteristic (Hartnoll 1974). *Ocypode quadrata* is no exception, nevertheless the difference in growth between sexes had not been previously documented. There is a slight difference, which

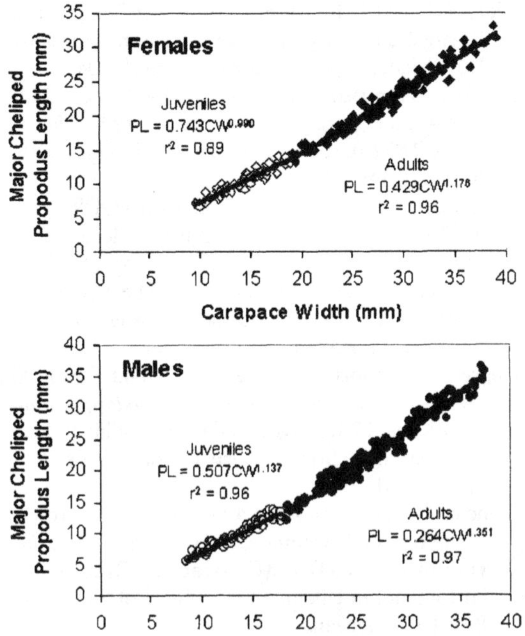

Figure 4. *Scatter diagram and adjusted curve for the relationship CW vs. PL in **Ocypode quadrata**, upper graphic for females and lower graphic for males (□○ = juveniles; ■ ● = adults).*

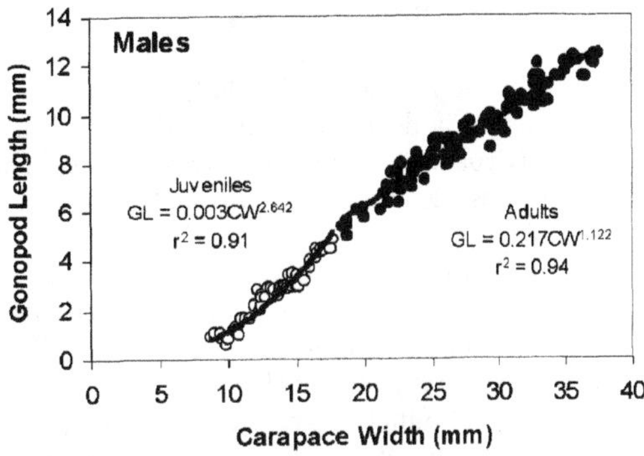

Figure 5. *Scatter diagram and adjusted curve for the relationship CW vs. GL in **Ocypode quadrata** males (O = juveniles; ● = adults).*

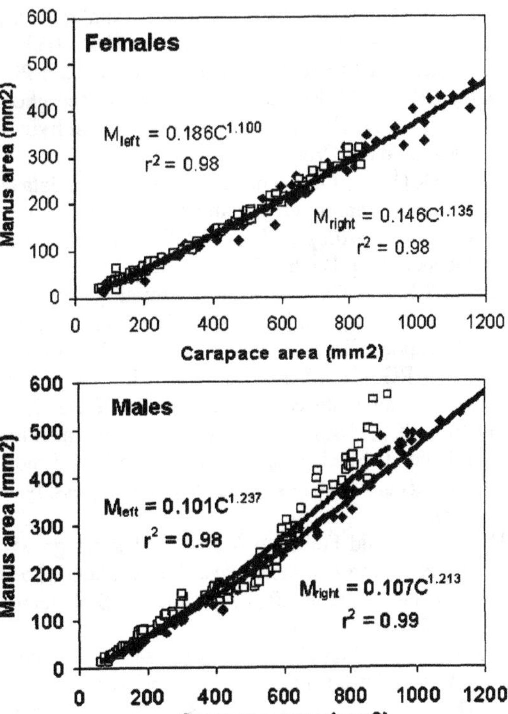

Figure 6. *Ocypode quadrata.* *Comparative growth of the cheliped using manus area (PL . PH) vs. carapace area (CL . CW), upper graphic for females, lower graphic for males (□ = left cheliped major; ■ = right cheliped major).*

was only detected by an statistical test. As already pointed by Tushida and Fujikura (2000), for other brachyuran species, the role of the cheliped may not be exclusively related to reproduction. In this sense, the growth pattern might be related to feeding or defense purposes. Thus, a more detailed investigation is needed for this species, in order to determine what are the more important cheliped functions.

ACKNOWLEDGEMENTS

The authors are grateful to "The State of São Paulo Research Foundation – FAPESP" (#94/4878-8; #95/08520-3) and "Foundation for Develoment of the São Paulo State University – FUNDUNESP" which provided financial support. We are also thankful to MSc. Rogério Caetano da Costa for his help during field samplings.

REFERENCES

Alberto RMF and Fontoura NF (1999) Distribuição e estrutura etária de *Ocypode quadrata* (Fabricius, 1787) (Crustacea, Decapoda, Ocypodidae) em praia arenosa do litoral sul do Brasil. Rev Bras Biol 59:95-108

Cobo VJ and Fransozo A (1998) Relative growth of *Goniopsis cruentata* (Crustacea, Brachyura, Grapsidae), on the Ubatuba region, São Paulo, Brazil. Iheringia, Ser Zool 84:21-28

Crane J (1941) On the growth and ecology of brachyuran crabs of the genus *Ocypode*. Zoologica 26:297-310

Góes JM and Fransozo A (1997) Relative growth of *Eriphia gonagra* (Fabricius, 1781) (Crustacea, Decapoda, Xanthidae) in Ubatuba, State of São Paulo, Brazil. Náuplius 5:85-98

Haley SR (1969) Relative growth and sexual maturity of the Texas ghost crab, *Ocypode quadrata* (Fabr.) (Brachyura, Ocypodidae). Crustaceana 17:285-297

Haley SR (1972) Reproductive cycling in the ghost crab, *Ocypode quadrata* (Fabr.) (Brachyura, Ocypodidae). Crustaceana 23:1-11

Haley SR (1973) On the use of morphometric data as a guide to reproductive maturity in the ghost crab, *Ocypode ceratophtalmus* (Pallas) (Brachyura, Ocypodidae). Pac Sci 27:350-362

Hartnoll RG (1974) Variation in growth pattern between some secondary sexual characters in crabs (Decapoda, Brachyura). Crustaceana 27:130-136

Hartnoll RG (1978) The determination of relative growth in Crustacea. Crustaceana 34:282-293

Hartnoll RG (1982) Growth (pp 111-196) In: Bliss, DE (ed) The biology of Crustacea, embriology, morphology and genetics. Vol.2, Academic Press, New York

Hiyodo CM and Fransozo A (1995) Relative growth of spider crab *Acanthonyx scutiformis* (Dana, 1851) (Crustacea, Decapoda, Majidae). Arq Biol Tecnol 38:969-981

Mantelatto FLM and Fransozo A (1994) Crescimento relativo e dimorfismo sexual de *Hepatus pudibundus* (Herbst, 1785) (Decapoda, Brachyura) no litoral paulista. Pap Avulsos Zool 39:33-48

Mantelatto FLM and Fransozo A (1999) Relative growth of the crab *Sesarma rectum* Randall 1840 (Decapoda, Brachyura, Grapsidae) from Bertioga, São Paulo, Brazil. Pak J Mar Biol 5:11-21

Melo GAS (1996) Manual de identificação dos Brachyura (caranguejos e siris) do litoral brasileiro. Ed. Plêiade, Fapesp, São Paulo, Brazil, 604 p

Negreiros-Fransozo ML, Fransozo A and Reigada ALD (1994) Biologia populacional de *Epialtus brasiliensis* Dana, 1852 (Crustacea, Decapoda, Majidae). Rev Bras Biol 54:173-180

Negreiros-Fransozo ML, Nakagaki JM and Reigada ALD (1999) Seasonal occurrence of decapods in shallow waters of a subtropical area. (pp. 351-361) In: Klein JCV and Schram FR (eds) The Biodiversity crisis and Crustacea. Crustacean Issues. Vol. 12, AA Balkema, Rotterdam, Brookfield

Negreiros-Fransozo ML, Fransozo A and Bertini G (2002) Reproductive cycle of *Ocypode quadrata* (Fabricius, 1787) (Decapoda, Ocypodidae) on a sand beach in Southeastern Brazil. J Crust Biol 22:157-161

Pinheiro MAA and Fransozo A (1993) Relative growth of the speckled swimming crab *Arenaeus cribrarius* (Lamarck, 1818) (Crustacea, Brachyura, Portunidae) in Ubatuba Coast, State of São Paulo, Brazil. Crustaceana 65:377-389

Pires AMS (1992) Structure and dynamics of benthic megafauna on the continental shelf offshore of Ubatuba, southeastern Brazil. Mar Ecol Prog Ser 86:63-76

Sandon H (1937) Differential growth in the crab *Ocypoda*. Proc Zool Soc London, (A) 107:397-414

Santos S, Negreiros-Fransozo ML and Fransozo A (1995) Morphometric relationship and maturation in *Portunus spinimanus* Latreille, 1819 (Crustacea, Brachyura, Portunidae). Rev Bras Biol 55:545-553

Sokal RR and Rohlf FJ (1995) Biometry. 3rd edition (Third printing). Freeman WH and Co, New York

Somerton D (1980a) Fitting a straight line to Hiatt growth diagrams: a re-evaluation. J Cons Int Explor Mer 39:15-19

Somerton D (1980b) A computer technique for estimating the size of sexual maturity in crabs. Can J Fish Aquatic Sci 37:1488-1494

Teissier G (1960) Relative growth (pp 537-560) In: Watermann TH (ed) The Physiology of Crustacea, Vol 1, Academic Press, New York

Tushida S and Fujikura K (2000) Heterochely, relative growth, and gonopod morphology in the Bythograeid crab, *Austinograea williamsi* (Decapoda, Brachyura). J Crust Biol 20:407-414

Vannini V and Gherardi F (1988) Studies on the pebble crab, *Eriphia smithi* MacLeay, 1838 (Xanthoidea, Menippidae): patterns of relative growth and populations structure. Trop Zool 1:203-206

Warner GF (1970) Behaviour of two species of grapsid crab during intraspecific encounters. Behaviour 36:9-19

Zar JH (1996) Bioestatistical Analysis. Prentice-Hall, Upper Saddle River, 662 p

SHALLOW-WATER BENTHIC DECAPOD CRUSTACEANS OF CHANKANAAB PARK, COZUMEL ISLAND, MEXICO

Patricia Briones-Fourzán and Enrique Lozano-Álvarez

Instituto de Ciencias del Mar y Limnología, Unidad Académica Puerto Morelos, Universidad Nacional Autónoma de México, Ap. Postal 1152, Cancún, Q. R., 77500 México

briones@mar.icmyl.unam.mx

ABSTRACT

Chankanaab Park is a protected area in Cozumel Island, Mexico, that includes a portion of the coral reef marine environment, and a small inland lagoon connected to the sea by a narrow underground tunnel, 60 m long. The present study is the first addressing the benthic decapod fauna in this Park. Because of the protected status and the small area of the marine zone of the Park (~7 ha), collections were made only by hand using SCUBA diving. Seventy-three decapod species belonging to 22 families were identified in the Park area. Twenty-one species occurred in both habitats (marine zone and lagoon), indicating that there is faunal exchange between both sites through the tunnel, either by means of larval phases, or by crawling or walking individuals. In fact, nine of these 21 species were actually recorded in the tunnel. Of the remaining species, 49 were restricted to the marine zone, and one was found solely in the lagoon. The higher species richness in the marine zone seems related to the wider variety of habitats in that area, especially living substrates such as erect sponges, octocorals, seagrasses, and anemones, which harbored a number of associated decapod species but were completely absent from the lagoon. Although no human activities are allowed in the lagoon, and fishing is forbidden in the whole Park area, the Park receives a daily average of 1000 visiting tourists, most of which conduct aquatic or underwater activities in the marine zone. Further protection of the live substrates in the marine zone is recommended, to warrant the preservation of the rich decapod fauna in the area.

I. INTRODUCTION

Coral reefs are heterogeneous environments that include many types of habitats. In coral reefs, many living species are closely associated with others or with a specific type of substrate (Bruce 1976, 1977). Crustaceans are among the dominant members of the coral reef community (McCloskey 1970), and in many coralline environments the majority of crustacean species and individuals are decapods (Coull and Bell 1983). For example, of 2,000 species of crabs reported in the Pacific, more than 500 occur in coral reefs (Castro 1976). In *Pocillopora* heads, decapods represent between 80 and 96% of all individuals, and from 76 to 89% of all species (Abele 1979). The complex and intricate structure of coral reefs provides a wide variety of substrates that serve as shelter, feeding sites, and nutritional sources for benthic decapods. In addition, many decapod species make a differential use of reef resources (Abele 1974), increasing the possibilities for coexistence.

Coral reefs abound along the Caribbean coast of Mexico, i.e. the eastern margin of the Yucatán Peninsula (state of Quintana Roo) and in the islands off this coast. However, studies on decapods in the Mexican Caribbean are scarce, and the majority are checklists based on samples taken from disparate places —including Cozumel Island— and from different kinds of marine habitats (Chace 1972, Kensley 1988, Markham et al. 1990).

In 1995-1997, the authorities of Chankanaab Park, a protected area located on the western margin

E. Escobar-Briones & F. Alvarez Eds.
MODERN APPROACHES TO THE STUDY OF CRUSTACEA
PP. 197-204

of Cozumel Island that bears a rich coral reef assemblage, funded a research program on the marine system of the Park. The study of the decapods in the Park area was addressed as part of this research program. In this paper, we provide the first characterization of the benthic decapod fauna and its relationship with the diversity of habitats found in Chankanaab Park.

II. MATERIALS AND METHODS

Study area

Cozumel Island, located 20 km from mainland Quintana Roo (Fig. 1a), is a karstic platform with virtually no soil, where rain-water percolates and exits to the sea through underground rivers. Chankanaab Park is located on the western margin of Cozumel, and includes a portion of land and a portion of the adjacent marine zone. The shoreline of the Park is a rocky, karstified fossil reef of Pleistocene origin (Spaw 1978) with numerous caves and fissures, which descends rather abruptly to an almost horizontal sand terrace (Fig. 1b). A coral reef community colonizes the hard substrate, from the surface to a depth of around 8 m, that conforms the rocky shore . This community includes 27 species of hard corals -mostly of small size- and 21 species of octocorals (Jordán-Dahlgren and Rodríguez-Martínez 1998), as well as numerous erect sponges and anemones. Within the boundaries of the park, the extension of the hard substrate varies according to its angle, from < 10 to 40 m. The sand terrace then extends for about 500 m to a depth of 25 m, where the insular slope abruptly begins, and falls almost vertically to depths in excess of 400 m. The first tens of meters of the sand terrace are virtually denuded, with a few isolated large coral heads (Fig. 1c). Further seawards, there are sparse seagrass beds, composed mainly of *Syringodium filiforme*, with some *Thalassia testudinum* interspersed. The marine portion of the Park extends for approximately 200 m from the shoreline. The total marine portion of the Park is about 7 ha; hard substrates account for ~20% of the area (Guevara-Muñoz 1998), and the rest consists of bare sand or sparse seagrass (Fig. 1c). Water temperature ranges from 25 to 30° C (Ruiz-Rentería and Jordán-Dahlgren 1998).

Chankanaab is a Mayan word meaning "little sea". This name refers to a small lagoon 50 meters inland from the shoreline (Fig. 1c). The lagoon is actually a sinkhole, formed by the collapse of the roof of a large underground cavern. The calcareous, almost vertical walls of the lagoon are covered by filamentous green and blue-green algae. Most of the bottom of the lagoon is bare sand, with a few sparse patches of short red algae, and is < 2 m in depth. The lagoon is connected to the sea through an underwater tunnel, formed by a fissure of irregular shape, ~60 m long, 1.0-1.5 m high on average, and up to 12 m wide (Fig. 1b, 1c). A narrower tunnel, whose extension has not been determined, runs inland from the lagoon (Fig. 1b). Both tunnels and the former cave that gave rise to the lagoon were part of the same underground water flow system. The southern and eastern margin of the lagoon bear rocky outcrops, and depth reaches 3-4 m at the entrances to the tunnels.

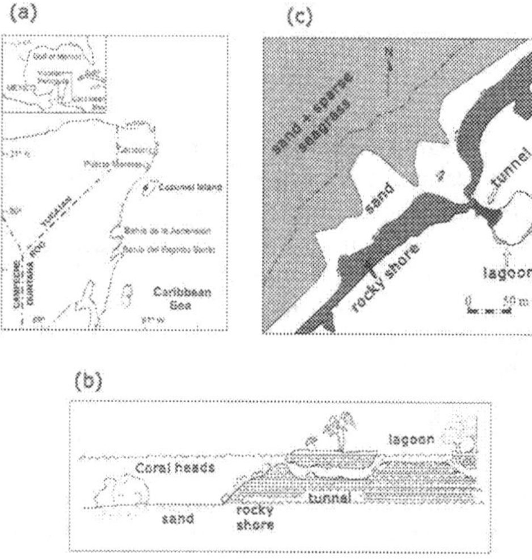

Figure 1. Study area. *(a) Cozumel Island; black circle on western margin of the island shows the location of Chankanaab Park. (b) Lateral schematic representation of coastal features in Chankanaab Park (not to scale). (c) Distribution of substrates in the studied portion of the marine zone, and location of the lagoon and tunnel in the Park. Broken line indicates the marine boundary of the Park area (drawn to scale).*

Decapod sampling

On 12 different occasions between February 1995 and June 1997, decapods were collected in the marine zone of the Park, from the waterline to a depth of approximately 10 m, and throughout the entire lagoon. Because the protected status of the Park forbids the use of any sampling, collecting, or fishing devices either in the lagoon or at the adjacent marine zone, most decapods were collected by hand, using SCUBA diving. However, the most conspicuous and/or easily identifiable underwater (such as palinurids, scyllarids, large portunids, the majids *Mithrax spinossisimus* and *Stenorhynchus seticornis,* and the stenopodid *Stenopus hispidus*) were recorded and/or photographed but not necessarily collected. Collections were made between 18:00 and 21:00 h, and occasionally between 06:00 and 08:00 h. Whenever a specimen was collected, photographed or observed, the divers recorded the type of habitat it was occupying. In the marine zone, these habitats were: hard substrates (rock + dead and live hard corals), octocorals, erect sponges, anemones, bare sand, seagrass, and isolated algal patches. In the lagoon there were fewer types of habitats: hard substrates, bare sand, algal patches, and a few anemones. On two occasions, the divers crossed the tunnel and recorded the visible species of decapods.

Specimens were introduced in numbered ziplock plastic bags and taken to land, where they were changed to glass jars, preserved in 70% ethanol, labeled and transported to the laboratory to be identified. The percentage of species occurrence by habitat type was analyzed in the two sampling sites (marine zone and lagoon), and a cluster analysis was applied to these results, based on presence/ absence of species by site and type of habitat.

III. RESULTS

A total of 73 species of decapods belonging to six infraorders and 22 families were identified in the whole Park area (Table 1). The infraorder Brachyura was the most diverse, with 31 species, of which 17 belonged to the family Majidae.

Seventy-two species occurred in the marine zone and 22 in the lagoon (Table 1). Of the latter 22 species, 21 also occurred in the marine zone, suggesting that despite its connection to the underground freshwater flow system of the island, the lagoon is a fully marine environment. This was confirmed by a concurrent study on the hydrology and hydrodynamics of the Chankanaab system (Ruiz-Rentería and Jordán-Dahlgren 1998), which showed that the salinity within the lagoon is virtually the same as in the marine zone (35.1-37.0 ppt) throughout the year. Therefore, some decapods can cross freely between the marine zone and the lagoon through the tunnel, either as larvae or as crawling or swimming benthic stages (e.g., the nine species observed in the tunnel, Table 1).

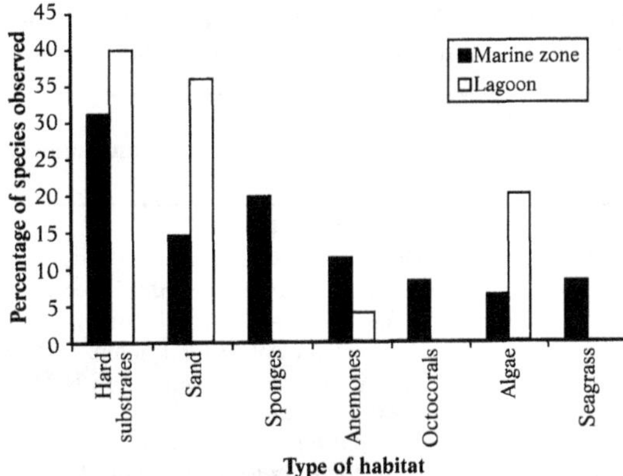

Figure 2. *Percentage of decapod species recorded in different types of habitats in the marine zone and in the lagoon, Chankanaab Park, Cozumel Island.*

The majority of decapod species (30) in the marine zone were collected from the hard subtrates (rock + live hard corals) (Fig. 2). Nineteen species were found inside erect sponges, and 14 in sand bottoms, whereas 11 species occurred on or beneath anemones. Some species found in anemones were also collected from gorgonians (8), and most species found among seagrass (8) also occurred among algae (6). In the lagoon, 10 species were collected from hard substrates, and the sandy bottom (9) and algal patches (5) yielded similar species. The cluster analysis (Fig. 3) reflected this distribution of species by habitat type.

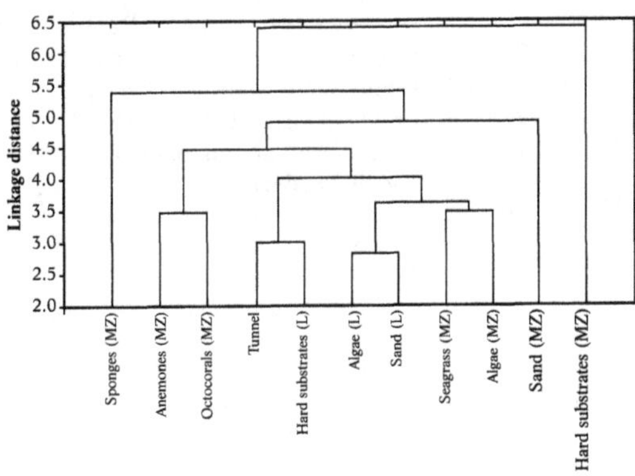

Figure 3. *Cluster (complete linkage, euclidean distances) of decapod species (based on presence/absence) by habitat type and site (MZ: marine zone; L: lagoon), Chankanaab Park, Cozumel Island.*

Table 1. *Distribution of decapod species by site (marine zone, lagoon, or tunnel) in Chankanaab Park, Cozumel Island, Mexico. (X present; - absent; *new record for Quintana Roo; ^ new record for Cozumel Island).*

Species number	Species name and authority	Site of occurrence		
		Marine zone	Lagoon	Tunnel
	Family Penaeidae			
1	*Farfantepenaeus brasiliensis* (Latreille, 1817)	-	X	-
2	* ^ *Metapenaeopsis gerardoi* Pérez-Farfante, 1971	X	X	-
3	^ *Metapenaeopsis goodei* (Smith, 1885)	X	-	-
4	^ *Metapenaeopsis smithi* Schmitt, 1924	X	X	-
5	^ *Sicyonia laevigata* Stimpson, 1871	X	-	-
	Family Stenopodidae			
6	*Stenopus hispidus* (Olivier, 1811)	X	-	-
7	^ *Stenopus scutellatus* Rankin, 1898	X	-	-
	Family Rhynchocinetidae			
8	* ^ *Rhynchocinetes rigens* Gordon, 1936	X	X	X
	Family Palaemonidae			
9	^ *Brachycarpus biunguiculatus* (Lucas, 1849)	X	-	-
10	*Periclimenes americanus* (Kingsley, 1878)	X	-	-
11	*Periclimenes pedersoni* Chace, 1958	X	-	-
12	*Periclimenes yucatanicus* (Ives, 1891)	X	-	-
	Family Alpheidae			
13	*Alpheus armatus* Rathbun, 1901	X	-	-
14	*Alpheus armillatus* H. Milne Edwards, 1837	X	-	-
15	^ *Alpheus heterochaelis* Say, 1818	X	-	-
16	^ *Alpheus normanni* Kingsley, 1878	X	-	-
17	*Synalpheus minus* (Say, 1818)	X	-	-
	Family Hippolytidae			
18	* ^ *Lysmata grabhami* (Gordon, 1935)	X	-	-
19	*Thor amboinensis* (De Man, 1888)	X	-	-
20	* ^ *Tozeuma cornutum* A. Milne-Edwards, 1881	X	-	-
21	^ *Trachycaris restrictus* A. Milne-Edwards, 1878	X	-	-
	Family Processidae			
22	* ^ *Processa bermudensis* (Rankin, 1900)	X	-	-
23	*Processa fimbriata* (Manning and Chace, 1971)	X	-	-
	Family Palinuridae			
24	*Panulirus argus* (Latreille, 1804)	X	X	X
25	^ *Panulirus guttatus* (Latreille, 1804)	X	X	X
	Family Scyllaridae			
26	* ^ *Parribacus antarcticus* (Lund, 1793)	X	-	X
27	* ^ *Scyllarides aequinoctialis* (Lund, 1793)	X	-	-
	Family Synaxidae			
28	* ^ *Palinurellus gundlachi gundlachi* Von Martens, 1878	X	-	-
	Family Coenobitidae			
29	*Coenobita clypeatus* (Herbst, 1791)	X	X	-
	Family Diogenidae			
30	*Calcinus tibicen* (Herbst, 1791)	X	X	-
31	*Clibanarius tricolor* (Gibbes, 1850)	X	X	-
32	^ *Dardanus venosus* (H. Milne Edwards, 1848)	X	X	X
33	*Paguristes cadenati* Forest, 1954	X	X	-
34	^ *Paguristes puncticeps* Benedict, 1901	X	X	-
35	* ^ *Paguristes tortugae* Schmitt, 1933	X	-	-
36	*Petrochirus diogenes* (Linnaeus, 1758)	X	X	-

Table 1. *(Continued)*

Species number	Species name and authority	Site of occurrence		
		Marine zone	Lagoon	Tunnel
	Family Paguridae			
37	*^Iridopagurus caribbensis* (A. Milne-Edwards and Bouvier, 1893)	X	X	-
38	*^Pagurus annulipes* (Stimpson, 1860)	X	X	-
39	*^Pagurus brevidactylus* (Stimpson, 1859)	X	-	-
40	*^Phimochirus holthuisi* (Provenzano, 1961)	X	X	-
	Family Galatheidae			
41	*^Munida pusilla* Benedict, 1902	X	-	-
	Family Porcellanidae			
42	*Petrolisthes galathinus* (Bosc, 1802)	X	-	-
	Family Calappidae			
43	*Calappa gallus* (Herbst, 1803)	X	-	-
44	*^Calappa ocellata* Holthuis, 1958	X	X	-
	Family Leucosiidae			
45	*^Speloeophorus nodosus* (Bell, 1855)	X	-	-
	Family Majidae			
46	*^Batrachonotus fragosus* Stimpson, 1871	X	-	-
47	*^Macrocoeloma trispinosum* (Latreille, 1852)	X	-	-
48	*Microphrys bicornutus* (Latreille, 1825)	X	-	-
49	*Mithraculus cinctimanus* Stimpson, 1960	X	-	-
50	*Mithraculus coryphe* (Herbst, 1801)	X	-	-
51	*^Mithraculus forceps* A. Milne-Edwards, 1875	X	-	-
52	*^Mithraculus ruber* Stimpson, 1871	X	-	-
53	*Mithraculus sculptus* (Lamarck, 1818)	X	X	-
54	*^Mithrax hemphilli* Rathbun, 1892	X	-	-
55	*^Mithrax hispidus* (Herbst, 1790)	X	-	-
56	*^Mithrax holderi* Stimpson, 1871	X	-	-
57	*^Mithrax pilosus* Rathbun, 1892	X	-	-
58	*Mithrax spinosissimus* (Lamarck, 1818)	X	X	X
59	*^Pitho mirabilis* Herbst, 1794	X	-	-
60	*Podochela* sp.	X	-	-
61	*Stenorhynchus seticornis* (Herbst, 1788)	X	X	X
62	*^Teleophrys ornatus* Rathbun, 1900	X	-	-
	Family Portunidae			
63	*Portunus ordwayi* (Stimpson, 1860)	X	-	X
64	*^Portunus sebae* H. Milne-Edwards, 1834	X	X	X
	Family Xanthidae			
65	*^Melybia thalamita* Stimpson, 1871	X	-	-
66	*^Micropanope nuttingi* (Rathbun, 1898)	X	-	-
67	*^Pilumnus dasypodus* Kingsley, 1879	X	-	-
68	*Pilumnus longleyi* Rathbun, 1930	X	-	-
69	*^Platypodiella spectabilis* Herbst, 1794	X	-	-
	Family Goneplacidae			
70	*^Euryplax nitida* Stimpson, 1859	X	-	-
	Family Grapsidae			
71	*Grapsus grapsus* (Linnaeus, 1758)	X	X	-
72	*Pachygrapsus transversus* (Gibbes, 1850)	X	-	-
73	*Percnon gibbesi* H. Milne Edwards, 1853	X	-	-

Alpheus armillatus, A. heterochaelis, Macrocoeloma trispinosum, Melybia thalamita, Microphrys bicornutus, Mithrax pilosus, Pagurus brevidactylus, Petrolisthes galathinus, Pilumnus sayi, Sicyonia laevigata, Synalpheus minus and *Teleophrys ornatus* were found only in sponges. Species associated with anemones included *A. armatus, Percnon gibbesi* and *Thor amboinensis*. Decapods found solely in sand bottoms were *Calappa gallus, C. ocellata* and *Pachygrapsus transversus*. In contrast, species found in at least four different types of substrates included *Panulirus argus, Metapenaeopsis smithi, Mithrax hispidus, Paguristes puncticeps, Pagurus brevidactylus, Portunus ordwayi* and *Stenorhynchus seticornis*.

IV. DISCUSSION

Despite the small area of Chankanaab Park, and the limitations of our visual collecting methods, the number of benthic decapod species recorded was high. Because of the non-destructive collecting method employed, cryptic and burrowing decapods (e.g., alpheids, xanthids, and thalassinids) were either very scarce or altogether absent from our samples. Moreover, areas with relatively stable water temperature tend to bear large numbers of decapods, as Lemaitre (1980) found in reef environments along the Colombian Caribbean. Therefore, the decapod fauna in Chankanaab Park must be richer, owing to its relatively stable temperatures and the variety of as yet unsearched habitats, and remains to be more thoroughly investigated.

Most of the species recorded in the Park area were typically collected from hard substrates, which have numerous holes and fissures that shelter many decapods (Provenzano 1959, Chace 1972, Williams 1984, Markham et al. 1990). Erect sponges (such as *Cribrochalina vasculum, Callispongia fallax, C. vaginalis, Niphates digitalis, Ircinia strobilina*) and a variety of anemones (e.g., *Bartholomea annulata, Condylactis gigantea, Stichodactyla helianthus*), were hosts to many decapod species in the marine zone of the Park, a common association found in coral reefs throughout the world (Chace 1972, Bruce 1976, Williams 1984).

The richest family was the Majidae, with 17 species in the marine zone, but most of these species were absent from the lagoon. In Atlantic oculinid reefs, majids are nonsymbiotic omnivores which utilize the coral as a refuge but not as a source of food (Reed et al.

1982). In contrast, in Indopacific reefs majids are among the most abundant decapods and some species are commensals of corals and other anthozoans (Patton 1976). In Chankanaab Park, *Mithraculus cinctimanus, M. coryphe, M. ruber, Mithrax spinosissimus* and *Pitho mirabilis*, were observed only on hard substrates, where they probably feed on the filamentous algae attached to the rock. Other species were observed in sponges, and the rest in several types of habitats.

In coral reef habitats, the primary determinant of species richness of decapods is the available surface, but the effect of habitat heterogeneity cannot be dismissed (Coles 1980). Therefore, the lower species richness found in the lagoon compared to the marine zone is probably related both to its small size and to its lower diversity of habitats. For example, erect sponges, octocorals, and seagrasses were completely absent in the lagoon. Anemones, although present, were very scarce, and the only decapod species found associated with anemones in the lagoon was *Mithraculus sculptus*.

Many of the decapod species in Chankanaab Park are tropical and occur throughout the Caribbean Sea. Of the 73 decapod species from Chankanaab, 51 (69.9%) have also been reported from Venezuela (Rodríguez 1980); 46 (63.0%) from Cuba (Gómez-Hernández and Martínez-Iglesias 1986, Martínez-Iglesias and Gómez-Hernández 1986, Martínez-Iglesias and García-Raso 1999); 42 (57.5%) from Bermuda (Markham and McDermott 1980); 42 (57.5%) from the Mexican coasts of the Gulf of Mexico (Hernández-Aguilera et al. 1996); 41 (56.2%) from Maine to Florida, USA (Wenner and Read 1982, Williams 1984), and 20 (27.4%) from Alacrán Reef (Martínez-Guzmán and Hernández-Aguilera 1993). Of the 31 species of brachyurans, 20 (64.5%) were reported from the Caribbean coast of Colombia (Lemaitre 1981), and 11 (35.5%) from the Bahamas (Garth 1978). Therefore, the decapod fauna of Chankanaab exhibits affinities with the following zoogeographic regions: Caribbean, Floridian, Carolinian, Bermudian, and Southern Gulf of Mexico. However, these apparent affinities should be viewed with caution, because in the cited studies decapods have been collected from a wide variety of habitats and depths, and by means of many different collecting techniques.

Markham et al. (1990) published the most comprehensive list of shallow-water crustaceans of the Caribbean coast of Quintana Roo to date. In addition to collecting their own material, these authors reviewed most of the previously published literature on crustaceans from the Mexican Caribbean. Markham et al. (1990) reported 233 decapod species along the coast of Quintana Roo. Of the 73 species recorded in the present paper from Chankanaab Park, 53 (72.6%) had already been reported by Markham et al. (1990) from Quintana Roo. The remaining 20 species are new records for Quintana Roo, and a total of 43 are new records for Cozumel Island (see Table 1). Considering the scarcity of cryptic and burrowing species in our collections, this underlines the decapod richness of Chankanaab Park.

The present study was not quantitative, but many decapod species (particularly penaeids, diogenids, pagurids, and majids) were observed in large quantities, and must be important in the food webs in Chankanaab Park, where the main attraction is the rich and varied tropical fish fauna. A large proportion of the fish species recorded in a concurrent study in the marine area of the Park (Guevara-Muñoz 1998) are members of the families Lutjanidae, Haemulidae, Scorpenidae, Carangidae, Tetraodontidae, Balistidae, Serranidae and Holocentridae, which prey either preferentially or secondarily on decapods (Randall 1967).

Although no human activities are allowed in the lagoon, and fishing is forbidden in the whole Park area, the Park receives a daily average of 1000 visiting tourists, most of which conduct some sort of aquatic or underwater activity in the shallow grounds of the relatively small marine portion of the Park. It has been proposed that species richness and diversity in coralline habitats is increased rather than decreased by periodical physical disturbances (Coles 1980), and in western Cozumel the damaging effects of the very strong Hurricane Gilbert (September 1988) on corals and sponges were surprisingly small (Fenner 1991). However, the long-term cumulative effect of the daily activities of a large number of humans in Chankanaab Park may be deleterious for the coral reef community, especially for fragile living substrates such as hard corals, octocorals and erect sponges (Jordán-Dahlgren and Rodríguez-Martínez 1998), which harbor a considerable number of associated decapod species. Further protection of the living substrates in the marine zone is recommended, to warrant the preservation of the rich decapod fauna in Chankanaab Park.

ACKNOWLEDGEMENTS

We greatly acknowledge Fernando Negrete-Soto for his invaluable help in the planning and conduction of field activities, and Cecilia Barradas-Ortiz for aiding in the laboratory. Erick Cadena and Verónica Monroy identified most of the decapod species. Verónica Castañeda, Erick Cadena, Jaime Estrada, Surya Garza, Patricia Rangel and Daniella Guevara collaborated in the field work. This study was funded by Fundación de Parques y Museos de Cozumel, through an agreement with Universidad Nacional Autónoma de México.

REFERENCES

Abele LG (1974) Species diversity of decapod crustaceans in marine habitats. Ecology 55:156-161

Abele LG (1979) The community structure of coral associated decapod crustaceans in variable environments. (pp 265-290) In: Livingston RJ (ed) Ecological Processes in Coastal and Marine Systems. Plenum, New York

Bruce AJ (1976) Shrimps and prawns of coral reefs, with special reference to commensalism. (pp 37-94) In: Jones OA and Endean R (eds) Biology and Geology of Coral Reefs, Vol. 3. Academic Press, New York

Bruce AJ (1977) Shrimps that live on corals. Oceans 1:70-75

Castro P (1976) Brachyuran crabs symbiotic with scleractinian corals: A review of their biology. Micronesica 12:99-110

Chace FA Jr (1972) The shrimps of the Smithsonian-Bredin Caribbean expeditions, with a summary of the West-Indian shallow-water species (Crustacea: Decapoda: Natantia). Smithson Contrib Zool 98:1-79

Coles SL (1980) Species diversity of decapods associated with living and dead reef coral *Pocillopora meandrina*. Mar Ecol Progr Ser 2:281-291

Coull BC and Bell SS (1983) Biotic assemblages: Populations and communities. (pp 283-319) In: Vernberg FJ and Vernberg WB (eds) The Biology of Crustacea, Vol. 7 Academic Press, New York

Fenner DP (1991) Effects of Hurricane Gilbert on coral reefs, fishes and sponges at Cozumel, Mexico. Bull Mar Sci 48:719-730

Garth JS (1978) Marine biological investigations in the Bahamas 19. Decapoda Brachyura. Sarsia 63:317-333

Gómez-Hernández O and Martínez-Iglesias JC (1986) Nueva lista de pagúridos cubanos (Crustacea, Decapoda, Anomura, Paguroidea) Rev Invest Mar 7:21-29

Guevara-Muñoz MD (1998) Análisis de la estructura comunitaria de los peces arrecifales del Parque Marino Chankanaab, Cozumel, Quintana Roo. Tesis Profesional Univ Nal Autón México

Hernández-Aguilera JL, Toral-Almazán RE and Ruiz-Nuño JA (1996) Especies Catalogadas de Crustáceos Estomatópodos y Decápodos para el Golfo de México, Río Bravo, Tamps. a Progreso, Yuc. Secretaría de Marina y CONABIO, México

Jordán-Dahlgren E and Rodríguez-Martínez R (1998) Caracterización de la comunidad coralina del Parque Chankanaab. (pp 1-64) In: Jordán-Dahlgren E (ed) Ecología del Ambiente marino del Parque Chankanaab Informe Final Fund Parques Museos Cozumel / Inst Cien Mar Limnol, Univ Nal Autón México, Puerto Morelos

Kensley B (1988) New species and records of cave shrimps from the Yucatan Peninsula Decapoda: Agostocarididae and Hippolytidae. J Crust Biol 8:688-699

Lemaitre R (1981) Shallow-water crabs (Decapoda, Brachyura) collected in the southern Caribbean near Cartagena, Colombia. Bull Mar Sci 31:234-266

Markham JC and McDermott JJ (1980) A tabulation of the Crustacea Decapoda of Bermuda. Proc Biol Soc Wash 93:1266-1276

Markham JC, Donath-Hernández FE, Villalobos-Hiriart JL and Cantú Díaz-Barriga A (1990) Notes on shallow-water marine Crustacea of the Caribbean coast of Quintana Roo, Mexico. An Inst Biol, Univ Nal Autón México Ser Zool 61:405-446

Martínez-Guzmán LA and Hernández-Aguilera JL (1993) Crustáceos estomatópodos y decápodos del Arrecife Alacrán, Yucatán. (pp 609-629) In: Salazar-Vallejo SI and González NE (eds) Biodiversidad Marina y Costera de México CONABIO-CIQRO, México

Martínez-Iglesias JC and Gómez-Hernández O (1986) Los crustáceos decápodos del Golfo de Batabanó: Brachyura. Poeyana 332:1-91

Martínez-Iglesias JC and García-Raso JE (1999) The crustacean decapod communities of three coral reefs from the southwestern Caribbean sea of Cuba: species composition, abundance and structure of the communities. Bull Mar Sci 65:539-557

McCloskey LR (1970) The dynamics of the community associated with a marine Scleractinian coral. Int Rev Gesamten Hydrobiol 55:13-82

Patton WK (1976) Animal associates of living reef corals. (pp 1-36) In: Jones OA and Endean R (eds) Biology and Geology of Coral Reefs, Vol. 3. Academic Press, New York

Provenzano AJ Jr (1959) The shallow-water hermit crabs of Florida. Bull Mar Sci Gulf Caribb 9: 349-420

Randall JE (1967) Food habits of reef fishes of the West Indies. Stud Trop Oceanogr 5:655-847

Reed JK, Gore RH, Scotto LE and Wilson KA (1982) Community composition, structure, areal and trophic relationships of decapods associated with shallow- and deep-water *Oculina varicosa* coral reefs. Bull Mar Sci 32:761-786

Rodríguez G (1980) Los crustáceos decápodos de Venezuela. Instituto Venezolano de Investigaciones Científicas, Caracas

Ruiz-Rentería F and Jordán-Dahlgren E (1998) Hidrología e hidrodinámica en el sistema marino-lagunar del Parque Chankanaab. (pp 1-41) In: Jordán-Dahlgren E (ed) Ecología del Ambiente marino del Parque Chankanaab Informe Final Fund Parques Museos Cozumel / Inst Ciencias del Mar y Limnol Univ Nal Autón México, Puerto Morelos

Spaw RH (1978) Late Pleistocene carbonate bank deposition: Cozumel Island, Quintana Roo, Mexico. Gulf Coast Assoc Geol Soc Trans 28:601-620

Wenner EL and Read TH (1982) Seasonal composition and abundance of decapod crustacean assemblages from the South Atlantic Bight, USA. Bull Mar Sci 32:181-206

Williams AB (1984) Shrimps, lobsters and crabs of the Atlantic coast of the eastern United States, Maine to Florida. Smithsonian Institution Press, Washington

SYMBIOSIS BETWEEN *PORTUNUS SPINIMANUS* LATREILLE, 1819 (DECAPODA, PORTUNIDAE) AND *OCTOLASMIS LOWEI* (DARWIN, 1852) (THORACICA, POECILASMATIDAE) FROM UBATUBA, SAO PAULO, BRAZIL

Sandro Santos

Departamento de Biologia, Universidade Federal de Santa Maria, Santa Maria, Rio Grande do Sul 97105-900, Brazil

ssantos@ccne.ufsm.br

ABSTRACT

From May 1991 to May 1993, 1798 *Portunus spinimanus* were captured by monthly samplings in the Ubatuba region, São Paulo, Brazil. *Octolasmis lowei* was present on the gills of 183 crabs. The interaction between the two species was analyzed, focusing on the monthly presence of *O. lowei* on the gills of *P. spinimanus*, and its relationship with sex, molt cycle, stage of gonad development, and weight of the host crab. *Octolasmis lowei* was registered in all months except in July, August, September 1991, February and April 1992 . The infestation rates in male and female hosts were similar, 11% of adults and 1.1% of juveniles had *O. lowei*. Among infested crabs, 91.8% were in the intermolt stage; 55.9% of females were in an initial stage of gonad development and 53.6% of males presented fully developed gonads. The weight-carapace width relationship is described as follows: crabs infested with *O. lowei*, $W = 3.9 \times 10^{-4} CW^{2.87}$ and uninfested crabs, $W = 56 \times 10^{-4} CW^{3.31}$. The obtained data suggest that *O. lowei* infests adult crabs preferentially, and it doesn't interfere with the gonad development of *P. spinimanus*. However, it seems to negatively affect the weight of infested crabs.

I. INTRODUCTION

The swimming crab *Portunus spinimanus* Latreille is distributed in the west Atlantic, from New Jersey in the USA to Rio Grande do Sul in Brazil (Melo 1996). This species is one of the most valuable resources in the region of Ubatuba, State of São Paulo, as a result of its size and taste; it is also one of the most abundant species in the area (Fransozo et al. 1991, Santos et al. 1996). During previous studies on the reproductive biology of this species (Santos 1998, Santos and Negreiros-Fransozo 1999), it was observed that some animals hosted from one to many barnacles (*Octolasmis lowei* Darwin) on their gills.

The adults of the genus *Octolasmis* live attached by a basal disc to a variety of hosts that include corals, echinoderms, mollusks, and within the Crustacea, crabs, lobsters, and isopods. The barnacles depend on their hosts for a substratum, and perhaps also for protection and nutrition (Voris and Jeffries 1997).

Gannon and Wheatley (1992, 1995) studied the physiological effects of *Octolasmis muelleri* on the swimming crab *Callinectes sapidus*. According to these authors, infested animals maintain a similar oxygen uptake as control animals, but the heart rate and the scaphognathite rate increased 1.4 and 1.8 times, respectively. Moderately infested crabs subjected to exercise could compensate for the presence of barnacles; however, heavily infested animals, with more than 50 specimens in the gill chambers did not survive the experimental stress. Hudson and Lester (1994) while evaluating what parasites and symbionts could have a negative impact on cultures of *Scylla serrata*, determined that crabs hosting *Octolasmis* were more stressed due to a decreased respiratory efficiency.

E. Escobar-Briones & F. Alvarez Eds.
MODERN APPROACHES TO THE STUDY OF CRUSTACEA
PP. 205-209

Table 1. *Characterization of stages of gonad development in males and females of the species* **P. spinimanus** *(adapted of Santos and Negreiros-Fransozo 1999).*

	Characteristics	
Stages	**Males**	**Females**
Immature (IM)	Undifferentiated gonads. Impossible to recognize testis and vas deferens.	Undifferentiated gonads. Stage of gonad development coincides with stage of morphological maturation.
Rudimentary (RU)	Gonads recognizable only with magnification. Vas deferens can be observed behind the stomach. Gonad looks filamentous and pale-yellow.	Undeveloped gonads with filamentous appearance, thin and pale yellow.
Developing (DG)	Gonads are visible. Vas deferens can be divided in two different regions, the anterior one being thinner. Gonad/hepatopancreas ratio is 1/4.	Beginning of ovary maturation. Gonad/hepatopancreas ratio is 1/8. Yellow color, connection between right and left trumpets visible.
Developed (DD)	Vas deferens clearly divided into anterior, median, and posterior sections. Vessels look entwined. Gonad/hepatopancreas ratio is 1/2.	Ovary fills almost the whole thoraxic cavity. Bright orange Lobes are evident

Shields (1992) studied the parasites and symbionts of *Portunus pelagicus* in Moreton Bay, Australia, and observed the presence of the barnacle *Octolasmis,* which had a higher prevalence in female crabs. Jeffries et al. (1992) investigated the age of *Scylla serrata* at colonization by *Octolasmis* and observed that the percentage of crabs hosting barnacles increased as the crab approached sexual maturity and the magnitude of the infestation on individual crabs increased with their size.

The present study aimed to investigate the seasonal frequency of *O. lowei* infestation on the gills of *P. spinimanus* and the relationship between overall barnacle infestation and host sex, molt stage, gonad development and weight.

II. MATERIAL AND METHODS

Monthly collections were taken for two years (May 1991 to May 1993) in Ubatuba, on the northern coast of the State of São Paulo, Brazil. Sampling was done using an otter-trawl (side mesh = 15 mm, cod end mesh = 10 mm) towed for one hour.

In the laboratory, the following data were recorded for each crab: sex, stage of maturity (based on external morphology), weight, carapace width, and molt stage (Santos et al. 1995, Santos 1998). After dissection of each individual crab, the stages of gonad development were verified (Table 1) and the presence or absence of *O. lowei* on the gills was recorded.

Chi-square tests (Sokal and Rohlf 1995, Zar 1996) and analysis of multinomial proportions (Goodman 1964, 1965) applied to contingency tables were used to compare the numbers of collected animals among months. To analyze the influence of *Octolasmis* on the molt cycle, sexual maturity and gonad development of host crabs, chi-square tests were applied to contingency tables (Sokal and Rohlf, 1995, Zar 1996). For the analysis of the weight-carapace width relationship, the allometric coefficients obtained for infested and uninfested animals were compared to the isometric value with a t-test.

III. RESULTS

During the study period, the presence of *O. lowei* was registered on the gills of *P. spinimanus* in all months, except in July, August and September 1991, and February and April 1992 (Table 2). The highest frequency of infested animals was recorded in March,

1992. A small but significant variation in the number of infested animals was observed among months ($P < 0.05$). The number of barnacles on the gills of the host varied from 1 to 29.

Adult crabs were more infested than young ones ($P < 0.05$). Among adults, 11% of collected crabs were infested, with a similar proportion for males and females ($P > 0.05$). Regarding young animals, 1.1% of males and 1.4% of females had barnacles on their gills (Fig. 1).

In relation to the molt cycle, infested and uninfested crabs were in intermolt (stage C - 91.80 and 72.90%, respectively), advanced post-molt (stage B - 4.92 and 22.22%, respectively), and pre-molt (stage D - 3.28 and 2.85 %, respectively). The highest number of crabs, in the two groups, was collected in intermolt stage (Fig. 2). However, the percentage of animals in intermolt was significantly greater for infested than for uninfested crabs ($P < 0.05$).

The sample of *Portunus spinimanus* analyzed included crabs with developed (DD), developing (DG), rudimentary (RU), and immature (IM) gonads for both groups (Fig. 3). Among infested crabs, the percentage of animals with developed or developing gonads was significantly higher than those for the other stages ($P < 0.05$).

The weight-carapace width relationship exhibited a positive allometric growth for uninfested animals, with a coefficient of 3.31. Infested animals exhibited an isometric growth, with an allometric coefficient of 2.87, which was significantly different from that of uninfested crabs ($P < 0.05$). Size range was greater for uninfested than for infested crabs, 24.2 – 99.6 mm CW and 36.0 – 91.2 mm CW, respectively (Table 3); accordingly, the weight range was also greater for uninfested crabs.

Table 2. *Sample size/number of infested crabs (adult males, juvenile males, adult females, and juvenile females) in each month.*

Month	Year	Adult Male	Juvenile Male	Adult Female	Juvenile Female	Total
May	1991	21 / 06	00 / 00	65 / 08	00 / 00	86 / 14
June	1991	13 / 01	00 / 00	26 / 02	01 / 00	40 / 03
July	1991	20 / 00	00 / 00	41 / 00	00 / 00	61 / 00
August	1991	07 / 00	01 / 00	46 / 00	01 / 00	55 / 00
September	1991	17 / 00	01 / 00	34 / 00	04 / 00	56 / 00
October	1991	13 / 00	03 / 00	40 / 03	01 / 00	57 / 03
November	1991	11 / 03	00 / 00	45 / 10	01 / 00	57 / 13
December	1991	15 / 01	11 / 00	18 / 00	10 / 00	54 / 01
January	1992	12 / 01	10 / 00	50 / 10	20 / 01	92 / 12
February	1992	58 / 00	25 / 00	44 / 00	16 / 00	143 / 00
March	1992	29 / 02	18 / 00	38 / 00	12 / 00	97 / 02
April	1992	18 / 00	09 / 00	35 / 00	04 / 00	66 / 00
May	1992	38 / 5	03 / 01	54 / 08	00 / 00	95 / 14
June	1992	23 / 2	00 / 00	43 / 03	01 / 00	67 / 05
July	1992	21 /02	01 / 00	38 / 00	00 / 00	60 / 02
August	1992	16 / 00	00 / 00	43 / 02	00 / 00	59 / 02
September	1992	23 / 01	00 / 00	61 / 04	00 / 00	84 / 05
October	1992	19 / 03	01 / 00	65 / 09	00 / 00	85 / 12
November	1992	29 / 01	01 / 00	85 / 01	00 / 00	115 / 02
December	1992	28 / 08	02 / 00	32 / 07	01 / 00	63 / 15
January	1993	16 / 01	01 / 00	32 / 11	00 / 00	49 / 12
February	1993	08 / 04	00 / 00	27 / 07	00 / 00	35 / 11
March	1993	24 / 08	00 / 00	52 / 18	01 / 00	77 / 26
April	1993	14 / 05	01 / 00	64 / 11	01 / 00	80 / 16
May	1993	03 / 01	00 / 00	62 / 12	00 / 00	65 / 13

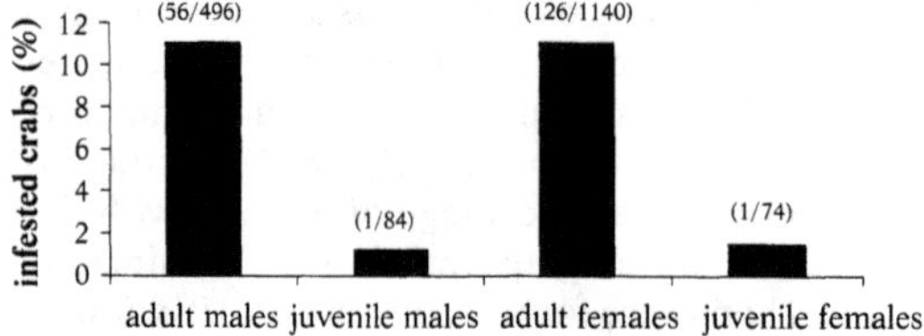

Figure 1. *Percentage of infestation of adult and juvenile male and female* **Portunus spinimanus** *by* **Octolasmis lowei**.

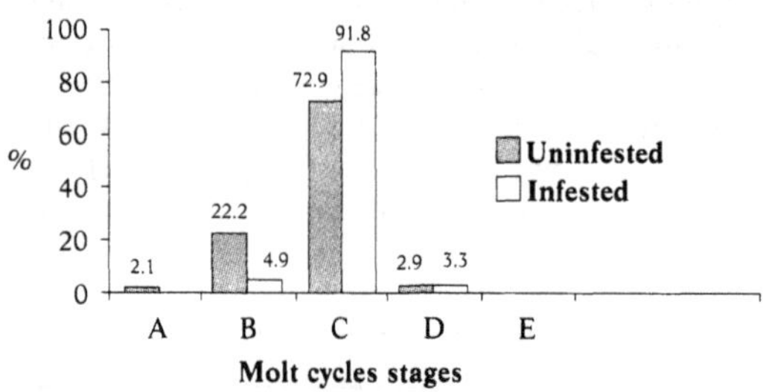

Figure 2. *Frequency of* **Portunus spinimanus** *in initial post-molt (stage A), advanced post-molt (stage B), intermolt (stage C), pre-molt (stage D), and ecdysis (stage E), for infested and uninfested crabs.*

IV. DISCUSSION

The continuous presence of *O. lowei* on the gills of *P. spinimanus* for most of the year, except for some months suggests a possible synchronization between the life cycle of the barnacle and the adult period of *P. spinimanus*.

The fact that the highest number of infested crabs were adults, suggests that the infestation occurs preferentially in late juveniles which are approaching maturity. The smallest infested crab captured was 36.0 mm CW, near 42 mm CW, the estimated size of the first reproduction for *P. spinimanus* (Santos et al. 1995, Santos and Negreiros-Fransozo 1996). Jeffries et al. (1992) also reported a low number of young *Scylla serrata* infested with *Octolasmis*. The presence of *O. lowei* does not seem to interfere with the molt cycle of *P. spinimanus*, since infested crabs in pre-molt and post-molt stages were observed. However, the higher percentage of infested animals in intermolt suggests that crabs in this stage are more suitable as barnacle substrata.

Barnacles of the genus *Octolasmis* find the necessary nutrients and protection in the branchial chambers of *P. spinimanus* and other crabs to complete their development. Considering that a crab is safer in intermolt, symbionts like *Octolasmis* should preferentially select hosts in this stage. Intermolt periods tend to increase after sexual maturity is reached, a fact that explains the greater frequency of barnacles in adult crabs.

Host gonad development does not seem to be affected by the presence of *O. lowei*, because the highest number of infested crabs, including both males and females, presented developing gonads (stages DG and DD for males and females).

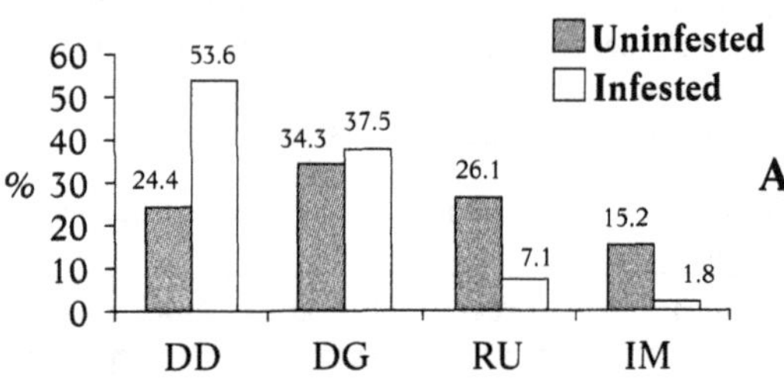

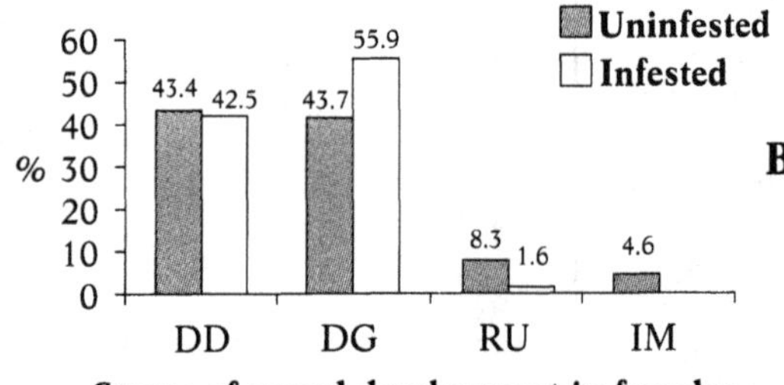

Figure 3. *Frequency of infested and uninfested male (3A) and female (3B)* **Portunus spinimanus** *with immature (IM), rudimentary (RU), developing (DG), and developed gonads (DD).*

Although *O. lowei* does not seem to interfere with molting and the gonad development of *P. spinimanus*, it may be affecting the weight gain of infested crabs. The data presented for infested crabs shows an isometric response for the weight-carapace width relationship, while uninfested animals presented a positive allometric response. These results suggest that in this relationship there is a negative effect for the host, hence, it is an example of parasitism.

Table 3. *Allometric equations for the weight-carapace width relationship and maximum and minimum weight and carapace width values, for infested and uninfested crabs.*

	Y = aXb	r2	Alom.	Max.CW Min.CW	Max.W Min.W
Uninfested	W = 56 x 10⁻⁴CW³·³¹	0.93	+	99.6 mm	201.3g
				24.2 mm	2.2g
Infested	W = 3,9 x 10⁻⁴CW²·⁸⁷	0.82	o	91.2 mm	193.7 g
				36.0 mm	9.4 g

ACKNOWLEDGEMENTS

I would like to thank Dr. Paulo Young (Museu Nacional-Universidade Federal do Rio de Janeiro, Brazil) for the identification of *Octolasmis lowei* (Darwin, 1852).

REFERENCES

Fransozo A, Negreiros-Fransozo ML, Mantelatto FLM, Pinheiro MAA and Santos S (1991) Composição e distribuição dos Brachyura (Crustacea, Decapoda) do sublitoral não consolidado da enseada da Fortaleza, Ubatuba, SP. Rev Bras Biol 52:667-675

Gannon AT and Wheatly MG (1992) Physiological effects of an ectocommensal gill barnacle, *Octolasmis muelleri*, on gas exchange in the blue crab *Callinectes sapidus*. J Crust Biol 12:11-18

Gannon AT and Wheatly MG (1995) Physiological effects of a gill barnacle on host blue crabs during short-term exercise and recovery. Mar Behav Physiol 24:215-225

Goodman LA (1964) Simultaneous confidence intervals for contrasts among multinomiae populations. An Math Stat 35:716-725

Goodman LA (1965) On simultaneous confidence intervals for multinomiae proportions. Technometrics 7:247-254

Hudson DA and Lester RJG (1994) Parasites and symbionts of the wild mud crab *Scylla serrata* (Forskal) of potential sgnificance in aquaculture. Aquaculture 120:183-199

Jeffries WB, Voris HK and Poovachiranon S (1992) Age of the mangrove crab *Scylla serrata* at colonization by stalked barnacles of the genus *Octolasmis*. Biol Bull 182:188-194

Melo GAS (1996) Manual de identificação dos Brachyura (caranguejos e siris) do litoral brasileiro. São Paulo: Editora Plêiade, 603 p

Santos S (1998) Moult cycle in the swimming crab *Portunus spinimanus* (Brachyura, Portunidae from Ubatuba, São Paulo, Brazil. Iheringia, Sér Zool 85:51-57

Santos S and Negreiros-Fransozo ML (1999) Reproductive cycle of the swimming crab *Portunus spinimanus* Latreille (Crustacea, Decapoda, Brachyura). Rev Bras Zool 16:1183-1193

Santos S, Negreiros-Fransozo ML and Fransozo A (1995) Morphometric relationships and maturation in *Portunus spinimanus* Latreille, 1819 (Crustacea, Brachyura, Portunidae). Rev Bras Biol 55:545-553

Santos S, Negreiros-Fransozo ML and Fransozo A (1996) Maturidade fisiológica em *Portunus spinimanus* Latreille, 1819 (Crustacea, Brachyura, Portunidae). Papéis Avulsos de Zoologia 39:365-377

Shields JD (1992) Parasites and symbionts of the crab *Portunus pelagicus* from Moreton bay, eastern Australia. J Crust Biol 12:94-100

Sokal RR and Rohlf FJ (1995) Biometry. WH Freeman, New York, 887 p

Voris HK and Jeffries WB (1997) Size, distribution, and significance of capitular plates in *Octolasmis* (Cirripedia: Poecilasmatidae). J Crust Biol 17:217-226

Zar JH (1996) Bioestatistical analysis. Prentice-Hall, Upper Saddle River, 662 p

HABITAT SPECIALIZATION AND ITS RELATION TO CONSERVATION POLICY IN CRUSTACEA

Marjorie L. Reaka-Kudla

Department of Biology
The University of Maryland
College Park, Maryland 20742, USA

mr9@umail.umd.edu

ABSTRACT

This work provides a survey of habitat diversity in all of the major groups of fossil and extant Crustacea. Analysis of the data suggests, however, that the hypothesis that habitat specialization leads to increased risk of extinction must be rejected. This conclusion has significant implications for conservation biology, where emphasis on conserving taxa that occupy unique, restricted, or specialized habitats has been one of the cornerstones of policy. Instead, we must focus on other attributes of these lineages, namely their taxonomic diversity and distributional ecology, in order to target organisms and regions for priority conservation attention.

I. INTRODUCTION

Since its inception about 20 years ago (Norse and McManus 1980, Lovejoy 1980, Wilson and Peters 1988, Wilson 1992, Reaka-Kudla et al. 1997), the field of "biodiversity" has come to represent our knowledge of the range of variation among all of the different levels and types of living organisms on Earth, including the processes that generate and maintain diversity. The levels of biodiversity often are conceptualized as molecular and cellular diversity, organismal diversity (variety among individual organisms), ecological diversity, and taxonomic diversity (variety among species and higher lineages). Of these, ecological diversity possibly has received the least attention. Yet ecological diversity is the sector of biodiversity that most closely relates to the field of conservation biology.

"Ecological diversity" includes (a) the range and types of habitats occupied by a taxon; (b) the ecological role of the organism (particularly its abundance, position in the food web, and impact on the chemical or structural environment); and (c) the life historical and distributional ecology of the organism (Reaka-Kudla 2000). Seeking to stem the extermination of populations and habitats, modern conservation biology (Soule and Wilcox 1980, Soule 1986) has identified as particularly vulnerable those species or assemblages that occupy highly specialized, unique or restricted habitats; have low population sizes; or are large in body size, long lived, or highly active. Top predators or species that play critical roles in their communities also are considered to be important for maintenance of diversity.

Because of space limitations, this paper will focus on only one aspect of ecological diversity, habitat diversity, in the major groups of extinct and extant crustaceans. The primary hypothesis to be tested is whether or not living in a highly specialized, unique or restricted habitat renders a species vulnerable to extinction.

II. METHODS

Although the need for brevity constrained this analysis to patterns of habitat use of the major taxonomic groups (orders, suborders, infraorders) of Crustacea, future studies should include analyses of these patterns at lower taxonomic levels as well. Greater ecological variability undoubtedly occurs than is represented for these taxa. However, it is hoped that these necessarily brief summaries encompass the

E. Escobar-Briones & F. Alvarez Eds.
MODERN APPROACHES TO THE STUDY OF CRUSTACEA
PP. 211-221

major trends for the groups treated. While attempts to categorize the "typical" habitats of a taxon are bound to be somewhat subjective, the summary table institutes some approximate boundaries for judging whether or not taxa are considered to occupy specialized and unique habitats. Although crude, these are sufficiently consistent to allow testing of the general idea in this paper. Sources included, among others, Barnes (1980, 1998), Abele (1982a, 1982b), Bowman and Abele (1982), Schram (1983, 1986), Gore and Heck (1986), Pearse et al. (1987), Brusca and Brusca (1990), Alvarez et al. (2000), Dworschak (2000), and the author's own experience. Special thanks are given to Todd Haney and Marilyn Schotte for sharing recent unpublished information on the habitats of leptostracans and isopods.

III. RESULTS

Class Remipedia. Probably an ancient, basal group with a few living relicts, the remipedes include 1 extinct order (Enantiopoda) and the contemporary Order Nectiopoda. The extinct enantiopodans are known only from the Lower Pennsylvanian of Texas (warm coastal water, probably epibenthic). The nectiopodans live (often at low oxygen concentrations) deep within fresh-brackish water caves in the Bahamas, Turk and Caicos Islands and Yucatan Peninsula, and in lava tubes in the Canary Islands.

Class Cephalocarida. There are no fossils within the living order, but the extinct Devonian lipostracans likely are allied to this group. Extant cephalocarids inhabit benthic marine habitats from 0-1500m. Of species known from more than 1 site, their depth distribution is moderately wide despite lack of lateral distribution.

Order Lipostraca. An extinct monospecific group that likely is related to the modern cephalocarids or perhaps to the anostracan branchiopods, the lipostracans are known only from the Middle Devonian of Scotland (tropical-subtropical at the time). They achieved high population densities in hot springs.

Class Branchiopoda, Order Anostraca. Although the Devonian lipostracans may be related, anostracan fossils are known from the Eocene to the Recent. The "fairy shrimps", though, are probably much older than their fossil record indicates. Contemporary anostracans inhabit ephemeral pools (including melt water, flood catchments, brine pools), hypersaline lakes and marine lagoons.

Order Notostraca. Paleozoic "tadpole shrimps" (incompletely known) are indistinguishable from modern living taxa. Notostracans occupy inland waters of all salinities, including temporary pools wherever such pools occur (although 1 species inhabits permanent ponds).

Order Kazacharthra. This extinct order, known only from the Jurassic of Kazachstan, is related to and may belong within the Notostraca. These shrimps probably lived in warm shallow benthic environments.

Order Cladocera. "Water fleas" do not have an extensive fossil record (Oligocene to Recent). Most cladocerans inhabit inland waters, usually in semi-permanent or permanent bodies of water, but several genera are marine. Most crawl in benthic habitats but some swim with the antennae; 1 species is ectoparasitic on *Hydra*.

Order Conchostraca. An old group known from the Devonian to the Recent, "clam shrimps" live on the bottom of permanent and ephemeral fresh water habitats. Most are benthic, but they may swim during reproductive periods.

Class Maxillopoda, Subclass Ostracoda. One of the oldest and most successful groups of crustaceans, the "seed shrimps" have existed since the Cambrian. Ostracods inhabit marine (to 7000m), brackish, and fresh water environments. A few are supralittoral on sandy beaches, and a few occupy moist terrestrial moss or humus habitats. Some taxa are commensal on echinoderms or other crustaceans. Most taxa are predominantly benthic crawlers or burrowers (though they may swim during reproductive phases), but a number are pelagic.

Subclass Mystacocarida. This interstitial group leaves no fossil record. Within any particular sandy habitat, mystacocarids are patchily distributed, implying either specializations for existence at particular grain sizes, disturbance regimes, salinities, or temperatures, or perhaps sporadic colonization due to disturbance events by small, highly local populations.

Subclass Copepoda. Although 1 of the most diverse and important crustacean groups, the fossil record of copepods includes only a parasite from the gill of a Lower Cretaceous fish and other taxa from the Miocene through the Recent. Copepods occupy fresh, brackish, and marine waters as well as moist terrestrial (moss, humus, leaves) habitats. Some groups inhabit primarily the deep sea (misophrioids, spinocalanoids), but others inhabit broad categories of habitats. Five orders are primarily planktonic (most calanoids, most cyclopoids, misophrioids, mormonilloids, monstrilloids); 1 order is primarily benthic (most harpacticoids). Two orders (poecilostomatoids, siphonostomatoids) are exclusively endo- and ectoparasitic on invertebrates, marine and fresh water fishes, and occasionally other hosts; some members of other orders are parasitic (e.g., larval stages of monstrilloids are endoparasitic in invertebrates).

Subclass Branchiura. No fossils of "whale lice" are known, but their distribution suggests that branchiurans are an ancient group. Branchiurans are ectoparasitic on the skin and gills of marine, brackish, and fresh water fishes and occasionally amphibians. Tolerant of salinity changes and usually not host specific, they may leave the host to find mates, lay eggs, or locate new hosts.

Subclass Tantulocarida. No fossils of this rare (only 5 species), deep water, recently discovered ectoparasitic group are known. Relatively host specific, 4 species each inhabit only 1 host species (a harpacticoid copepod, a benthic ostracod, and 2 species of isopods) and 1 species lives on 3 species of hosts, all harpacticoid copepods. All inhabit the deep ocean (400-1400, 2000-5000, 3000, 3300 m).

Order Ascothoracida (recently considered to be outside of the Cirripedia). Although no fossils exist, the morphology of these marine ecto- and endoparasites on soft and hard corals and echinoderms suggests that the group is primitive relative to the cirripedes; they likely are very old. Cretaceous octocorals and irregular sea urchins contain possible trace fossils of ascothoracidans. Except for 2 species that are freeswimming throughout their life cycle (intermittently ectoparasitic), all are obligate marine parasites that reside permanently on a host. Some live in galls on corals. Most attach and suck fluids from the host but some live inside the host tissue. In 1 group, the female is endoparasitic while the male is ectoparasitic on an anthozoan host.

Subclass Cirripedia, Order Acrothoracica. The "boring barnacles" have a modest fossil record (mostly of burrows) from the Devonian onward. Exclusively marine, the group is best represented in tropical and subtropical waters (though 1 species inhabits cold Antartic environments) and generally is restricted to shallow water (although 2 species inhabit the deep sea). All burrow into calcareous substrates (corals and mollusc shells, including hermit crab shells) except 1 species that excavates burrows in soft siliceous clay.

Order Thoracica. The "goose", "wart", and "acorn barnacles" comprise 3 living suborders (lepadomorphs, verrucomorphs, balanomorphs). Although a possible lepadomorph fossil is recognized in the Middle Cambrian Burgess shale, barnacles are known definitively from the Silurian (attached to a sea scorpion) to the present. All thoracicans are marine, inhabiting hard surfaces in all world oceans, including Antarctica and the deep sea. Some species of both goose and acorn barnacles live on floating or swimming objects (logs, seaweed, pumice, whales, sea turtles) or may settle on benthic motile objects such as crabs or other organisms. Some species of lepadomorphs are specialized for living only on sessile coronulid balanomorphs which in turn attach to certain species of whales, and others inhabit the gill chambers of crabs. Some thoracicans are obligate symbionts in soft corals or other hosts.

Order Rhizocephala. Leaving no fossil trail, this group (whose taxonomy is controversial) is endoparasitic in other crustaceans (especially decapods, but also other thoracican barnacles, isopods, and cumaceans), where they interfere with the host's reproduction. The rhizocephalans inhabit shallow marine and brackish habitats, particularly in tropical-subtropical waters.

Class Malacostraca, Order Belotelsonidea. Of uncertain affinity within the Malacostraca, the extinct belotelsonideans are documented only from the Carboniferous. They inhabited tropical-subtropical shallow nearshore marine environments. Abundant, they were among the more prominent crustaceans in the benthic fauna.

Order Waterstonellidea. Also of uncertain affinities, these extinct malacostracans are known from the Carboniferous to the Permian. They were relatively abundant inhabitants of tropical-subtropical marine habitats. Morphological features suggest that the waterstonellideans may have been pelagic; they are preserved in gregarious associations, perhaps schools.

Order Eocaridacea. This heterogeneous collection of poorly preserved extinct malacostracans may not represent a natural group. These taxa are recorded from the Carboniferous to the Permian. Their habitat was tropical-subtropical and probably benthic marine. Some morphological features of the eocaridids resemble the aeschronectid hoplocarids, and some characteristics of the anthracophausiids and essoidiids are somewhat similar to those of burrowing and tube-dwelling peracarids.

Subclass Phyllocarida, Order Hymenostraca. Few fossils document this extinct order from the Lower-Middle Cambrian to the Lower Ordovician. The group inhabited tropical-subtropical, marine benthic environments.

Order Canadaspida. This extinct phyllocarid is known only from Cambrian Burgess Shale. These relatively abundant crustaceans probably were epibenthic in tropical-subtropical marine sediments.

Order Archaeostraca. These extremely successful but now extinct phyllocarids are known from the Cambrian until the Permian. They were world-wide, abundant, and occupied a great variety of marine habitats. Some heavily armored forms may have inhabited benthic reefs. Thin shelled, possibly burrowing forms likely inhabited benthic coastal/lagoonal habitats. Some forms appear to have been pelagic.

Order Hoplostraca. This extinct phyllocarid is recorded only from the Carboniferous. They were benthic, probably lie-in-wait predators in coastal fine grained sediments.

Order Leptostraca. The only phyllocarid still extant, the leptostracans are known from the Permian to the Recent. All living forms are marine. Most are epibenthic, often associated with muds of low oxygen content. Some occur with decaying algae, seagrass, mangroves, coral and sponges, and 1 species is known from marine caves. Although usually inhabiting relatively shallow (0-400m) level bottoms, several genera, including those with pelagic habits, live in deep water (to 6000m); 1 species lives inside deep sea hydrothermal vents (East Pacific).

Subclass Hoplocarida (some workers include the hoplocarids and all subsequent malacostracan orders within the Subclass Eumalacostraca), Order Aeschronectida. This extinct group is known only from the Carboniferous. Extremely successful, they dominated their tropical near-shore shallow marine communities in numbers. Most species probably lived on the surface of sediment, although 1 apparently was pelagic.

Order Palaeostomatopoda. The extinct palaeostomatopods persisted from the Late Devonian to the Late Mississippian in shallow tropical near-shore marine waters. Most species probably were benthic, surface-dwelling predators.

Order Stomatopoda. The "mantis shrimps" comprise 2 suborders, the extinct archaeostomatopods (1 family, 2 species) and the living unipeltids. The Archaeostomatopoda (family Tyrannophontidae)—documented from the Middle Mississippian to Upper Pennsylvanian—link the extinct Order Palaeostomatopoda to the living unipeltids. The unipeltid Superfamily Bathysquilloidea usually includes the extinct Sculdidae (Upper Jurassic-Cretaceous, possibly Eocene), which probably inhabited warm coastal marine benthic environments, and the extant Bathysquillidae, known only from deep continental shelf habitats in the Recent. Other modern unipeltid superfamilies diversified during the Mesozoic (Cretaceous onward). Stomatopods are entirely marine, though some taxa tolerate hypersaline and estuarine environments. Most inhabit shallow tropical and subtropical habitats, although a few exist in cold or deep waters (bathysquillids, some pseudosquilloids, squilloids). The gonodactyloids (and smallest lineages of squilloids and lysiosquilloids) live in preformed holes and crevices in coral and rock (small pseudosquilloids live in cavities in coral and larger individuals and species may excavate burrows under coral), while most squilloids and lysiosquilloids excavate burrows in mud and sand, respectively. The deep sea bathysquilloids live on muddy bottoms (burrowing habits not known).

Subclass Syncarida, Order Palaeocaridacea. This extinct order lived only during the Carboniferous and Permian. Inhabiting brackish to fresh waters, they often occurred in deltas and swampy lagoonal habitats with aeschronectid hoplocaridans and mysidaceans. They were very abundant components of their tropical-subtropical epibenthic communities.

Order Anaspidacea. These living syncarids, probably derived from the fossil Order Palaeocaridacea, are known from the Triassic to the present, although the lineage probably is older. The Suborder Anaspidinea includes the single fossil representative, recorded from southeastern Australia; living members of this suborder occur only in surface fresh water in Tasmania. Some genera inhabit streams, lakes, and caves, and others are found in mats on lake bottoms, small ponds and crayfish burrows in grass swamps, or deep within semi-terrestrial crayfish burrows. Occurrence is erratic and patchy (though populations are relatively dense when found). However, 2 species of *Anaspides* are fairly abundant and well distributed in streams. Members of the other suborder, Stygocaridinea, are known only from interstitial ground water and isolated fresh water springs in the southern hemisphere.

Order Bathynellacea. No fossils of this syncarid group exist, although their distibution implies that the order is very old. Tiny organisms that crawl over and around sand grains, most bathynellaceans are interstitial ground water forms collected from wells, but 2 species inhabit Lake Baikal (20-1440m). The group usually inhabits fresh water, though some species have been collected in brackish conditions and 1 species is found in warm springs (to 55°C).

Subclass Mysidacea (some workers include the mysidaceans within the Pericarida), Order Pygocephalomorpha. The extinct pygocephalomorphs have a good fossil record from the Carboniferous through the Permian. Abundant, they were among the most important components of their Late Paleozoic tropical-subtropical marine benthic communities.

Order Mysida. Fossil "opossum shrimps" are rare and relationships uncertain, but likely mysidan fossils are recorded from the Jurassic; otherwise all taxa are Recent. The group inhabits a range of aquatic environments, especially marine environments down to 7210m. A few are pelagic but most live on, near, or in the substrate, where they sometimes can be abundant. On coral reefs, they frequently form swarming swimming aggregations in the lee of reef structures such as corals and sometimes are found in cryptic reef habitats. Some species are intertidal, where they burrow in sand. A number of species live in brackish or fresh water environments, and several groups inhabit marine, brackish and fresh water caves.

Order Lophogastrida. The fossil record of this mysidacean group extends from the Pennsylvanian (extinct Peachocarididae) to the present. All marine, most contemporary taxa are active pelagic swimmers, but some are benthic.

Subclass Pancarida, Order Thermosbaenacea (some experts place this group within the Peracarida). No fossils exist for this rare and unique taxon. These shrimps inhabit a variety of benthic environments, including marine habitats, inland salt ponds, fresh or salty ground waters and caves; 1 species inhabits hot springs (> 40°C) in North Africa that were repopulated (apparently from thermal ground water) after sterilization.

Subclass Peracarida, Order Tanaidacea. This order is known from the lower Carboniferous to the present. Inhabiting all latitudes, most species live in or on benthic marine substrates (though some inhabit floating algal mats) from the intertidal to the deep sea, but a few inhabit brackish and fresh water environments.

Tanaids construct tubes or burrows or live in holes, often defending them tenaciously; 1 species lives in minute empty gastropod shells similar to hermit crabs.

Order Spelaeogriphacea. Probably allied to the tanaids, fossils of this exceedingly rare group of peracarids are documented only from the Mississippian (Acadiocarididae). The 2 living species (Spelaeogriphidae) are recorded only from relatively acidic stream water in a cave at Table Mountain, South Africa, and from a cave in Brazil.

Order Cumacea. Cumacean-like forms, whose morphology suggests an early phase of evolution of the order, are known from the Pennsylvanian and Permian, but all other taxa are known only from the Recent. The vast majority of cumaceans are marine, but a few occupy brackish or fresh water habitats. Primarily benthic, they typically remain buried in bottom sediments during the day (and also are found in crevices in coral and rubble), but they swim into the water column at night, especially during reproductive phases. Some species are abundant in nocturnal plankton tows. Distributions of species are patchy and correlated with particle size of the substrate. Although cumaceans occur from intertidal to abyssal depths, they are most abundant and diverse in the deep sea.

Order Mictacea. This recently discovered order contains only 3 species. No fossils are known. One species is pelagic in the water of several brackish marine caves (Bermuda), and 2 benthic deep sea species each are known from a single site (1000m off northeastern South America; 1500m off south Australia). Their distributon suggests an old group.

Order Isopoda. All infraorders of this peracarid group are still extant after a Paleozoic origin. The oldest fossils (the extinct Palaeophreatoicidae) occurred in tropical-subtropical benthic marine habitats during the Pennsylvanian but invaded fresh water in the Permian. The marine flabelliferans are recorded from the Triassic onward, and the extinct urdids from the Jurassic-Cretaceous; most other groups appear in the Tertiary. Marine parasites (epicarideans) are known from the last half of the Mesozoic by the swollen gill chambers of their crustacean hosts. Isopods are common in almost all environments, from marine (supratidal to deep sea, tropical to polar), brackish, fresh water (including caves and hot springs), to terrestrial (even arid) habitats. With reproduction freed from water, the Oniscoidea are the most successful terrestrial crustaceans. Some phreatoicideans also are terrestrial. The

Infraorders Phreatoicidea and Asellota are particularly successful in fresh water and the deep sea, respectively; some asellotans and microcerberideans (commonly from shallow marine habitats) also inhabit fresh water. Some groups are exclusively or partly ectoparasitic on marine or fresh water fishes (aegid and cymothoid flabelliferans, juvenile gnathiids) or ecto- and endoparasitic on other crustaceans (bopyrid, dajid, entoniscid, and liriopsid epicarideans). Some parasites are host-specific, and exhibit specializations for a particular microhabitat such as crustacean branchial chambers or fish gill slits or tongues. Many higher taxa occupy broad categories of habitats (terrestrial-fresh water, fresh water-marine, shallow-deep sea), though individual species often have specialized habitat requirements (some terrestrial species live only in certain humidity conditions in desert microhabitats, some sphaeromatids are found only where they can bore into muddy shale, some exhibit morphological specializations for living in caves).

Order Amphipoda. Probably closely related to the isopods, these peracarids occur in Late Eocene-Early Oligocene Baltic amber (Crangonyctidae of the Infraorder Gammaridea), but probably are much older. Amphipods inhabit most marine (supratidal to deep sea, tropical to polar), brackish, and fresh water environments, and can be very abundant. Most aquatic groups are benthic, but some are pelagic, often in deep oceanic waters. Some pelagic species associate with other zooplankton such as cnidarian medusae, comb jellies, or salps; different taxa eat the host, use it for transport or as a nursery for the young. Within the Infraorder Caprellidea, the Caprellidae are highly modified for clinging to other organisms and the Cyamidae are parasitic on whales and dolphins. Amphipods are very successful in fresh water, and certain isolated sites such as Lake Titicaca and Lake Baikal are centers of rapid diversification; Baikal hosts over 40 genera and 300 species (25% of the world's fresh water amphipods). Some lineages inhabit fresh or brackish caves and other ground water (ingolfiellideans; gammaridean crangonyctids, hadziids, and bogidiellids; these groups also often include marine species). A number of amphipod taxa are interstitial, living among sand grains. A few (e.g., gammaridean talitrids) are terrestrial in moist leaf litter or vegetation on supratidal sandy beaches, but none are as successful or specialized for this habitat as are the terrestrial isopods. Many amphipods camouflage themselves in or make domiciles of plant matter; some cut and fashion algae into a protec-

tive bivalve "pod". A few build tubes similar to those of tanaids; 1 group spins a silk tube inside a tiny gastropod shell and carries it like a hermit crab, mimicing the color, shape and posture of the mollusc. Using a gut symbiont to digest cellulose, 1 family (Limnoridae) bores in wood.

Subclass Eucarida, Order Euphausiacea. There are no definitive fossils of "krill". Inhabiting oceanic waters from 0-5000m, euphausiaceans occur at all latitudes but are most abundant in cold (temperate, boreal) conditions. Almost all euphausiaceans are pelagic. Over 2/3 of Pacific species occur only in the epipelagic, about 1/4 are mesopelagic, and a few are bathypelagic. Although species are rarely benthic, a few stir up bottom sediment while feeding on suspended particles. Gregarious, krill occur in huge schools (> 1,000 individuals/m^3).

Order Amphionidacea. This exceedingly rare, unique group contains only 1 variable oceanic species (35°N to 35°S). Fossils are lacking. Exclusively pelagic, these eucarids occur down to at least 2000m. Although only 3 males are known, most females are taken at depth, but a few have been collected near the surface. Larvae are found in the upper 30m (0-10m at night), probably indicating extensive dispersal.

Order Decapoda (higher classification within the decapods is controversial), Suborder Dendrobranchiata. Although no fossils are known from the Paleozoic, a diverse assemblage of Penaeoidea was present from the Triassic-Jurassic on. The Sergestoidea are known only from the Recent. Extant penaeoideans include both pelagic and benthic species; most are marine but some inhabit brackish estuaries. Sergestoideans are all marine and pelagic.

Suborder Caridea (including the Procarididae, which some experts place in a unique suborder, Procaridea). Relatively few fossils are known, including the extinct family Udorellidae from the Upper Jurassic and several other taxa from the Jurassic and Cretaceous. The procaridids—known only from brackish tide pools on Ascension Island and brackish water caves in Bermuda and the Hawaiian Islands (Maui, Hawaii)—may be allied to the extinct udorellids. Carideans inhabit marine, brackish and fresh water environments. Some are pelagic but most are epibenthic; some lie buried in sediment. Possibly due to their hemolymph chemistry, carideans do well in cold environments, and are well represented in the Antarctic. The group occurs at all depths, including the abyssal plain; relatively large white carideans have been observed around deep sea vents. Many species inhabit cryptic environments (e.g., crevices and holes in coral, rock) or are specialized for commensal life in, on, or with sponges, cnidarians (anemones, coral), kelp, sargassum, or other organisms (e.g., echinoderms, molluscs, fishes).

Suborder Stenopodidea (termed Euzygida by some authors to include the fossil uncinideans). The fossil record of this group of "cleaner shrimps" extends back to the Jurassic. Contemporary stenopodideans inhabit mostly tropical and warm temperate, marine to brackish coastal and lagoonal environments. Most are benthic and associated with hard bottoms, especially coral reefs. Many are commensal, including species that are associated with sponges or cnidarians as well as the conspicuously colored "cleaner shrimps" that remove parasites from fishes at cleaning stations using complicated interspecific communication systems. Stenopodideans usually occupy shallow waters, but some live encased as pairs in glass sponges in deep habitats (to 1480m). Generally, stenopodids are characterized by specific habitat requirements (temperature, substrates, symbiotic associations).

Suborder Reptantia (including all the remaining groups of decapods, this taxonomic category is controversial), Infraorder Thalassinidea. The "mud" and "ghost shrimps" are known from the Jurassic onward. Although they extend from about 70°N-50°S and a few species are known from abyssal and hadal depths, the group is best represented in tropical-temperate latitudes and relatively shallow depths (often in very shallow [<2m] environments). As a whole, thalassinideans inhabit benthic marine and estuarine coastal environments and lagoons (mass migrations have been observed in African rivers). One species recently has been reported from marine caves. Most of these shrimps construct large, permanent U-shaped burrows or large (sometimes >3-4m), highly complex, changing burrows in mud, sand, shell, among rubble, or occasionally in moist "firm ground". Some species inhabit pre-formed cavities in coral rubble or hard calcareous substrate (e.g., upogebiids, callianassids, axiids), including vertical canyon walls (axiids); a few upogebiids actually bore into calcareous substrate.

Infraorder Astacidea. This group, including the "crayfish" and "clawed lobsters", extends back at least to the Permian (Erymidae) and possibly to the Devonian. The Devonian *Palaepalaemon* may be allied to Astacidea or Palinura, but probably was pelagic rather than benthic. All extant astacideans are benthic. Most

crayfish inhabit fresh and brackish waters (ponds, bayous, streams, rivers, caves), but some excavate complex burrow systems in damp terrestrial environments. Nephropids (Maine and Norway lobsters) inhabit temperate marine environments where they live under rocks and excavate shallow shelters, ranging extensively on level bottoms to feed at night.

Infraorder Palinura. Probably inhabiting a warm benthic coastal environment, the Devonian *Palaepalaemon* possibly is related to this group, the "spiney" and "slipper lobsters". Other fossil representatives (*Protoclytiopsis*, probably a glypheid) appeared in the Carboniferous and the group diversified in the Mesozoic. All extant taxa inhabit benthic, tropical-subtropical marine environments. Most are associated with coral reefs or inter-reef areas. They often aggregate in cavernous shelters in reef or coarse substrate, but rove extensively over level bottoms at night to feed.

Infraorder Anomura (termed Anomala by some experts). This group, including the "hermit crabs", "porcelain crabs", "mole" or "sand crabs", "squat lobsters", and "lobster krill", is known from the Jurassic onward. Most species are marine, but some species occupy fresh or brackish water and terrestrial habitats (those in the latter must return to the sea to reproduce). The hermit crab life style has arisen independently several times in the anomurans (as well as in the tanaids and amphipods). Most anomuran hermit crabs occupy vacant gastropod shells (or hollow twigs, bamboo, tusk shells, even bottle caps) for protection, employing complicated communication and intense aggressive behavior to commandeer preferred shells. Commensals also inhabit the shell (e.g., hydroids, bryozoans, foraminiferans, sponges, anemones, seaweeds, tunicates, barnacles); some hermit crabs actively "plant" anemones on their shell and may steal anemones from other hermit crabs. Some lithodids camouflage themselves with sponges. A few inhabit stationary worm or vermetid mollusc tubes or holes in coral; the coconut crab (a hermit with no gastropod shell) avoids predators by its large size, digging a burrow, and inhabiting isolated islands. Many squat lobsters are small and occupy cryptic or commensal benthic habitats (e.g., crevices in rocks and coral, on or under cnidarians), but a number of galatheideans are large and benthic in deep environments; a relatively large white species has been observed around deep sea vents. Several groups (e.g., lobster krill) are exclusively pelagic (species of *Pleuroncodes* occur in vast floating aggregations). Streamlined egg-shaped mole and sand crabs live in the sandy intertidal, burrowing backward after each wave. Anomurans are particularly successful in cold environments, being one of the few reptantians moderately well represented in the Arctic (some hermit crabs, lithodids, galatheideans).

Infraorder Brachyura. Possibly appearing as early as the Carboniferous, when a probable dromiacean fossil is recorded from the Mississippian, the "true crabs" diversified in successive radiations during the Mesozoic (Jurassic-Cretaceous), Tertiary (especially Eocene) and Recent. Brachyurans occupy mostly marine and brackish aquatic habitats in shallow to deep (continental shelf, deep sea vents) environments, but freshwater and semi-terrestrial groups (mostly limited to tropical and subtropical regions) are well represented. Some fresh water species return to saline waters to release young, but others are isolated in fresh water lakes or caves, and a few species in fresh water families are truly terrestrial (inhabiting leaf litter where embryos develop directly into juvenile crabs that cling to the mother's body). "Land crabs" (gecarcinids, oxypodids) and intertidal species (e.g., grapsids) must return to the ocean to release young. Most crabs are benthic, although a few portunids are exclusively pelagic. Several groups are well adapted for commensal life in other species' burrows or mantle cavities of molluscs (e.g., pinnotherids in the mantle cavity of oysters or burrows of thalassinideans and other organisms) or in, on, or under other organisms such as anemones (majids), corals (hapalo-carcinids and some xanthids are obligate commensals in corals; the xanthid *Trapezia* defends coral), or kelp (the majid *Pugettia*). Some dromiids carry clam shells and others culture sponges on their back. Many crabs live cryptic lifestyles in crevices in coral and rock; xanthids and majids are particularly well represented in these tiny habitats on coral reefs. Some large species such as xanthid coral and stone crabs live in larger cavernous refuges among coral, rocks, or oyster shell, and emerge onto adjacent coarse or level bottoms at night to feed. Calappids and portunids bury themselves in sediment. Majid snow and king crabs inhabit level bottoms in deeper coastal waters, and geryonids inhabit deep continental shelf environments. A large white cancrid species lives near deep sea hydrothermal vents.

IV. DISCUSSION

Table 1 summarizes the data presented above, using + as a signal of risk due to occupation of unique, restricted, or highly specialized habitats. Although habitat specialization has been considered to be important by the conservation biology community in assessing vulnerability to extinction, the paleontological community has reported, instead, that taxonomic diversity and restricted geographic distributions are important factors in increasing the risk of extinction.

Table 1 shows that habitat specialization does not enhance the likelihood of extinction in Crustacea: lineages that became extinct most often inhabited generalized benthic coastal marine habitats, while taxa that survived are more specialized for restricted types of habitats (average habitat specialization/taxon 0.2 and 1.3 in extinct and extant groups, respectively). Of the extinct taxa, only the lipostracans (7% of extinct taxa) occupied unique, highly specialized habitats (++). Only the palaeocaridaceans (7% of extinct taxa) inhabited moderately specialized brackish and fresh water habitats (+), and the waterstonellideans may have been somewhat specialized (±) for a pelagic existence (7% of extinct taxa). All 3 of these groups likely succumbed to the great, relatively nonselective mass extinction at the end of the Paleozoic rather than to processes of background extinction. While only 14% of the extinct taxa could be considered to have extremely or moderately specialized habitat requirements, 87% of the living groups exhibited habitat specialization. The nectiopodans, anostracans, notostracans, mystacocaridans, tantulocaridans, ascothoracidans, rhizocephalans, anaspidaceans, bathynellaceans, thermosbaenaceans, spelaeogriphaceans, mictaceans, and stenopodideans (34% of extant taxa) were categorized as primarily occupying highly specialized habitats (++). At least some components of another 20 living groups (53% of living taxa) inhabited moderately specialized environments (+).

In other studies of crustacean biodiversity (Reaka 2001), I have considered the relevance of other factors (population size, life history ecology, trophic position and ecological role, taxonomic diversity, breadth of species ranges, and global extent of the entire group's distribution) to extinction. Of these factors, taxonomic diversity, sizes of species ranges, and extent of the group's distribution are most significantly related to extinction vulnerability. Ranking of these 3 risk factors among all living crustaceans indicates that the nectiopodans, cephalocaridans, tantulocaridans, anaspidaceans, lophogastridans, mictaceans, and especially the spelaeogriphaceans merit priority conservation attention.

Table 1. Habitat specialization in extinct and extant crustacean taxa. Within taxa, + + indicates that the majority of species inhabit highly specialized or unique habitats, + means that most taxa occupy somewhat specialized habitats or that a significant minority of taxa live in specialized or restricted environments, - indicates that most taxa do not inhabit specialized habitats but usually occur in coastal benthic marine environments. For extinct taxa, L,M,U refer to Lower, Middle, Upper, and * indicates that their disappearance may have coincided with a mass extinction. FW = fresh water. To compute average + signals/taxon, ? were not counted at all, ± and +? were counted as 0.5+.

Taxon	Unique, specialized or restricted habitat
EXTINCT GROUPS	
Remipedia,	
Enantiopoda	
(Pennsylvanian)	-
Cephalocarida?	
Lipostraca	
(Devonian*)	+ + (hot springs, abundant)
Branchiopoda	
Kazarthra	
(Jurassic*)	-
Malacostraca	
Belotesonida	
(Carboniferous)	- (abundant)
Waterstonellida	
(Carboniferous-Permian*)	± (pelagic schools?, abundant)
Eocaridacea	
(Carboniferous-Permian*)	-
Hymenostraca	
(LM Cambrian-L Ordovician)	-
Canadaspida	
(Cambrian)	- (abundant)
Archaeostraca	
(Cambrian-Permian*)	- (extremely abundant)
Hoplostraca	
(Carboniferous)	-
Aeschronectica	
(Carboniferous)	- (extremely abundant)
Palaeostomatopoda	
(U Devonian-U Mississippian)	-
Stomatopoda (Archaeostomatopoda)	
(M Mississippian-U Pennsylvanian)	-
Palaeocaridacea	
(Carboniferous-Permian*)	+ (brackish-FW, abundant)
Pygocephalomorpha	
(L Carboniferous-Permian*)	- (abundant)
Average + signals/taxon for extinct groups	**3.5/15 = 0.2**
EXTANT GROUPS	
Remipedia	
Nectiopoda	+ + (brackish caves)
Cephalocarida	-
Branchiopoda	
Anostraca	+ + (ephemeral FW)
Notostraca	+ + (ephemeral FW)
Cladocera	+ (FW)
Conchostraca	+ (FW)

Maxillopoda
  Ostracoda — + (some deep sea, FW, beach, semi-terrestrial, commensal, pelagic)
  Mystacocarida — + + (interstitial)
  Copepoda — + (most pelagic, some deep sea, parasitic)
  Branchiura — + (ectoparasitic)
  Tantulocarida — + + (parasitic)
  Ascothoracida — + + (parasitic)
  Acrothoracica — + (borers)
  Thoracica — + (some intertidal, commensal-parasitic)
  Rhizocephala — + + (endoparasitic)
Malacostraca
  Leptostraca — + (a few commensal, marine caves, deep sea & vents)
  Stomatopoda (unipeltids) — + (some cryptic reefs, some deep sea)
  Anaspidacea — + + (FW, ground water)
  Bathynellacea — + + (interstitial ground water, deep lakes, warm springs)
  Mysida — + (some deep sea, FW, caves, intertidal)
  Lophogastrida — ± (pelagic)
  Thermosbaenacea — + (salt ponds, ground water, hot springs)
  Tanaidacea — + (some deep sea, FW)
  Spelaeogriphacea — + + (FW caves)
  Cumacea — + (deep sea, a few FW, patchy distributions)
  Mictacea — + + (brackish caves, deep sea)
  Isopoda — + (some truly terrestrial, deep sea, pelagic, FW, hot springs, caves, parasitic, most with specific habitat)
  Amphipoda — + (a few truly terrestrial, some deep sea, pelagic, FW, caves, ground water, interstitial, commensal-parasitic, many with specific habitat)
  Euphausiacea — ± (pelagic, a few deep sea)
  Amphionidacea — + (pelagic, usually deep sea)
  Dendrobranchiata — ± (many pelagic, some estuarine)
  Caridea — + (some FW, cave, cryptic, commensal-mutualistic, Antarctic, some deep sea & vents)
  Stenopodidea — + + (most commensal-mutualistic, a few deep sea)
  Thalassinidea — + (some intertidal, many estuarine, a few into FW, 1 marine cave, a few deep sea, some cryptic in coral)
  Astacidea — + (many FW, caves, semi-terrestrial)
  Palinura — ± (coral reefs)
  Anomura — + (some FW, many inter-tidal, some terrestrial, many shell & cryptic habitats, some commensal associations, some deep sea & vents, a few pelagic, some polar)
  Brachyura — + (many FW, some caves, some deep sea & vents, many intertidal, some terrestrial, many commensal, many cryptic, a few pelagic)

Average + signals/taxon for living groups   48/38 = 1.3

REFERENCES

Abele LG (ed.) (1982a) The Biology of Crustacea. Vol. 1 Systematics, the Fossil Record, and Biogeography. Academic Press, New York

Abele LG (ed.) (1982b) The Biology of Crustacea. Vol. 2 Embryology, Morphology, and Genetics. Academic Press, New York

Alvarez F, Villalobos JL and Iliffe TM (2000) *Naushonia manningi*, new species (Decapoda: Thalassinidea: Laomedidae), from Aklins Island, Bahamas. J Crust Biol 20 (Special No 2):192-198

Barnes RD (1980) Invertebrate Zoology, 4th ed. Saunders, Philadelphia

Barnes RSK (ed.) (1998) The Diversity of Living Organisms. Blackwell Scientific, London

Bowman TE and Abele LG (1982) Classification of the Recent Crustacea. In: Abele L G (ed.) The Biology of Crustacea. Vol. 1 Systematics, the Fossil Record, and Biogeography (pp. 1-27) Academic Press, New York

Brusca RC and Brusca GJ (1990) Invertebrates. Sinauer Associates, Sunderland, Massachusetts

Dworschak PC (2000) Global diversity in the Thalassinidea (Decapoda). J Crust Biol 20 (Special No 2):238-245

Gore RH and Heck KL (eds.) 1986. Crustacean Biogeography. Balkema Press, Rotterdam

Jablonski D (1986) Background and mass extinctions: The alternation of macroevolutionary regimes. Science 231:129-133

Jablonski D (1991) Extinctions: A paleontological perspective. Science 253:754-757

Lovejoy TE (1980) Changes in biological diversity. In: Barney GO (ed.) The Global 2000 Report to the President 2 (pp. 327-332) Harmondsworth, Penguin Books, New York

Norse EA and McManus RE (1980) Ecology and living resources: Biological diversity. In: Environmental Quality 1980: The Eleventh Annual Report of the Council on Environmental Quality (pp. 31-80) Council on Environmental Quality, Washington, DC

Pearse V, Pearse J, Buschbaum M and Buschbaum R (1987) Living Invertebrates. Blackwell Scientific, Palo Alto

Reaka-Kudla ML (2001) Crustaceans. In: Levin SA (ed.) Encyclopedia of Biodiversity. Vol. 1:915-943 Academic Press, San Diego

Reaka-Kudla ML, Wilson DE and Wilson EO (1997) Santa Rosalia, the turning of the century, and a new age of exploration. In: Reaka-Kudla ML, Wilson DE and Wilson EO (eds.) Biodiversity II: Understanding and Protecting Our Natural Resources (pp. 507-524) Joseph Henry/National Academy Press, Washington, DC

Schram FR (ed.) (1983) Crustacean Phylogeny. Crustacean Issues. Vol. 1. Balkema Press, Rotterdam

Schram FR (1986) Crustacea. Oxford University Press, Oxford

Soule ME (ed.) (1986) Conservation Biology: The Science of Scarcity and Diversity. Sinauer Associates, Sunderland, Massachusetts

Soule ME and Wilcox BA (eds.) (1980) Conservation Biology: An Evolutionary-Ecological Perspective. Sinauer Associates, Sunderland, Massachusetts

Wilson EO (1992) The Diversity of Life. Belknapp Press, Cambridge, Massachusetts

Wilson EO and Peters FM (eds.) (1988) BioDiversity. National Academy Press, Washington, DC

TEMPORARY WATER CRUSTACEANS: BIODIVERSITY AND HABITAT LOSS

D. Dudley Williams

Division of Life Sciences, University of Toronto at Scarborough, 1265 Military Trail, Scarborough, Ontario, Canada M1C 1A4

williamsdd@utsc.utoronto.ca

ABSTRACT

Temporary waters are those that exhibit recurrent dry phases that, for specific waterbodies, are oftentimes predictable both in their time of onset and their duration. They include coastal marine, inland saline, and freshwater bodies, but this discussion focuses on the latter two types. Physicochemical features strongly influence the faunas present, but biological factors may be important also especially with increased duration of the aquatic phase. Insects and crustaceans dominate the fauna and comparisons between these two groups are made. The latter are well represented by micro- and macro-forms from the following major taxonomic groups: Branchiopoda, Ostracoda, Copepoda, Decapoda, Peracarida, and many rare species are present. Highest species richness appears to be associated with a hydroperiod of between 150-250 days per year. Crustacean biodiversity may be extremely high in some regions; for example, over 90 species of branchiopods and copepods have been recorded in wetland ponds on the coastal plain of Atlantic North America. This diversity is under threat from a variety of human activities, especially agriculture and land development, and also is likely to be affected by global climate change. Case histories are examined, as are management practices that may negatively impact crustacean biodiversity. Maximum diversity is likely to be best achieved by ensuring a range of natural drought regimes in temporary waters across wide geographical, physiographical, and climate conditions.

I. RATIONALE

The cyclical nature of temporary waters creates habitats that are structurally and functionally different from those found in permanent waterbodies. Their properties result in communities with distinctive, and oftentimes unique, elements, which are still poorly known - yet both habitats and faunas are increasingly under threat from human activities. The purposes of this paper are to briefly review the nature of these habitats and their communities - specifically for crustaceans but using insect analogues where appropriate, to examine global and regional threats to their survival, and to highlight management issues that may provide some protection.

II. TEMPORARY WATERS AS HABITATS

Temporary water are to be found in most regions of the world and are defined as bodies of water that experience a recurrent dry phase of varying length that is sometimes predictable in both its time of onset and duration (Williams 1996). They include both static and running waters, and may be subdivided into: *intermittent waters* - which contain water on a cyclical basis or are dry at times of the year that are more or less predictable; and *episodic waters* - which contain water on a more or less unpredictable basis (Comin and Williams 1994). Temporary waters include coastal marine,

Kluwer Academic/Plenum Publishers

inland saline, and freshwater bodies, but this paper will be largely confined to the latter two types. The physical and chemical nature of the temporary water environment typically fluctuates more than that of adjacent, permanent water bodies, and as the aquatic phase draws to a close, the habitat cyclically assumes semi-terrestrial and, finally, fully terrestrial characteristics. Many temporary waters eventually fill in as a result of excess production and accumulation of organic matter, and they thus represent important stages in the hydroseral succession of wetlands to more terrestrial habitats.

Such changes require the biotas to be specially adapted if they are to take advantage of the rich resources, freedom from fish predation, and reduced competition that temporary waters often provide. Adaptations to the loss of water include physiological tolerance (primarily through dormancy), migration, and life history modification. Although physicochemical factors are extremely influential, biological interactions may be important also, but they have been less studied.

Foremost amongst the physicochemical factors are the length of the aquatic phase, the pattern of disappearance of the water, and whether the latter is predictable or not. Linked to water loss are decrease in habitat volume, and increase in insolation with subsequent affects on water temperature, dissolved oxygen, primary production, pH, and water chemistry (Williams 1996). The physicochemical environment also will influence community structure through trophic processes, such as via the effects of dissolved nutrients, temperature, etc. on the qualitative and quantitative nature of a pond's phytoplankton community and the microbial and aquatic fungal processing of detritus on the bed (Bärlocher et al. 1978). Ironically, it has been suggested that temporary ponds represent the most persistent of freshwater habitat types throughout geological time, and that this long association has enabled invertebrates to become well adapted to even the most extreme of temporary waters (Fryer 1996).

Features of the biological environment that are likely to influence community structure include the degree of inter/intraspecific competition and predation (Morin et al. 1988, Johansson 1993) - both of which are likely to be affected by the succession of species that is a characteristic feature of temporary waters in general (Williams 1983, Jeffries 1994), and the seasonal influx of aerial colonisers (Fernando 1958, Nilsson and Svensson 1994). A number of other factors, termed genotype-environment interactions, may affect individual organisms, for example modification of growth, morphology and life history (Harrison 1980, Landin 1980, Juliano and Stoffregen 1994).

III. PROPERTIES OF THE COMMUNITIES

The faunas of temporary waters tend to be dominated by species with opportunistic and pioneering traits (*r*-selected), particularly insects and crustaceans. Within a given temporary water type, community composition appears to be quite similar across the globe. For example, temporary freshwater ponds from Britain, northeastern North America and northwestern Australia, despite having wide geographic separation and large differences in climate (maritime, temperate/continental, and wet-dry tropics, respectively), all characteristically contain anostracans, notostracans, microcrustaceans, mites, collembolans, insects (odonates, dipterans, hemipterans, and beetles), and snails (Williams 1997). The insect and crustacean fractions of temporary water communities both contain species with broad (e.g., holarctic) distributions, some of which are known from permanent waters (e.g., *Daphnia pulex*; but see Korinek and Hebert 1996), and species that occur only, or predominantly, in temporary waters (e.g., *Triops cancriformis*). Limited dispersal powers in some species of crustaceans may lead to a greater incidence of endemism than in insects, although transport of eggs via wind, bird's feet, and human activities has produced the cosmopolitan distributions seen in many species.

Although insects and crustaceans coexist, and have come to interact, in temporary waters, their evolutionary history is quite different, arising as they do from terrestrial and aquatic ancestors, respectively. It might be expected, therefore, that each demonstrate different adaptations for surviving the dry phase. Table 1 shows that whereas active migration to a suitable resting site is seen in some of the higher crustaceans (isopods, amphipods, decapods), it is more common in insects, by virtue of their powers of flight. However, a "wait it out" strategy involving diapause, particularly in the egg stage, is common to both groups. Diapause is an especially effective adaptation allowing organisms to synchronize their development with a particular phase in their habitat (e.g., a seasonal food resource), and it has been suggested that it is a relatively simple adaptation to acquire, having evolved in arthropods and non-arthropods several times (Farner 1961, Danks 1987). Particularly relevant to temporary water survival strategies is the suggestion that diapause may be primed by the same few dominant hormones that promote migration (Rankin 1974).

Table 1. *Comparison of the known mechanisms used to survive drought seen in crustaceans and insects (based on literature cited in Williams, 1987).*

Taxon	Drought survival mechanism/stage in life cycle
Crustacea	
Anostraca	Diapausing eggs
Notostraca	Diapausing eggs
Conchostraca	Diapausing eggs
Cladocera	Diapausing eggs; adults may survive in moist substrate
Ostracoda	Diapausing eggs; juveniles in torpor in moist substrate
Copepoda	Diapausing eggs; late copepodite stages and adults in cysts
Isopoda	Juveniles near the groundwater table
Amphipoda	Juveniles near the groundwater table
Decapoda	Juveniles and adults in burrows at the groundwater table
Insecta	
Plecoptera	Diapausing early instars
Ephemeroptera	Diapausing eggs
Odonata	Recolonising adults; drought-resistant eggs and nymphs
Hemiptera	Recolonising adults
Trichoptera	Recolonising adults; diapausing eggs; gelatinous egg mass; larvae deep in moist substrate; terrestrial pupae
Coleoptera	Recolonising adults; burrowing adults; semiterrestrial pupae; ?diapausing eggs
Diptera	
Chironomidae	Diapausing eggs; drought-resistant late instars, sometimes in cocoons of silk or mucus
Culicidae	Diapausing eggs; recolonising adults; drought-resistant late instar larvae and pupae
Other	Diapausing eggs, larvae and pupae; recolonising adults

The fact that crustaceans complete their entire life cycles in temporary waters whereas many insects have to recolonise each year may lead to some differences in the timing of occurrence of these two groups and these, in turn, may contribute to the strong seasonal succession so characteristic of the fauna. The greater diversity in development rates seen among temporary water crustaceans has been interpreted as an evolutionary consequence of their *permanent* association with the habitat, resulting in a greater variety of life history strategies than in temporary water insects (Lahr et al. 1999).

A tendency towards dominance of early pond stages by crustaceans has been noted in many localities, for example, the American Southwest (Sublette and Sublette 1967), Louisiana (Moore, 1970), and South and West Africa (Meintjes 1996, Lahr et al. 1999). However, in contrast, in temporary ponds in northeastern North America (e.g., Michigan: Kenk 1949, Ontario: Wiggins et al. 1980, Williams 1983), early phases in the hydroperiod have been shown to be characterized by both crustaceans (e.g., branchiopods, copepods) and insects (e.g., coleopterans, hemipterans, midges, mosquitoes, caddisflies), with insects becoming even more abundant in the later phases. Perhaps the differences in crustacean dominance between the northeastern North American and the more southerly located ponds are related to the more rapid recolonization of aquatic habitats by insects after Pleistocene glaciation. The predominance of insects in other freshwater habitats in previously glaciated regions, over the approximately 15,000 years since the ice retreated, is known (Williams and Williams 1998). None of the other locations cited above was impacted as much by ice (Guenther 1970, Matthews 1979).

IV. CRUSTACEANS SPECIFICALLY

In an attempt to begin to understand aspects of the ecology of temporary water crustaceans, I have compiled data on the distribution of taxa in relation to one of the most influential environmental variables in these habitats - the length of the hydroperiod (Table 2). These data represent some 46 studies from across North America, spanning intermittent water bodies that dry for only a few weeks each year to highly episodic ones that contain water for 10 days or less. The habitats vary considerably in their physical (and sometimes chemical) characteristics and, as is inevitable in such a compilation, the source studies vary in their degree of taxonomic resolution (from order to genus). Nevertheless, some patterns are evident. Notably, cumulatively, the highest taxon richness seems to occur in habitats that contain water from 150 to 250 days of the year (41 taxa), compared with 23 taxa in habitats that more approach permanent waters (250-330 days of water). The latter total was closer to that found (24) in habitats containing water from 70 to 150 days each year. Habitats containing water for between 40 and 70 days supported only 6 crustacean taxa, those wet for between 10 and 40 days 4 taxa, and only 2 taxa were

Table 2. *Occurrence of crustaceans in North American temporary waters in relation to length of the hydroperiod (based on data in Batzer et al., 1999).*

Taxon	Genus	<10	10-40	40-70	70-150	150-250	250-330	Location and habitat diversity
Mean Length of Hydroperiod (days)								**Location and habitat diversity**
Number of studies done		2	2	4	10	11	17	
Taxa:								
Branchiopoda								
Anostraca	Artemina					*	*	PPW[2], HPW[2]
	Chirocephalopsis				A			OVWP
	Branchinecta					*		PPW[1], SPP, SUT
	Eubranchipus		*	*	*	*		WSP, CB, PPW[2]
	Streptocephalus					C		CB, SPP, SUT
	Thamnocephalus					*		SPP
Notostraca	Triops					*		SPP
Conchostraca	Caenestheriella					*		SPP
	Eocyzicus					*		SPP
	Eulimnida					*		SUT
	Leptestheria					*		SPP
	Limnadia					C		CB
	Lynceus			*	A	*	*	WSP, CB, PPW[2], SPP, OVWP
Cladocera	Alona					C		CB, TFW[2]
	Alonella				*	*		WSP, CB
	Bosmina					R		BPW, PPW[2], TFW[2]
	Chydorus				C	*	*	WSP, BPW, SAW, CB, TFW[2], OVWP
	Chydoridae				C			OVWP
	Ceriodaphnia				C			OVWP
	Daphnia		*	*	*	C	*	WSP, BPW, SAW, CB, SUT, TFW[2], PPW[2], HPW[2]
	Diaphanosoma					C		CB
	Macrothrix					C	C	BPW, TFW[2]
	Moina					C		OVWP
	Pleuroxus						*	BPW, TFW[2]
	Polyphemus					C		CB
	Pseudosida					C		CB
	Scapholeberis	*	*		*	C	*	WSP, CB
	Sididae						R	BPW
	Simocephalus			R	C	C	*	BPW, SAW, CB, OVWP
Ostracoda	Candona				A	*		SPP, TFW[2], OVWP
	Cypria				A			OVWP
	Cypricercus				C			OVWP
	Cypridopsis						*	BPW, CB, TFW[2]
	Cyprinotus					*		SPP
	Cyclocypris					*		PPW[2], SPP
	Eucypris				C			OVWP
	Limnocythere					*		SPP
	Megalocypris					C		SPP
	Pleocypris					*		SPP
	Potamocypris					*		SPP
Copepoda								
Calanoida					*	*	*	WSP, CB, PPW[2], USPP, HPW[2]
	Diaptomus				C			OVWP
	Leptodiaptomus			C	C	C		SAW, OVWP
Harpacticoida						*	*	BPW, CB, TFW[2]
	Canthocamptus				A			OVWP
Cyclopoida		*	*	*	*	*	*	WSP, SAW, CB, USPP, SUT, TFW[2], HPW[2]
	Acanthocyclops				C			OVWP
	Cyclops				A			OVWP
Siphonostomatoida	Argulus					*	*	FM
Malacostraca								
Isopoda	Caecidotea				C	*	*	AWP, BPW, FM, TFW[2], OVWP
	Oniscus				R			OVWP
Amphipoda	Crangonyx					*	C	AWP, BPW, FLW, FM, CB
	Gammarus						C	BPW, CB, PPW, TFW, HPW[2]
	Hyallela					*	C	AWP, BPW, FM, CB, PPW, DM, TFW[2], HPW[2], FLW
Decapoda	Palaemonetes					C	*	FM, CB, USPP, TFW[2]
	Cambarus				C	*	R	BPW, FLW, OVWP
	Procambarus					*	C	BPW, FM, CB

Figure 2. *Key*

AWP	Autumnal Woodland Ponds (dry during most of summer and early fall)		CB	Carolina Bays
WSP	Wisconsin Snowmelt Ponds		PPW	Prairie Pothole Wetlands - some may be hypersaline at low water levels
OVWP	Ontario Vernal Woodland Ponds (dry from mid-summer until snowmelt)		HPW	High Plains Wetlands (S.E. Wyoming)
DM	Delta Marsh (Manitoba)		TFW	Temporarily Flooded Wetland (Missouri)
SAW	Subalpine Wetlands (Colorado)		A	Abundant
SPP	Southern Plains Playas		C	Common
BPW	Beaver Pond Wetlands (S.E. Coastal Plain & N.W. Pennsylvania)		R	Rare
USPP	Urban Southern Plains Playas		[1]	present at initial stages of wet/dry cycle, cannot tolerate high salt concentrations
FLW	Forested Limesink Wetlands (S.W. Georgia)		*	Present (not quantified)
SUT	Southeastern Utah Tinajas		[2]	can be present in highly saline wetlands
FM	Florida Marshes (Cypress Domes, Riparian Mashes, Everglades etc.)			

recorded in habitats wet for less than 10 days. Significantly, each of the last three hydroperiods was based on less than 5 surveys and this is likely to have had some impact on the low taxon richness found. The crustaceans recorded from this short hydroperiod group included an anostracan (*Eubranchipus*), a conchostracan (*Lynceus*), three cladocerans (*Daphnia, Scapholeberis* and *Simocephalus*), and calanoid (*Leptodiaptomus*) and cyclopoid copepods. All of these taxa were found in habitats with much longer hydroperiods but clearly have superior survival abilities. Collectively, habitats with hydroperiods in excess of 70 days supported a greater diversity of anostracans, a single notostracan (*Triops*), conchostracans ($\geq$150 days), cladocerans, ostracods, calanoid, harpacticoid, cyclopoid and siphonostomatoid copepods, isopods, amphipods, and decapods. Clearly, then, bearing the limitations of this compilation in mind, several conclusions may be drawn:

(1) maximum taxon richness in these waters occurs with a hydroperiod of between 150 and 250 days;

(2) some additional taxa occur in habitats with hydroperiods of between 70 and 150 days, but no additions occur (to the genus level) in habitats wet for less than 70 days;

(3) richness of the classic, three, large temporary water branchiopods (Anostraca, Notostraca and Conchostraca) falls dramatically in habitats containing water for more than 250 days per year;

(4) there is an urgent need for further study of short hydroperiod habitats (< 70 days).

V. GLOBAL THREATS TO TEMPORARY WATER CRUSTACEANS

Comparison of the traits of temporary water crustaceans with those of insects, begun in the preceeding paragraphs, is both inevitable - as they share many of the same habitats - and desirable - as, in general, many more studies have been conducted on insect populations and the knowledge base is therefore larger. For example, on a global scale it is clear that, amongst the insects, species diversity is dropping (between 100,000 and 500,000 species ($\sim$ 20%) are predicted to be lost over the next 300 years, Mawdsley and Stork 1995) and the world fauna is becoming more homogeneous. Factors contributing to this include increased abundance and spread of a small number of native species that thrive in habitats disturbed by human activity; the attraction of invading exotic species to native ecosystems; human-induced species extinctions and population declines; and the overarching yet poorly understood synergistic effects of global change (Samways 1996). Fossil evidence from the Quaternary indicates that, in the past, it has been possible for insects to track climatic changes, gradually altering their geographical ranges and thus preventing species extinctions (Coope 1995). However, the rate of human-induced climate change is far faster (e.g., mean global air temperature is predicted to increase by 1.5-4.5°C over the next 20 years; IPCC 1990) than even insects appear to be able to cope with. Presumably, the effects on freshwater crustaceans with their, typically, lower dispersal abilities are likely to be far more severe. Not only will crustacean extinctions occur, but also the population characteristics of surviving species are likely to be altered. Virtually no studies have been done to assess such changes in freshwater crustaceans - although a large-scale field manipulation to mimic global warming showed an increase in growth, smaller size at maturity, and precocial breeding in a population of the amphipod *Hyalella azteca* (Hogg and Williams 1996). Assessing such differences/evolutionary change among/within invertebrate populations typically has involved use of morphological and ecological characters. However, it has recently been shown (Müller et al. 2000) for *Gammarus fossarum*, in Europe, that morphological traits were 10-times less effective as genetic

Table 3. *Examples of environmental impacts likely to be associated with mankind's attempts to conserve water resources at risk from global climate change (based on Covich et al., 1997; Meyer et al., 1999).*

Impact	Likely to affect temporary water crustaceans	
	Positively	Negatively
Changed magnitude and seasonality of runoff regimes	?	✓
Accompanying altered nutrient loading, e.g., release from shoreline soils due to increased water level change	?	✓
Accompanying limited habitat availability at low flow		✓
Reduction in phytoplankton diversity, e.g., increased phosphorus loading + higher temperatures promote competitively superior cyanobacteria over diatoms		✓
Loss of wetlands due to water extraction and tillage		✓
Disconnection of groundwater table from shallow basins	?	✓
Increase in amount of emergent vegetation due to lowered water table	?	?
Increasing distance among habitats due to landscape fragmentation		✓
Reduced availability of waterfowl as dispersal agents		✓
Increasing salinity due to evaporation & reduced runoff	?	✓
Shifts in riparian vegetation (qualitative & quantitative) e.g., increased temperature favours emergence of weed species from wetland seed banks; also fire-resistant forms predominate because of increasingly dry conditions	?	?
Creation of dams for increased water retention		✓
Increased irrigation networks for agriculture	✓	?
Increased proximity of agricultural chemicals		✓

characters (enzyme loci) in revealing population variance. These findings suggest that molecular techniques should be applied when the status of threatened temporary water species is being evaluated, as observed morphological stasis may belie the level of genetic differentiation, which perhaps may lead to an alternative conservation stratagem.

Quite apart from the direct effects of global climate change on crustacean populations, there will be many serious effects as a consequence of how mankind responds to the problem of reduced water retention on the planet surface as a result of elevated temperatures. Table 3 provides some examples of these impacts and indicates whether they are likely to have a positive or negative effect on temporary water crustaceans. Most of the effects are seen, intuitively, as being negative although manipulation experiments are required for confirmation. Only one impact, that of an increase in irrigation networks needed for agriculture under a warmer climate, is likely to be beneficial (see Caspers and Heckman 1981).

By way of underlining our lack of knowledge of precisely how temporary water crustaceans are likely to respond to global changes, Table 4 compares what is known for insect populations. The few studies (cited in the legend) that address, or come close to addressing, the likely responses by crustaceans all point to an escalation of the listed variables as a result of human-induced changes to temporary waters - in fact pretty much paralleling the insect responses. The severity and speed of response are, however, largely unknown and require considerable research and managerial input.

VI. REGIONAL THREATS AND SOLUTIONS

Temporary ponds frequently support high levels of crustacean biodiversity, for example, ponds on the upper coastal plain of South Carolina support exceptionally rich zooplankton communities, including 44 species of cladoceran and seven species of calanoid copepod (Mahoney et al. 1990). In a single temporary pond in western Morocco, Thiery (1991) found six species of Anostraca, two of Notostraca and two of Conchostraca (Spinicaudata). Petrov and Cvetkovic-Dragana (1997) similarly found up to seven species of large branchiopod in a single pond in Yugoslavia. Southern Africa is particularly rich in anostracans, with 80% (38 species) being endemic. However, the fauna is under threat from agricultural practices, urbanization, pollution and pesticides (Hamer and Brendonck 1997). For example, in the Sahel, Senegal,

Table 4. *Comparison of the responses of insects and crustaceans to human-induced changes to ecosystems (insect responses in part based on Samways, 1996).*

	Insects		Crustaceans	
Variable	Natural	Human-induced	Natural	Human-induced
Population surges	Common	Common	Common	Increasing[1]
Range increase	Rare	Common	Rare	Increasing[2]
Population fragmentation (via landscape fragmentation, habitat loss, etc.)	Rare	Very common	Rare	Increasing[3]
Population crashes	Fairly common	Common	Fairly common	Increasing[4]
Species extinctions	Occasional	Increasingly more common & widespread	Occasional	Strong potential[5] + actual?
Loss/reduction of keystone species	Rare	Increasing	Rare	Strong potential[6]
Shifts in life history traits	Occasional	Increasing	Occasional	Increasing[7]
Competition/ predation from invading species	Occasional	Increasing	Occasional	Increasing[8]
Loss of genetic variability	Rare	Increasing	Rare	Strong potential[9]

(Examples, but not necessarily from temporary water: [1]Patalas 1975; [2]Maude 1998; Mills *et al*. 1993; [3]Goettle 2000; [4]Hobbs & Hall 1974; [5]Collinson *et al*., 1995; [6]Neckles *et al*. 1990; [7]Abdullahi 1990; Gallaway & Hummon 1991; Hogg & Williams 1996; [8]Berrill 1978; Garvey & Stein 1993; Lehman & Caceres 1993; [9]Korpelainen 1986)

downwind drift from routine application of a variety of insecticides used to control Desert Locusts has been shown to be detrimental to anostracans and cladocerans living in local ponds (Lahr 1998).

Temporary waters, in the U.K. for example, are known to be custodians of more rare species than permanent waters (Collinson et al. 1995). Further, the highest degree of genetic differentiation for some crustacean species (e.g., *Daphnia magna*, in Rhode Island, USA) has been shown to occur in small temporary ponds (Korpelainen 1986), suggesting that such populations are vital to the gene pool of certain species. However, via transplant experiments, Hairston and Olds (1984) showed that populations of the copepod *Diaptomus sanguineus* are adapted to the conditions of specific isolated ponds rather than to a broader geographical region. Transplanted females were not able to sense a change in pond type nor were able to adjust egg production (subitaneous versus diapausing eggs) accordingly, hence establishment in new ponds, should their home ponds be destroyed, cannot be taken for granted. In contrast to the cladoceran example above, the genetic variability exhibited by the anostracan *Branchinecta sandiegonensis* (endemic to temporary pools along the coast of San Diego County, USA) is very low - perhaps a consequence of low gene flow and founder effects resulting from habitat fragmentation and a lack of potential vectors for dispersal of their cysts (Davies et al. 1997). Such species are especially at risk from human activities - in the case of *B. sandiegonensis*, San Diego County already has lost 90-95% of its historic vernal pool habitat (Goettle 2000). Paradoxically, some so-called "pond-management programmes", such as are maintained by volunteer organizations in the U.K., dredge and remove silt from shallow temporary waters in an attempt to bring them back to permanent water status (Biggs et al. 1994). Although well-meaning, such practices are detrimental to the survival of drought-adapted species.

Protective legislation would seem to be an obvious solution, however another Californian example serves as a warning. *Lepidurus packardi*, the vernal pool tadpole shrimp, is known only from California's Central Valley and from a few locations in the San Francisco Bay area. Because loss of vernal pool habitat in California's Central Valley is estimated at between 65 and 90% of its former extent - due to urbanization of flat lands (slope ≤ 3-4%) close to metropolitan areas - in 1994 this species was designated as endangered and protected by federal law (Goettle 2000). In 1996 agricultural and business groups asked for delisting of two of the four species of endangered vernal pool shrimp, and are currently in the process of suing the US Fish and Wildlife Service over the matter claiming that protection of the shrimp is a surrogate for land-use control (Evans 2000). Another example of lack of public sympathy for the fate of notostracans is the fact that in California rice fields, *Triops* are looked on as pest species as they are reported to uproot and eat young rice plants (Fry and Mulla 1992).

In some cases, judgement based on habitat size rather than importance has failed to protect species. For example, small, isolated wetlands, such as are found in the southeastern Coastal Plain of the US, lack the legal protection given to riparian and lacustrine wetlands. Indeed, the US Army Corps of Engineers permits wetlands less than 0.12 ha to be filled on an *ad hoc* basis, and requires only a minimal review of circumstances if they do not exceed 1.2 ha (Federal Register 1996). As a result of uncontrolled agricultural conversion and regional development, loss of these habitats and their unique faunas is escalating (Kirkman et al. 1999).

The following are some management issues that should be considered wherever and whenever human activity and temporary water communities come into conflict:

(1) It has to be understood that temporary waters, be they ponds, pools, streams or wetlands are not "wasted" areas of land; they are natural features of the environment representing distinct and unique habitats for many species - some that are not found elsewhere, others that reach their maximum abundance there. Sufficient examples exist to show that drainage of these habitats results in the loss of both aquatic and terrestrial species, sometimes on an enormous scale;

(2) There is a strong likelihood that temporary waters contribute to maximizing the gene pool of certain species that occur in both temporary and permanent waters and that this increased diversity may be crucial to the survival of species facing future changes to global environments. It is important, therefore, that scientists, managers and developers consider the *evolutionary* as well as the *ecological* consequences of habitat destruction or alteration;

(3) That a diversity of temporary water habitats exists in nature, particularly with respect to the length of the hydroperiod and size of basin, and that maximum species diversity is best achieved by ensuring a range of natural drought regimes in a variety of environments across wide geographical, physio-graphical, and climate conditions;

(4) That the current state of knowledge of the biology and ecology of temporary water crustaceans is meager, as is how species will react to contemporary changes in global climate. Regional development issues therefore require some indulgence, even if this means enactment of overly protective legislation, to allow the knowledge base to catch up so that species and their environmental role are not lost before their importance is understood.

Alongside these issues is the need, as recently emphasized by Everard et al. (1999), for an intensified "gentle education" of the public about pond environments so that an awareness of the collective ownership of ponds takes hold. In this way, a "bottom-up" process may develop that will temper the ravages that current "top-down" directives have wrought upon these habitats and their biotas.

ACKNOWLEDGEMENTS

I thank Lucie Sliva, University of Toronto, for technical and library assistance, and Dr. Karen Lee, University of Pittsburgh, for inviting me to present this paper. Funding was provided by the Natural Sciences and Engineering Research Council of Canada.

REFERENCES

Abdullahi BA (1990) The effect of temperature on reproduction in three species of cyclopoid copepods. Hydrobiologia 196:101-109

Bärlocher F, Mackay RJ and Wiggins GB (1978) Detritus processing in a temporary vernal pool in southern Ontario. Arch Hydrobiol 81:269-295

Batzer DP, Rader RB and Wissinger SA (eds) (1999) Invertebrates in Freshwater Wetlands of North America. John Wiley & Sons, New York

Berrill M (1978) Distribution and ecology of crayfish in the Kawartha Lakes region of southern Ontario. Can J Zool 56:166-177

Biggs J, Corfield A, Walker, D, Whitfield M and Williams PJ (1994) New approaches to the management of ponds. British Wildl 5:273-287

Caspers H and Heckman CW (1981) Ecology of orchard drainage ditches along the freshwater section of the Elbe Estuary. Biotic succession and influences of changing agricultural methods. Arch Hydrobiol Suppl 43:347-486

Collinson NH, Biggs J, Corfield A, Hodson MJ, Walker D, Whitfield M and Williams PJ (1995) Temporary and permanent ponds: an assessment of the effects of drying out on the conservation value of aquatic macroinvertebrate communities. Biol Conserv 74:125-133

Comin FA and Williams WD (1994) Parched continents: our common future? (pp 473-527) In: Margalef R (ed) Limnology Now: a Paradigm of Planetary Problems. Elsevier, Amsterdam

Coope GR (1995) The effects of Quaternary climatic changes in insect populations: lessons from the past. (pp 29-48) In: Harrington R and Stork NE (eds) Insects and a Changing Environment. Academic Press, London

Covich AP, Fritz SC, Lamb PJ, Marzolf RD, Matthews WJ, Poiani KA, Prepas EE, Richman MB and Winter TC (1997) Potential effects of climate change on aquatic ecosystems of the Great Plains of North America. Hydrol Process 11:993-1021

Danks HV (1987) Insect dormancy: an ecological perspective. Biol Surv Canada Monogr Ser 1, Ottawa

Davies CP, Simovich MA and Hathaway SA (1997) Population genetic structure of a California endemic branchiopod, *Branchinecta sandiegonensis*. Hydrobiologia 359:149-158

Evans RR (2000) California Farm Bureau Federation, ag groups: fairy shrimp not threatened. Agriculture Alert http://www.cfbf.com/archive/aa-1029c.htm

Everard M, Blackham R, Rouen K, Watson W, Angell, A and Hull A (1999) How do we raise the profile of ponds? Freshwat Forum 12:32-43

Farner DS (1961) Comparative physiology: photoperiodicity. Ann Rev Physiol 23:71-96

Federal Register (1996) Final notice of issuance, reissuance and modification of nationwide permits. Army Corps of Engineers, DOD, Vol 61, p 65916

Fernando CH (1958) The colonisation of small freshwater habitats by aquatic insects. I. General discussion, methods and colonisation in the aquatic Coleoptera. Ceylon J Sci (Biol Sci) 1:117-154

Fry LL and Mulla MS (1992) Effect of drying period and soil moisture on egg hatch of the tadpole shrimp (Notostraca: Triopsidae). J Econ Entomol 85:6569

Fryer G (1996) Diapause, a potent force in the evolution of freshwater crustaceans. Hydrobiologia 320: 1-14

Gallaway MS and Hummon WD (1991) Adaptations of *Cambarus bartoni cavatus* (Hay) (Decapoda,

Cambaridae) to acid mine polluted water. Ohio J Sci 91:167-171

Garvey JE and Stein RA (1993) Evaluating how chela size influences the invasion potential of an introduced crayfish (*Orconectes rusticus*). Amer Midl Nat 129:172-181

Goettle B (2000) A "living fossil" in the San Francisco Bay Area? http:/ www.r1.fws.gov/sfbnwr/ tadpole.html

Guenther EW (1970) Die pleistozäne Eiszeit und ihre Vorläufer. Kiel Naturwisse Ver Schleswig-Holst Schrift. 40 pp.

Hairston NG and Olds EJ (1984) Population differences in the timing of diapause: adaptations in a spatially heterogeneous environment. Oecologia 61:42-48

Hamer ML and Brendonck L (1997) Distribution, diversity and conservation of Anostraca (Crustacea: Branchiopoda) in southern Africa. Hydrobiologia 359:1-12

Harrison RG (1980) Dispersal polymorphisms in insects. Ann Rev Ecol Syst 11:95-118

Hobbs HH and Hall ET (1974) Crayfishes (Decapoda: Astacidae). (pp 195-241) In: Hart, CW and

Fuller SLH (eds) Pollution Ecology of Freshwater Invertebrates. Academic Press, New York

Hogg ID and Williams DD (1996) Response of stream invertebrates to a global-warming thermal regime: an ecosystem-level manipulation. Ecology 77:395-407

IPCC [Intergovernmental Panel on Climate Change] (1990) Scientific assessment of climate change. Report of Working Group 1. World Meteorological Organisation, Geneva, Switzerland

Jeffries MJ (1994) Invertebrate communities and turnover in wetland ponds affected by drought. Freshwat Biol 32:603-612

Johansson F (1993) Intraguild predation and cannibalism in odonate larvae: effects of foraging behaviour and zooplankton availability. Oikos 66:80-87

Juliano SA and Stoffregen TL (1994) Effects of habitat drying on size at, and time to, metamorphosis in the tree hole mosquito *Aedes triseriatus*. Oecologia 97:369-376

Kenk R (1949) The animal life of temporary and permanent ponds in southern Michigan. Misc Publs Mus Zool Univ Mich 71:1-66

Kirkman LK, Golladay SW, Laclaire L and Sutter R (1999) Biodiversity in southeastern, seasonally ponded, isolated wetlands: management and policy perspectives for research and conservation. J N Am Benthol Soc 18:553-562

Korinek V and Hebert PDN (1996) A new species complex of *Daphnia* (Crustacea, Cladocera) from the Pacific northwest of the United States. Can J Zool 74:1379-1393

Korpelainen H (1986) Genetic variation and evolutionary relationships within four species of *Daphnia* (Crustacea: Cladocera). Hereditas 105:245-254 Lahr J (1998) An ecological assessment of the hazard of eight insecticides used in Desert Locust control, to invertebrates in temporary ponds in the Sahel. Aquat Ecol 32:153-162

Lahr J, Diallo AO, Ndour KB, Badji A and Diouf PS (1999) Phenology of invertebrates living in a sahelian temporary pond. Hydrobiologia 405:189-205

Landin J (1980) Habitats, life histories, migration, and dispersal by flight of two water beetles, *Helophorus brevipalpis* and *H. stigifrons* (Hydrophilidae). Holarct Ecol 3:190-201

Lehman JT and Caceres CE (1993) Food-web responses to species invasion by a predatory invertebrate: *Bythotrephes* in Lake Michigan. Limnol Oceanogr 38:879-891

Mahoney DL, Mort MA and Taylor BE (1990) Species richness of calanoid copepods, cladocerans and other branchiopods in Carolina Bay temporary ponds (South Carolina, USA). Am Midl Nat 123:244-258

Matthews JV (1979) Tertiary and Quaternary environments: historical background for an analysis of the Canadian insect fauna. (pp 31-86) In: Danks HV (ed) Canada and its Insects. Mem Ent Soc Can, Ent Soc Canada, Ottawa

Maude SH (1998) A new Ontario locality record for the crayfish *Orconectes rusticus* from West Duffins Creek, Durham Region Municipality. Can Field-Nat 102:66-67

Mawdsley NA and Stork NE (1995) Species extinctions in insects: ecological and biogeographical considerations. (pp 321-369) In: Harrington R and Stork NE (eds) Insects in a Changing Environment. Academic Press, London

Meintjes S (1996) Seasonal changes in the invertebrate community of small shallow ephemeral pans at Bain's Vlei, South Africa. Hydrobiologia 317:51 64

Meyer JL, Sale MJ, Mulholland PJ and Poff NL (1999) Impacts of climate change on aquatic ecosystem functioning and health. J Amer Wat Res Assn 35:1373-1386

Mills EL, Leach JH, Carlton JT and Secor CL (1993) Exotic species in the Great Lakes: a history of biotic crises and anthropogenic introductions. J Great Lakes Res 19:1-54

Moore WG (1970) Limnological studies of temporary ponds in southeastern Louisiana. S West Nat 15:83110

Morin PJ, Lawler Sp and Johnson EA (1988) Competition between aquatic insects and vertebrates: interaction strength and higher order interactions. Ecology 69:1401-1409

Müller J, Partsch E and Link A (2000) Differentiation in morphology and habitat partitioning of genetically characterised *Gammarus fossarum* forms (Amphipoda) across a contact zone. Biol J Linn Soc 69:41-53

Neckles HA, Murkin HR and Cooper JA (1990) Influences of seasonal flooding on macroinvertebrate abundance in wetland habitats. Freshwat Biol 23:311-322

Nilsson AN and Svensson BW (1994) Dytiscid predators and culicid prey in two boreal snowmelt pools differing in temperature and duration. Ann Zool Fenn 31:365-376

Patalas K (1975) The crustacean plankton communities of 14 North American Great Lakes. Verh Internat Verein Limnol 19:504-511

Petrov B and Cvetkovic-Dragana M (1997) Community structure of branchiopds (Anostraca, Notostraca and Conchostraca) in the Banat Province in Yugoslavia. Hydrobiologia 359:23-28

Rankin MA (1974) The hormonal control of flight in the milkweed bug *Oncopeltus fasciatus*. (pp 175-182) In: Barton Browne L (ed) Experimental Analysis of Insect Behaviour. Springer, New York

Samways MJ (1996) Insects on the brink of a major discontinuity. Biodiv Conserv 5:1047-1058

Sublette JE and Sublette MS (1967) The limnology of playa lakes on the Llano Estacado, New Mexico and Texas. S West Nat 12:369-406

Thiery A (1991) Multispecies coexistence of branchiopods (Anostraca, Notostraca and Spinicaudata) in temporary ponds of Chaouia Plain (western Morocco): sympatry or syntopy between usually allopatric species. Hydrobiologia 212:117-136

Wiggins GB, Mackay RJ and Smith IM (1980) Evolutionary and ecologival strategies in annual temporary pools. Arch Hydrobiol Suppl 58:97-206

Williams DD (1983) The natural history of a nearctic temporary pond in Ontario with remarks on continental variation in such habitats. Int Rev ges Hydrobiol 6:239-254

Williams DD (1987) The Ecology of Temporary Waters. Croom Helm, London

Williams DD (1996) Environmental constraints in temporary fresh waters and their consequences for the insect fauna. J N Am Benthol Soc 15:634-650

Williams DD (1997) Temporary ponds and their invertebrate communities. Aquatic Conserv: Mar Freshw Ecosyst 7:105-117

Williams DD and Williams NE (1998) Invertebrate communities from freshwater springs: what can they contribute to pure and applied ecology? (pp 251-261) In: Botosaneanu L (ed) Studies in Crenobiology. Backhuys, Leiden

DISTRIBUTION OF CONTINENTAL SHELF DECAPOD CRUSTACEANS ALONG THE AMERICAN PACIFIC COAST

Enrique E. Boschi

Consejo Nacional de Investigaciones Científicas y Técnicas (CONICET)
Instituto Nacional de Investigación y Desarrollo Pesquero (INIDEP)
Facultad de Ciencias Exactas y Naturales (UBA)
Paseo Victoria Ocampo N° 1 CP: B7605AVN Mar del Plata, Argentina

eboschi@inidep.edu.ar

I. INTRODUCTION

This contribution deals with the distribution and abundance of decapod crustacean species in the continental shelves of the Eastern Pacific Ocean. In recent years, the large number of publications available on the tropical, subtropical and temperate seas have allowed to develop a list of species in this region and to include them in the zoogeographical provinces and subprovinces of the American Pacific Ocean. The complete list of species registered in the sea shelves, which includes the American Atlantic and the Arctic Sea coasts, is given in Boschi (in press). Constant findings of new species make the list very dynamic and it can be supposed that the number will increase in the next years, especially in the tropical and subtropical regions, thus requiring a constant updating of results.

II. MATERIALS AND METHODS

The study area extends from Nunivack Island (60°N) in the Alaska Peninsula, to Cabo de Hornos (56°S) in southern South America. According to Boschi's (2000) criterion and Boschi (2000b), who established the limits of the zoogeographic provinces on the basis of marine currents and water temperature in each region, the distribution and diversity of species, can be classified in 6 provinces and 2 subprovinces. The continental shelf areas considered comprise from the highest tide to depths of 200-300 m.

For the division of the littoral zone into zoogeographical provinces and subprovinces, the pioneer work of scientists such as Ekman (1953), Balech (1954), Hedgpeth (1957), Boschi (1966, 1979), Briggs (1974), and Brusca and Wallerstein (1979) were taken into account. In this analysis not only decapods, but also other groups of organisms present in the continental shelves were considered (Brattström and Johanssen 1983, Bernard et al. 1991).

The following zoogeographical provinces and subprovinces were considered: Aleutian, Oregonian, Californian, Cortes (subprovince), Panamic, Galapagos (subprovince), Peru-Chilean, and Magellanic. Only the Pacific portion of the Magellanic province was considered in this study (Fig. 1).

The information on the distribution and species richness of decapod crustaceans in the study area was obtained from numerous papers that considered marine and brackish waters of the continental shelves of the Eastern Pacific Ocean, from the exchange of opinions with several specialists from different institutions, as well as from my own investigations.

The complete list of the 2,472 decapod species, the oceanographic characteristics of the 14 provinces and 2 subprovinces of the continental shelves of the Americas, and the species distribution will be published elsewhere (Boschi 2000b). For that reason, no reference to the habitat conditions of the species is made here.

Kluwer Academic/Plenum Publishers

E. Escobar-Briones & F. Alvarez Eds.
MODERN APPROACHES TO THE STUDY OF CRUSTACEA
PP. 235-239

Due to new taxonomic reviews and the constant updating of species distribution, the number of species and the records per province have changed during the course of this study. This is why the numbers mentioned by the author in previous papers are slightly different from the ones appearing in this study.

Figure 1. *Zoogeographical provinces and subprovinces of the Americas and their limits: Arctic (1), between Nunivak Island to Strait of Belle Isle; Aleutian (2), Nunivak Island to Puget Sound; Oregonian (3), Puget Sound to Point Concepción; Californian (4), Point Concepción to Bahia Magdalena; Cortes (5) subprovince, from Tiburon and Angel de la Guarda Islands (Gulf of California) to the northern limit of the Gulf; Panamic (6), Bahia Magdalena to Gulf of Guayaquil/Bahia Sechura; Galapagos (7) subprovince, (0° 40' N, 91° 50' W to 1° 30' S, 89° 20' W); Peru-Chilean (8), Bahia Sechura to the north Chiloé Island; Magellanic (9), Chiloe Island to 35° S southwest Atlantic; Argentinian (10), 43/44° S to Cabo Frío; Brazilian (11), Cabo Frio to Delta Rio Orinoco; Caribbean (12), Delta Rio Orinoco to Cabo Rojo, Gulf of Mexico, Caribbean Islands and Cape Romano to Cape Canaveral, Florida Peninsula, Island of Bermuda; Texan (13), Cabo Rojo to Cape Romano, Gulf of Mexico; Carolinian (14), Cape Canaveral to Cape Hatteras; Virginian (15), Cape Hatteras to Cape Cod; Boreal (16), Cape Cod to Strait of Belle Isle (Boschi 2000a).*

III. RESULTS

The total number of species in the Eastern Pacific Ocean shelves, for the provinces and subprovinces described above (Fig. 1), is estimated at 1,279 (Table 1). Species richness in these regions, that corresponds to the addition of species found in each province or subprovince, amounts to 2,293 (Table 2). Of these, 627 is the figure for endemic species (Table 3).

Table 1. *Total Decapod species per group in the continental shelf in the American Pacific coast.*

Group	Number of Species
Anomura	279
Astacidea	1
Brachyura	567
Caridea	305
Palinura	15
Penaeoidea	47
Sergestoidea	12
Stenopodidea	4
Thalassinidea	49
Total	**1279**

The highest number of records (825) corresponds to the tropical region contained in the Panamic Province (Table 2). While the most species rich group is the Brachyura, with 567 species in the whole study region (Table 1).

Of the 62 species found on the Pacific side of the Magellanic Province, 31 are also present in the Atlantic coast. Ekman (1953) considered the Magellanic Province and other southern regions of the world's oceans Antiboreal. Nevertheless, it is recommended to call them Notal (Boschi 1976) which refers precisely to the south.

Table 2. *Total number of species per province/subprovince and per group in the American Pacific coast.*

Group	Prov2	Prov3	Prov4	Sprov5	Prov6	Sprov7	Prov8	Prov9	Total
Anomura	56	50	67	47	165	35	43	18	481
Astacidea								1	1
Brachyura	32	60	126	139	411	129	103	23	1028
Caridea	85	67	74	50	161	67	41	12	557
Palinura	1	2	2	3	7	3	7	1	26
Penaeoidea	2	7	20	14	37	11	12		103
Sergestoidea	1	2	4	3	7	2	3	1	23
Stenopodidea				1	3	2			6
Thalassinidea	5	5	8	8	34	4	3	1	68
Total	**182**	**193**	**301**	**265**	**825**	**253**	**212**	**62**	**2293**

When comparing the number of species on both coasts of the continent, the highest diversity of species (1318) is found in the Atlantic Ocean shelves (Table 4).

The distribution of 2,472 known species along all the American coasts was compared (Fig. 2). Two well-defined groups were obtained. The first one groups the species that correspond to the zoogeographic provinces in the Atlantic and the Arctic Oceans (Arctic, Boreal, Virginian, Carolinian, Texan, Caribbean, Brazilian, Argentinian, and Magellanic) (Table 5). The second one covers the Pacific Ocean provinces and subprovinces (Aleutian, Oregonian, Californian, Cortes, Panamic, Galapagos, Peru-Chilean, and Magellanic).

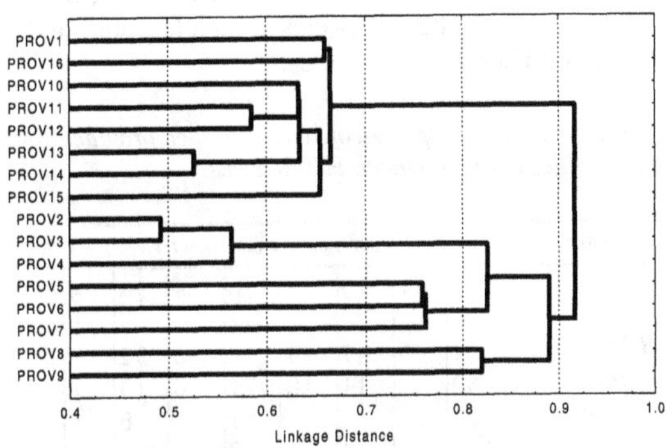

Figure 2. *Dendogram using the Jaccard index of the total number of Decapod species of the Arctic, Pacific and Atlantic oceans (N=2472).*

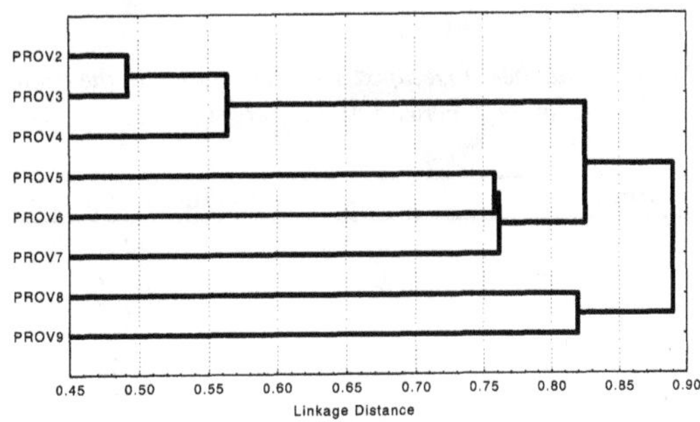

Figure 3. *Dendogram using the Jaccard index of the total number of Decapod species of the east Pacific coast (N=1279).*

The analysis of the American Pacific species (Fig. 3, Table 2), suggests a high level of similarity among the temperate northern provinces (Aleutian, Oregonian, and Californian). It is also apparent a moderate affinity among the warm water Cortes, Panamic and Galapagos provinces and a minor degree of affinity among the temperate southern provinces (Peru-Chilean and Magellanic).

The distribution and diversity patterns of decapods in the studied region show a latitudinal gradient with number of species decreasing towards high latitudes (Table 2, 5). This trend is more pronounced for Brachyura than for Caridea. This coincides with results obtained for all decapods in the Americas (Boschi 2000b). In a recent paper (Dworschak 2000) on the global diversity of the Thalassinidea, the author highlights the presence of a latitudinal distribution gradient with a lower number of species at high latitudes and a higher number at low latitudes.

Table 3. *Endemic species of decapods per province and subprovince in the American Pacific coast.*

Group	Prov2	Prov3	Prov4	Sprov5	Prov6	Sprov7	Prov8	Prov9	Total
Anomura	18	1	12	1	70	4	24	8	138
Astacidea								1	1
Brachyura	4	5	17	15	168	30	32		271
Caridea	32		11	10	71	6	13	7	150
Palinura					1		6	1	8
Penaeoidea			4	1	12				17
Sergestoidea			1		3	1	1		6
Stenopodidea					2	1			3
Thalassinidea	1	1	4		23	1	2	1	33
Total	**55**	**7**	**49**	**27**	**350**	**43**	**78**	**18**	**627**

Table 4. *Number of Decapod species per group in the continental shelf in the American Atlantic coast.*

Group	Number of Species
Anomura	251
Astacidea	7
Brachyura	558
Caridea	283
Palinura	22
Penaeoidea	55
Sergestoidea	19
Stenopodidea	9
Thalassinidea	114
Total	**1318**

Table 5. *Total number of records for species along the Pacific and Atlantic coasts of the Americas.*

ATLANTIC								
Province	9	10	11	12	13	14	15	16
Total	48	330	572	1058	422	386	158	77	3051

PACIFIC								
Province	2	3	4	5	6	7	8	9
Total	182	193	301	265	825	253	212	62	2293

Total records = 5344

IV. CONCLUSIONS

1. The number of species so far described for the zoogeographical provinces and subprovinces of the continental shelves of the Eastern Pacific Ocean is estimated at 1,279. Species richness for the whole region reaches 2,293 records, endemic species amount to 627.

2. The distribution of decapods shows a latitudinal gradient with a higher number of species in tropical and subtropical areas and a lower number in temperate and cold regions.

3. The Brachyura is the group with the highest number of species, while Astacidea accounts for the lowest.

4. The similarity analysis shows a cluster formed by species from different provinces, being observed that the main environmental condition that influences this pattern is water temperature.

5. It is possible to expect a considerable increase in the number of new species in the future, especially coming from tropical and subtropical waters.

REFERENCES

Balech E (1954) División zoogeográfica del litoral Sudamericano. Revista de Biología Marina. Estación de Biología Marina de la Universidad de Chile 4:184-195

Bernard FR, McKinnell SM and Jamienson GS (1991) Distribution and zoogeography of the bivalvia of the eastern Pacific Ocean. Can Spec Pub Fish Aquat Sci 112:1-60

Boschi EE (1966) Preliminary note on the geographic distribution of the decapod crustaceans of the marine waters of Argentina (South-west Atlantic Ocean). Proceedings of the Symposium on Crustacea, India 1:449-456

Boschi EE (1979) Geographic distribution of Argentinian marine Decapod Crustaceans. Bull Biol Soc Wash 3:134-143

Boschi EE (2000a) Biodiversity of marine decapod brachyurans of the Americas. J Crust Biol 20:337-342

Boschi EE (2000b). Species of decapod crustaceans and their distribution in the American marine zoogeographic provinces. Rev Inv Des Pes, Mar del Plata, Argentina N°13:1-136

Brattström H and Johanssen A (1983) Ecological and regional zoogeography of the marine benthic fauna of Chile. Sarsia 68:289-339

Briggs JC (1974) Marine zoogeography. McGraw- Hill Co, New York

Brusca RC and Wallerstein BR (1979) Zoogeographic patterns of idoteid isopods in the northeast Pacific, with a review of shallow water zoogeography of the area. Bull Biol Soc Wash 3:67-105

Dworschak PC (2000) Global diversity in the Thalassinidea (Decapoda). J Crust Biol 20 (special number 2):238-245

Ekman S (1953) Zoogeography of the Sea. Sidgwick and Jackson Ltd, London

Hedgpeth JW (1957) Marine Biogeography. In: Hedgpeth JW (ed) Treatise on marine ecology and paleoecology. Vol 1. Ecology. Memoir 67 (pp 359-382) The Geological Society of America

DISTRIBUTION OF INTERTIDAL NON-BRACHYURAN DECAPODS FROM THE GULF OF CALIFORNIA ISLANDS AND ITS BIOGEOGRAPHICAL IMPLICATIONS

José Luis Villalobos and Fernando Alvarez

Colección Nacional de Crustáceos, Instituto de Biología, Universidad Nacional Autónoma de México.
Apartado Postal 70-153, México 04510, D.F., México

hiriart@servidor.unam.mx

ABSTRACT

The systematics and spatio-temporal distribution of 6525 specimens of non-brachyuran decapods, collected in the islands of the Gulf of California were analyzed. Samples were collected between May 1985 and May 1987, in the intertidal zone of 23 island. The 82 species recognized belong to the superfamily Penaeoidea and to the infraorders Caridea, Palinura, Thalassinidea and Anomura. The anomurans were the most abundant, diverse and frequently collected crustaceans in the islands, representing six families, 17 genera and 43 species. Porcellanidae, with 28 species, was the most diverse family while Alpheidae was the second with 21 species. The genus *Petrolisthes* had the highest diversity with sixteen species. The hermit crab *Clibanarius digueti* was the most abundant species with 1171 organisms, occurring in 91% of the samples. Tiburon and Angel de la Guarda islands in the upper gulf, and San Jose, Espiritu Santo and Cerralvo islands in the southern portion, had the highest specific richness. A slightly higher abundance of organisms was detected in the cold season. Most of the species classified as rare had a tropical affinity, while dominant species were mostly gulf endemics. The results obtained support the idea that the Gulf of California must be considered as a distinct biogeographical unit.

I. INTRODUCTION

The Gulf of California is considered a unique region in Mexico and in the tropical eastern Pacific due to its biotic richness and high number of endemic forms (Villalobos 2000). Only a few groups of crustaceans that inhabit the Sea of Cortez have been studied with some detail to evaluate their diversity, characterize their populations, and study their spatial and temporal distribution. The order Decapoda stands out among these groups, because of its economic, biological and ecological importance. The information that we have now on the Gulf of California decapods is taken as the most complete, among the marine invertebrates. Around 600 species of decapods have been registered, making the gulf the highest diversity area inside the tropical eastern Pacific (Hendrickx 1996), and the second in comparison with other regions of the western Atlantic, such as the Florida peninsula.

Although the Gulf of California is considered as one of the better known regions in Mexico, still many questions remain unanswered about the diversity and distribution of most of the invertebrate groups. Regarding the crustacean fauna, recent samples continue to bring new records of species previously unknown from inside the gulf, as well as new, undescribed genera and species. The wealth of endemic decapod crustaceans, together with the hundreds of other decapod species present inside the gulf, should become a conservation priority and a constant research subject (Alvarez et al. 1996b, Campos et al. 1998, Hendrickx and Esparza-Haro 1997, Hendrickx 1998, Villalobos 2000).

The present study focuses on the spatial and temporal distribution of 83 species and 6,525 specimens of non-brachyuran decapods, collected as a part of the project "Conservación de las islas en un mar en el desierto, Golfo de California", which had as its pri-

Kluwer Academic/Plenum Publishers

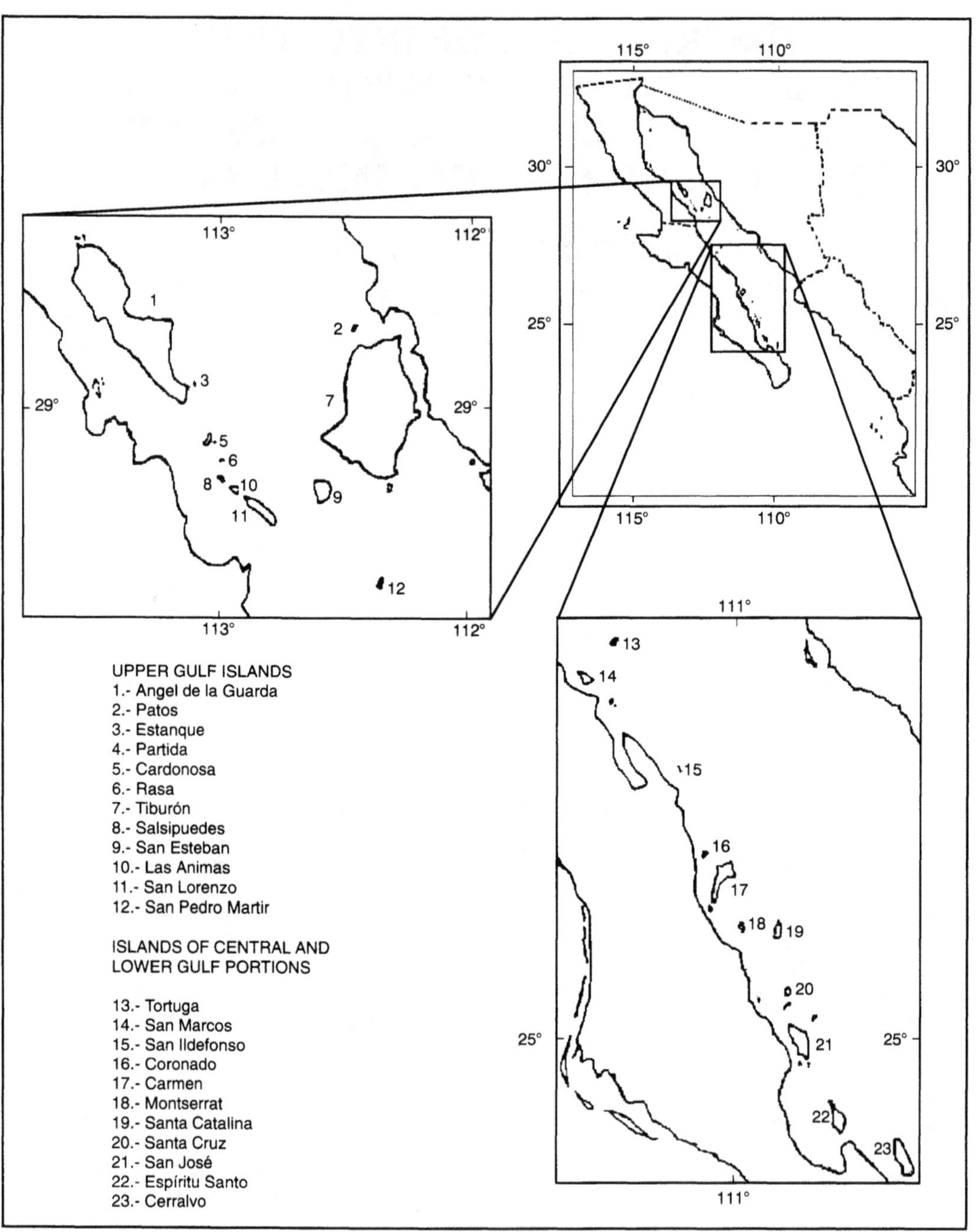

Figure 1. *Islands sampled in the Gulf of California.*

mary objective to build an inventory of as many plant and animal species present in the islands of the Gulf of California, as possible.

II. MATERIALS AND METHODS

Between May 1985 and May 1987, the intertidal decapod carcinofauna of 23 islands of the Sea of Cortez was sampled (Fig. 1). Eleven of these islands are located in the upper gulf (insular belt or midriff area) and the remaining twelve are distributed along the east coast of the Baja California Peninsula, through the central and lower portion of the gulf.

Eight trips were organized to sample the islands, each one of them with a 20-day duration. They were scheduled to cover the two seasons, cold (November to March) and warm (April to October), that have been registered in the Gulf of California (Maluf 1983, Cano 1991). The sampling localities were situated in the protected zones of the islands and visited systematically through the two-year period. Crustaceans were collected by hand in the inter- and subtidal zones to a 5 m depth, in which case scuba was used to sample sponges, rocks, or coral rubble. These materials were brought out of the water and fragmented until all the crustaceans present were obtained. The specimens captured were separated, classified to the family level, and preserved in 70% ethanol. In the laboratory, all specimens were identified, registered, and deposited in the Colección Nacional de Crustáceos (CNCR), Instituto de Biología, UNAM.

Data analysis included cluster analysis based on species presence/absence data to group islands. A simple single linkage, euclidean distance procedure, was chosen and performed with the statistical software Statistica. The Olmstead-Tukey technique was used to classify species according to their abundance and frequency of appearance.

III. RESULTS

Species diversity and abundance

The 83 species recognized (Tables 2 and 3) belong to the superfamily Penaeoidea and to the infraorders Caridea, Palinura, Thalassinidea, and Anomura. The group of anomurans was the most abundant, diverse and frequently collected in the islands, representing 53% of the species total, and 85% of all the specimens studied (Fig. 2). The 5,575 anomurans

recognized were classified into six families, 17 genera, and 43 species. Carideans were next in abundance with four families, 15 genera, 31 species, and 887 organisms; the thalassinids with 4, 5, 6 and 40; the penaeid shrimps with 1, 1, 1 and 15, and the palinurids with 1, 1, 1 and 1, respectively, followed.

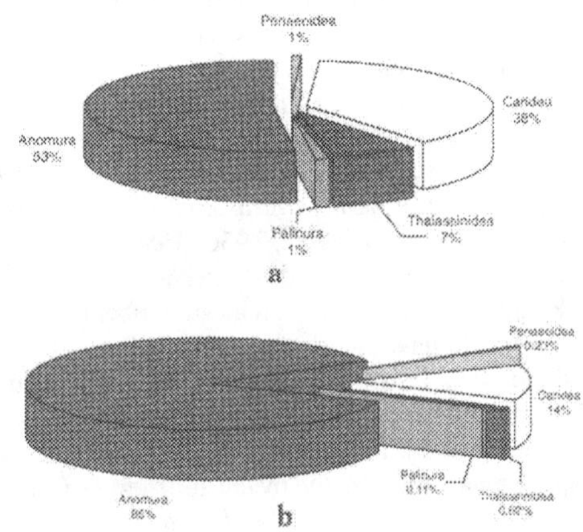

Figure 2. *Higher taxa represented by number of species (a) and specimens (b).*

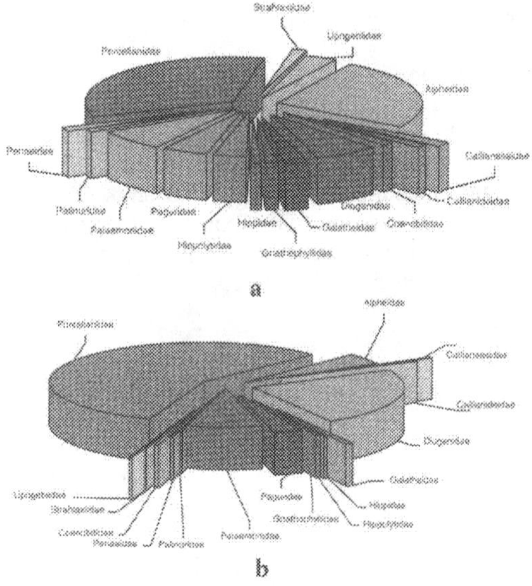

Figure 3. *Families represented by number of species (a) and specimens (b).*

Sixteen families were identified with Porcellanidae as the best represented with 28 species and 3,650 specimens. In second place was Alpheidae with 21 species and 317 specimens (Fig. 3). The remaining 14 families had fewer species, between one and seven. Within this group the family Diogenidae had only six species, but it was second in abundance, with 1,550 specimens. The family Palaemonidae showed a similar pattern, with seven species and 550 specimens, being third in abundance.

The genus *Petrolisthes* had the highest diversity with 16 species. Porcellanid crabs like *Petrolisthes gracilis* Stimpson, 1860, *P. galapagensis* Haig, 1960, *P. hirtipes* Lockington, 1878, *P. hirtispinosus* Lockington, 1878, *P. tiburonensis* Glassell, 1936, *Megalobrachium smithi* (Glassell, 1936), and *Pachycheles setimanus* (Lockington, 1878), had the greatest number of specimens in this family. They occurred in 40% to 70% of the islands. The genus *Alpheus* with 13 species was the second in importance. Nevertheless, this relevance was not reflected in the number of organisms, because all the species were represented by few individuals (one to six), with the exception of *A. hyeyoungae* and *A. sulcatus*, that had 44 and 31 specimens, respectively.

Within the genus *Clibanarius*, *C. digueti* was the most abundant species with 1,171 specimens and the most frequently captured through the islands. It was registered at 91% of the collecting sites.

Species composition per island

The results show that the larger the island the higher the number of species found. This was the case of Tiburon and Angel de la Guarda islands, in the upper gulf and San Jose, Espiritu Santo, and Cerralvo, to the south near of the gulf´s mouth. At the smaller islands we found mainly endemic species, representing approximately 50% of all the species (Fig. 4).

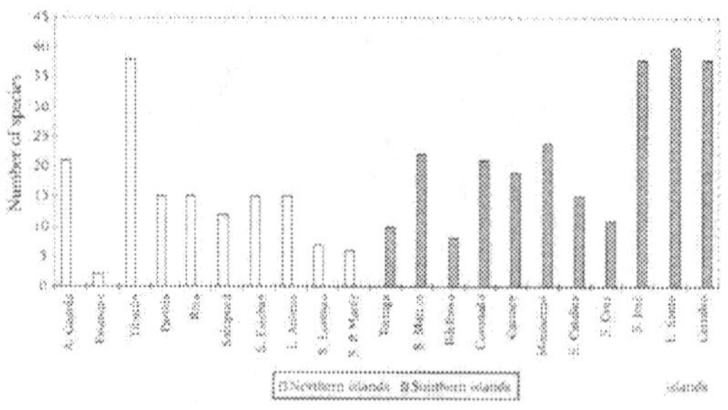

Figure 4. *Number of species per island.*

Two groups of islands were defined according to the number of shared species. The first group was composed by the islands that lie closest to the gulf's mouth, Cerralvo-Espiritu Santo-San Jose, which showed great affinity for tropical species. The three islands share more than 60% of the species, of which 65% are tropical, 30% endemic and 5% eurythermal. The second group included the remaining islands, which formed other secondary groups, such as Angel de la Guarda-San Marcos-Coronado-Carmen-Montserrat. In spite of being far from each other, Angel de la Guarda with San Marcos and Coronado, and Carmen and Montserrat, they share between 45% and 72 % of the species, with the endemic component representing between 61% and 73% (Fig. 5).

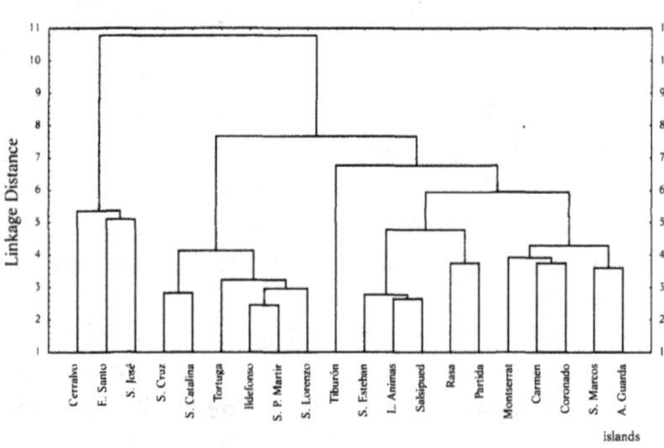

Figure 5. *Similarity of the Gulf of California islands by species composition.*

The intermediate position of Tiburon in the second group is interesting since it appears more related to the southern group than to the northern one. This is probably the result of the great spatial heterogeneity found in Tiburon which can house a greater diversity of species. Tiburon shares from 39% to 52% of the species found in the islands closest to the gulf's mouth (Fig. 5).

Seasonal species diversity and abundance

In 50% of the islands more species were collected during the warm season, especially in the southern section. In 10 other islands the opposite pattern was found, while in the rest there were no changes (Fig. 6). With respect to abundance, there was a slight preference for the cold season, with 55% of the islands registering higher numbers of organisms. Northern gulf islands, like Tiburon and Salsipuedes, had an evident variation in number of specimens through the two seasons, more than twice the specimens were collected during the cold months. In the southern islands only Coronado had a noticeable difference, with four times as many organisms in the cold season than in the warm one (Fig. 7).

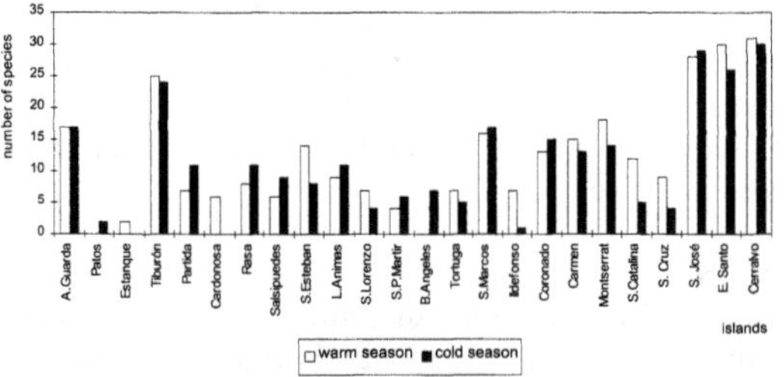

Figure 6. *Number of species per island by sampling season.*

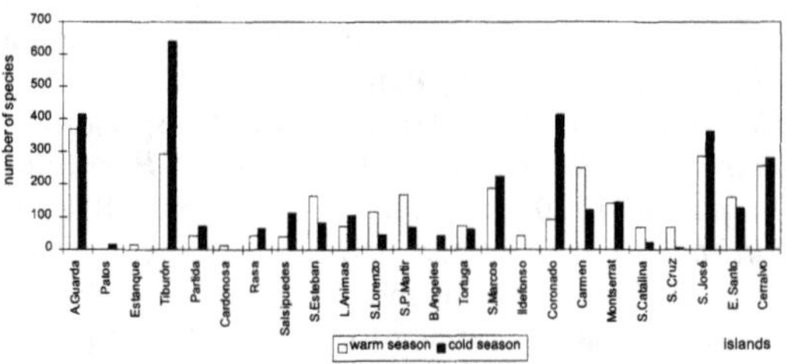

Figure 7. *Number of specimens per island by sampling season.*

Biogeographical analysis of species composition

According to geographical distribution and biogeographical affinity, the species studied were divided into four faunistic components (tropical, endemic, eurythermic, and warm-temperate). By number of species, the tropical component was the most important with 69% of the species, followed by the endemic with 19%, eurythermic with 10%, and warm-temperate with 2% (Fig. 9a). The majority of tropical species (71%) belong in the Panamic province, showing the strong influence of the eastern tropical Pacific carcinofauna inside the Gulf of California. The other 29% was composed of species related to other tropical regions (western and eastern Atlantic, Caribbean, and Indo-west Pacific).

The number of specimens by faunistic component, shows that the endemic group is dominant with 62.7% of the organisms collected, followed by the tropical (35%), the eurythermic (2%), and the warm-temperate (0.3%) components (Fig. 8b).

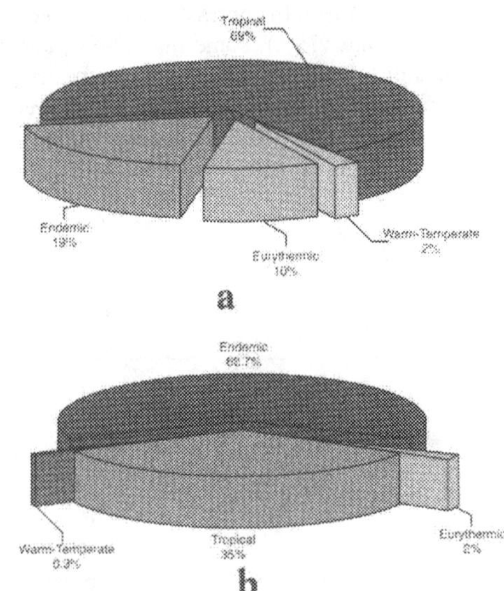

Figure 8. *Faunistic components represented by number of species (a) and specimens (b).*

The presence of the different faunistic components by island and number of species showed a dominance of tropical forms particularly through the southern islands, such as Espiritu Santo (30 species), Cerralvo (29), and San Jose (28). In the upper gulf islands, Tiburon stands out with 28 tropical species. The endemic species appear in approximately the same proportion in almost all the islands sampled. Angel de la Guarda, San Marcos and Espiritu Santo with 12 endemic species were the most important, followed by Tiburon island with 11 and Coronado and Montserrat with 10 (Fig. 9). The group of endemic species was the most abundant in most of the islands except for Rasa, San Pedro Martir, and Espiritu Santo. The eurythermic and warm-temperate components had very low numbers of organisms, appearing the former one in Montserrat, San Jose, and Cerralvo, and the latter one in Bahia de los Angeles only (Fig. 10).

Finally, 58% of tropical species were classified as rare (low frequency and abundance), 23% as dominant (high frequency and abundance), 12% as indicator (low frequency and high abundance), and 7% as common (high frequency and low abundance). Of the tropical dominant species, more than half of them (62%) appear with abundances that range between 31-102 organisms, and a frequency of less than 50% of the islands (30-43%). In contrast, endemic species are usually dominant (75%), some rare (24%), and very few indicator (1%). Most of the dominant species (9 out of 12) appear with very high abundances between 219 and 1082 organisms, at 12 of the 22 islands (Fig. 11).

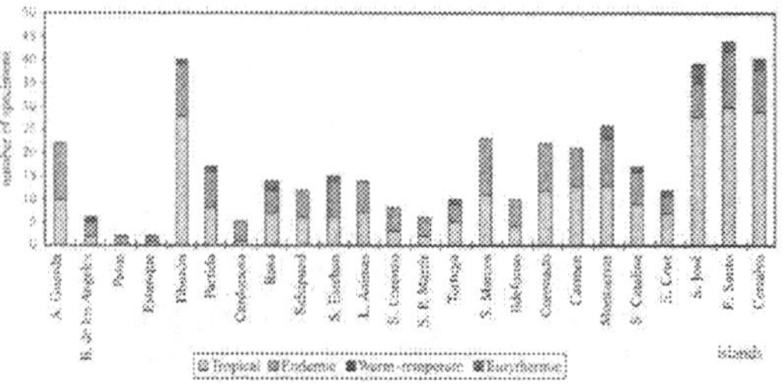

Figure 9. *Faunistic components per island represented by number of species.*

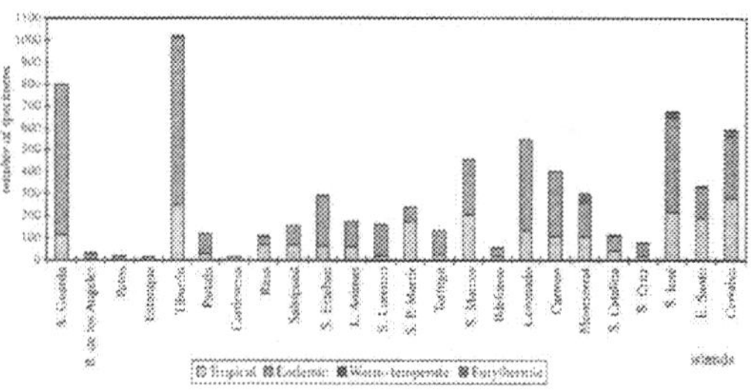

Figure 10. *Faunistic components per island represented by number of specimens.*

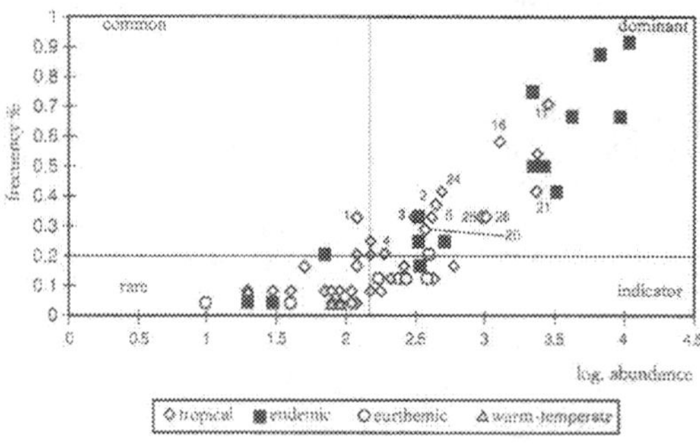

Figure 11. *Olmstead-tukey analisys to the four faunistic components registered in the Gulf of California islands. The numbers indicate some of tropical species that must be considered part of the eurythermic component (see Table 1).*

IV. DISCUSSION

The results obtained offer interesting new information about the presence and distribution pattern of the shallow water non-brachyuran species of the islands of the Gulf of California. For the first time 23 islands were systematically sampled over a two year period with the goal of registering as many species of decapods as possible. Prior to this study only the information from isolated samples from some of the islands was available.

Species abundance

The infraorder Anomura was the most important group recorded, particularly the family Porcellanidae which had a high diversity of species and, in most cases, a high number of organisms per species. The dominance of porcellanids in the intertidal zone of the Gulf of California has already been noted (Westervelt 1967, Carvacho 1980, Ramírez 1983, Romero and Carvacho 1987, Villalobos et al. 1992). Species like *Petrolisthes galapagensis* Haig, 1960, *P. gracilis* Stimpson, 1859, *P. hirtipes* Lockington, 1878, *P. hirtispinosus* Lockington, 1878, and *P. tiburonensis* Glassell, 1936, were the most important porcellanids in the total sample; except for the first two (the first one is a tropical species and the second one has two extralimital records outside the Gulf of California), the remaining three species are gulf endemics.

Other numerically important species were the diogenids *Clibanarius digueti* Bouvier, 1898, *Calcinus californiensis* Bouvier, 1898, and *Paguristes anahuacus* Glassell, 1938. *Clibanarius digueti* was the most abundant and frequently collected species throughout the study, coinciding with Snyder-Conn's (1980) observation.

Caridea resulted the second most diverse group, however, most species were represented by only a few organisms. The fact that the northern limit of the distribution range of many tropical caridean species lie in the Gulf of California, may explain why although they occur inside the gulf they do it in very low numbers (Brusca and Wallerstein 1979, Brusca 1980). One exception is *Palaemon ritteri* Holmes, 1895, which was the most abundant and frequently collected caridean species in this study; the same observation was made by Wicksten (1983) for continental coastal areas inside the gulf.

The remaining infraorders, Palinura and Thalassinidea, were scarcely represented, both in terms

Table 1. *Species considered as members of the tropical component collected in the upper gulf islands during the cold season, in temperatures under 20 °C.*

	Species	Island	Material Examined
1	*Alpheus galapagensis*	Tiburon, Las Animas	2♂, 1♀
2	*Alpheus hyeyoungae*	Tiburon, Partida, San Esteban, Rasa, Las Animas	6♂, 7♀, 3♀ ov.
3	*Alpheus sulcatus*	Salsipuedes, San Esteban, Las Animas	7♂, 8♀, 1♀ ov, 2 juv.
4	*Alpheus tenuis*	Angel de la Guarda, Tiburon	3♂, 1♀, 2♀ ov.
5	*Alpheus umbo*	Tiburon	1♂
6	*Alpheus villus*	Angel de la Guarda, Rasa, Tiburon	2♂, 1♀
7	*Callianidea l. occidentalis*	Cardonosa	1♂
8	*Clibanarius panamensis*	Tiburón	3♂, 1♀
9	*Dardanus sinistripes*	Tiburón	1♂
10	*Gnathophyllum panamense*	Partida	1♀ ov.
11	*Megalobrachium smithi*	Angel de la Guarda, Tiburon, San Esteban, Partida, Salsipuedes, Las Animas, Bahia de los Angeles	29♂, 16♀, 11♀ ov, 2 juv.
12	*Megalobrachium tuberculipes*	Angel de la Guarda, Tiburon	3♂, 1♀, 1♀ ov.
13	*Neaxius vivesi*	Tiburon, Rasa	5♂, 2♀, 1♀ ov, 2 specimens
14	*Pachycheles calculosus*	Tiburon	1♂
15	*Pachycheles panamensis*	Tiburon	6♀, 6♀ ov.
16	*Pagurus lepidus*	Tiburon, Angel de la Guarda, Partida, San Lorenzo	16♂, 7♀, 8♀ ov, 11 specimens
17	*Panulirus inflatus*	Angel de la Guarda, San Pedro Martir	2♂, 2♀
18	*Petrochirus californiensis*	Tiburon	1♂
19	*Petrolisthes armatus*	Tiburon	1♂, 3♀, 4♀ ov.
20	*Petrolisthes edwardsii*	San Esteban	1♂
21	*Petrolisthes galapagensis*	Estanque, Partida, Rasa, Salsipuedes, San Lorenzo, San Pedro Martir	103♂, 114♀, 12♀ ov.
22	*Petrolisthes lewisi*	Tiburon	8♂, 3♀, 8♀ ov.
23	*Petrolisthes ortmanni*	Tiburon	1♀, 1♀ ov.
24	*Phimochirus roseus*	Tiburon, Angel de la Guarda, Partida, Bahia de los Angeles	5♂, 11♀, 8♀ ov, 10 specimens
25	*Pontonia margarita*	Tiburon, Las Animas	2♂, 1♀, 1♀ ov.
26	*Pontonia pinnae*	Salsipuedes, Las Animas	24♂, 36♀, 2♀ ov.
27	*Porcellana cancrisocialis*	Tiburon	6♀, 2♀ ov.
28	*Porcellana paguriconviva*	Tiburon	2♂, 2♀
29	*Salmoneus ortmanni*	Tiburon	2 specimens
30	*Synalpheus digueti*	Tiburon	1♀
31	*Ulloaia perpusillia*	Tiburon	1♀ ov.
32	*Upogebia dawsoni*	Tiburon	2♂, 1♀

Table 2. *Specimens and species collected in the upper Gulf of California islands.*

	species/islands	A.Guarda		Partida		Rasa		Salsipuedes		L. Animas		S.Lorenzo		S.P.Martir		Tiburón		S.Esteban		B.Angeles		Cardonosa		Patos		Estanque		TOTAL
	season	warm	cold	warm	cold	warm	cold	warm	cold	warm	cold	warm	cold	warm	cold	warm	cold	warm	cold	warm	cold	warm	cold	warm	cold	warm	cold	TOTAL
1	Alpheus galapagensis									1							2											3
4	Alpheus felgenhaueri				1																							1
5	Alpheus hyeyoungae				1		1			11						2		1										16
9	Alpheus sulcatus						1		1	13									3									18
10	Alpheus tenuis	2					1										6											9
11	Alpheus umbo															1												1
12	Alpheus villus	1					1									1												3
15	Betaeus longidactylus																				8							8
17	Calcinus califomiensis		2													1												3
18	Callianidea l. occidentalis																					1						1
19	Clibanarius digueti	110	130	10	8	10	1	35	11	41		4	19			63	137	10			9	5				11		614
20	Clibanarius panamensis																4											4
21	Coenobita compressus						1																					1
22	Dardanus sinistripes															1												1
23	Gnathophyllum panamense				1																							1
27	Lysmata califomica				1													1										2
29	Megalobrachium sinuimanus	10	1	2												1	85	6	41									146
30	Megalobrachium smithi	1	19	4						1		1	3			24		2	3			1						59
31	Megalobrachium tuberculipes	1	1					4											3									9
32	Munida tenella										3																	3
33	Neaxius vivesi					3										3	5											11
35	Pachycheles calculosus															1												1
36	Pachycheles panamensis																12											12
37	Pachycheles setimanus		2													3	7											12
38	Paguristes anahuacus	38	25	1	11			19	3			1		2		9	5	2	15			5						136
40	Pagurus lepidus	11	5	1									3	1		1	21											43
41	Palaemon ritteri	16	21			11	47	7		13	12	5				11	5	10										158
43	Panulirus inflatus	3													1													4
44	Farfantepenaeus califomiensis				1											1	9											11
46	Petrochirus califomiensis															1												1
47	Petrolisthes armatus															8	1											9
48	Petrolisthes crenulatus	1	5														1											7
49	Petrolisthes edwardsii																	1										1
50	Petrolisthes galapagensis			19		3	1	1				9		134	40												2	209
51	Petrolisthes gracilis	121	83	3			10	5	1	4						135	300	3	8		13					7	13	706
54	Petrolisthes hirtipes	23	55	2	30	2			1	7	1	27	16	27	4	2		63	3			1						264
55	Petrolisthes hirtispinosus		1							2	1	1				3	4	1	8									21
56	Petrolisthes lewisi																		19									19
57	Petrolisthes nigrunguiculatus	1	2							6					3			1										13
58	Petrolisthes ortmanni															2												2
60	Petrolisthes sanfelipensis																	2										2
61	Petrolisthes schmitti		4	3					1							8						2						18
62	Petrolisthes tiburonensis	19	49	6	11	4	10	4	12	28	55	27	27					1	64		2	1						320
64	Phimochirus roseus	10	11	2															9			2						34
65	Phimochirus venustus		3	2																								5
67	Polyonyx quadriungulatus																				9							9
70	Pontonia margarita			2							2					2												6
71	Pontonia pinnae							58			4																	62
72	Porcellana cancrisocialis															7	1											8
73	Porcellana paguriconviva															4												4
74	Salmoneus ortmanni																	2										2
77	Synalpheus digueti																	1										1
78	Synalpheus t. mexicanus	1									3																	4
80	Ulloaia perpusillia															1												1
81	Upogebia dawsoni																		3									3
	TOTAL	369	419	44	76	44	66	40	114	72	106	115	46	168	71	293	641	165	83	0	43	14	0	0	0	18	15	3022

of species and number of organisms per species. Although this result may be attributable to the type of sampling used, three thalassinids were collected inside the gulf for the first time:

1.Callianidea laevicauda occidentalis Schmitt, 1939.- From some point in the western coast of Baja California Peninsula (Hendrickx 1995) to Cardonosa Island, Baja California, in the Gulf of California, Mexico.

2.Biffarius debilis Hernández-Aguilera, 1998.- From Azufre Bay in Clarion Island, Revillagigedo Archipielago, Colima to San Jose Island, Baja California Sur, in the Gulf of California, Mexico.

3.Pomatogebia cocosia (Williams, 1986).- From Cocos Island, Costa Rica to San Jose Island, Baja California Sur, in the Gulf of California, Mexico.

Island species composition

The species composition in each island was different; some of the factors affecting what species occur in what islands may be: island size, latitude, and position within the gulf (Abele 1976, Brusca 1980, Ricketts et al. 1985). Island size may determine the number of different habitats and thus the number of species present; evidence of this is the high specific richness in larger islands (100 - 780 km^2), such as Tiburon and Angel de la Guarda in the upper gulf, and San Jose, Espiritu Santo, and Cerralvo, near the gulf's mouth. In the smaller islands (0.25 to 50 km^2), spatial heterogeneity decreases considerably, with rocky shores, composed mainly of large and smooth boulders, dominating the coastline. In these intertidal habitats endemic species compose about 50% of the community, suggesting that endemic species are highly adapted to this particular environment.

Seasonal species composition

The comparison of the accumulative number of species by island present in the warm vs the cold season showed that there is very little overall variation in species richness throughout the year (289 vs 264 spp.). This stability in species richness may be the result of two factors combined: first, the endemic species group is one that, as expected, has a constant presence in most of the islands; and second, many species previously recognized as tropical (Table 1), appear well established in the northern gulf, withstanding winter temperatures below 20°C. The strictly tropical forms reach their northern distribution limit near Coronado island (26° 07' N), where they are replaced by more resistant species that coexist with the endemic forms which are adapted to the extreme environmental conditions of the northern gulf. In turn, south from Coronado tropical species diversity increases evidently, although the number of organisms per species is very low.

Endemic species

Endemic species have a constant presence in most of the islands with higher population numbers than any other zoogeographical component, even in those islands bordering the gulf's mouth. The results of this study showed that 58% of the tropical species were classified as rare and only 23% were dominant, while 75% of the endemic species were dominant components.

From the zoogeographical point of view, two different patterns can be recognized inside the gulf depending on the interpretation given to the available data. If a species presence/absence data set is used, then it becomes clear that the most important zoogeographical affinity of the Gulf of California is with the tropical component. This would be particularly true for San Jose, Espiritu Santo, and Cerralvo islands, which have been considered as part of the Mexican Province (Hendrickx 1992, 1993). However, if species abundance and frequency data are considered, then the endemic species become the most important element of the crustacean fauna in the Gulf of California. The relevance of the endemic species supports the idea of considering the Gulf of California as a distinct biogeographical entity within the tropical eastern Pacific.

Island associations

The clustering of islands, based on the presence of species, allowed for the recognition of a minimum of six groups that may be shaping the distribution pattern of the species in the gulf. In the group composed by Angel de la Guarda, San Marcos, Coronado, Carmen, and Montserrat, it was evident that the shared endemic species supported this cluster. Angel de la Guarda island, with a large size and a high proportion of endemic species, may be acting as a dispersal center for these species' larvae. Once released, the larvae may reach the southern gulf, as far south as Montserrat island, transported by the strong surface currents that cross the Ballenas channel (Alvarez and Schwartzlose 1979, Maluf 1983).

In contrast, Tiburon island, appeared in the analysis with an intermediate position due to a large

Table 3. *Specimens and species collected in the central and lower Gulf of California islands.*

#	species/islands	Tortuga		S. Marcos		Ildefonso		Coronado		Carmen		Montserrat		S. Catalina		S. Cruz		S. José		E. Santo		Cerralvo		TOTAL
	season	warm	cold	warm	cold	warm	cold	warm	cold	warm	cold	warm	cold	warm	cold	warm	cold	warm	cold	warm	cold	warm	cold	TOTAL
1	Alpheus galapagensis			3					1	2											1		1		1	9
2	Alpheus cristulifrons																					1				1
3	Alpheus cylindricus																								2	2
4	Alpheus felgenhaueri				6				24															1	2	33
5	Alpheus hyeyoungae			3															11	2	2	9		1	28
6	Alpheus lottini																		2	3	4	16	6	16	47
7	Alpheus normanii																		5	1		1			7
8	Alpheus paracrinitus																						8		8
9	Alpheus sulcatus				2				4										1				3	3	13
10	Alpheus tenuis											2							1	2		1			6
11	Alpheus umbo			2	1			2	14			4	1			1				4	4	3	1	3	40
12	Alpheus villus									1	1														2
13	Alpheus websteri																					3	8		11
14	Automate dolichognatha																				1	1			2
16	Biffarius debilis																			2					2
17	Calcinus californiensis				11	11		3	26	24	1	5		23		2				26	6	7	44	30	219
18	Callianidea l. occidentalis										1														1
19	Clibanarius digueti			52			1	14	69	30	71	16		1		1			16	131	21	18	1	26	468
20	Clibanarius panamensis									11															11
21	Coenobita compressus											3	9						1	3	2		8	12	38
22	Dardanus sinistripes																		5	2					7
23	Gnathophyllum panamense												1						6			3	1		11
24	Harpiliopsis depressa															2			1	15	3	18	22	4	65
25	Hippa pacifica									1		2	8									1			12
26	Hippolyte williamsi											4													4
28	Megalobrachium garthi																			2					2
29	Megalobrachium sinuimanus				40	2			38		1	6		1								1		5	94
30	Megalobrachium smithi	3			50	76		3	9	16		17	26	3		2			1	8		1	5	4	224
31	Megalobrachium tuberculipes			1	2																				3
34	Pachycheles biocellatus																						3		3
37	Pachycheles setimanus				2					33		2										2			39
38	Paguristes anahuacus	35			1	3		12		20	1	3		2					1	1	2		2		83
39	Pagurus benedicti																		4	3					7
40	Pagurus lepidus	2				5		3		3		18		1		7			20	14	1	1	9	1	85
41	Palaemon ritteri			17	7				8	18	9	3							18		7	8			95
42	Palaemonella holmesi																		3	9	2	2	6	2	24
43	Panulirus inflatus						1					1									1				3
44	Farfantepenaeus californiensis																		6						6
45	Periclimenes infraspinis																			1					1
48	Petrolisthes crenulatus			1	6							5										1		13	26
49	Petrolisthes edwardsii	1			1										1	1			1				1	30	36
50	Petrolisthes galapagensis		18														3			1			2		24
51	Petrolisthes gracilis			10	19				17	52	26	1	6						30	24	30	25			240
52	Petrolisthes haigae																			1			6	2	9
53	Petrolisthes hians		1																				7		8
54	Petrolisthes hirtipes	23	17	22	3	7		15	6	40	4	39	11	12	8	45	2	84	9	2			27	25	401
55	Petrolisthes hirtispinosus	9	1	8	33	14		5	107	5	2	9	18	3	5	10	1	20	51	14	2	24	57	398
56	Petrolisthes lewisi			5	1																			2	8
57	Petrolisthes nigrunguiculatus		25	26	27			8	57	9	3	9	21					27	8			3	1	25	249
58	Petrolisthes ortmanni							1						1								1	5	3	11
59	Petrolisthes polymitus																						15	2	17
61	Petrolisthes schmitti							1						11							2	1			15
62	Petrolisthes tiburonensis			2	3																				5
63	Phimochirus californiensis	1																							1
64	Phimochirus roseus									2		3		5		1			1	1	2				15
65	Phimochirus venustus															2			1						3
66	Pleuroncodes planipes												28		8	2									38
68	Pomagnathus corallinus																				1	7			8
69	Pomatogebia cocosia																						2		2
70	Pontonia margarita									21	24	2	14						14	25	19	6	5	2	132
71	Pontonia pinnae							22		2			1						2	8	1		3	1	40
75	Synalpheus charon																						9	2	11
76	Synalpheus nobilii																					1	6	1	8
77	Synalpheus digueti																			2	7	1	11	4	25
78	Synalpheus t. mexicanus				1														1		1		1		4
79	Thor algicola							3											3	5	13	3	4	3	34
81	Upogebia dawsoni									17									1						18
82	Upogebia galapagensis																				1				1
	TOTAL	74	62	204	240	43	1	92	414	271	146	144	146	71	23	73	8	288	362	166	136	255	284	3503

number of tropical species combined with a moderate number of endemic species. It is more similar to the islands near the gulf's mouth, rather than to the islands in the middle and upper gulf, which are closer. The presence of many tropical species in Tiburon island may be the result of larval transport, from south to north, in a superficial, warm current along the continental coastline (Alvarez and Schwartzlose 1979, Cano 1991); this current may have important fluctuations during El Niño events, reaching the northernmost part of the gulf. The incoming larvae of tropical species combined with the high habitat heterogeneity present in Tiburon island create a favourable environment for the settlement of this species and a stepping stone for the dispersal towards the upper gulf.

REFERENCES

Abele LG (1976) Comparative species composition and relative abundance of decapod crustaceans in marine habitats of Panama. Mar Biol 38:263-278

Alvarez F, Camacho ME and Villalobos JL (1996) The first species of *Prionalpheus* from the eastern Pacific, and new records of caridean shrimps (Crustacea: Decapoda: Caridea) from the western coast of Mexico. Proc Biol Soc Wash 109:715-724

Alvarez S and Schwartzlose RA (1979) Masas de agua del golfo de California. Cien Mar 6:19-63

Brusca RC (1980) Common intertidal invertebrates of the Gulf of California. University of Arizona Press, Tucson, Arizona, 2nd Ed, 513 p

Brusca RC and Wallerstein BR (1979) Zoogeographic patterns of Idoteid isopods in the northeast Pacific, with a review of shallow water zoogeography of the area. Bull Biol Soc Wash 3:67-105

Campos E, Díaz EV and Gamboa-Contreras JA (1998) Notes on distribution and taxonomy of five poorly known species of Pinnotherid crabs from the eastern Pacific (Crustacea: Brachyura: Pinnotheridae). Proc Biol Soc Wash 111:372-381

Cano FA (1991) Golfo de California. Oceanografía física (pp. 453-495, 505-511) In: de la Lanza G. (comp) Oceanografía de mares mexicanos. AGT Editor, México, DF 569 p

Carvacho A (1980) Los porcelánidos del Pacífico americano: Un análisis biogeográfico (Crustacea: Decapoda). An Cen Cien Mar Limnol, UNAM 7:249-258

Hendrickx ME (1992) Distribution and zoogeographic affinities of Decapod Crustaceans of the Gulf of California, Mexico. Proc San Diego Soc Nat Hist 20:1-12

Hendrickx ME (1993) Crustáceos decápodos del Pacífico mexicano (pp 271-318) In: Salazar-Vallejo SI y González NE (eds) Biodiversidad Marina y Costera de Mexico. CONABIO-CIQRO, 865 p

Hendrickx ME (1996) Habitats and biodiversity of decapod crustaceans in the SE Gulf of California, Mexico. Rev Biol Trop 44:603-617

Hendrickx ME (1998) A new genus and species of "goneplacid-like" brachyuran crab (Crustacea: Decapoda) from the Gulf of California, Mexico, and a proposal for the use of the Family Pseudorhombilidae Alcock, 1900. Proc Biol Soc Wash 111:634-644

Hendrickx ME and Esparza-Haro JA (1997) A new species of *Clibanarius* (Crustacea, Anomura, Diogenidae) from the eastern tropical Pacific. Zoosystema 19:111-119

Maluf LY (1983) The Physical Oceanography (pp 26-45) In: Case TJ and Cody ML (eds) Island Biogeography in the Sea of Cortez. University of California Press, Berkeley-Los Angeles-London, 508 p

McClure MR and MK Wicksten (2000). Taxonomic studies of snapping shrimp of the *Alpheus* "Edwardsii" group from the Galápagos Islands (Decapoda: Caridea: Alpheidae). Proc Biol Soc Wash 113:964-973

Ramírez PA (1983) Sistemática, Ecología y Biogeografía de los Crustáceos Decápodos Anomuros de Bahía Concepción, B. C. S. Tesis de Licenciatura. Facultad de Ciencias Biológicas, Universidad Autónoma de Nuevo León, 132 p

Ricketts EF, Calvin J, Hedgpeth JW and Phillips DW (1985) Between Pacific Tides. 5th Edition. Stanford University Press, Stanford, California, 652 p

Romero C and Carvacho A (1987) Estudios ecológicos en laguna Percebú, alto golfo de California. I. Crustáceos Decápodos: Anomuros, sistématica, ecología, biogeografía y claves de identificación. Cien Mar 13:59-88

Snyder-Conn E (1980) Arthropoda: Crustacea Paguroidea and Coenobitoidea (Hermit Crabs) (pp 275-285) In: Brusca RC (ed) Common intertidal invertebrates of the Gulf of California. University of Arizona Press, Tucson, Arizona. 2nd ed, 513 p

Villalobos JL, Cantú A, Valle MD, Flores P, Lira E and Nates JC (1992) Distribución espacial y consideraciones zoogeográficas de los crustáceos decápodos intermareales de las islas del golfo de California. Proc San Diego Soc Nat Hist 11:1-13

Villalobos JL (2000) Estudio monográfico de los decápodos no braquiuros de la zona intermareal de las islas del golfo de California, México. Tesis de maestría. Facultad de Ciencias, Universidad Nacional Autónoma de México, 312 p

Westervelt CA (1967) The littoral Anomuran Decapod Crustacean fauna of the Punta Peñasco-Bahia La Cholla, area in Sonora, Mexico. Diss. Abstracts, 27B: 4183-b, University of Arizona (Ph. D. Thesis), University of Arizona, Tucson, 143 p

Wicksten MK (1983) A monograph on the shallow water caridean shrimps of the Gulf of California, Mexico. Allan Hancock Found Mon Mar Biol 13:1-59

CONSERVATION OF CONTINENTAL COPEPOD CRUSTACEANS

Janet W. Reid, Ian A. E. Bayly, Giuseppe L. Pesce, Nancy A. Rayner,
Y. Ranja Reddy, Carlos E. F. Rocha, Eduardo Suárez-Morales
and Hiroshi Ueda

JWR, Department of Systematic Biology, National Museum of Natural History, Smithsonian Institution,
Washington, D.C. 20560-0163, USA; Address for correspondence: 1100 Cherokee Court,
Martinsville, Virginia 24112, USA
reid.janet@nmnh.si.edu

IAEB, 501 Killiekrankie Road, Flinders Island, Tasmania 7255, Australia

GLP, Dipartimento di Scienze Ambientali, Università de L'Aquila, Via Vetoio, Località Coppito, 67100 L'Aquila, Italy

NAR, Zoology Department, University of Durban-Westville, P. Bag X54001, Durban 4000, South Africa

YRR, Department of Zoology, Nagarjuna University, Nagarjunanagar 522510, India

CEFR, Departamento de Zoologia/IB, Universidade de São Paulo, C.P. 11461, São Paulo, SP 05422-970, Brazil

ESM, El Colegio de la Frontera Sur (ECOSUR) - Unidad Chetumal, A. P. 424, Chetumal, Q. Roo 77000, Mexico

HU, Center for Marine Environmental Studies, Ehime University, Bunkyo-cho 3, Matsuyama, Ehime 790-8577, Japan

ABSTRACT

A discussion on the conservation problems of continental copepods from several parts of the world is presented, based on the experience of local experts. The status of some copepod species of concern for the IUCN Red List is evaluated.

I. INTRODUCTION

Freshwater copepods have not attracted much, if any, specific concern from the general conservationist community, nor have many copepodologists yet dealt directly with this aspect of their field. Most of those who have are Europeans: lists of small crustaceans, including copepods, were furnished for parts of Germany by Blab et al. (1977), Herbst (1982), Flössner (1993), Maier (1998), and Maier et al. (1998), for Slovenia by Sket and Brancelj (1992), and for Italy by Pesce (2000). Some species of concern in Europe are large calanoids, inhabitants of temporary ponds, and

most of the remainder are stygobiont species with extremely limited known distributions. Outside Europe, lists exist for Australia (Horwitz 1990) and the state of North Carolina, USA (Clamp et al. 1999). Fryer (1972) mentioned the *Schizopera* species-flock in his argument for protection of Lake Tanganyika and the other African Great Lakes. Bayly contributed largely to the first IUCN Red List for copepods, and also (1992a, 1997, 1999) argued strongly in both the scientific and popular literature for preservation of distinctive ephemeral freshwater habitats, including gnammas (solution holes in rock outcrops) in Australia. Danielopol (1992, 1998) argued equally strongly for preservation of groundwater organisms in general. Reid (1994a, 1998) called attention to threats to endemic copepods in the central plateau (cerrado) of Brazil. The 1996 Red List (IUCN 1996) included 118 species of copepods.

In the following sections, several of us discuss conservation problems and some individual species in our respective parts of the world (Bayly, Australia; Pesce, Italy; Rayner, southern Africa; Reid, USA; Rocha, Brazil; Suárez-Morales, Mexico; Ueda, East Asia). All of us

E. Escobar-Briones & F. Alvarez Eds.
MODERN APPROACHES TO THE STUDY OF CRUSTACEA
PP. 253-261

have contributed to the ongoing assessment of the statuses of copepod species of concern for the IUCN Red List.

A. Salinization: A major threat to the conservation of Australian freshwater Crustacea

Salinization of inland waters has emerged as a major factor in the disappearance of habitat for Australian freshwater crustaceans. This is most striking in the southwest of the continent, especially in the Wheatbelt of Western Australia (Yencken and Wilkinson 2000). This is a region in which all but a small fraction of the original vegetation has been removed. With the clearing of deep-rooted native vegetation, ground-water levels rise and salt is brought to the surface. This is referred to as dryland salinity.

Pinder et al. (2000) drew attention to the way in which fresh waters associated with granite outcrops may play an increasingly important conservation role as the Wheatbelt becomes more saline over the next few decades. Among the Copepoda, salinization in Western Australia may be regarded as a threat to species as *Boeckella geniculata* (vulnerable) and *B. shieli* (critically endangered) (IUCN 1996).

The Murray-Darling Basin in eastern Australia is another region that is beset with salinization. Here, there are both dryland and irrigation-induced salinity problems (Yencken and Wilkinson 2000). Among the copepods, species such as *Calamoecia australica*, which is known from only four localities in western Victoria (Bayly 1992b), are threatened by rising salinity.

The conservation status of Australian freshwater Crustacea in general, was well summarized by Horwitz (1990), who also dealt with threatening processes other than increased salinity.

Draining of coastal lagoons and swamps may threaten certain species such as the vulnerable calanoids *Boeckella nyoraensis* and *B. propinqua*, for which this type of habitat is important. *Boeckella nyoraensis* is known only from coastal localities in Victoria and Tasmania (Bayly 1992b).

B. Threatened groundwater copepods of Italy

In recent years, pollution of subterranean aquifers has become a serious problem in several Italian regions and, in spite of their remarkable scientific and economical importance, underground fresh and coastal biological resources of the country are declining at a very fast rate. Several forces are swiftly producing impoverishment or disappearance of aquatic communities with great economic and scientific value. These include the intrusive proliferation of agriculture and pastoral activities, chemical manufacturers, and petroleum refineries which discharge their waste into large and small aquatic reservoirs of the country. There are also high levels of pollution along the Mediterranean coasts, particularly those related to smaller basins such as the Adriatic and Ionian Seas, as well as pollution of coastal caves, including anchialine caves.

At present, the most endangered aquifers are in the Po, Arno, and Tevere river basins, the northeastern karst region, and the karst of southern Italy (Salentine Peninsula), Sicily, and Sardinia. In these regions the karstic biota, phreatic habitats, springs, and caves are particularly threatened, and detailed conservation measures are necessary for their protection and for assessing preservation priorities: "Buso della Rana" (Vicenza, northern Italy); the Mugnone River basin (Tuscany); the Pescara springs (L'Aquila, central Italy); "Grotta di Stiffe" Cave (L'Aquila, central Italy); "Grotta dell'Arco" Cave (Rome, central Italy); "Grotta di Pertosa" and "Grotta di Castelcività" Caves (Salerno, southern Italy); "Grotta di S. Pietro", "Grotta di Nettuno," and "Grotta su Coloru" Caves (Sardinia); anchialine systems of Porto Palo (Sicily); "Zinzulusa," "Buco dei Diavoli," and "L'Abisso" Caves (southern Italy); and anchialine systems along the Adriatic coast (southern Italy).

These ecosystems are variously threatened by tourism, agriculture and urbanization, proximity to hydro-electric construction, invasion of alien species, and the carelessness of cavers who often are unaware of the presence of rare, often the only known population of particular species, in these environments. In some cases, the impact of stochastic events on habitat can affect populations or subpopulations with a limited distribution in the same areas.

Locally, the tunnels recently excavated under Gran Sasso Mountain in the central Apennines are causing the lowering of a rich phreatic watertable complex and consequently the impoverishment of the surrounding small springs and their remarkable fauna, which includes several interesting rare and endemic species. Unevaluated and extensive sedimentation and landslips in northwestern and southern Italy, resulting from dams and hydroelectric reservoirs, have disturbed and continue to disturb large areas as well as the local ecosystems and fauna.

As an example of present threats to the Italian subterranean ecosystems and fauna, Zinzulusa Cave has been recently included in the 'Ten Most Endangered Karst Communities for 1999' by the Karst Water Institute. Although this cave probably harbors the most rich and diverse cave animal fauna in Europe (more than 60 stygobitic or stygophilic taxa, most of them endemic and of ancient lineage), it is at present seriously polluted from urban discharge. Water also threatens the cave, as do tourists who litter and destroy natural formations, mainly near the cave entrance. Moreover, guano mining in the past removed an already limited energy resource for the communities living in this cave. The fauna of this extraordinary cave is certainly vulnerable, and its protection requires habitat preservation as well as conservation of the associated epigean habitat, since, as in most karstic situations, protection of surface environments may be the key to the conservation of groundwater ecosystems.

Unfortunately, most subterranean fauna is not taken into account in the existing laws affecting environmental protection in Italy. Moreover, the relevant Red Lists are in great need of updating and revision.

As regards copepods, of the more than 150 subterranean species and subspecies presently known from Italy, only 37 could be classified as threatened in that country. Most of these are stygobitic or eustygophilic taxa, of remarkable scientific interest, and belong to different IUCN categories; most, on account of the restriction of their area or the few known localities, fit Red List category VU D (Vulnerable, restricted area). But this number is certainly a minimum estimate, since more taxa could be added to the list in the categories "Data deficient" or "Near Threatened", since there are no adequate data to assess their risk of extinction, or they do not qualify as conservation-dependent, but are close to qualifying as vulnerable. Therefore, it is imperative to know the present status of these taxa, obtain more information to determine their appropriate listing, and to implement necessary and opportune studies.

C. Some ephemeral and semi-terrestrial wetlands and their copepod communities in southern Africa

South Africa occupies a vast area stretching from the western Cape Province in the south to Namibia in the west and subtropical Kwa Zulu-Natal on the East Coast. In the north it is bordered by Botswana, Zimbabwe, and Mozambique. Very little is known regarding the freshwater invertebrate fauna of these neighboring countries. There are two landlocked countries within the borders of South Africa, the mountain kingdom of Lesotho, and Swaziland. Except for the eastern border of Lesotho, which it shares with South Africa, nothing is known about the water bodies and freshwater fauna of these two countries. Lesotho has become more accessible with the construction of the very large Katse Dam, which was built as a source of water for South Africa and power for Lesotho. Water is a critical resource in the subcontinent of southern Africa, where rivers flow erratically and natural lakes are few. Most of South Africa lies in the arid latitudes of the southern hemisphere where rainfall is seasonal, the mean annual figure being 475 mm compared with a global average of 860 mm. There is grave concern about the availability of water in the next decades, and it has been predicted that there will be severe water shortages, despite the fact that more than 100 major impoundments have been built in 23 drainage basins (Noble and Hemens 1978). Many of the wetlands have been drained for agriculture and polluted by mining, industry, and informal settlements. There is a diversity of wetlands in southern Africa, such as marshes, vleis, pans, temporary pools, mountain tarns, coastal lakes, and in Namibia, wetlands with some unusual local names such as "oshana" (a shallow watercourse which may become a flat expanse of sand in the dry season) and "ondombe" (a deep, large pool which may retain water year-round) (Clarke and Rayner 1999).

The information on cyclopoids and harpacticoids in southern Africa is limited. There are no records of subterranean species. Most species were described by Brady, Sars, or Kiefer from 1904 to 1934, and no research has been done since then. Three of the better known cyclopoids are *Mesocyclops major, Thermocylops oblongatus,* and *Afrocyclops gibsoni.* These species are usually collected from permanent waters such as man-made impoundments and reservoirs. Except for coastal lakes which are estuarine in origin, South Africa has no natural lakes (Lake Chrissie is a large shallow pan). Species such as *Thermocyclops schuurmanae* and *Microcyclops crassipes* have been recorded from the vleis of the Western Cape.

Information on arid-adapted species of Diaptomidae is available in Rayner and Heeg (1994) and Rayner (1999). Species which are adapted to temporary waters belong to the subfamily Paradiaptominae, and it seems likely that they are obligate temporary pool dwellers. All these species will have resting eggs and this will be a safeguard against extinction despite environmental pressures. The six calanoid species which occur in the Western Cape Province of South Africa (*Lovenula simplex, Paradiaptomus lamellatus, P. peninsularis, P. hameri, Metadiaptomus capensis, M. purcelli*) are probably the most vulnerable. They are all endemic, occurring in low-lying vleis and temporary pools. This area of the Cape has been isolated by the uplift of the Cape Fold Mountains in the middle Tertiary, and some species occupy a very small geographical area. Any change in sea level would be a threat, especially to *Paradiaptomus peninsularis,* which is confined to the narrow Cape Peninsula. The western and central areas of South Africa are semi-arid, with hundreds of temporary pools and pans. Two species that are very common and widespread in semi-arid areas are *Lovenula falcifera* and *Metadiaptomus meridianus.* They often co-occur and are adapted to wide variations in altitude, water temperature, and salinity. In contrast, *Lovenula excellens* has been recorded only from the Lake Chrissie area of Mpumalanga, where its ancestor (probably *L. falcifera)* appears to have been trapped in an ancient river system. The fourth *Lovenula* species, *L. africana,* has been recorded from waters of high conductivity from Ethiopia through East Africa, with the southernmost record being Makgadikdadi Pan in Botswana. Its affinity for saline waters is confirmed by its presence in this pan, where birds' legs are often encrusted with salt deposits. With the exception of the rare *Paradiaptomus natalensis,* all the other known African species of *Paradiaptomus* (*lamellatus, schultzei, similis, rex, hameri, peninsularis, warreni*) occur in temporary waters. Both *P. schultzei* and *P. similis* extend their range into North Africa, and *P. rex* is known from Senegal, Upper Volta, and Chad. *Metadiaptomus meridianus* and *M. colonialis* occur extensively in permanent waters, although *M. colonialis* is a subtropical species. The other known African species (*chevreuxi, capensis, purcelli, aethiopicus, mauretanicus, gauthieri*) are temporary pool dwellers.

Species of *Tropodiaptomus* and *Thermodiaptomus* of the subfamily Diaptominae are found mainly in permanent, warm waters. *Tropodiaptomus bhangazii* has been recorded from one small lake, Bhangazi South on the KwaZulu-Natal Coast, and is endangered especially from pollution in an area of controversial dune mining for titanium.

From the days of colonization of South Africa in the 17th century, the land and its fauna and flora have been ravaged. There is a great deal of information on the loss and threatened extinction of the vertebrate fauna, but little attention has been paid to the invertebrates and especially freshwater Crustacea. Wetlands have been drained for agriculture, and overgrazing by cattle has caused serious erosion. The African people judge their wealth by head of cattle, and this has serious implications. In the major cities, pollution of rivers and ground water is serious, especially in the mining areas. Informal settlements, which are always established near rivers or streams, are a serious source of pollution, not to mention the health problems of malaria and bilharzia. More than a quarter of the population is poor and illiterate, and there is an ongoing conflict between conservation of resources and impoverished people. The hope lies in the education of the youth, and every effort is being made to encourage them to understand what conservation means. Environmental education has a high priority in schools and universities.

With regard to the freshwater Copepoda, some species are threatened and some may have already become extinct, especially in marshy areas that have been drained. The positive aspect lies in the vastness and semi-aridity of the land, as well as the inaccessibility of some of the mountainous areas. Although water is scarce and rainfall unpredictable in the arid areas, copepod populations have evolved survival techniques. Despite the harsh conditions, it seems doubtful that extinctions will occur in arid-adapted species. Species in the Western Cape that occur in low-lying vleis are probably at risk, and it is possible that some of them have become extinct. As always, there is a great need for more research, but this will not happen until funds are available.

D. Stream degradation and surface wetlands loss in the USA

Aquatic ecosystems in the USA have been strongly affected, in some cases destroyed, by dams or channelization, water-level control, contaminants and reduced water quality, and introductions of exotic species (Mac et al. 1995). Recent data for several major watersheds indicates that only a tiny proportion of each is protected from development, with much original forest cover lost (e.g., Alabama/Tombigbee, 0% protected and 25% forest lost; Mississippi, 2 and 52%; Hudson, 2 and 9%) (Revenga et al. 1998). Only 2% of the 5.1 million kilometers of streams remain free flowing and undeveloped (Abell et al. 2000).

The rapid pace of development in the past century has led to destruction of an unknown number of ephemeral water bodies and to silting of streams. Because of the poor knowledge of most groups of aquatic invertebrates including copepods in North America, it is impossible accurately to assess the main threats to their existence (Strayer 2000). A case in point is an endemic genus of cyclopoids, *Rheocyclops*, several species of which were recently described from streambeds and a cave in the southeastern USA (Reid et al. 1999). Although most of the habitats were considered relatively unaffected by human activities, only 29 specimens of the most numerous species were collected, and of some species only 1 or 2 individuals were found. The extent to which silting or pollution might affect these populations is unknown.

Although calanoid copepods are relatively large and easily collected, and some are brightly colored, the distributions of many, especially those inhabiting small and/or ephemeral waterbodies, are known only fragmentarily. For instance, *Aglaodiaptomus kingsburyae* was described from a roadside ditch in Oklahoma and a small pool and a pond in Texas in 1975, and *A. marshianus* was described in 1953 from Lake Jackson, northern Florida. *Hesperodiaptomus augustaensis* was first found in a temporary pond in Georgia in 1910 and was reported in 1938 from a floodplain pond, which had already been destroyed, in North Carolina. None of these species has been reported since. *Hesperodiaptomus californiensis* was described from three vernal pools in northern California in 1996. *Skistodiaptomus carolinensis* is known only from two large artificial lakes in western North Carolina.

As in most of the rest of the world, thorough local surveys are necessary to underpin estimates of conservation needs of the microcrustacean faunas. Although individual states and private conservation agencies sponsor faunal inventory programs, there is as yet no national survey or central information source for all faunal groups.

E. Conservation concerns regarding Brazilian copepods

Knowledge of the copepod fauna of Brazilian fresh waters is quite disparate if we consider the pelagic and benthic habitats separately. Although Brazil is poor in natural lakes, the number of artificial lakes formed by damming medium and large rivers is large and continues to grow with the construction of new hydroelectric dams in the north and central-west. The zooplankton of these waterbodies has been extensively and intensively studied, and the copepod diversity is well known. It appears that this fauna may be threatened by species recently introduced into reservoirs in the Brazilian southeast, such as the Afro-Asian cyclopoid *Mesocyclops ogunnus* (Reid and Pinto-Coelho 1994). We cannot ignore the possibility that the native species of *Mesocyclops* (*M. brasilianus, M. meridianus*) and perhaps other genera may be supplanted by introduced species.

In addition, the damming of rivers has led to the disappearance of many floodplain lakes, and it is possible that certain copepod species will not survive in the new environmental conditions. The harpacticoid genus *Potamocaris* is widely distributed in the Paraná and Paraguay River basins and in the small rivers of the Brazilian east coast, where its members live in the interstitial water of sandy riverbeds. The damming and growing pollution of the rivers by domestic and industrial wastes, as well as agricultural and mining activities may bring changes in the granulometry and oxygen that threaten the survival of these harpacticoids. For now, data are lacking to evaluate the extinction risk of these species.

Knowledge of the diversity of copepods in benthic and ephemeral habitats is very fragmentary because of the great extent of the country, the enormous diversity of aquatic habitats, the varying methodologies of the few ecological studies (most of which treated the macrofauna), and the lack of a national project of faunistic studies. Existing data are spotty and result from individual initiatives by a small number of investigators. The distribution of most benthic copepod species reflects more the distribution of research institutions where there are or have been researchers interested in the group, than the true distribution of the animals. A pioneering initiative is the BIOTASP Project (Biota of São Paulo), sponsored by FAPESP (Fundação de Amparo à Pesquisa do Estado do São Paulo), which aims mainly to finance projects on the biodiversity of the state of São Paulo. Mosses and other microhabitats of the Atlantic Forest biome are inhabited by cyclopoid species which can be considered in the Red List category of Lower Risk (conservation-dependent). *Muscocyclops operculatus, Bryocyclops caroli, B. campaneri, Hesperocyclops herbsti,* and *Metacyclops oraemaris* are all known only from remnants of the Atlantic Forest, which since 1500 has been gradually and continuously devastated for farming, ranching, and residential purposes. The uppermost layers of the substrate in the restinga (coastal dune forest) of the coastal plain of the state of São Paulo are made up of decomposing fallen leaves, sediment, and plant roots, forming a soft substrate in which *Allocyclops silvaticus* and *Metacyclops hirsutus* live. *Metacyclops paludicola* was described by Herbst (1959) from *Sphagnum* mats in this coastal plain and never found again, in spite of collecting efforts in different seasons about 70 km to the north in the same type of habitat (Rocha, unpublished data). Tiny restinga pools are inhabited by an undescribed, planktonic species of the cyclopoid genus *Neutrocyclops*. All these species of the restinga forest are also conservation-dependent, since the area that they occupy is under intense human pressure for tourism development. Much of the forest has been razed, with no effective measures by governmental authorities to preserve representative areas of this ecosystem.

Semiterrestrial habitats of the cerrado (savanna) of central Brazil, including permanent marshes, wet campo marshes, and ephemerally to permanently moist organic soils, harbor a highly diverse copepod fauna, which is up to now known only through studies in a small area. Of the 79 copepod taxa recorded in such habitats by Reid (1994b), 44 endemic species are threatened by the continual expansion of pasture and farmland, changes in drainage patterns, and input of fertilizers and pesticides. These factors may be highly damaging and cause the elimination of species about which we know nearly nothing regarding their geographical distribution, biology, and ecological preferences.

F. Conservation concerns regarding Mexican copepods

Only three neotropical diaptomid species were included as Threatened in the 1996 Red List, and four species were listed in the DD (Data Deficient) section. Recent surveys have provided new information to propose additional species to be included in the list, and also to exclude at least one. *Microdiaptomus cokeri*, recently redescribed by Elías-Gutiérrez and Suárez-Morales (1998), is the only true troglobitic diaptomid copepod known in the Americas. It is known exclusively from a cave system in northeastern Mexico (Osorio-Tafall 1942, Elías-Gutiérrez and Suárez-Morales 1998), and its local distribution in the cave remains unknown. These caves are increasingly used to provide water for human activities (mainly watering cattle), and the system is connected to the heavily polluted hydrological basin of the Panuco River. Hence, as recommended by Elías-Gutiérrez and Suárez-Morales (1998), this species should be included in the next IUCN list as endangered.

Two species, *Mastigodiaptomus reidae* and *M. maya*, were recently described from a single small ephemeral pond within the limits of a protected area (Calakmul) in southeastern Mexico (Suárez-Morales and Elías-Gutiérrez 2000). However, increasing human pressure is expected over the entire area; the type locality is within an archaeological site that will be progressively developed for tourism. The distribution of this species is still unknown, but it is probable that, as is common in diaptomid copepods, it is restricted in both time and space (Dussart and Defaye 1995), and should be considered vulnerable.

Leptodiaptomus dodsoni was described recently (Elías-Gutiérrez et al. 1999) from a single ephemeral pond near the large Lake Chapala in northwestern Mexico; two individuals were also collected in Lake Chapala. The entire perimeter of this lake is under human pressure due to agricultural development and water extraction; in particular, the type locality has adjacent rural communities that use water from the pond.

Leptodiaptomus mexicanus was described in 1901 from an undetermined locality near Mexico City. It was not observed thereafter, until it was collected recently in a small (less than 1 km²) temporary pond located between the lanes of a highway. The male was described based on this material (Grimaldo-Ortega et al. 1998). The species appeared in this pond over a period of only one month. Only one Mexican calanoid species has been removed from the 1996 Red List.

Mastigodiaptomus montezumae was included in the DD (Data Deficient) category, but was recorded in several localities in central Mexico (Santos-Silva et al. 1996). Several species of cyclopoid copepods were described recently from isolated populations in sinkholes and associated cave systems in the karst plains of the Yucatan Peninsula. Some occur in only a single system, with apparently very low population numbers. Many sinkholes all over the Yucatan are being used for touristic, recreational, and water supply purposes. The isolated populations of some of these species (*Mesocyclops chaci*, a well and a cave; *M. yutsil*, a few isolated wells and cenotes; *Diacyclops puuc*, a semi-enclosed cenote; *D. chakan*, several isolated sinkholes; see Fiers et al. 1996, Suárez-Morales et al. 1996, Suárez-Morales and Rivera-Arriaga 2000) seem to indicate that they are vulnerable.

G. Copepods in East Asia

In the 1996 Red List, only one copepod, *Neutrodiaptomus formosus*, was listed from East Asia (Japan). This is not because there are few species to be listed in East Asia, but because there was no person who could cover East Asian copepods among the writers of the 1996 List. It is needless to say that there are as many species described from East Asia, especially from China and Japan, as from Europe and other geographic regions, and that some species in East Asia are recorded from very limited localities (sometimes only from type localities). These could be ranked as "VU D2" because of recent general environmental changes in their habitats due to global warming, acid rain, artificial land improvements, and so on. In the new Red List, we will add such rare Chinese and Japanese copepods as were recorded only from the type localities, mainly according to Shen (1979) and Mizuno and Miura (1984).

II. SUMMARY

The number of copepod species of concern to be entered in the next IUCN Red List will much exceed the 118 species listed in the 1996 version. The number of species considered recently extinct has increased from 2 to 6; more than 220 species are now considered under some degree of threat. The increased number of species resulted directly from information provided by several contributors who did not participate in composing the former version; on the other hand, as several of us have noted, many more species could be added. As expressed in various ways by all the contributors, the basic taxonomy, preferred habitats, and composition of local continental copepod faunas are under active investigation in many parts of the world. There are, however, some rays of hope. For example, Maier (1998) and Maier et al. (1998) called attention to the fact that, while natural temporary ponds in riverine floodplains are disappearing in Germany, temporary ponds that are artificially maintained in early successional stages by vehicular activity, mainly in military reserves, continue to support rare crustaceans, in-

cluding 4 species of copepods. Reid (1994a) suggested that because many copepod taxa are highly endemic, strategies aimed at conserving many small areas might be appropriate; and the increased concerns of local authorities for conserving green strips, and riparian wetlands, erosion control, and groundwater pollution abatement will go far to preserve suitable habitats for microcrustaceans. Specialists are beginning to include assessments of the conservation status of copepods in their publications (e.g., Elías-Gutiérrez and Suárez-Morales 1998, Boxshall and Jaume 2000, for anchialine caves). In recent decades a heartening degree of attention has been given to preserving special kinds of surface wetlands, and to problems with groundwater pollution in many countries.

REFERENCES

Abell RA (ed) (2000) Freshwater Ecoregions of North America: A Conservation Assessment. World Wildlife Fund, Washington, DC, 319 p

Bayly IAE (1992a) Freshwater havens. Landscope 7:49-53

Bayly IAE (1992b) The Non-marine Centropagidae (Copepoda: Calanoida) of the World. Guides to the Identification of the Microinvertebrates of the Continental Waters of the World 2:1-30

Bayly IAE (1997) Invertebrates of temporary waters in gnammas on granite outcrops in Western Australia. J R Soc West Austr 80:167-172

Bayly IAE (1999) Rock of Ages: Human Use and Natural History of Australian Granites. University of Western Australia Press, Nedlands, 144 p

Blab J, Nowak E, Sukopp H and Trautmann W (eds) (1977) Rote Liste der gefährdeten Tiere und Pflanzen in der Bundesrepublik Deutschland. Naturschutz Aktuell 1:1-67

Boxshall GA and Jaume D (2000) Discoveries of cave misophrioids (Crustacea: Copepoda) shed new light on the origin of anchialine faunas. Zool Anz 239:1-19

Clamp JC (ed) (1999) A Report on the Conservation Status of North Carolina's Freshwater and Terrestrial Crustacean Fauna. Report of the Scientific Council on Freshwater and Terrestrial Crustaceans. Raleigh, North Carolina, 90 p

Clarke NV and Rayner NA (1999) Freshwater Crustacea (Ostracoda, Copepoda, Branchiopoda, Cladocera) of the Cuvelai wetlands in northern Namibia. Cimbebasia 15:117-126

Danielopol DL (1992) New perspectives in ecological research of groundwater organisms (pp 15-22) In: Stanford JA and Simons JJ (eds) Proceedings of the First International Conference on Ground Water Ecology. American Water Resources Association, Bethesda, Maryland, 419 p

Danielopol DL (1998) Conservation and protection of the biota of karst: assimilation of scientific ideas through artistic perception. J Cave Karst Stud 60:67

Dussart BH and Defaye D (1995) Introduction to the Copepoda. Guides to the Identification of the Microinvertebrates of the Continental Waters of the World 7:1-277

Elías-Gutiérrez M and Suárez-Morales E (1998) Redescription of *Microdiaptomus cokeri* (Crustacea: Copepoda: Diaptomidae) from caves in central Mexico, with the description of a new diaptomid subfamily. Proc Biol Soc Wash 111:199-208.

Elías-Gutiérrez M, Suárez-Morales E and Romano B (1999) A new species of *Leptodiaptomus* (Copepoda, Diaptomidae) from northwestern Mexico with comments on the distribution of the genus. J Plankton Res 21:603-614

Fiers F, Reid JW, Iliffe TM and Suárez-Morales E (1996) New hypogean cyclopoid copepods (Crustacea) from the Yucatan Peninsula, Mexico. Contrib Zool 66:65-102

Flössner D (1993) Rote Liste der Süsswasserkrebse (Branchiopoda et Copepoda) Thuringens. I. Fassung. Stand 1992. Naturschutzreport 5:49-51

Fryer G (1972) Conservation of the Great Lakes of East Africa: a lesson and a warning. Biol Conserv 4:256-262

Grimaldo-Ortega D, Elías-Gutiérrez M, Camacho-Lemus M and Ciros-Pérez J (1998) Additions to Mexican freshwater copepods with the description of the female *Leptodiaptomus mexicanus* (Marsh). J Mar Syst 15:382-390

Herbst HV (1959) Brasilianische Süsswassercyclopoiden (Crustacea, Copepoda). Gewäss Abwäss 24:49-73

Herbst HV (1982) Deutsche existenzbedrohte Branchiopoda und Copepoda (Crustacea). Arch Hydrobiol 95:107-114

Horwitz P (1990) The conservation status of Australian freshwater Crustacea (With a provisional list of threatened species, habitats and potentially threatening). Australian National Parks & Wildlife Service Report Series No. 14

IUCN (1996) 1996 IUCN Red List of Threatened Animals. IUCN, Gland, Switzerland, 368 p

Mac MJ (1995) Aquatic ecosystems (pp 233-258) In: LaRoe ET, Farris GS, Puckett CE, Doran PD and Mac MJ (eds) Our Living Resources: A Report to the Nation on the Distribution, Abundance, and Health of US Plants, Animals, and Ecosystems. US Department of the Interior, National Biological Service, Washington, DC, 530 p

Maier G (1998) The status of large branchiopods (Anostraca, Notostraca, Conchostraca) in Germany. Limnologica 28:223-228

Maier G, Hössler J and Tessenow U (1998) Succession of physical and chemical conditions and of crustacean communities in some small, man made water bodies. Internat Rev Ges Hydrobiol 83:405-418

Mizuno T and Miura Y (1984) Freshwater Copepoda in Japan (pp 471-646) In: Shen CJ and Mizuno Y (eds) Chinese/Japanese Freshwater Copepoda. Tatara Shobo, Yonago, 650 p (in Japanese)

Noble RG and Hemens J (1978) Inland water ecosystems in South Africa: A review of research needs. South Africa National Scientific Programmes Report No 34

Osorio-Tafall B (1942) *Diaptomus (Microdiaptomus) cokeri*, nuevos subgénero y especie de diaptómido de las cuevas de la región de Valles (San Luis Potosí, México) (Copep., Calan.). Ciencia 3:206-210

Pesce GL (2000) Copepods Web Portal: Copepod Red List. Threatened Species in Italy According to IUCN Red List Categories. University of L'Aquila, Italy. http://copepods.interfree.it/red.htm

Pinder AM, Halse SA, Shiel RJ and McRae JM (2000) Granite outcrop pools in south-westem Australia: foci of diversification and refugia for aquatic invertebrates. J R Soc West Austr 83:149-161

Rayner NA (1999) Copepoda: Calanoida. Diaptomidae, Paradiaptominae. Guides to the Identification of the Microinvertebrates of the Continental Waters of the World 15:1-126

Rayner NA and Heeg J (1994) Distribution patterns of the Diaptomidae (Calanoida: Copepoda) in southern Africa. Hydrobiologia 272:47-75

Reid JW (1994a) Latitudinal diversity patterns of continental benthic copepod species assemblages in the Americas. Hydrobiologia 292/293:341-349

Reid JW (1994b) The harpacticoid and cyclopoid copepod fauna in the Cerrado region of Central Brazil. 1. Species composition, habitats, and zoogeography. Acta Limnol Bras 6:5-68

Reid JW (1998) Copepod crustaceans of the Brazilian Cerrado and Pantanal: different systems, different assemblages. World Wide Web: Homepage of Workshop on "Priorities for Conservation of the Cerrado and Pantanal," Brasilia, March 1998. http://www.bdt.org.br/bdt/workshop/cerrado/br/contribuicao/crustaceos

Reid JW and Pinto-Coelho RM (1994) An Afro-Asian continental copepod, *Mesocyclops ogunnus*, found in Brazil; with a key to the species of *Mesocyclops* in South America and a review of intercontinental introductions of copepods. Limnologica 24:359-368

Reid JW, Strayer DL, McIntosh N, Stibbe S and Lewis JJ (1999) *Rheocyclops*, a new genus of copepods from the southeastern and central U.S.A. (Copepoda: Cyclopoida: Cyclopidae). J Crust Biol 19:384-396

Revenga C, Murray S, Abramovitz J and Hammond A (1998) Watersheds of the World: Ecological Value and Vulnerability. World Resources and Worldwatch Institutes, Washington, DC, 164 p

Santos-Silva EN, Elías-Gutiérrez M and Silva-Briano M (1996) Redescription and distribution of *Mastigodiaptomus montezumae* (Copepoda, Calanoida, Diaptomidae) in Mexico. Hydrobiologia 328:207-213

Shen CJ (ed) (1979) Fauna Sinica, Crustacea, Freshwater Copepoda. Science Press, Beijing, 450 p

Sket B and Brancelj A (1992) The Red List of freshwater "Entomostraca" (Anostraca, Cladocera, Copepoda, Ostracoda) in Slovenia. Varstvo Narave 17:165-172 (In Slovenian; English abstract)

Strayer DL (2000) North America's freshwater invertebrates: a research priority (p 104) In: Abell RA (ed) Freshwater Ecoregions of North America: A Conservation Assessment. World Wildlife Fund, Washington, DC, 319 p

Suárez-Morales E and Elías-Gutiérrez M (2000) Two new *Mastigodiaptomus* (Copepoda, Diaptomidae) from southeastem Mexico, with a key for the identification of the known species of the genus. J Nat Hist 34:693-708

Suárez-Morales E, Reid JW, Iliffe TM and Fiers F (1996) Catálogo de los Copépodos Continentales (Crustacea) de la Península de Yucatán. ECOSUR/CONABIO, Mexico City, 292 p

Suárez-Morales E and Rivera-Arriaga E (2000) The aquatic fauna of karstic environments in the Yucatan Peninsula, Mexico: an updated overview. In: Munawar M, Lawrence S, Munawar IF and Malley D (eds) Aquatic Ecosystems of Mexico: Status and Scope. ECOVISION World Monograph Series. Backhuys Publishers, Leiden

Yencken D and Wilkinson D (2000) Resetting the Compass: Australia's Journey Towards Sustainability. CSIRO Publishing, Melbourne

OSTRACODA (CRUSTACEA) FROM THE LOWLAND FLOODPLAIN OF THE RIVER PRIPYAT

Liubov Nagorskaya and Jolande de Jonge

LN, Institute of Zoology NAS B, 27, Academitcheskaya st., 220072, Minsk, Belarus

zoo467@biobel.bas-net.by

JdJ, Riza, Postbus 17, 8200 AA, Lelystad, the Netherlands

ABSTRACT

Ostracoda species richness was compared for various patches of Pripyat river catchment (so called Polesye), which covered about 30% of the territory of Belarus. Internationally the Pripyat is one of the least disturbed lowland rivers in Europe and its floodplain (to 10-15 km wide) from the past is still present in nearly natural condition. The total Ostracoda list for 4 regions of the Pripyat catchment supports 51 species including rare species, which are more than 30% of the total species composition.

The investigations of abundance and distribution of Ostracoda assemblages along the floodplain water body's lateral gradient was performed both in May-June and August of 1999 (in the middle river course). Species richness was equal to 42 and about 20% of the Ostracoda assemblage composition were rare species. Species number and rare species indices were increased at water bodies both with the biotope diversity and the underground water upwelling influence made out with hydrochemical data.

The analyses of the abundance and ecological patterns of Ostracoda species demonstrated the ecological groups, which inhabited various water bodies in the gradient from a riverbed to the floodplain margin.

I. INTRODUCTION

A well-known fact is that most European lowland rivers and floodplains have been changed in quality, morphology and hydrology during last century because of human impact. A reduction in species diversity and number occurs when habitats are affected by a decline in water quality and water level fluctuations (Wiggins et al. 1980, Neckles et al. 1990, Gilliam 1994, Svobodová 1994, Vranovský 1997). Siltation is a big problem. A very few lowland rivers are unaffected (Amoros and Roux 1988, Middelkoop and van Hansen 1999). One of them is the Belarussian Pripyat River because of agrarian management is still extensive and landscape morphology, hydrology and water quality is still virtually unimpacted. Internationally the Pripyat is one of the least influenced lowland rivers in Europe (De Jonge et al. 1999).

In less disturbed and more vulnerable regions like the Pripyat one might expect to find the sensitive species. Since the parameters of the environment define the composition and abundance of populations, the specific species complexes could be used as an indicator of the trophic status of river systems.

Study area

The Polesye lowland surface is a plain (altitude 100-130 m) with occasional small hills. The region is characterized by both high water table and poor drainage. Valley fens and boggy soils covered with pine and deciduous forests are common. The Pripyat is a tributary of Dnepr River. The catchment of the Pripyat is 121 000 km² and its length is 761 km.

Considering the characteristics of landscapes in the floodplain of the Pripyat river, it is clear that different flow regimes could be possible on the different landscape patches along a transverse gradient: from the main stream to the margin of the floodplain. They are precipitation (rain and snow), soil and underground water and the level of the river during flooding as well as the run-off from nearby forests and bogs (Fig. 2).

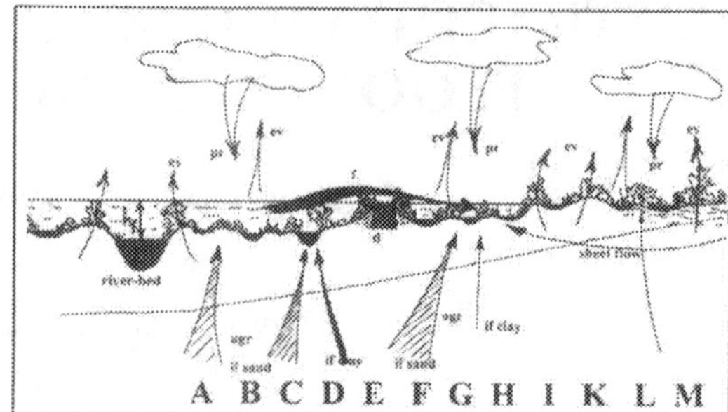

Figure 2. *Scheme of different water sources of floodplain water bodies (ugr - underground waters, pr - precipitation (rain, snow), ev - evaporation, f - flooding, h_f - flooding height over the normal water level, d - dam/road).*

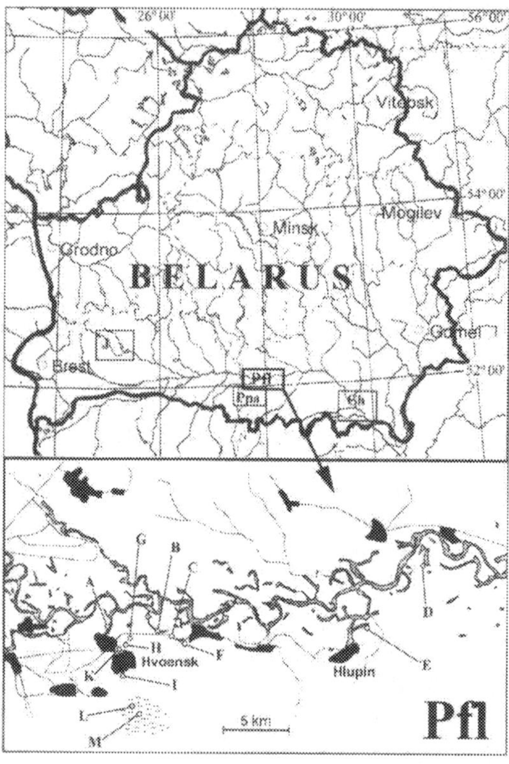

Figure 1. *The map of Pripyat in area of the investigations and the sampling sites.*

The study area was located in the middle part of the river, 15-40 km east of the town of Turov (Fig. 1). At this point the floodplain of the river is approximately 5 km wide. Human influence in this region is limited. There are no dams or heavy industries upstream. The course of the river has not been altered by canalization and boat traffic is almost absent. The natural morphology of the streambed is largely intact. Agricultural activity is minimal. Drainage is small, resulting in a very high water table. Vast expanses of marsh and

forest in the watershed are pristine and uninhabited. In short, the system is representative of the situation lost in most of Europe many centuries ago.

Pripyat floodplain waters are characterised by a considerable amount of organic (humic) matter as well as iron (3.5-7.5 mg Fe.l^{-1}). Because of their boggy catchment the organic humic matters were in all water bodies especially after a flooding. The typical feature of waters at this area is a deficit of oxygen saturation (40-85%) (Kagan and Gelfer 1956).

II. MATERIALS AND METHODS

Data were collected during two sampling periods – spring (May—June) and summer (August) 1999 at 12 water bodies (47 biotopes) situated along a lateral Pripyat floodplain gradient. Data from 3 other patches of the river catchment at the Polesye lowland (Fig. 1) were used for comparison of the Ostracoda fauna composition.

Ostracoda species richness and abundance as one part of the whole community were estimated along a lateral gradient of the Pripyat floodplain (Fig. 2). Water bodies investigated were ordered by Pripyat River influence and titled conditionally with letters from A to M. The lateral continuum of floodplain habitat includes: backwater connected to main stream at downstream end only (A); backwaters without permanent connection to river but strongly influenced by river with flooding (B, C, D); rarely inundated backwaters (oxbows) (E, F, G), occasionally inundated wet woodlands (H, I, K), and fens and marshes which are situated in the margin of a floodplain (L, M).

The measurement of hydrochemical parameters was performed in situ during the sampling with the Hydrochemical express laboratory (Merck [R]) (hardness total and carbonate, NH_4^+, NO_2^-, NO_3^-, PO_4^{3-}), Viscor test kit Alkalinity AL 7 (alkalinity) and Corning [R] Checkmate System (pH, temperature, oxygen saturation and conductivity). Hydrochemical parameters correlation was tested.

The benthos was sampled by a hand dredge net with an 0.031 m² opening. This hand dredge

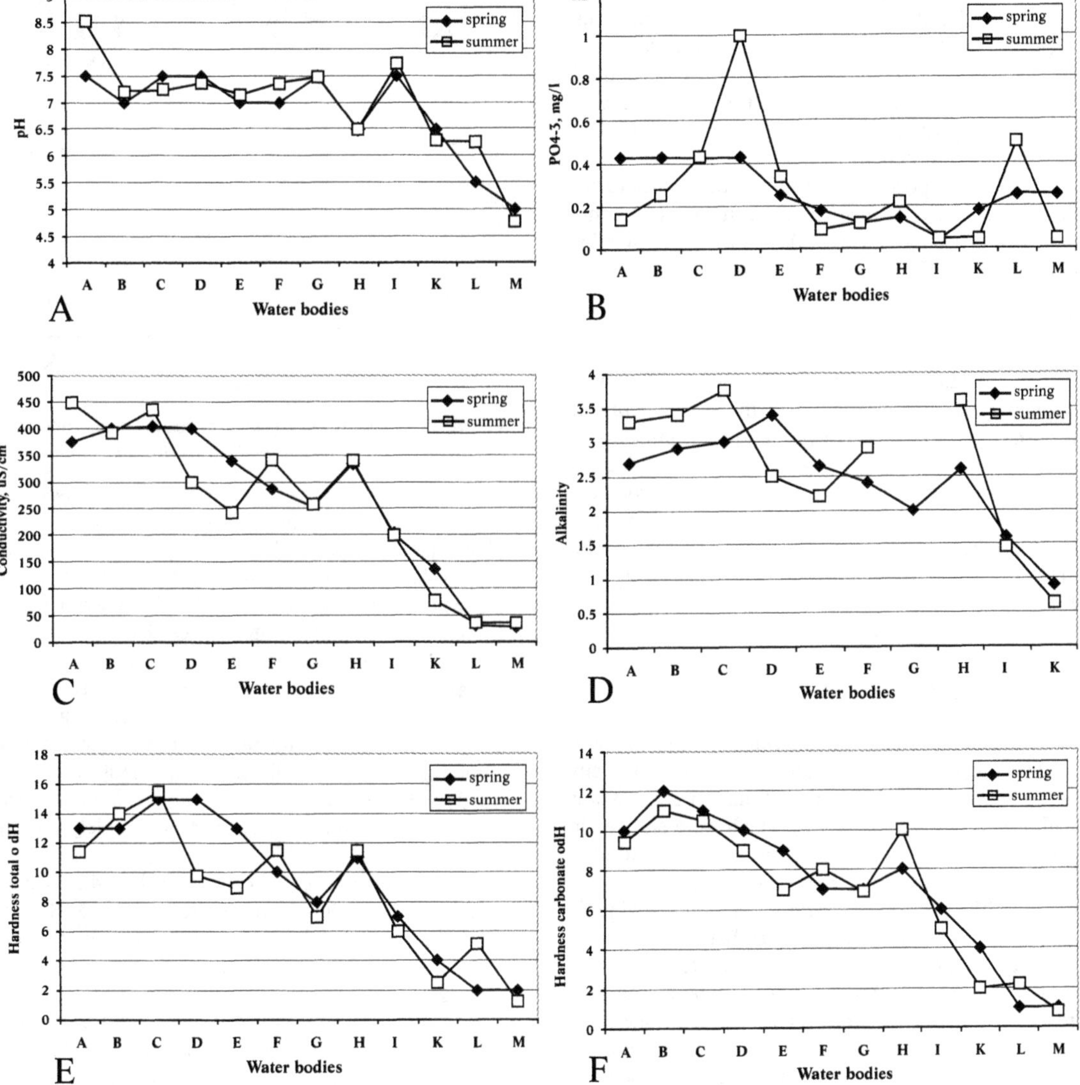

Figure 3. *The hydrochemical parameters dynamics along the water bodies gradient in two seasons (A - pH; B – orthophosphates PO_4^{3-}, mg.l⁻¹; C – conductivity, iS.cm⁻¹; D – alkalinity; E – total hardness; F – carbonate hardness °dH.*

was dragged over the bottom surface to a depth of 3-5 cm. Each replicate was 1 m long by 25 cm wide (an area 0.25 m²). From 2 to 8 replicates were taken for every sample, in order to encompass each habitat type. The replicate number depended on the sediment type. Then the total sample (all replicates together) was washed through the sieve set with holes 1.0, 0.5 and 0.25 mm. First and second fractions were selected to taxa in the field. Separate taxa and the fraction < 0.25 mm were fixed with 70% alcohol in tubes. Ostracoda was one of numerous taxa collected. We identified Ostracoda species and estimated their population density.

We proposed three indices for species rarity that reflect the relationship between rare (r), common (c), cosmopolitan (csm) and total (T) species number of the Ostracoda assemblages. They are:

1). $I_{(T\text{-csm})/T} = (T - csm) / T = (r + c) / T$;

2). $I_{r/T} = r / T$;

3). $I_{rc} = r / c$;

In addition, the Ostracoda fauna similarity was estimated for different regions of the river catchment (Pripyat protected area, Pripyat floodplain, Jaselda and Chernobyl protected zone) (Fig. 1). The similarity indices of Szekanowski-S?rensen used for fauna comparison were calculated as: I ss = 2a / (a+b) + (a+c), where a - species are in the both lists; b - species in list 1 only; c - species in list 2 only; (a+b) - total list 1; (a+c) - total list 2.

III. RESULTS

Pripyat lowland pool waters are mildly alkaline and mildly mineralized (Fig. 3). They belonged to hydrocarbonate-calcium type II. Only the bog pools have really acidic waters. Ammonium was not found in any water bodies while nitrite and nitrate ion content was high exclusively for the points, which were situated in the area with visible anthropogenic impact and used as watering places by herds from villages (E, F).

The trend of decreasing pH, conductivity, alkalinity and both total and carbonate hardness occurred along the water body gradient from river to inner water bodies (Fig. 3). There were a few exceptions from that tendency for pools, which had anthropogenic alteration in the past (the water bodies C, F, H). These pools demonstrated the high values, which are close to oxbows by the Pripyat River. These values are high because of the increase in underground water upwelling through the disturbed bottom. The orthophosphate values confirmed

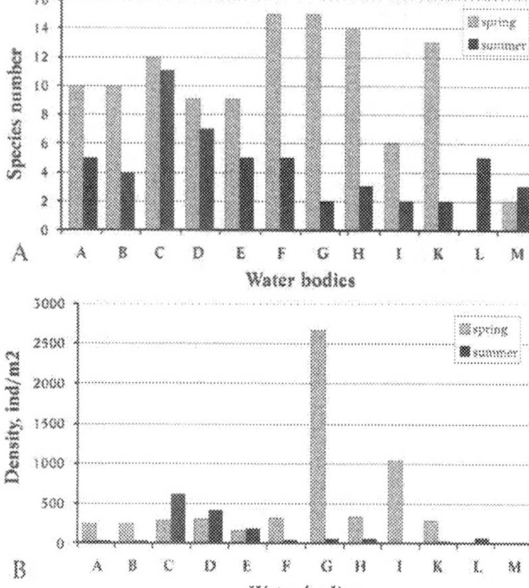

Figure 4. *A - Ostracoda species number, B - population density along the water bodies gradient (spring and summer).*

that most water bodies were eutrophic. The oxbows situated by a main river channel (A-C) had high orthophosphate values in spring after flooding, while in summer these values dropped down to 0.15-0.25 mgP.l⁻¹. The inner pools had 0.05-0.25 mg P.l⁻¹. The increase of orthophosphate values in water bodies D and L in August was probably due to both the destruction of macrophytes organic remains and seepage waters from the neighbouring forest bank vault (D) and peatbog (L).

The temperature ranged from 17-25°C in both sampling seasons. The dissolved oxygen saturation was 40-85%. Most ostracods inhabiting the floodplain water bodies were presumably tolerant of widely varying oxygen levels.

The total list of Ostracoda had 42 species. (Tables 1-2). A significant difference (t = 3.21, p = 0.0041) was found the mean species number in spring ($\bar{x}$ = 9.6, sd = 4.85, n =36) and in summer ($\bar{x}$ = 4.5, sd = 2.58, n = 21).

The species number per pool in spring ranges from 2 to 15 (Fig. 4A). The richest pools were oxbows with underground water upwelling and pools with a diversity of biotopes (C, F, G, H, K). In summer the species number per pool decreased along a gradient from riverbed to inner water bodies.

Table 1. *Species abundance of Ostracoda assemblages from floodplain water bodies (spring).*

Species	A	B	C	D	E	F	G	H	I	K	L	M
						cosmopolitan species						
Candona candida		0,3	0,75									
Candona neglecta		4,5				4,8						
Cyclocypris ovum		60,8		175	73		6,7	6,3	6,0	1,7		
Cypria ophtalmica	0,8	2,0	0,3	22,0	6,8	2,5	1338,7	100,8	6,5	159,7		
Cypridopsis vidua	76,3	3,0	7,0	18,0	22,3	18,0	274,7	0,3	1003,0	7,0		
Heterocypis incongruens						0,1						
Pseudocandona compressa	3,8	0,5	12,0	5,0	3,8	117,0	403,7	10,2	0,5	6,7		
						rare species						
Cypria lata						0,1						
Fabaeformiscandona acuminata		1,8								0,7		
Fabaeformiscandona holzkampfi	0,3	0,8				45,0	132,0	1,3		2,0		
Paralimnocythere relicta							0,7					
Pseudocandona hartwigi						14,5		27,5		0,7		
Pseudocandona rostrata				14,0	0,5			5,7				
Pseudocandona semicognita							20,0					
Trajancypris clavata						0,5						
						common species						
Bradleystrandesia reticulata			1,8			26	4,7					
Candonopsis kingsleii				1,0						1,7		
Candonopsis scourfieldi								0,7				
Cyclocypris laevis	5,7		33,0		6,5	11,1	37,7	131,0	3,0	4,7		4,0
Cypria exculpta				5,0								
Cypridopsis elongata						1,3	0,7					
Cypridopsis parva		1,3			8,8				1,0	3,3		
Cypris pubera	49,3	42,0	0,8	2,0	12,3	36,6	322,0	0,7				
Dolerocypris fasciata								0,2		3,3		
Fabaeformiscandona fabaeformis		0,3					17,3			0,3		
Fabaeformiscandona fragilis	2,0	1,0	4,3							14,7		
Fabaeformiscandona protzi			2,3									
Ilyocypris decipens	2,0			2,0								
Limnocythere inopinata	18,3											
Metacypris cordata							77,7					
Notodromas monacha							4,7	10,0				
Physocypria kraepelini	0,3							0,3				
Plesiocypridopsis newtoni							13,3					
Pseudocandona insculpta			1,3			20,5		2,3				
Pseudocandona marchica					1,0							
Pseudocandona pratensis							3,3					4,0
Candona juvenile	68,2	104,0	209,0	28,0	21,0	8,6		20,0	4,5	47,7		
Fabaeformiscandona juvenile		23,0		22,0		0,8		11,0		23,3		
Pseudocandona juvenile				5,0	6,5	1,8	3,0	0,7	6,0			

Table 2. *Species abundance of Ostracoda assemblages from floodplain water bodies (summer).*

Species	A	B	C	D	E	F	G	H	I	K	L	M
						cosmopolitan species						
Candona candida	0,5	3,5	98,0	41,0	8,0	0,5						
Candona neglecta	0,5	1,5	4,0						2,5			
Cypria ophtalmica	17,3	7,0	17,0	31,0	0,5	11,0	0,5	57,0	13,0			
Cypridopsis vidua	10,8		26,0	4,5		26,5	61,3			11,0	6,0	
Heterocypis incongruens											24,0	6,0
						rare species						
Candona lindneri	0,75							2,0				
Cypria curvifurcata				8,5								
Fabaeformiscandona acuminata			4,0									
Fabaeformiscandona caudata		22,0								9,5		
Pseudocandona insculpta			1,0									
Trajancypris clavata											16,0	
						common species						
Candona weltneri					40,0							
Cyclocypris laevis			3,0									
Cypria exculpta				44,0							8,0	4,0
Fabaeformiscandona fabaeformis						1,0		1,0				
Fabaeformiscandona fragilis			27,0	168,0	115,0							
Fabaeformiscandona hyalina			48,0									
Fabaeformiscandona protzi			3,0									
Ilyocypris gibba											24,0	6,0
Physocypria kraepelini				76,0	5,5							
Pseudocandona compressa			8,0			0,5						
Candona juvenile	5,3		310,0	33,0	21,0					1,0	4,0	
Pseudocandona juvenile	0,25		8,0	3,0								
Fabaeformiscandona juvenile		0,5	96,0	2,5	0,5							

The population density of Ostracoda assemblages was characterized by relatively low values (Fig. 4B). In spring in most water bodies population density was 200-400 ind/m^2, with the exception of two pools that were much higher density (>1000 ind/m^2 in I and 2600 ind/m^2 in G pools) because of the spring generation of a few cosmopolitan ostracod (*Cypria ophtalmica, Cypridopsis vidua*) and the spatial heterogeneity of the oxbows. In summer the population density of Ostracoda assemblages was in 4-10 time less that in spring. The maximal values were > 600 ind/m^2 in C, 411 ind/m^2 in D and 190 ind/m^2 in E water bodies.

The abundance and distribution of Ostracoda both in spring and in summer is presented in Tables 1-2. In most cases Ostracoda species inhabited various water bodies without preference. Nevertheless, four ostracod species were found only in pools influenced by the Pripyat River (A, B, C and D) *(Cypria curvifurcata, Ilyocypris decipens, Physocypria kraepelini, Pseudocandona insculpta)*. These species were found in lotic water systems very often, although they inhabit lakes and reservoirs also (Martens and Dumont 1984, Semyonova 1993).

In the very soft acidic boggy pools (L, M) 5 ostracod species were found *(Cypria exculpta, Cypridopsis vidua, Heterocypris incongruens, Ilyocypris gibba and Trajancypris clavata)* and Candona juvenile as well. Most of them are tolerant the wide range of environmental factors.

The widely distributed and abundant group was conditionally cosmopolitan Ostracoda species characterized by high tolerance levels and inhabited freshwater in general (no specialization) according to

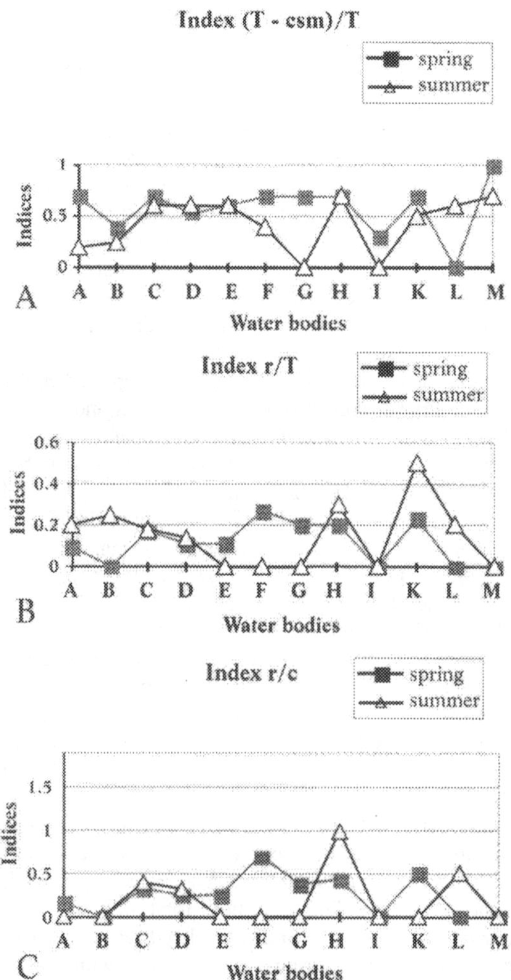

Figure 5. *Indices are been offered for an estimation of the water ecosystem level disturbance.* $A - [I_{(T-csm)/T} = (T - csm) / T = (r + c) / T]$; $B - [I_{r/T} = r / T]$; $C - [I_{r/c} = r / c]$, *where T - total species number; csm - cosmopolitan species, r - rare species and c - common species number.*

modern classification (Limnofauna Europea 1978). It was represented by 7 species (*Cypria ophtalmica, Pseudocandona compressa, Heterocypris incongruens, Cypridopsis vidua, Cyclocypris ovum, Candona candida* and *C. neglecta*) (Tables 1, 2). They were abundant in most floodplain water bodies. Cosmopolitans seem to be the most resistant taxa to environment fluctuations.

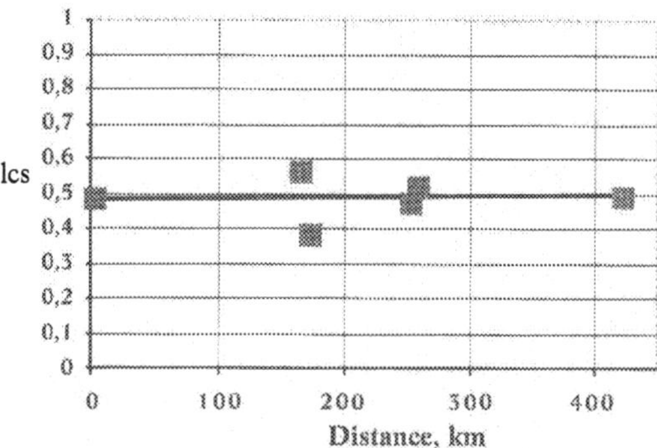

Figure 6. *Species assemblages similarity (Czekanovski-Sørensen indices) vs. linear distance between investigated area.*

The second group included 10 species (24%) of Ostracoda, which were collected in single biotopes and could be considered rare for Belarus as well as for some European countries (Sywula 1974, Meisch 2000). These species were scattered throughout floodplain water bodies with a low abundance (Tables 1, 2).

The third group was common species which are widespread and also occur in diverse eutrophic and mesotrophic water bodies as lakes, ponds, temporary water bodies, wells, ditches and rivers (Tables 1, 2). They were not of highest abundance, but sometimes were moderately abundant as *Cypris pubera, Cyclocypris laevis* and *Fabaeformiscandona fragilis.*

It is evident that the floodplain water bodies investigated are characterised by a large proportion of relatively rare species. We used three proposed rarity indices mentioned above for the estimation of rare species relative number along the water bodies' gradient. All three indices together describe the relationships between groups and their dynamic in spatial and temporal aspects (Fig. 5). It is obvious that rare species were found mainly at the water bodies with underground water upwelling (C, F, H) and an isolated one (K).

The Ostracoda species composition was compared from various patches of the catchment of the Pripyat River (Fig. 1). The total Ostracoda list for 4 regions of the Pripyat catchment supports 51 species including relatively rare species, which account for 30 % of the total species composition (Table 3). Figure 6 presents similarity indices of Szekanowski-Sørensen counted for this area vs. patches some distance away (see Fig. 1). Results suggest that Ostracoda species composition of these patches have about 50% fauna similarity.

L. Nagorskaya & J. de Jonge

Table 3. Ostracoda from Pripyat catchment: J - Jaselda river; Pfl - Pripyat floodplain; Ch - Chernobyl zone; Ppa - Pripyat protected area (*-rare species).

Species	J	Pfl	Ch	Ppa
Bradleystrandesia fuscata			X	
Bradleystrandesia reticulata	X	X	X	
Candona balatonica*			X	
Candona candida	X	X	X	X
Candona lindneri*		X	X	
Candona muelleri*			X	
Candona neglecta		X		X
Candona weltneri		X	X	
Candonopsis kingsleii		X		
Candonopsis scourfieldi		X		
Cyclocypris globosa			X	
Cyclocypris laevis		X	X	
Cyclocypris ovum		X	X	X
Cypria curvifurcata*		X		
Cypria exculpta	X	X	X	X
Cypria lata*		X	X	
Cypria ophtalmica	X	X	X	X
Cypridopsis elongata		X		
Cypridopsis parva	X	X		X
Cypridopsis vidua	X	X	X	X
Cypris pubera	X	X	X	X
Darwinula stevensoni	X			
Dolerocypris fasciata	X	X		
Eucypris virens			X	
Fabaeformiscandona acuminata*	X	X		
Fabaeformiscandona caudata*		X		
Fabaeformiscandona fabaeformis	X	X	X	
Fabaeformiscandona fragilis		X		
Fabaeformiscandona holzkampfi*		X		
Fabaeformiscandona hyalina*		X	X	
Fabaeformiscandona protzi	X	X		
Heterocypris incongruens		X	X	X
Ilyocypris gibba		X		
Ilyocypris decipiens		X		X
Limnocythere inopinata		X		
Metacypris cordata		X		
Notodromas monacha	X	X	X	
Paralimnocythere relicta*		X		
Physocypria kraepelini		X		
Plesiocypridopsis newtoni		X		
Pseudocandona albicans			X	
Pseudocandona compressa	X	X	X	
Pseudocandona hartwigi*		X		X
Pseudocandona insculpta	X	X	X	
Pseudocandona marchica		X		
Pseudocandona pratensis	X	X	X	X
Pseudocandona rostrata*		X	X	
Pseudocandona semicognita*		X	X	
Pseudocandona sucki*			X	
Trajancypris crassa*			X	
Trajancypris clavata*		X		
Total 51 species	16	42	28	13

IV. DISCUSSION

Twelve Pripyat floodplain pools had 42 ostracod species (2-15 species per pool). This represents 5% of the total fauna list for the investigated floodplain water bodies, which exceeds 840 species, and 37% of the crustacean species. It is interesting that a similar proportion was reported for the community from different water bodies' types. In northern California 67 crustacean species occurred in the 44 vernal pools, including 24 ostracod species (36%). Ostracod species number per pool varied from 1 to 12 (King et al. 1996).

There is little information on ostracods from the Pripyat catchment. Only 13 species were reported earlier for the Pripyat protected area (Nagorskaya and Keyser 1999). For Russian and Moldavian river systems sampled annually for many years the Ostracoda species number did not exceed 70 (Semyonova 1993, Keyser and Nagorskaya 1998). The list of Ostracoda in East European lowlands has about 84 species (Limnofauna Europea 1978). The Pripyat catchment is rich and diverse in ostracod species despite species lists based on our local sampling which are almost certainly incomplete.

Undoubtedly, both the water feeding type and the soil character as well as hydrochemical characteristics determine the species composition and fauna richness of the wetland pools. Also the hydrological and hydrochemical characteristics are strongly reflected in the Ostracoda assemblages (Hartmann and Hiller 1977, Meisch 2000). The species number (Fig. 4a) decreased in summer along the lateral water bodies' gradient in accordance with the hydrochemical values (Fig. 3). pH, conductivity, alkalinity and both total and carbonate hardness were significantly correlated ($P < 0.01$) to each other both in spring and in summer, while phosphates were highly variable. The strong correlation between conductivity, alkalinity and TDS was reported earlier (King et al. 1996). High species richness could not be explained by hydrochemical patterns only. In spring the morphologically diverse water bodies with various biotopes (F, G, H, K) were the richest in species number. It is necessary to emphasize that in particular the landscape diversity is one of the reasons of the biotope diversity in water bodies and thus, the species richness (Pickett and Cadenasso 1995, Oertli 1995). Flooding as a natural disturbance factor is a necessary element of river system function. The natural disturbance sustains habitat type diversity, spatio-temporal heterogeneity and supports a high level of biodiversity and species richness.

270

The water bodies with acidic soft waters (L, M) were relatively poor in species (Tables 1, 2). Ostracods inhabiting these pools have adaptations to the wide range of factors and inhabit different reservoirs (Martens and Dumont 1984, Meisch 2000). Species from patches L, M with low pH values (*Heterocypris incongruens and Trajancypris clavata*) were reported from Karelian small pools with pH 5—6 together with *B. reticulata, Cyclocypris ovum, C. serena, C. globosa, P. rostrata* and *L. inopinata* (Akatova and Filimonova 1975).

There are striking seasonal differences between ostracod assemblages. Thirty six ostracod species were present in spring samples and 21 in summer ones. (Tables 1, 2). Among them 15 ostracod species were present both in spring and summer samples. There were 6 cosmopolitan species (see Tables 1, 2), 4 of the genus *Fabaeformiscandona* as well as *Cypria exculpta, Cyclocypris laevis, Physocypria kraepelini, Pseudocandona insculpta*, and *Trajancypris clavata*.

In contrast to 15 species mentioned above, 21 species in spring and 6 in summer were distinctly seasonal in occurrence. In May—June samples there were 3 characteristic spring species (*B. reticulata, C. pubera and P. relicta*), cosmopolitan *C. ovum*, 6 relatively rare species (Table 1), in two species, genus *Candonopsis* and *Cypridopsis* and *Dolerocypris fasciata, Ilyocypris decipens Limnocythere inopinata, Metacypris cordata, Notodromas monacha*, and *Plesiocypris newtoni*. In August 6 rare species (Table 2) and the common *Ilyocypris gibba* were found alone. One of them was *Cypria curvifurcata*, which was not previously reported for Belarus but was a common species for many pools of the former USSR (Semyonova 1993).

Ostracod abundance in the Pripyat floodplain water bodies was within the range of values reported for other wetland systems (Leeper and Taylor 1998). Nevertheless it was much less than in other kinds of pools (Semyonova 1993).

Probably both a characteristic abiotic condition of the floodplain water bodies as well as dormancy period in different ostracod species and biotic relationships in the community (competition for space and food, predation) defined the species growth and reproduction (Hairston and de Stasio 1988, Williams 1987). The variability of ostracods ecological patterns seems to be based on clonal predominance at different seasons. Different clones have genetically based adaptations to environmental factors (Geiger et al. 1998, Rossi et al. 1998).

The area of investigation has a curiously large proportion of relatively rare species. Species were considered relatively rare if they have only single pool or single site occurrence (King et al. 1996). We used indices for relative rarity estimation along the water body gradient from the riverbed to the floodplain margin (Fig. 5). The inner pools (I, L, M) as well as water bodies by the riverbed (B) had no rare species in spring whereas the vernal pools, after widespread flooding, (F, G, H, K) had a rich and diverse species range. In summer, when water level dropped down and the water surface reduced, the highest species richness was in pools by the riverbed (A, B, C, and D). The water bodies with underground water upwelling (F and H) supported predominantly cosmopolitan widespread species.

It was shown for various taxa that widespread species should more often be found in disturbed or strongly seasonal habitats then relatively rare (Glazier 1986).

Typically rare species are ignored for many cases and the dominant complexes are mainly used for the community structure analyses (King et al. 1996). However, rare species were used for ostracod assemblages' estimation (Slack et al. 2000).

The similarity coefficients of Szekanowski-Sørensen calculated for this area vs. patches some distance away (Fig. 6) indicated that ostracod assemblages were characterized by comparable patterns (0.39-0.57 on average 0.49). This means that Ostracoda species composition of various areas of the Pripyat catchment had about 50% fauna similarity. The Pripyat catchment area investigated (Jaselda river floodplain, Chernobyl zone, Pripyat protected area and Pripyat floodplain) has ostracod species, which belong to the common fauna composition (Table 6).

It was shown that, for pools of the same kind, geographic proximity was not correlated with species assemblage similarity indices. In spite of the fact that indices varied from 0.1 to 0.75 for the same distances, the relationship between Jaccard's index of similarity and distance between 44 temporary pools was not clear. Species assemblages of crustaceans were strongly related to habitat type (King et al. 1996).

Jaccard's index of similarity for microcrustacean assemblages from 88 ponds changed from 0.5 (nearby patches) to 0.25 (distant ones) at average value 0.36 (Taylor et al. 1999). A weak decrease in the index value occurred with an increase in distance.

Such varied disposition of similarity indices vs. distance demonstrates a patchwork pattern of crustacean occurrence (King et al. 1996, Mahoney et al. 1990).

Ostracod complexes as well as nutrient maintenance in the Pripyat floodplain water bodies point out eutrophic types of these pools, particularly in spring. Ostracoda is one of the taxa that could be used as a tool for environmental impact assessment, monitoring observations at the river floodplain and a framework for more effective management of river ecosystems. Conservation activity should use this taxa for complex ecosystem assessment. To provide reliable information on species richness is a major information challenge that must be met in order to assess priorities in conservation biology.

ACKNOWLEDGEMENTS

This work was fulfilled thanks the grant from RIZA, the Netherlands (contract N R1-2692) with the financial support from Beijerinck Popping Fund. The data used are also from projects supported by RSPB BYE99/003 UK and by NASB. We appreciate to Dr. C. Meisch (Luxembourg), Dr. K. Martens (Brussels) and an anonymous reviewer for helpful remarks and MS criticizing. We greatly thank to As. Prof. K. Lee (Johnstown, Pennsylvania) for improving the MS text.

REFERENCES

Akatova NA and Filimonova ZI (1975) O faune ostracod (Ostracoda, Crustacea) malych vodojemov Karelii. (pp110-116) In: Vodnye resursi Karelii i ich ispolzovanie. Pietrozavodsk (in Russian)

Amoros C and Roux AL (1988) Interaction between water bodies within the floodplains of large rivers: function and development of connectivity. Münstersche geogr. Arbeiten 29:125-130

De Jonge J, Kerkum FCM and Van Schie K (1999) Pripyat: reference for the lower part of the Rhine (The Netherlands). (pp 50-56) In: Biologicheskoe raznoobrazie Natsionalnogo Parka "Pripyatski" (Biodiversity of the National Park "Pripyatski"),Turov-MozyrGeiger W, Otero M and Rossi V (1998) Clonal ecological diversity. (pp 243-256) In: Martens K (ed.) Sex and Parthenogenesis, Evolutionary Ecology of Reproductive Modes in Non-Marine Ostracods. Backhuys Publishers, Leiden

Gilliam JW (1994) Riparian wetlands and water quality. J Environ Qual 23:896-900

Glazier DS (1986) Temporal variability of abundance and the distribution of species. Oikos 47:309-314

Hairston NDJr and De Stasio BTJr (1988) Rate of evolution slowed by dormant propagule pool. Nature 336:239-242

Hartmann G and Hiller D (1977) Beitrag zur Kenntnis der Ostracodenfauna des Harzes und seines nördlichen Vorlandes (unter besonderer Berücksichtigung des Männchens von *Candona candida*). Naturw Verein Goslar 125:99-116

Kagan CA and Gelfer EA (1956) Characteristica gumusovych veschestv necotorych vodoemov Polesya. (pp 69-93) In: Vinberg GG (ed) Trudy complexnoi expeditsii po izucheniu vodijemov Polesya Izd. BGU, Minsk (in Russian)

Keyser D and Nagorskaya L (1998) Ostracods in the vicinity of Minsk, Belarus. Mitt hamb zool Mus Inst 95:115-131

King JL, Simovich MA and Brusca RS (1996) Species richness, endemism and ecology of crustacean assemblages in northern California vernal pools. Hydrobiologia 328:85-116

Leeper DA and Taylor BE (1998) Abundance, biomass and production of aquatic invertebrates in Rainbow Bay, a temporary wetland in South Carolina, USA. Arch Hydrobiol 143:335-363

Limnofauna Europea (1978) In: Illies J (ed) Gustav Fischer Verlag Stuttgart, New York, Swets and Zeitlinger BV, Amsterdam

Mahoney DL, Mort MA and Taylor BE (1990) Species richness of calanoid copepods, cladocerans and other branchiopods in Carolina bay temporary ponds. Am Midl Nat 123:244-258

Martens K and Dumont HJ (1984) The ostracod fauna (Crustacea, Ostracoda) of Lake Donk (Flanderes): a comparison between two surveys 20 years apart. Biol Jb Dodonaea 52:94-111

Meisch C (2000) Freshwater Ostracoda of Western and Central Europe. In: Suesswasserfauna von Mitteleuropa, Band 8/3 Spectrum Academischer Verlag, Germany

Middelkoop H and van Hansen (eds) (1999) Twice a River. Rhine and Meuse in the Netherlands RIZA report no 99.003 Arnhem: RIZA

Nagorskaya L and Keyser D (1999) Rakushkovye raki (Ostracoda, Crustacea) iz vodoemov Natsionalnogo Parka "Pripyatski". (pp 50-56) In: Biologicheskoe raznoobrazie Natsionalnogo Parka "Pripyatski" (Biodiversity of the National Park "Pripyatski",Turov—Mozyr, in Russian)

Neckles HA, Murkin HR and Coorer JA (1990) Influences of seasonal flooding on macroinvertebrate abundance in wetland habitats. Freshwat Biol 23:311-322

Oertli B (1995) Spatial and temporal distribution of the zoobenthos community in woodland pond (Switzerland) Hydrobiologia 310:189-196

Pickett STA and Cadenasso ML (1995) Landscape ecology: spatial heterogeneity in ecological systems. Science 269:331-334

Rossi V, Schön I, Butlin RK and Menozzi P (1998) Clonal genetic diversity. (pp 257-274) In: Martens K (ed) Sex and Parthenogenesis Evolutionary Ecology of Reproductive Modes in Non-Marine Ostracods. Backhuys Publishers, Leiden

Schneider DW and Frost TM (1996) Habitat duration and community structure in temporary ponds. J North Am Benth Soc 15:64-86

Semyonova LM (1993) Rakushkove ratchki (Ostracoda) basseina Volgi (Musselshrimps (Ostracoda) of the Volga Basin) Trudy Rosijskoy Akad Nauk 68:109-118 (in Russian)

Slack JM, Kaesler RL and Kontrovitz M (2000) Trend, signal and noise in the ecology of Ostracoda: information from rare species in low-diversity assemblages. Hydrobiologia 419:181-189

Svobodová A (1994) Evolution of arm systems and their functional typology (example of the Slovak-Hungarian Danube river reach). Ekológia (Bratislava) 13:369-383

Sywula T (1974) Malzoraczki (Ostracoda). Fauna slodkowodna Polski 24:1-315 (in Polish)

Taylor BE, Leeper DA, McClure MA and DeBiase AD (1999) Ecology of aquatic invertebrates and perspectives on conservation. (pp 167-196) In: Batzer D (ed) Invertebrates in freshwater wetlands of North America: ecology and management

Vranovský M (1997) Impact of the Gabèikovo hydropower plant operation on planktonic copepods assemblages in the River Danube and its floodplain downstream of Bratislava. Hydrobiologia 347:41-49

Wiggins GB, Mackay RJ and Smith IM (1980) Evolutionary and ecological strategies of animals in annual temporary pools. Arch Hydrobiol Suppl 58:97-206

Williams DD (1987) The Ecology of Temporary Waters. Timber Press, Portland, Oregon

EFFECTS OF ABANDONED MINE DRAINAGE ON CRAYFISH DISTRIBUTION ALONG A PH GRADIENT

Beth A. Dillon and Karen T. Lee

*Deptartment of Biology, University of Pittsburgh at Johnstown,
Johnstown, PA 15904 USA*

ktlee@pitt.edu

ABSTRACT

The Kiski-Conemaugh river basin in western Pennsylvania is heavily impacted by drainage from abandoned coal mines. Invertebrate diversity is decreased, with only acid tolerant species inhabiting the most affected streams. Crayfish, which are not acid tolerant, disappear as streams approach pH = 6.0. Little, however, is known about the distribution of crayfish in streams of intermediate pH. We sampled water quality and crayfish and invertebrate abundance at five sites along the pH gradient of Little Paint Creek, near Johnstown, PA. The first site (pH > 8.0) is an unimpacted area upstream from a coal refuse pile. The second site (7.0 < pH < 8.0), at the upstream edge of the pile, has been impacted at some time, but is not impacted now. The remaining sites (pH < 7.0), downstream from the mine, are heavily impacted year round. Crayfish abundance as determined by kick net and hand collecting, declines downstream, from site 1 to site 2. Crayfish were not found at the downstream sites.

I. INTRODUCTION

Abandoned Mine Drainage (AMD) from coal mines is the largest non-point source of pollution of rivers and streams in the Appalachian region of the United States. The Environmental Protection Agency estimates that there are over 6,000 stream miles impacted by AMD in Pennsylvania and West Virginia. The Kiski-Conemaugh Watershed, extending from central to western Pennsylvania, is severely impacted by abandoned mine drainage. It is estimated that over 90% of the surface waters in this watershed are contaminated by AMD (Williams 1991). The Pennsylvania Department of Environmental Protection has estimated the cost of remediation in Pennsylvania alone at $5 billion (Griner 1996).

AMD is the result of water coming into contact with the iron and sulfur rich overburden associated with coal beds in the Appalachians. Chemical reactions take place that release sulfuric acid into the water. The subsequent drop in pH increases the solubility of metal ions including iron, manganese, and aluminum. When the affected water comes into contact with oxygen the metals precipitate in the oxidized forms causing the stream bed and rocks to turn the characteristic red/orange color. In moderate concentrations AMD is toxic to fish and aquatic invertebrates. In high concentrations AMD is lethal.

The general result of the acid inputs is a reduction in species richness in the streams. As species intolerant to acid die off, other acid-tolerant forms may invade and fill the vacant niches (Griffith et al. 1996, Kobuszewski and Perry 1993). Therefore, the benthic macroinvertebrate community is a good indicator of the health of freshwater systems. Research further suggests that the health, abundance, and distribution of crayfish, in particular, could serve as an important biological indicator of the early stages of acidification (France 1984, 1993, Kobuszewski and Perry 1993).

![Kluwer logo] Kluwer Academic/Plenum Publishers

E. Escobar-Briones & F. Alvarez Eds.
MODERN APPROACHES TO THE STUDY OF CRUSTACEA
PP. 275-279

Table 1. *Summary of water quality and aquatic invertebrate data from Little Paint Creek.*

	Site 1	Site 2	Discharge*	Site 3	Site 4	Site 5
Field pH	8.6	8.2	/	6.9	6.3	4.3
Lab pH	8.74	8.44	2.7-3.2	6.73	6.53	4.34
Iron (ppm)	0.32	0.32	800-2700	13.92	8.56	15.76
Sulfate (ppm)	15	16	/	24	24	276
Aluminum (ppm)	0.04	0.14	65-160	0.64	0.24	Over Range
Alkalinity (ppm)	50	45	/	40	35	Below Detection
Copper (ppm)	0.01	Below Detection	/	0.18	0.14	0.13
Turbidity (FTU)	8	9	/	33	41	34
DO (ppm)	8.4	9	/	7.7	7.5	6.1
Temperature (?C)	24	15	/	24.5	18.5	24
Crayfish Density (/m²)	10	0	/	0	0	0
Crayfish Abundance (CPU)	13	3	/	0	0	0
Total # Invertebrates	386	49	/	21	22	27
Invertbrate Families	24	13	/	13	11	8
Invertebrate Density (/m²)	42.9	5.4	/	2.4	2.4	3

* *Discharge Data provided by PA DEP, Pamela Milavec Personal Communication*

Several studies have demonstrated that the distribution of crayfish is correlated with lake pH. In lakes with pH less than 6.0 crayfish abundance is decreased (France 1993). River systems, such as the Cheat River Watershed of West Virginia, often contain no crayfish in stream areas with pH lower than 6.0 (Schwartz and Meredith 1962). However, if crayfish have adapted over the long term to natural additions of acid due to seepage from exposed coal seems, as found in Central and Western Pennsylvania, we can expect to find crayfish populations in streams with low to intermediate levels of mine drainage (Gallaway and Hummon 1991). Some species are more tolerant of low pH than others. The *Cambarus* genus has been found in streams that range from pH of 4.6-4.9. *Cambarus* juveniles will molt and survive at pH 4.0, but the *Orconectes* genus fails to survive in pH of 5.0 or less (Berrill et al. 1985). Crayfish cannot acclimate to sudden decreases in pH. If the pH of a stream dropped to below 5.5 on a regular basis, an *Orconectes* population would go extinct due to mortality of the young (Berrill et al. 1985, France 1984, Gallaway and Hummon 1991). One can assume *Cambarus* would react similarly to lower pH values. For instance, the central portion of the Cheat River Watershed of West Virginia is heavily impacted by acid mine wastes that keep the pH at 4.0 for six to eight months of the year. No crayfish or other aquatic life exist in that zone (Schwartz and Meredith 1962).

Field studies have established the relationship between the abundance of crayfish and other macroinvertebrates and lake pH. However, there is some question whether the effects of pH can be distinguished from other related factors such as water hardness and metal concentrations (Lonergan and Rassmussen 1996).

In addition, most stream studies report crayfish abundance on a presence/absence basis assuming that pH is a threshold variable (Clark and Scruton 1997, Lodge and Hill 1994, Schwartz and Meredith 1962). However, no one thus far has studied the abundance of crayfish in moderately impacted streams, or along a pH gradient within the same stream. Knowledge of the distribution of crayfish in sites of intermediate impact may help to answer the questions about their use as indicator species in acidified water systems.

II. MATERIALS AND METHODS

Site Selection

Little Paint Creek, a headwater stream of the Stonycreek River watershed near the city of Johnstown in West-Central Pennsylvania, provides a perfect location for studying the effects of gradual pH change on crayfish abundance. Little Paint is uncontaminated, with a pH range of 7.4-8.1, minimal amounts of iron (.03- .05 mg/L) and sulfate (7-13 mg/L) and abundant crayfish (Lee et al. unpublished data) until it reaches a large refuse pile left over from Eureka Mine 40 coal mine. At the pile a small, highly acidic discharge (0-45 gpm) enters the stream. The pH of this discharge ranges from 2.7 – 3.2. It also contains high levels of iron (800-

2700 mg/L) and aluminum (65-160 mg/L) (Table 1). The stream pH downstream from the discharge decreases from 8.0 to 7.0. The pH continues to decrease further downstream and pH as low as 5.0 has been recorded (Pam Milavec, DEP, personal communication).

Five sample sites (Table 1) were selected using pH change and visual inspection as an indicator of pollution level. Site 1 has no obvious precipitate of iron, manganese or aluminum and a pH of 8.6. Site 2, which lies just above the principle discharge, contains rocks stained a light orange and an increased level of sedimentation indicating that the area has been impacted by AMD though the pH, 8.2, is only slightly lower than site 1. This represents a moderately impacted site. Site 3, downstream from the discharge, has a considerably lower pH, 6.9, than sites 1 and 2 and the rocks and sediment are dark orange. Site 4 lies downstream from site 3 and is similar in appearance and pH. Site 5 lies approximately one mile downstream of site 4. It is the most heavily impacted site with a pH of 4.3 and extensive sedimentation due to the iron and other heavy metal precipitates.

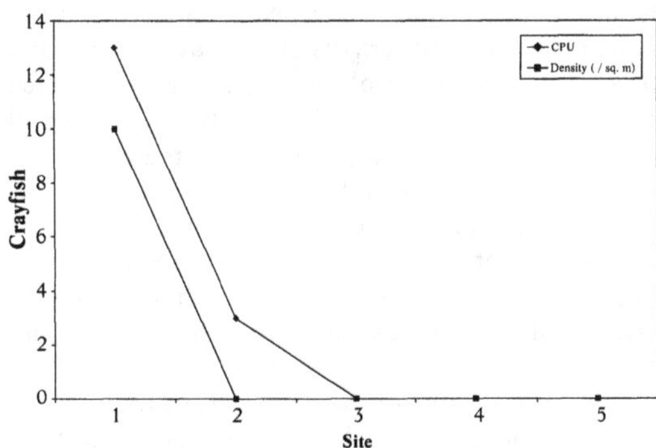

Figure 1. *Crayfish abundance (measured as catch per unit effort) and density at five sites along Little Paint Creek.*

Invertebrate Sampling

All water chemistries and invertebrate sampling were performed during low water conditions in August of 1999. Crayfish and invertebrates were collected using a kick-net and hand collecting. Three 3m² areas were sampled by kick-net at each site: two riffles and one pool. The contents of each kick-net were transferred into storage jars, taken to the lab and preserved in alcohol for later sorting and identi-

fication. The stream invertebrates were identified to the family level. Only intact specimens of 1mm or larger were identified and counted. Data are reported as density of crayfish per square meter, invertebrate density per square meter, number of invertebrate families, and total number of invertebrates.

In addition, crayfish were captured by hand during careful searching under rock and debris, by the same investigator at each site. Each site was carefully searched for 0.5 hours. These data are reported as catch per unit effort.

Water Quality

Water chemistries were measured simultaneous to invertebrate sampling. Tests for nitrates, phosphates, sulfates, iron, aluminum, turbidity, field and lab pH, dissolved oxygen, and total alkalinity were performed using a LaMotte portable SMART colorimeter, an Orion hand-held pH meter, and a hand titration kit for alkalinity (Titrets brand).

III. RESULTS

Crayfish abundance varies from site to site (Table 1, Fig. 1). Crayfish were abundant at site 1. At site 2 few crayfish were collected by hand and none were collected by kick-net. No crayfish were collected by either method at sites 3, 4, or 5. All crayfish collected were *Orconectes obscurus* with the exception of one *Cambarus bartonii* captured at site 1.

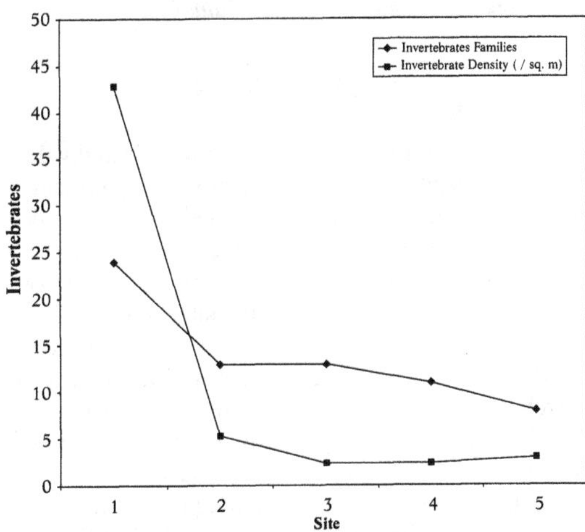

Figure 2. *Invertebrate density at five sites along Little Paint Creek.*

A similar trend is evident in aquatic invertebrate distribution. Aquatic invertebrates, mostly insects, were more numerous, of a higher density, and more diverse (more families) at site 1, somewhat less abundant and less diverse at site 2, and considerably less abundant at sites 3, 4, and 5 (Table 1, Fig. 2).

It is clear from the water chemistry data (Table 1, Fig. 3) that site 1, with relatively high pH and alkalinity and relatively low iron, sulfate, aluminum, copper and turbidity, is a healthy section of stream. Site 2, with a slightly decreased pH and alkalinity and slightly higher sulfate, aluminum, and turbidity is moderately impacted. Sites 3, 4, and 5 are severely impacted with decreased pH and alkalinity and considerably increased levels of iron, sulfate, aluminum, and copper, and higher turbidity.

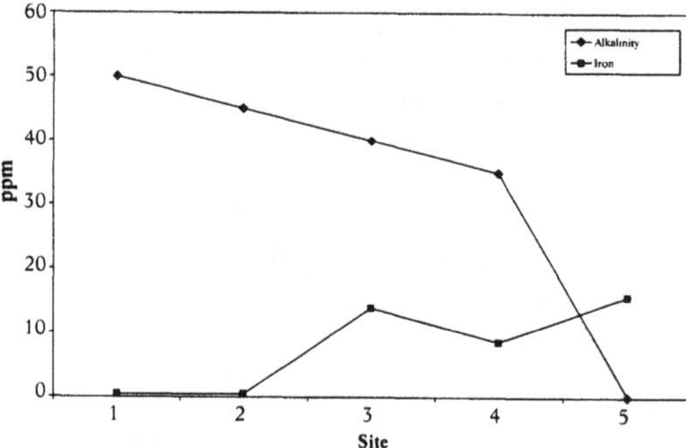

Figure 3. *Selected water chemistry (Alkalinity and Iron concentrations) at five sites along Little Paint Creek.*

IV. DISCUSSION

Sites on Little Paint Creek were initially designated as healthy, intermediate or severely impacted based on visual inspection and field pH. Detailed chemical measurements, and invertebrate and crayfish sampling from sites 1, 3, 4 and 5 support our initial site designations. Healthy sites in central and western Pennsylvania typically support abundant crayfish populations and diverse and abundant invertebrate populations. Sites severely impacted by AMD often cannot support any invertebrates except the most acid tolerant. We found the same phenomenon along a pH gradient in a single stream.

Site 2, however, is not typical. Crayfish and invertebrate abundance and diversity are lower there than at site 1, though water chemistry data suggests that the site could support abundant invertebrates. Invertebrate abundance must be affected by some aspect that our study did not measure. Visual inspection of site 2 indicates that this area has been impacted at some time by AMD, though not impacted during the two weeks of our study.

There are several possible mechanisms that could keep crayfish abundance low at site 2. One possibility is seasonal changes in water quality. In August at low water the chemistries indicate that a crayfish population would be sustainable; however, site 2 may be impacted severely at high water in the spring. Any crayfish living there at that time could be killed or forced to move to another part of the stream. As water levels decrease in the summer, only a few crayfish may repopulate the site.

Another possibility is that crayfish avoid site 2 either because of the precipitate or because they can sense the decreased pH of the water. Crayfish burrow into the streambed and take shelter under rocks and debris. Perhaps the iron and other heavy metal precipitate makes site 2 a less desirable place to live due to excess sedimentation, or the heavy metal precipitate may be detrimental to crayfish health. Whether crayfish would avoid site 2 due to slightly decreased pH is unknown, but France (1985) suggests that crayfish are able to detect and actively avoid acidified portions of streams.

A final possibility is decreased food availability. The total number of invertebrates and their diversity is appreciably decreased at site 2, decreasing the available food for crayfish. In addition decreased pH interferes with crayfish chemoreceptors (Tierney and Atema 1986, Allison et al. 1992), making it difficult for crayfish to find food.

One of the problems with using crayfish as biological indicators has been the lack of clear data on crayfish populations at sites that are moderately impacted by abandoned mine drainage, but not destroyed. In Little Paint Creek, the intermediate site is also intermediate in abundance and diversity of aquatic invertebrates, but the mechanism for the change in the invertebrate community is not clear. Future research will explore the possibility of seasonal differences in water quality and its effects on stream health and crayfish distribution and investigate other streams with intermediate AMD sites.

ACKNOWLEDGEMENTS

This project was funded by grants from the Central Research Development Fund of the University of Pittsburgh (K.L.), and the University of Pittsburgh at Johnstown (UPJ) Undergraduate Research Initiative (B.D.). Additional support was provided by the Biology and Geology Departments at UPJ. Rick Harckom and Tony Sossong provided access to field sites on private property.

Proof correction: Due to a typographical error, the data for the crayfish density at site one (Table 1) is incorrectly reported as 10. The correct figure is 1.1. This error was carried over into figure 1 as well. Please note, however that this change in the data does not ultimately change the trends reported in this paper, but only the magnitude of changes in crayfish density. For corrected table and figure please contact the author.

REFERENCES

Allison V, Dunham DW and Harvey HH (1992) Low pH alters response to food in the crayfish *Cambarus bartoni*. Can J Zool 70:2416-2420

Berrill M, Hollett L, Margosian A and Hudson J (1985) Variation in tolerance to low environmental pH by the crayfish *Orconectes rusticus, O. propinquus* and *Cambarus robustus*. Can J Zool 63:2586-2589

Clarke KD and Scruton DA (1997) The benthic community of stream riffles in Newfoundland, Canada and its relationship to selected physical and chemical parameters. J Fresh Ecol 12:113-121

France R (1993) Influence of lake pH on the distribution, abundance, and health of crayfish in Canadian shield lakes. Hydrobiologia 271:65-70

France R (1985) Low pH avoidance by crayfish (*Orconectes virilis*): Evidence for sensory conditioning. Can J Zool 63:258-262

France R (1984) Comparative tolerance to low pH of three life stages of the crayfish *Orconectes virilis*. Can J Zool 62:2360-2363

Gallaway MS and Hummon WD (1991) Adaptation of *Cambarus bartonii cavatus* to acid mine polluted waters. Ohio Jour Sci 91:167-171

Griffith MB, Wolcott LT and Perry S (1996) Production of the crayfish *Cambarus bartonii* in an acidic Applachian stream. Crustaceana 69:974-984

Griner B (1996) A Conjoint analysis of water quality enhancements and degradations in a Western Pennsylvania watershed. http://info.pitt.edu/~griner/ov7.html

Kobuszewski D and Perry S (1993) Aquatic insect community structure in an acidic andcircumneutral stream in the Appalachian Mountains of West Virginia. J Freshwat Ecol 8: 37-45

Lodge DM and Hill AM (1994) Factors governing species composition, population size, and productivity of cool-water crayfishes. Nord J Fresh Res 69:111-136

Lonergan SP and Rasmussen JB (1996) A multi-taxonomic indicator of acidification: Isolating the effects of pH from other water chemistry variables. Can J Fish Aquat Sci 53:1778-1787

Schwartz FJ and Meredith WG (1962) Crayfishes of the Cheat River Watershed in West Virginia and Pennsylvania part II: Observations upon ecological factors relating to distribution. Ohio Jour Sci 62:260-273

Tierney AJ and Atema J (1986) Effects of acidification on the behavioral response of crayfishes (*Orconectes virilis* and *Procambarus acutus*) to chemical stimuli. Aquat Toxicol 9:1-11

Williams BT (1991) Coal Dust in Their Blood: The Work and Lives of Underground Coal Miners. AMS Press, NY

LARGE FRESHWATER BRANCHIOPODS IN AUSTRIA: DIVERSITY, THREATS AND CONSERVATIONAL STATUS

Erich Eder and Walter Hödl

Institute of Zoology, University of Vienna, Althanstr. 14, A-1090 Vienna, Austria

Erich.Eder@univie.ac.at

ABSTRACT

For the first time, Anostraca, Notostraca and Conchostraca (Laevicaudata, Spinicaudata) will be listed in the Austrian Red Data Book of Endangered Species. According to the 1999 IUCN draft criteria, eight out of 15 large branchiopod species recorded between 1994 and 1999 are considered as critically endangered, three of which (*Chirocephalus shadini, Eoleptestheria ticinensis, Streptocephalus torvicornis*) occur at only one site each. Five species are considered as endangered, two of them (*Branchinecta orientalis, Triops cancriformis*) showing a statistically significant decrease of sites. *Eubranchipus grubii* and *Lepidurus apus*, both found abundantly mainly in the flood plains along the rivers Thaya and Morava, are near threatened. *Lynceus brachyurus* is extinct in Austria. Large branchiopods are mainly threatened by agricultural activities and hydrological/hydrochemical changes. Presently, three Austrian locations are protected exclusively on the basis of large branchiopod occurrence; three additional habitats are in the process of obtaining official protection. Several large populations are situated in the WWF nature reserve "March-Auen", and in the National Parks "Donau-Auen" and "Neusiedler See - Seewinkel", respectively.

I. INTRODUCTION

The Austrian share of the Pannonian region, of the Danube as well as of the Morava, the westernmost steppe river of Europe, results in a remarkable diversity of large branchiopods within a small area. Ranges of species with holarctic, Eurasian, European, Pannonian, and southern European distribution overlap in Austria due to its specific geographic position (Eder et al. 1997). Sixteen large branchiopod species belonging to 14 genera have been reported from Austria (Vornatscher 1968, Löffler 1993). Large branchiopods are considered to be endangered throughout Europe (Alonso 1985, Brendonck 1989, Mura 1993, Petrov and Petrov 1997, Defaye et al. 1998, Maier 1998). The primary threats are comparable to those threatening aquatic insects (Polhemus 1993). Physical destruction due to agricultural development, changes of hydrologic conditions and urbanisation play a major role (Rieder 1989, Löffler 1993, Hödl and Eder 1996a). "Invertebrates traditionally attract very little conservation funding in relation to vertebrates, although they may play more subtle and significant ecological roles and monitor environmental change better" (New 1993). Large branchiopods have proven to attract public interest as "primeval shrimps" (Hödl and Eder 1996a) as well as due to their extreme ecology, and thus may help to propagate the relevance of invertebrates for conservation biology (Eder and Hödl 1996a). Recent public discussion on the biodiversity crisis and public awareness for the protection of endangered species favour effective conservational measures. Conservation authorities, however, increasingly demand quantitative data for valid assessments of extinction risks as basis of Red Data Books and legal actions (Müller-Motzfeld 1990, Gruttke et al. 1999). The aim of this study is to provide a quantitative evaluation of the status, threats

E. Escobar-Briones & F. Alvarez Eds.
MODERN APPROACHES TO THE STUDY OF CRUSTACEA
PP. 281-289

and conservation of Anostraca, Notostraca and Conchostraca, as a prerequisite for the planned Red Data Book of Endangered Species in Austria (Zulka et al. 2000). Main threatening factors are discussed, multispecies large branchiopod sites are ranked according to their relevance for conservational purposes, and the status of large branchiopod habitats and recent conservation activities by the authors are presented.

II. METHODS

Single pools, a group of water bodies within a radius of 100 m or connected at least once during a sampling visit are treated as one geographical unity (="site"). To quantify the status, threats, and causes of potential local extinction, all undoubtedly identifiable locations historically reported in Austria until 1990 (117) were revisited by us at least once in the ten year observation period (1991 - 2000), primarily between 1994 and 1996 (cf. Hödl and Eder 1996a). Eleven sites were insufficiently described and could not be inspected. In order to assess the current status for each species, a thorough search for potential other large branchiopod sites was undertaken, including appeals to the public through printed media and local exhibitions. Our field sampling focused on the Lower Austria and Burgenland provinces, where 288 (90,8%) large branchiopod occurrences were reported up to 1990. For each species, the number of entirely destroyed localities, intact sites, and locations under legal protection (National Park and nature reserves) was determined. The number of sites reported up to 1990 was compared for each species with the number of recent sites by means of Pearson c^2-test. The observed 95% significance level in all 16 tests was adjusted using Bonferroni correction (Sokal and Rohlf 1995, Milasowszky and Zulka 1998), i.e. $P=0.05/16=0.003125$.

III. RESULTS

To the best of our knowledge, a total of 318 reports of large branchiopod occurrences from 128 Austrian sites, according to the above definition, were known up to 1990. Seventy nine localities were reported in accessible publications (Brehm 1910, Puschnig 1918, Machura 1935, Pesta 1937, 1939, 1942, Kupka 1940, Heikertinger 1951, Nemec 1952, Vornatscher 1955, 1968, Löffler 1957, 1959, 1993, Sampl 1969, Metz and Forró 1987, Paar et al. 1993). Information about

Table 1. *Austrian large branchiopod sites, listed by species. Number of sites and protectional status. Sites protected by general conventions (such as Ramsar) are not included. * Statistically significant (P<0.003125) decrease, ** statistically significant increase of inhabited sites. (**Branchinecta** sp. site numbers up to 1990 are given in parentheses due to the lack of differentiation between **B. ferox** and **B. orientalis**. For statistical reasons, the records up to 1990 were split up according to the relation of still intact sites) Old: sites reported up to 1990, new: sites found 1991—2000.*

taxon	Old	intact	not vis.	new	recent	reason for local extinction	protected	aimed	% prot.
Anostraca									
Branchinecta ferox	(8)	3	0	0	3	hydrological changes / fish introduction	1	0	33
Branchinecta orientalis	(24)	9	0	1	10*	hydrological changes / fish introduction	7	0	70
Branchipus schaefferi	29	5	5	31	36	agriculture / road construction	7	1	22
Chirocephalus carnuntanus	2	0	0	2	2	hydrological changes / unknown	2	—	100
Chirocephalus shadini	1	1	0	0	1		1	—	100
Eubranchipus grubii	8	3	0	57	60**		9	0	15
Streptocephalus torvicornis	5	0	0	1	1	urbanisation / unknown	0	0	0
Tanymastix stagnalis	5	0	1	3	3	urbanisation / unknown	0	1	33
Notostraca									
Lepidurus apus	20	9	0	69	78**		16	2	23
Triops cancriformis	42	5	3	26	31*	agriculture / urbanisation	6	1	23
Conchostraca									
Cyzicus tetracerus	5	1	0	4	5	urbanisation / agriculture	3	1	80
Eoleptestheria ticinensis	1	0	0	1	1	urbanisation	1	—	100
Imnadia yeyetta	12	1	1	22	23	agriculture / hydrological changes	9	1	44
Leptestheria dahalacensis	18	3	1	11	14	agriculture / hydrological changes	3	1	29
Limnadia lenticularis	9	0	1	11	11	urbanisation / agriculture	2	0	18
Lynceus brachyurus	9	0	2	0	0*	unknown	0	—	—

Table 2. *Multispecies large branchiopod habitats(>2 species) in Austria.*

Habitat	province	Location	protectional status	species documented
"Blumengang" near Engelhartstetten	Lower Austria	Morava/Danube 48°10"42" N, 16°58"00" E	nature reserve	*C. tetracerus, E. ticinensis, I. yeyetta, L. dahalacensis, L. lenticularis, T. cancriformis*
"Triops-Senke" (Lange Lüsse)	Lower Austria	Lower Morava 48°14'24" N, 16°56'25" E	protection aimed	*C. tetracerus, I. yeyetta, L. apus, L. dahalacensis, T. cancriformis*
Pond near "Steinbrunner See"	Burgenland	Wiener Becken 47°50'04" N, 16°23'20" E	—	*B. schaefferi, S. torvicornis, L. dahalacensis, T. cancriformis*
"Kohler-Lacke" near Apetlon	Burgenland	Seewinkel 47°45'23" N, 16°50'50" E	national park	*B. schaefferi, I. yeyetta, L. dahalacensis, T. cancriformis*
"August-Lacke" (Lange Lüsse)	Lower Austria	Lower Morava 48°14'23" N, 16°56'12" E	—	*I. yeyetta, L. dahalacensis, L. lenticularis, T. cancriformis*
"T/L-Lacke" (Lange Lüsse)	Lower Austria	Lower Morava 48°13'11" N, 16°57'05" E	—	*I. yeyetta, L. apus, L. lenticularis, T. cancriformis*
"Alte Schanzen" near Parndorf	Burgenland	Parndorfer Platte 47°58'40" N, 16°51'20" E	protection aimed	*B. schaefferi, T. stagnalis, T. cancriformis*
"Pulverturm" ponds near Marchegg	Lower Austria	Lower Morava 48°16'30" N, 16°55'00" E	nature reserve	*C. shadini, I. yeyetta, L. apus*
Meadow near Wörten-Lacke	Burgenland	Seewinkel 47°46'32" N, 16°51'18" E	national park	*B. schaefferi, I. yeyetta, L. dahalacensis*
Gramatneusiedl	Lower Austria	„Wiener Becken" 48°01'15" N, 16°29'35" E	nature reserve	*B. schaefferi, C. tetracerus, T. cancriformis*
"Tanymastix-Lacke" near Wörten-Lacke	Burgenland	Seewinkel 47°46'40" N, 16°51'13" E	—	*I. yeyetta, T. stagnalis, T. cancriformis*
Field south of Markthof	Lower Austria	Lower Morava 48°11'16" N, 16°57'29" E	—	*B. schaefferi, I. yeyetta, T. cancriformis*
"Kreuzlacke" south of Markthof	Lower Austria	Lower Morava 48°11'21" N, 16°57'07" E	—	*B. schaefferi, I. yeyetta, T. cancriformis*
"Dammwiese" near Marchegg	Lower Austria	Lower Morava 48°16'45" N, 16°54'20" E	none (protected by a local agreement)	*I. yeyetta, L. apus, T. cancriformis*
"Loimersdorfer Wiesen" near Markthof	Lower Austria	Lower Morava 48°10'47" N, 16°57'10" E	—	*I. yeyetta, L. dahalacensis, T. cancriformis*
Field south of Engelhartstetten	Lower Austria	Lower Morava 48°10'26" N, 16°53'27" E	—	*I. yeyetta, L. dahalacensis, T. cancriformis*
"Schlosslacke" (Lange Lüsse)	Lower Austria	Lower Morava 48°13'05" N, 16°57'20" E	—	*I. yeyetta, L. lenticularis, T. cancriformis*
"Brachesenke" (Lange Lüsse)	Lower Austria	Lower Morava 48°13'10" N, 16°57'05" E	—	*I. yeyetta, L. lenticularis, T. cancriformis*
"Hoffnungslacke" (Lange Lüsse)	Lower Austria	Lower Morava 48°13"05" N, 16°57"30" E	—	*I. yeyetta, L. lenticularis, T. cancriformis*

the remaining 49 sites was obtained from unpublished reports (Marschitz and Käfel 1993), unpublished theses (Jungwirth 1973, Jahn 1981, Linder 1983, Lechthaler 1993), personal communications (E. Christian, Gartner, J. Gruber, R. Ille, B. Kohler, I.Korner, G. Lutschinger, M. Maslo, H. Palme, K. Schütz, U. Tessenow), and historical field notes from J. Vornatscher stored at the Natural History Museum of Vienna (NHMW). The species with most records were *Triops cancriformis* (Bosc, 1801), *Branchipus schaefferi* Fischer, 1834, *Branchinecta* sp. and *Lepidurus apus* (L., 1758) (Table 1).

Except *Lynceus brachyurus* Müller, 1776, all historically reported species were documented in Aus-

tria within the last ten years (Table 1, cf. Eder et al. 1997). Thus, almost one quarter of the 68 European large branchiopod species and two thirds of the 22 European genera (Brtek and Thiéry 1995) are present on less than 1% of the European land mass.

The highest numbers of recent sites were recorded for *L. apus* (78) and *Eubranchipus grubii* (Dybowski, 1869) (60), two species abundantly occurring in the Morava river flood plains. More than 20 recent sites were documented for *B. schaefferi, T. cancriformis*, and *Imnadia yeyetta* Hertzog, 1935, mainly in the lower Morava and Danube river flood plains, the Seewinkel region and the Wiener Becken depression. All other species occur at less than 15 sites, three of which are known from only a single site each (Table 1). Besides the extinct *L. brachyurus*, decrease of recent sites was statistically significant for two species, *Branchinecta orientalis* (G. O. Sars, 1901) and *T. cancriformis* (Table 1).

Almost one third of the Austrian large branchiopod sites is inhabited by more than one species, 18 sites by more than two (see Table 2). The most frequent co-occurrences are: *T. cancriformis* with either *I. yeyetta*, *Leptestheria dahalacensis* (Rüppell, 1837) or *Limnadia lenticularis* (L., 1761); *L. apus* with *E. grubii*; and *I. yeyetta* with either *L. dahalacensis* or *L. lenticularis* (Gottwald and Eder 1999).

Reasons for local extinction, as far as it is known, were physical destruction due to agricultural development, changes of hydrologic conditions (dikes, hydroelectric power plants), and urbanisation (including litter disposal, road constructions and recreational measures) (cf. Table 1). Until 1970, none of the Austrian large branchiopod occurrences was under legal protection. When the WWF bird sanctuaries "Marchauen-Marchegg" and "Breitensee" in the Morava inundation area were declared nature reserves in 1970, populations of the two most common Austrian large branchiopod species *E. grubii* and *L. apus* fell under local protection. In 1982, the "Pulverturm" ponds near the city of Marchegg, westernmost and single Austrian site of *Chirocephalus shadini* (Smirnov, 1928), became the world's first area declared as a nature reserve based solely on large branchiopod occurrence due to the initiative of one of the authors (Hödl 1994). In 1994, the first Austrian National Park meeting the IUCN criteria, "Neusiedler See - Seewinkel", was inaugurated. By this measure, both known Austrian sites of *Chirocephalus carnuntanus* (Brauer, 1877), occurrences of *Branchinecta ferox* (Milne-Edwards, 1840), *B. schaefferi*, *I. yeyetta*, *L. dahalacensis* and *T. cancriformis*, and most sites of *Branchinecta orientalis* G. O. Sars, 1901, came under protectional status (Eder et al. 1996). In 1996, the declaration of the National Park "Donau-Auen" (Manzano 2000) gave protection to two habitats with co-occurrences of *L. apus* with *L. lenticularis*, and *T. cancriformis* with *I. yeyetta*, respectively (Eder and Hödl 1996a).

In the course of our efforts for large branchiopod conservation, a broad public audience was addressed by a brochure (Hödl and Rieder 1993), more than 20 articles in local journals, and a popular science book (Hödl and Eder 1996b). The book, out of print

in 1998, was followed up by a revised and updated CD-ROM edition (Eder 1999a). Additionally, an exhibition ("primeval shrimps in Austria – living fossils in short-lived waters") was set up and shown in the museums of Marchegg (1996-1997) and Illmitz (1999), located in the main large branchiopod distribution areas of the Morava flood plains and Neusiedler See-Seewinkel, respectively. Media reactions followed with interviews for TV (7), radio programs (8), and newspapers (>30). A natural history film on large branchiopods worldwide, including scenes from Austrian populations, supported our conservational intentions (Bludszuweit et al. 1996). Public awareness caused the accelerated legal procedure regarding the protection of the most diverse Austrian large branchiopod site: on June 19, 1996, the 7.5 ha "Blumengang" site, habitat of six large branchiopod species, among them all spinicaudatan species known from Austria, was declared a nature reserve due to the initiative of the authors (Hödl and Eder 1996c). Conservation proposals for three additional large branchiopod sites (seven species) are presently treated by the responsible local authorities. If successful, with the exception of *Streptocephalus torvicornis* (Waga, 1842) and *L. brachyurus*, at least one locality of each Austrian large branchiopod species will face the strictest legal nature protection available (Table 1).

None of the large branchiopod species is listed in the presently valid Austrian Red Data Book of Endangered Species (Gepp 1994). Recently, large branchiopods were listed in local Red Data Books for the provinces Carinthia (Eder 1999b) and Lower Austria (Hödl and Eder 2000), and were recommended for the upcoming new version of the Austrian Red Data Book (Zulka et al. 2000).

IV. DISCUSSION

According to the statistical data, four different species groups are discussed:

(1) Two species, *E. grubii* and *L. apus*, show a remarkable increase of site records, which is due to the extensive quantitative sampling along the Morava river in the course of recent projects (Marschitz and Käfel 1993, Hödl and Rieder 1993). The abundant occurrence of these species along the Morava river was obvious to earlier investigators (J. Vornatscher †, pers. comm.) who did not go into detail documenting each encountered site (e.g., "Morava River", given in Vornatscher 1968).

(2) Species with no statistically significant increase or decrease of site numbers are *B. schaefferi, I. yeyetta, L. dahalacensis,* and *L. lenticularis.* These species are widespread in Europe (*B. schaefferi*), Eurasia (*L. dahalacensis*) or the Pannonian region (*I. yeyetta*), or show a holarctic distribution (*L. lenticularis*) (Dumont et al. 1995).

(3) *Branchinecta orientalis* and *T. cancriformis,* abundant along the Seewinkel region and the lower Morava river, respectively, show a statistically significant decrease of sites, while *L. brachyurus* is extinct. The decrease of *B. orientalis,* a species limited to steppic pools in central Spain, the Pannonian lowlands and Eastern Europe, is a result of the degradation of the alkaline pans in the Seewinkel region (Löffler 1993) which has stopped with the declaration of the National Park "Neusiedler See-Seewinkel" in 1994. For *T. cancriformis,* a well known species widely distributed all over Europe, the data reflect the general European trend of large branchiopod decrease. More than 80% of the historically reported sites for this species were destroyed, mainly by agriculture and urbanisation.

(4) Seven species occur at less than six sites (Table 1). With the exception of *Tanymastix stagnalis* (L., 1758), Austria lies within either their westernmost or northernmost distribution boundaries (Dumont et al. 1995). Due to the low numbers, the negative trend of *B. ferox* (cf. *B. orientalis*), *S. torvicornis* and *T. stagnalis* is not statistically significant.

In species producing long-lived permanent stages, cyst banks may remain in the soil for years or even decades, which complicates the evaluation for Red Data Books. Re-appearance of rare species such as *Cyzicus tetracerus* (Krynicki, 1830) at habitats where conditions have recently changed, could be either re-immigration (e.g., by birds), or hatching of cysts that have been present in the soil for years without developing due to suboptimal conditions. The irregular but widespread occurrence of *B. schaefferi* outside the main large branchiopod areas indicates the possibility of inactive cyst banks present at sites still unknown for large branchiopod occurrence in Austria.

The known ranges of smaller, less charismatic invertebrates often more accurately reflect the distribution of their experts than that of the animals themselves (Dumont et al. 1995), an aspect to be considered when comparing historic and recent distribution data. As the geographic range of historical and recent investigations are approximately the same, no bias is expected for the evaluation of large branchiopod status in eastern Austria. However, the necessity of additional faunistic research throughout the whole country is demonstrated by new reports from Upper Austria (W. Weißmair and R. Gottwald, unpubl.), Carinthia (Sampl and Fressner, unpubl.), and from Lower Austrian regions so far unknown for large branchiopod occurrence (T. Hochebner unpubl., R. Gottwald unpubl.).

The prospective evaluation mode of the Austrian Red Data Book (Zulka et al. 2000) follows the criteria by Schnittler et al. (1994), modified according to Gärdenfors et al. (1999) to provide IUCN-compatible results. Eight of the 16 large branchiopod species known from Austria are considered as critically endangered (CR), five species are endangered (EN), the two most common species near threatened (NT). One species is considered to be extinct in Austria (RE; EX according to IUCN 1994) (Table 3).

Most Austrian meadows result from long-term use as grassland for cattle. Traditional cultivation of these semi-natural habitats is needed to preserve fauna and flora diversity (cf. Rieder 1989). Due to changes in agricultural policy during the last decades, large meadow areas along the lower Morava river have become farm land. As a consequence, development led to the physical destruction of large branchiopod habitats through filling up or drainage of wetlands. Aiming the protection of private property as natural reserves leads to conflicts with land owners. Without adequate information strategies, farmers may refuse conservational measures (W. Suske, unpubl.). Since 1987, the "Verein zur Erhaltung und Förderung ländlicher Lebensräume (Distelverein)" tries to ensure "wise use" or renaturation of valuable anthropogenic habitats along the Morava river, including large branchiopod habitats, by direct payments to the local farmers (Schlederer 1999). Extirpation due to pollution has not been reported for large branchiopods in Austria. According to Owens et al. (2000), the effect of pesticides is lower for Anostraca (*Branchinecta* sp.) than for other aquatic organisms (Cladocera), but still needs a closer examination. In the Morava river flood plains, *Bacillus thuringensis israelensis* (BTI) is used against mosquitoes (B. Seidel, unpubl.). Neither experimental nor field data are available about possible effects of BTI on large branchiopods.

The storage lakes "Nove Mlyny" of the Thaya-River in Southern Moravia, built in 1989, reduce the dynamics of the Morava river's highwaters. Crossing the Morava river, the planned Danube-Oder-Elbe-Ca-

Table 3. *Evaluation of Austrian large freshwater branchiopod species according to the IUCN Red List draft Categories (Gärdenfors et al., 1999, and adaptation by Zulka et al., 2000), considering data until October 2000. VIE Vienna, LOA Lower Austria, BUR Burgenland, UPA Upper Austria, SAL Salzburg, STY Styria, CAR Carinthia. No large branchiopod records are known from the westernmost provinces Tyrol and Vorarlberg.*

Taxon	VIE	LOA	BUR	UPA	SAL	STY	CAR	Austria
Anostraca								
Branchinecta ferox (Milne-Edwards, 1840)	—	—	CR	—	—	—	—	CR
Branchinecta orientalis G.O.Sars, 1901	—	—	EN	—	—	—	—	EN
Branchipus schaefferi Fischer, 1834	CR	CR	EN	RE	RE	RE	—	EN
Chirocephalus carnuntanus (Brauer, 1877)	—	—	CR	—	—	—	—	CR
Chirocephalus shadini (Smirnov, 1928)	—	CR	—	—	—	—	—	CR
Eubranchipus grubii (Dybowski, 1860)	—	NT	CR	CR	—	—	CR	NT
Streptocephalus torvicornis (Waga, 1842)	RE	—	CR	—	—	—	—	CR
Tanymastix stagnalis (Linnaeus, 1758)	—	RE	CR	—	RE	—	—	CR
Notostraca								
Lepidurus apus (Linnaeus, 1758)	—	NT	CR	—	—	—	CR	NT
Triops cancriformis (Bosc, 1801)	CR	EN	EN	RE	RE	—	RE	EN
Conchostraca								
Cyzicus tetracerus (Krynicki, 1830)	RE	CR	RE	—	—	—	—	CR
Eoleptestheria ticinensis (Balsamo-Crivelli, 1859)	RE	CR	—	—	—	—	—	CR
Imnadia yeyetta Hertzog, 1935	—	EN	EN	CR	—	—	—	EN
Leptestheria dahalacensis (Rüppell, 1837)	RE	EN	EN	RE	—	—	—	EN
Limnadia lenticularis (Linnaeus, 1761)	RE	CR	RE	—	—	—	—	CR
Lynceus brachyurus Müller, 1776	—	RE	RE	—	—	—	—	RE

nal, 40 km on its Austrian course (Müller 2000), is considered a major threat to wetland habitats. Effects on water level and hydrologic dynamics of both Morava and Danube rivers are unknown.

To determine conservational priorities, large branchiopod occurrences were ranked using the main criteria listed by Usher and Erz (1994). Local diversity was the decisive factor, prior to species rareness. Coincidently, rarest Austrian species always occur at multispecies habitats, e.g., *Eoleptestheria ticinensis* (Balsamo-Crivelli, 1859) at the "Blumengang" depression (cf. Table 2). Five sites out of the 10 highest ranked habitats are protected (Table 2). The final decision about the aimed nature reserve "Triops-Senke" is expected within the next year, further conservational activities are planned.

ACKNOWLEDGEMENTS

We are grateful to all persons who supported our conservational activities, first of all to E. Kraus (Provincial Government of Lower Austria, dep. RU-5) for his engagement. D. Belk[†] (IUCN), W. Haas, W. Kaffarek, E. Neumeister, M. Pöckl and A. M. Sturm provided friendly and professional advice. G. Bieringer, R. Gottwald, H. Groß, E. Hable, T. Hochebner, E. Klotz, E. Kusel-Fetzmann, N. Milasowzky, G. Navara, I. Oberleitner-Fischer, L. Paulssen, R. Plöchl, E. Rieder, H. Sampl, T. Schlögl, N. Weißenböck, W. Weißmair, A. Welzl and T. Zuna-Kratky reported new large branchiopod occurrences in Austria. Finally, we appreciate the support in statistical analysis by N. Milasowzky and P. Zulka.

REFERENCES

Alonso M (1985) A survey of the Spanish Euphyllopoda. Misc Zool (Barcelona) 9:179-208

Bludszuweit G, Haft J and Riehl I (1996) "Heimische Urzeitkrebse". Zur Konzeption eines Dokumentarfilmes. Stapfia 42:159-165

Brehm V (1910) Seltene Phyllopoden von Pöchlarn in Niederösterreich. Arch Hydrob 6: 206-208

Brendonck L (1989) A review of the phyllopods (Crustacea: Anostraca, Notostraca, Conchostraca) of the Belgian fauna. Verh Symp "Invertebraten van Belgie":129-135

Brtek J and Thiéry A (1995) The geographical distribution of the European Branchiopods (Anostraca, Notostraca, Spinicaudata, Laevicaudata). Hydrobiologia 298:263-280

Defaye D, Rabet N and Thiéry A (1998) Atlas et bibliographie des crustacés branchiopodes (Anostraca, Notostraca, Spinicaudata) de France métropolitaine. MNHN, Paris

Dumont H, Mertens J and Maeda-Martinez AM (1995) Historical biogeography and morphological differentiation of *Streptocephalus torvicornis* (Waga) since the Würm III-glaciation. Hydrobiologia 298:281-286

Eder E (ed) (1999a) Urzeitkrebse - Lebende Fossilien. CD-ROM. Eigenverlag Eder-Steiner, Wien

Eder E (1999b) Rote Liste der Rückenschaler Kärntens (Crustacea: Branchiopoda: Notostraca). In: Rottenburg T, Wieser C, Mildner P and Holzinger WE (eds) Rote Listen gefährdeter Tiere Kärntens. Naturschutz in Kärnten 15:535-538

Eder E and Hödl W (1996a) Wozu "Urzeitkrebse"? Praktische Bedeutung der Groß-Branchiopoden für Wirtschaft, Naturschutz und Wissenschaft. Stapfia 42:149-158

Eder E and Hödl W (1996b). Die Groß-Branchiopoden der österreichischen Donau-Auen. Stapfia 42:85-92

Eder E, Hödl W and Milasowszky N (1996) Die Groß-Branchiopoden des Seewinkels. Stapfia 42:93-101

Eder E, Hödl W and Gottwald R (1997) Distribution and phenology of large branchiopods in Austria. Hydrobiologia 359:13-22

Gärdenfors U, Rodriguez JP, Hilton-Taylor C, Hyslop C, Mace G, Molur S and Poss S (1999) Draft guidelines for the application of IUCN Red List criteria at national and regional levels. Species 31/32:58-70

Gepp J (ed) (1994) Rote Listen gefährdeter Tiere Österreichs. Grüne Reihe des Bundesministeriums für Umwelt, Jugend und Familie, Band 2, Styria, Graz

Gottwald R and Eder E (1999) "Co-occurrence" - ein Beitrag zur Synökologie der Groß-Branchiopoden. Ann Naturhist Mus Wien 101B:465-473

Gruttke H, Ludwig G, Binot-Hafke M and Riecken U (1999) Perspektiven bundesweiter Roter Listen. Ergebnisse eines Symposiums des Bundesamtes für Naturschutz. Natur u Landschaft 74:281-284

Heikertinger F (1951) Erinnerungen an den Laaerberg von einst. Natur u Land 37:68

Hödl W (1994) A short review of the Anostraca, Notostraca, Laevicaudata and Spinicaudata of Austria. IUCN Anostraca News 2/1:2-3

Hödl W and Eder E (1996a) Rediscovery of *Leptestheria dahalacensis* and *Eoleptestheria ticinensis* (Crustacea: Branchiopoda: Spinicaudata): an overview on presence and conservation of clam shrimps in Austria. Hydrobiologia 318: 203-206

Hödl W and Eder E (eds) (1996b) Urzeitkrebse Österreichs. Stapfia 42, zugleich Kataloge des OÖ. Landesmuseums N. F. 100, Linz

Hödl W and Eder E (1996c) Die "Blumengang"-Senke: Chronologie eines Schutzgebietes für "Urzeitkrebse". Stapfia 42:71-84

Hödl W and Eder E (2000) "Urzeitkrebse" (Branchiopoda: Anostraca, Notostraca, Conchostraca). 1.Fassung 1999. (pp 4-33) In: Rote Listen ausgewählter Tiergruppen Niederösterreichs Amt d. NÖ. Landesregierung, St.Pölten

Hödl W and Rieder E (1993) Urzeitkrebse an der March. Verein zur Erhaltung und Förderung ländlicher Lebensräume (Distelverein), Orth/Donau

IUCN (1994) IUCN Red List categories. Gland, Switzerland

Jahn W (1981) Untersuchungen zur Entwicklungs- und Fortpflanzungsbiologie von *Chirocephalus grubei* Dyb.(1860) und *Chirocephalus shadini* Smirnov (1928). Unpublished Masters thesis, Univ Vienna

Jungwirth M (1973) Populationsdynamik und Populationsrate von *Branchinecta orientalis* (G.O.Sars) in der Birnbaumlacke (Seewinkel, Burgenland) unter besonderer Berücksichtigung der limnologischen Bedingungen des Gewässers. Unpublished PhD thesis, Univ Vienna

Kupka E (1940) Untersuchungen über die Schalenbildung bei den Eiern von *Branchipus schaefferi* Fischer. Zool Anz 132:130-139

Lechthaler W (1993) Gesellschaften epiphytischer Makroevertebraten in überschwemmten Wiesen an der March (Niederösterreich). Unpublished PhD thesis, Univ Vienna

Linder W (1983) Entwicklung und Biologie von *Lepidurus apus*. Unpublished Masters thesis, Univ Vienna

Löffler H (1957) Vergleichende limnologische Untersuchungen an den Gewässern des Seewinkels (Burgenland). Verh Zool Bot Ges Wien 97:27-52

Löffler H (1959) Zur Limnologie, Entomostraken- und Rotatorienfauna des Seewinkelgebietes (Burgenland, Österreich). Sber Österr Akad d Wiss, math nat Kl Abt I 168:315-362

Löffler H (1993) Anostraca, Notostraca, Laevicaudata and Spinicaudata of the Pannonian region and in its Austrian area. Hydrobiologia 264:169-174

Machura L (1935) Ökologische Studien im Salzlackengebiet des Neusiedlersees, mit besonderer Berücksichtigung der halophilen Koleopteren- und Rynchotenarten. Z wiss Zool 146:555-590

Maier G (1998) The status of large branchiopods (Anostraca, Notostraca, Conchostraca) in Germany. Limnologica 28:223-228

Manzano C (2000) Großräumiger Schutz von Feuchtgebieten im Nationalpark Donau-Auen. Stapfia 69:229-248

Marschitz G and Käfel G (1993) Über das Vorkommen anostraker und notostraker Krebse an den Flüssen Thaya und Obere March. Ergebnisse 1993. Unpublished report, Verein zur Erhaltung und Förderung ländlicher Lebensräume, Orth/Donau

Metz H and Forró L (1989) Contributions to the knowledge of the chemistry and crustacean zooplankton of sodic waters: the Seewinkel pans revisited. BFB-Bericht 70, Illmitz

Milasowzky N and Zulka P (1998) Habitat requirements and conservation of the "flagship species" *Lycosa singoriensis* (Laxmann 1770) (Araneae: Lycosidae) in the National Park Neusiedler See—Seewinkel (Austria). Z Ökologie u Naturschutz 7:111-119

Müller J (2000) Danube-Oder-Elbe-Canal: many open questions. Aqua Press International 4/2000:9-11

Müller-Motzfeld G (1990) Quantitative Ökofaunistik im Dienste des Insektenschutzes. Entomologische Nachrichten und Berichte 34:109-117

Mura G (1993) Italian Anostraca: distribution & status. IUCN Anostracan News 1:3

Nemec H (1952) Lebende Urzeittiere. N ill Wochenschau, Wien: 14.12.1952

New TR (1993) Angels on a pin: dimensions of the crisis in invertebrate protection. Amer Zool 33:623-630

Owens C, Ripley B and Simovich M (2000) Toxicity of malathion to the endangered vernal pool branchiopod *Branchinecta sandiegonensis*. TCS 2000 Summer Meeting Abstracts: 37

Paar M, Schramayr G, Tiefenbach M and Winkler I (1993) Burgenland, Niederösterreich, Wien. In: Tiefenbach M (ed.) Naturschutzgebiete Österreichs. Umweltbundesamt Monographien 38A, Wien

Pesta O (1937) Beiträge zur Kenntnis der Tierwelt (Entomostrakenfauna) des Zicklackengebietes am Ostufer des Neusiedlersees im Burgenland, Österreich. Zool Anz 118:177-192

Pesta O (1939) *Triops (Apus) cancriformis* aus dem Stadtgebiet von Wien. Ann Mus Wien 60:387-394

Pesta O (1942) Ein neuer Nachweis von *Triops (Apus) cancriformis* Bosc in Wien. Zool Anz 139:113-114

Petrov B and Petrov I (1997) The status of Anostraca, Notostraca and Conchostraca (Crustacea: Branchiopoda) in Yugoslavia. Hydrobiologia 359:29-35

Polhemus DA (1993) Conservation of Aquatic Insects: Worldwide Crisis or Localized Threats? Amer Zool 33:588-598

Puschnig R (1918) Vom Ausflußgebiete des Wörthersees. Carinthia II 108/28:136-141

Rieder N (1989) Veränderungen und neuere Entwicklungen im Gefährdungsstatus der Phyllopoden. Schr f Landschaftspflege u Naturschutz, Bonn-Bad Godesberg 29: 294-295

Sampl H (1969) Der Kiemenfuß *Lepidurus apus* (L.) (Phyllopoda, Crust.) erstmals in Kärnten nachgewiesen. Carinthia II 159:130-131

Schlederer R (1999) Das Ramsar-Konzept – Modell einer Konsensfindung. In: Kelemen J and Oberleitner I (eds.) Fließende Grenzen. Lebensraum March-Thaya-Auen (pp. 320—325) Umweltbundesamt, Wien

Schnittler M, Ludwig G, Pretscher P and Boye P (1994) Konzeption der Roten Listen der in Deutschland gefährdeten Tier- und Pflanzenarten – unter Berücksichtigung der neuen internationalen Kategorien. Natur und Landschaft 69:451-459

Sokal RR and Rohlf FJ (1995) Biometry. The principles and practice of statistics in biological research. Freeman, New York

Usher MB and Erz W (1994) Erfassen und Bewerten im Naturschutz. Probleme, Methoden, Beispiele. Uni-TB, Stuttgart

Vornatscher J (1955) Alte und neue Vorkommen von *Triops cancriformis* Bosc. (Apus) in Wien und Niederösterreich. Ann Naturhist Mus Wien 60:287-290

Vornatscher J (1968) Anostraca, Notostraca, Conchostraca. Catalogus Faunae Austriae VIIIaa. Österr Akad Wiss 5 p

Zulka P, Eder E, Höttinger H and Weigand E (2000) Fachliche Grundlagen zur Fortschreibung der Roten Listen gefährdeter Tiere Österreichs. Unpubl. Report, Federal Environmental Agency (UBA), Vienna

MICROSATELLITES IN SHRIMP SPECIES

John R. Scarbrough, David L. Cowles and Ronald L. Carter

Department of Natural Sciences
Loma Linda University
Loma Linda, California 92350, USA

rcarter@univ.llu.edu

ABSTRACT

A review of microsatellites in shrimp is presented. Microsatellites are abundant and polymorphic and should be useful for molecular studies, particularly those involving population genetic structure. Microsatellites are likely to become an increasingly important tool to help in the understanding of shrimp species with applications to aquaculture, ecology, conservation, and genetics.

I. INTRODUCTION

Many new classes of polymerase chain reaction (PCR)-based genetic markers have been developed over the last decade. One of these, microsatellites, is widely regarded as one of the most useful types of nuclear genetic markers to be identified so far (Bruford and Wayne 1993, Queller et al. 1993, Sunnucks 2000). In fact, microsatellites come close to being an optimal genetic marker (Queller et al. 1993, Jarne and Lagoda 1996). The utility of microsatellite markers derives from their abundance in the genome, single locus nature, simplicity of assay, high levels of allelic diversity, mendelian inheritance, codominance, and selective neutrality. The number of applications for microsatellite markers is impressive. They are currently used intensively for efforts involving genome mapping and they are also excellent markers for questions of behavioral ecology since they allow the determination of genetic identity, paternity, and kinship. In addition, microsatellites are becoming a marker of choice for the study of structure in natural populations and are therefore becoming increasingly important in the fields of population and conservation genetics. Microsatellite analysis can potentially be applied to a wide variety of ecological and evolutionary studies and the technique is allowing researchers to focus on questions that were previously difficult to address.

Microsatellites are defined as tandem arrays of short DNA sequence motifs of 1-6 bp in length (Schlötterer 1998) and are also referred to as simple sequence repeats (SSRs) or short tandem repeats (STRs). Microsatellite arrays are small. Total array lengths range from somewhere around a dozen up to a few hundred base pairs, with many being less than 100bp in total length. For example, a microsatellite locus may consist of a stretch of DNA with the base sequence of CA repeated 21 times in a row $[(CA)_{21}]$ or TGTA repeated 8 times $[(TGTA)_8]$. Microsatellite loci have been found in all organisms investigated so far and occur even in the smallest bacterial genomes (Field and Wills 1996, Hancock 1996). They are very abundant in eukaryotes (Tautz and Renz 1984) and are ubiquitously interspersed throughout the genome in a roughly uniform manner (Hancock 1999). Microsatellite loci can be found anywhere in the genome, both in protein coding and noncoding regions (Tóth et al. 2000). However, it has been demonstrated that the presence of microsatellite repeats is much higher in noncoding sequences (Hancock 1995). Microsatellite repeats in prokaryotes occur at significantly lower frequencies (Van Belkum et al. 1998).

Kluwer Academic/Plenum Publishers

E. Escobar-Briones & F. Alvarez Eds.
MODERN APPROACHES TO THE STUDY OF CRUSTACEA
PP. 291-299

Weber (1990) is responsible for much of the terminology associated with microsatellite sequences. Microsatellites can be described, based on the number of nucleotides in the repeat motif, with terms such as dinucleotide, trinucleotide, or tetranucleotide. These terms designate a pattern of two, three, or four nucleotides repeating. Microsatellites can also be described by the term perfect, which refers to an uninterrupted stretch of identical repeats, or the term imperfect, which refers to a repeat sequence in which there are one or more interruptions. Microsatellites are also referred to as compound, or composite, if made up of two or more adjacent tandem repeats. All these terms can be used in combination to describe various microsatellites.

. The widespread distribution of microsatellite repeats in the eukaryotic genome requires that most microsatellites be surrounded by single copy DNA— that is, DNA sequence which is unique in the genome. The small size of microsatellite arrays and the presence of single copy nuclear DNA in the flanking regions ensure that a microsatellite locus can be easily amplified by PCR in a specific repeatable manner resulting in PCR typing that is highly reproducible. This enables a precise scoring of alleles (see below) and takes advantage of the strengths of the PCR technique. These include a fast and relatively easy processing time, the ability to use minute amounts of tissue, and in the case of microsatellites, the potential use of low molecular weight DNA from partially degraded tissue samples.

Variability at microsatellite loci was originally demonstrated in several laboratories (Litt and Luty 1989, Tautz 1989, Weber and May 1989). This variability consists of length polymorphisms caused by changes in the number of repeat units in the microsatellite array. The changes are thought to be due to a mutation mechanism called DNA slippage which occurs during replication (Levinson and Gutman 1987). The most likely change is the gain or loss of a single repeat unit which suggests that microsatellite repeats are changing in a stepwise fashion. The observed mutation rates range from 10^{-3} to 10^{-6} events per locus per generation (Hancock 1999). Since rates of mutation at microsatellite loci are high, the result is a hypervariable marker locus with many alleles. The allelic state at a particular microsatellite locus can be visualized by the analysis of length variation seen in PCR products. Because the regions flanking the microsatellite are generally conserved within a species, primers in these flanking regions can be used for PCR

amplification and the products screened for size variation next to appropriate size standards on polyacrylamide gels (for examples, see Strassmann et al. 1996, Cregan and Quigley 1997, Schlötterer 1998). The determination of the absolute length of alleles allows inter-gel comparisons and makes possible the exchange of data produced in different laboratories. Note that microsatellite alleles are codominant and will be seen as one or two bands after PCR amplification representing homozygous or heterozygous states respectively.

A major drawback to the use of microsatellite markers is that it is necessary to know the sequence of DNA flanking the various tandem repeats in order to design appropriate PCR primers. Unless one is working with organisms for which DNA sequence is readily available (e.g., humans, common laboratory animals, or certain agricultural species), it is necessary to create and screen a genomic DNA library in order to locate microsatellite loci. The traditional strategy for this task typically involves using oligomeric probes to screen random short insert libraries for bacterial clones containing microsatellite repeats (Queller et al. 1993, Strassmann et al. 1996). After this, bacterial clones must be sequenced to obtain the information from the flanking regions needed to design PCR primers. Although microsatellite repeats can be abundant in the genome, the frequency of microsatellite containing clones in a random small insert library is low and it may become necessary to screen many bacterial clones. For this reason, new methods have been developed to enrich a genomic library for inserts with microsatellites (Ostrander et al. 1992, Karagyozov et al. 1993, Armour et al. 1994, Kandpal et al. 1994, Kijas et al. 1994, Edwards et al. 1996, Refseth et al. 1997, Schlötterer 1998, Hamilton et al. 1999). These enrichment strategies are particularly useful for finding trinucleotide and tetranucleotide repeats which occur more rarely in eukaryotic genomes. Library construction and screening may not be necessary if primer pairs for microsatellite markers have already been developed for closely related species. Priming sites in the flanking regions may be conserved among species and microsatellite primer pairs frequently amplify homologous loci in related species (e.g., Moore et al. 1991, Schlötterer et al. 1991, FitzSimmons et al. 1995, Tam and Kornfield 1996, Wimberger et al. 1999). Such cross species amplification has even been found to work across families (e.g., Valsecchi and Amos 1996, Jekielek and Strobeck 1999, Winker et al. 1999).

Although the procedures required for the development of microsatellites from poorly known species utilize standard techniques practiced in many molecular biology laboratories, microsatellite development usually requires considerable time and expense. The effort can be complicated by the fact that microsatellites are known to be present or absent to varying degrees in different taxonomic groups (Beaumont and Bruford 1999). Many plant species and birds, as well as several invertebrate groups such as lepidopterans (Meglécz and Solignac 1998, Nève and Meglécz 2000), have been difficult to work with, presumably because microsatellites are scarce. In addition, the relative abundance of different repeat motifs is taxon-dependent (Lagercrantz et al. 1993, Butcher et al. 2000, Tóth et al. 2000). Consideration of these relative abundances would be wise before beginning a potentially long and expensive procedure. Microsatellite studies using shrimp species are just now beginning to accumulate. The purpose of this paper is to review information about the nature of microsatellites in shrimp and to report, for the first time, the isolation of microsatellites from a shrimp species in the group Caridea.

Microsatellites in penaeid shrimp

Initial microsatellite studies involving shrimp have all been carried out in the family Penaeidae, with the focus mostly on species of economic importance (Table 1). The family Penaeidae includes several species that are very important to a large worldwide aquaculture industry and also species which make up important fisheries. Many of these initial studies were aimed at determining population genetic structure (Benzie 2000). Information regarding genetic structure is useful for determining stock structure in fisheries and the information is also important for aquaculture since wild populations are currently the primary source of broodstock. In the future, it is expected that microsatellite markers will greatly assist aquaculture research in Penaeidae by helping to select broodstock, establish pedigrees, monitor inbreeding levels in breeding programs, and develop genetic linkage maps (Wolfus et al. 1997, Moore et al. 1999, Xu et al. 1999, Bierne et al. 2000, Brooker et al. 2000).

Microsatellites in penaeid shrimps have been found to be abundant. All types of microsatellite sequence have been found including di-, tri-, tetra-, and hexanucleotide sequences (Table 2). Many repeat arrays were found to be of simple composition consisting of perfect uninterrupted stretches of identical repeats, however, imperfect and compound repeats are common. One of the most interesting characteristics of microsatellites in penaeid shrimps is their extreme length. Dinucleotide repeats in *Penaeus monodon* were found to be much longer than those in mammals and other species (Tassanakajon et al. 1998) and the occurrence of many compound microsatellites in penaeids serves to further increase the length of various repeat arrays. Some authors have reported the occurrence of long and complex microsatellite sequences with flanking sequence that is sometimes interspersed with smaller segments of repetitive DNA (Tassanakajon et al. 1998, Brooker et al. 2000). Most studies have reported high levels of polymorphism for microsatellite loci in penaeids with heterozygosities sometimes approaching 100 percent.

Table 1. *Shrimp species from which microsatellites have been isolated.*

Species	Reference	Type of repeat
PENAEIDAE		
Litopenaeus setiferus	Ball et al. 1998	dinucleotide
Litopenaeus stylirostris	Vonau et al. 1999	dinucleotide
Litopenaeus vannamei	Garcia and Alcivar-Warren 1996	dinucleotide
	Garcia et al. 1996	tri-, tetranucleotide
	Bagshaw and Buckholt 1997	pentanucleotide
Marsupenaeus japonicus	Moore et al. 1999	di-, trinucleotide
Penaeus monodon	Tassanakajon et al. 1998	dinucleotide
	Xu et al. 1999	di-, tri-, tetra-, hexanucleotide
	Brooker et al. 2000	dinucleotide
	Whan et al. 2000	tri-, pentanucleotide
CARIDEA		
Acanthephyra curtirostris	This paper	di-, tri-, tetra-, hexanucleotide

Table 2. *Common microsatellite repeat motifs isolated from shrimp.*

Length of repeat unit	*Penaeidae*	*Acanthephyra*
2	CA/GT CT/GA TA/AT	CA/GT TA/AT
3	CAA/GTT CTT/GAA TAT/ATA	CAC/GTG CTT/GAA
4	TGTA/ACAT TTTA/AAAT TTTC/AAAG	TCCG/AGGC TGTA/ACAT TGTC/ACAG

Microsatellites in caridean shrimp

We have isolated microsatellites from the deep-sea caridean shrimp *Acanthephyra curtirostris*. These are, to our knowledge, the first microsatellites isolated from the group Caridea. *Acanthephyra curtirostris* is a midwater species and is a member of the largest genus in the family Oplophoridae. This oplophorid species has an extremely large geographical range which is typical of many oplophorids and also deep oceanic species in general. *Acanthephyra curtirostris* is found at lower latitudes around the world and to 51 degrees north in the eastern Pacific. Studies of population genetic structure and genetic diversity within broadly ranging oceanic species such as *A. curtirostris* have been relatively few. The development of microsatellite markers for *A. curtirostris* should prove useful for the study of population genetic structure within the species. Studies of genetic subdivision and levels of gene flow in deep oceanic species are likely to increase not only our understanding of the nature of the species involved, but also have the potential of increasing our basic understanding of the marine environment. Primer pairs for microsatellite loci in *A. curtirostris* are currently being designed and used to examine population genetic structure.

Microsatellites were isolated from *A. curtirostris* using standard techniques (see Strassmann et al. 1996). Briefly, genomic DNA was extracted from muscle tissue using a conventional phenol-chloroform protocol (Sambrook et al. 1989). A partial genomic library was constructed from DNA of a single specimen. Genomic DNA 17(μg) was digested to completion with the restriction enzyme Sau3AI. Size selected fragments (500-1500 bp) were isolated from a poly-acrylamide gel by electroelution into a dialysis bag (Ausubel et al. 1999). Isolated fragments were ligated into the dephosphorylated BamHI site of the pUC19 cloning vector (Pharmacia). After transformation into DH5α competent cells, 3800 recombinant clones were transferred to nylon filters (Hybond N+, Amersham) and screened by hybridization using a mixture of radioactive oligonucleotide probes [$(TG)_{10}$ $(TC)_{10}$ $(AGG)_5AG$ $(CAC)_5CA$ $T(CCT)_5$ $CT(ATCT)_6$ $(TGTA)_6TG$]. Purified plasmid DNA from positive clones was sequenced on an ABI-377 automated sequencer.

Microsatellite sequence in *A. curtirostris* appears to be abundant with many dinucleotide, trinucleotide, and tetranucleotide sequences present (Table 2). Given a library of 3800 clones and based on the average size of the insert and the number of microsatellites isolated (around 28), a microsatellite containing sequence would be expected at least every 136 kb. The dinucleotide CA was the most commonly encountered microsatellite. Simple repeat sequences were found, but it was also found that most (64%) of the microsatellite arrays were composite consisting of adjacent runs of dinucleotide, trinucleotide, and tetranucleotide sequence. Some of the repeats were quite complex consisting of 3 or 4 different repeat sequences. Repeats tended to be pure, or perfect, even if they were composite. Long repeats were common, particularly in the case of the dinucleotide repeats—many of which had runs of 50-100 repeat units. There was also a strong suggestion that microsatellites may be clustered with two or more distinct loci occurring near each other in the same cloned insert. One hexanucleotide repeat was found $(CCCACA)_6$ which was part of a compound microsatellite sequence.

II. DISCUSSION

Abundant and varied microsatellites appear to be present in both penaeid and caridean shrimp. These microsatellites are similar to those found in other animal taxa studied, but microsatellite sequences in shrimps appear to be atypical in length and complexity with a high percentage of shrimp microsatellites being long and/or compound. Microsatellite sequence from decapod Crustacea other than shrimp is also beginning to accumulate. Microsatellites have been isolated from a lobster (Tam and Kornfield 1996), a crab (Urbani et al. 1998), and a few different crayfish species (Imgrund et al. 1997, Baker et al. 2000, Gouin et al. 2000). These decapods also show a tendency toward long and/or compound microsatellites.

The extreme length and complexity found in shrimp microsatellites may present challenges to their use—particularly for creating genetic linkage maps which require large numbers of marker loci (Tassanakajon et al. 1998, Moore et al. 1999). The lack of simple repeats makes usable markers technically difficult to obtain since many isolated positive clones will contain useful flanking sequence only on one side of a microsatellite repeat and this prevents the design of a primer pair. Such a problem may be overcome by cloning longer inserts which is a strategy that may be advisable when working with shrimp. However, the sequencing of longer inserts is more laborious and may require the design of internal primers. A long microsatellite may also present a problem if it results in PCR amplification products that are too large to be conveniently separated and sized by various polyacrylamide gel systems. Furthermore, complex compound loci may result in banding patterns that are difficult to score and interpret.

It is possible that the unusually high percentage of long microsatellite sequences reported from shrimp may be the result of a selection bias due to the method of isolation. During screening, high stringency conditions may lead to a preference for choosing clones with inserts that have long microsatellites. Xu et al. (1999) isolated microsatellites from *Penaeus monodon* genomic libraries by direct sequencing of random clones without screening. A large number of different microsatellite repeats were found revealing not only an abundance of microsatellite repeats, but also the presence of many shorter repeats with flanking sequence suitable for primer design. It is clear, however, that long and complex microsatellites are a major component in shrimp genomes.

An examination of the organization of microsatellite repeats in shrimp sequence reveals some interesting characteristics beyond length and complexity. Although most microsatellite arrays consist of dinucleotide, trinucleotide and tetranucleotide repeats, pentanucleotide, hexanucleotide, and extra long repeat units have also been observed. Pentanucleotide repeats have been found in *Penaeus monodon* and *Litopenaeus vannamei* and hexanucleotide repeats have been found in *P. monodon* and *A. curtirostris*. A microsatellite with an extra long repeat unit —an octanucleotide

$(ATTTATTC)_5$— was found in *P. monodon* (Xu et al. 1999). Such long repeats were found by chance and the frequency and usefulness of such repeats needs to be determined. Some authors have suggested that there may be an association or clustering of microsatellites in some genomic regions with microsatellite repeats having a non-independent distribution in the genome (Estoup et al. 1993, Fisher et al. 1996). Evidence of these types of clusters has been seen in shrimp species. Brooker et al. (2000) reported the isolation of two clones which contained two separate microsatellite arrays and such an arrangement of microsatellite repeats has also been observed in *A. curtirostris*.

The level of polymorphism in shrimp microsatellites is high. Heterozygosities are often greater than 90% and many alleles may be present. This is consistent with the current understanding that higher repeat unit counts appear to be associated with higher mutation rates (Weber 1990, Wierdl et al. 1997, Estoup and Cornuet 1999, Zhu et al. 2000). The range of microsatellites, both long and short, present in shrimp species make it possible to select loci for specific applications. Short microsatellite markers with fewer alleles can, if found, be well suited for population genetic studies. The longer, more variable microsatellites may be more appropriate for such things as genetic mapping, determination of parentage, and pedigree analysis. Imperfections and compound arrangements may also affect the level of polymorphism and thus the utility of the marker. Interrupting bases in imperfect microsatellites appear to decrease the mutation rate by reducing the likelihood of DNA slippage during replication (Estoup and Cornuet 1999). This results in lower levels of polymorphism than would be expected of a pure repeat with a similar number of repeats.

Information about microsatellites from previously unstudied taxonomic groups can be helpful in understanding microsatellite evolution. Amos (1999) has used a comparative taxonomic approach to examine microsatellite evolution. By comparing information about microsatellite loci found in different species, a model of microsatellite evolution is proposed in which microsatellites undergo a life cycle. The microsatellites are "born" and expand towards a length boundary (possibly a species specific boundary) at which point they become unstable and delete or degenerate into random sequence by accumulating point mutations. The life history of the microsatellite is proposed to occur faster with increasing mutability due to both greater heterozygosity and increased body temperature resulting in less time spent at longer lengths. Consequently, the model proposes that species with larger population sizes and/ or higher body temperatures will have fewer and shorter microsatellites. The results of many published studies seem to fit the proposed model (Amos 1999) and the model is useful to explain the decreased frequency of microsatellite loci in birds and insects. The data from shrimp species, however, do not appear to fit the model completely. While body temperatures are low, particularly in the deep-sea shrimp *A. curtirostris*, the population sizes for shrimp species are presumably quite large. A more complex model is likely needed to account for the frequency and long lengths of microsatellite loci in shrimp species.

The overall value of previously characterized microsatellite loci as a source of primer pairs for related shrimp species remains to be determined. Mitochondrial DNA analysis (Palumbi and Benzie 1991, Baldwin et al. 1998) suggests that morphological similarity in penaeids masks large genetic differences. Failure of microsatellite loci to amplify across species in penaeid shrimps may also point to large genetic divergence among taxa. Moore et al. (1999) tested microsatellite markers developed for *Marsupenaeus japonicus* in a number of penaeid species and found the markers to be species specific. Three microsatellite primer pairs developed for *Litopenaeus stylirostris* failed to amplify in *P. monodon* and *L. vannamei* (Vonau et al. 1999) and heterologous microsatellite loci from both *P. monodon* and *M. japonicus* failed to amplify products in *L. stylirostris* (Bierne et al. 2000). Nevertheless, cross species amplification has been reported. A primer pair originally developed for *L. vannamei* has proved useful in *L. stylirostris* (Bierne et al. 2000) and of six loci developed for *Litopenaeus setiferus*, four were found to amplify in at least one other penaeid species and two produced strong PCR products in *Farfantepenaeus aztecus, F. duorarum, L. vannamei*, and *L. stylirostris* (Ball et al. 1998). Two microsatellite primer pairs developed from coding regions in *P. monodon*, which are presumably more conserved, have been found to amplify products *in Penaeus esculentus, P. semisulcatus* and *Fenneropenaeus merguiensis* (Whan et al. 2000). Limited attempts at cross species amplification in the genus *Acanthephyra* using *A. curtirostris* primer sets have been moderately successful. Two primer pairs developed for *A. curtirostris* have produced PCR products in *A. prionota* and *A. acutifrons* but have failed to amplify in *A. eximia*.

III. CONCLUSIONS

The accumulated data from several studies has revealed that shrimp species are rich in microsatellite sequence. Microsatellites in shrimp are abundant and polymorphic and should be useful for molecular studies, particularly those involving population genetic structure. Microsatellites are likely to become an increasingly important tool to help in the understanding of shrimp species with applications to aquaculture, ecology, conservation, and genetics.

ACKNOWLEDGEMENTS

We are indebted to J.J. Childress for providing access to *Acanthephyra curtirostris* tissue and to J. Sands for providing technical advice relating to library construction and cloning.

Acanthephypa curtirostris tissue was collected during research cruises aboard the R/V New Horizon conducted under an NSF grant to JJ. Childrens OCE-9415543.

REFERENCES

Amos W (1999) A comparative approach to the study of microsatellite evolution. In: Goldstein DB and Schlötterer C (eds) Microsatellites: Evolution and Applications (pp 66-79) Oxford University Press, New York

Armour JAL, Neumann R, Gobert S and Jeffreys AJ (1994) Isolation of human simple sequence repeat loci by hybridization selection. Hum Mol Genet 3:599-605

Ausubel F, Brent R, Kingston RE, Moore DD, Seidman JG, Smith JA and Struhl K (1999) Short Protocols in Molecular Biology. 4th ed. Wiley, New York

Bagshaw JC and Buckholt MA (1997) A novel satellite/microsatellite combination in the genome of the marine shrimp *Penaeus vannamei*. Gene 184:211-214

Baker N, Byrne K, Moore S and Mather P (2000) Characterization of microsatellite loci in the redclaw crayfish, *Cherax quadricarinatus*. Mol Ecol 9:494-495

Baldwin JD, Bass AL, Bowen BW and Clark WH (1998) Molecular phylogeny and biogeography of the marine shrimp *Penaeus*. Molecular Phylogenetics and Evolution 10:399-407

Ball AO, Leonard S and Chapman RW (1998) Characterization of $(GT)_n$ microsatellites from native white shrimp (*Penaeus setiferus*). Mol Ecol 7:1251-1253

Beaumont MA and Bruford MW (1999) Microsatellites in conservation genetics. In: Goldstein DB and Schlötterer C (eds.) Microsatellites: Evolution and Applications (pp 165-182) Oxford University Press, New York

Benzie JAH (2000) Population genetic structure in penaeid prawns. Aquacult Res 31:95-119

Bierne N, Beuzart I, Vonau V, Bonhomme F and Bédier E (2000) Microsatellite-associated heterosis in hatchery-propagated stocks of the shrimp *Penaeus stylirostris*. Aquaculture 184:203-219

Brooker AL, Benzie JAH, Blair D and Versini J-J (2000) Population structure of the giant tiger prawn *Penaeus monodon* in Australian waters, determined using microsatellite markers. Mar Biol 136:149-157

Bruford MW and Wayne RK (1993) Microsatellites and their application to population genetic studies. Current Opinions in Genetics and Development 3:939-943

Butcher RDJ, Hubbard SF and Whitfield WGF (2000) Microsatellite frequency and size variation in the parthenogenetic parasitic wasp *Venturia canescens* (Gravenhorst) (Hymenoptera: Ichneumonidae). Insect Molecular Biology 9:375-384

Cregan PB and Quigley CV (1997) Simple sequence repeat DNA marker analysis. In: Caetano-Anollés G and Gresshoff PM (eds) DNA Markers: Protocols, Applications, and Overviews (pp 173–185) Wiley-Liss, New York

Edwards KJ, Barker JHA, Daly A, Jones C and Karp A (1996) Microsatellite libraries enriched for several microsatellite sequences in plants. BioTechniques 20:758-760

Estoup A, Solignac M, Harry M and Cornuet J-M (1993) Characterization of $(GT)_n$ and $(CT)_n$ microsatellites in two insect species:*Apis mellifera* and *Bombus terrestris*. Nucleic Acids Res 21:1427-1431

Estoup A and Cornuet J-M (1999) Microsatellite evolution: inferences from population data. In: Goldstein DB and Schlötterer C (eds) Microsatellites: Evolution and Applications (pp 49-65) Oxford University Press, New York

Field D and Wills C (1996) Long, polymorphic microsatellites in simple organisms. Proc R Soc Lond 263:209-215

Fisher PJ, Gardner RC and Richardson TE (1996) Single locus microsatellites isolated using 5' anchored PCR. Nucleic Acids Res 24:4369-4371

FitzSimmons NN, Moritz C and Moore SS (1995) Conservation and dynamics of microsatellite loci over 300 million years of marine turtle evolution. Mol Biol Evol 12: 432-440

Garcia DK and Alcivar-Warren A (1996) Identification and organization of microsatellites in *Penaeus vannamei* shrimp. Anim Genet 27 (Supplement 2):71

Garcia DK, Dhar AK and Alcivar-Warren A (1996) Molecular analysis of a RAPD marker (B20) reveals two microsatellites and differential mRNA expression in *Penaeus vannamei*. Molecular Marine Biology and Biotechnology 5:71-83

Gouin N, Grandjean F and Souty-Grosset C (2000) Characterization of microsatellite loci in the endangered freshwater crayfish*Austropotamobius pallipes* (Astacidae) and their potential use in other decapods. Mol Ecol 9:636-637

Hamilton MB, Pincus EL, Di Fiore A and Fleischer RC (1999) Universal linker and ligation procedures for construction of genomic DNA libraries enriched for microsatellites. BioTechniques 27:500-507

Hancock JM (1995) The contribution of slippage-like processes to genome evolution. J Mol Evol 41:1038-1047

Hancock JM (1996) Simple sequences in a "minimal" genome. Nat Genet 14:14-15

Hancock JM (1999) Microsatellites and other simple sequences: genomic context and mutational mechanisms. In: Goldstein DB and Schlötterer C (eds) Microsatellites: Evolution and Applications (pp 1-9) Oxford University Press, New York

Imgrund J, Groth D and Wetherall J (1997) Genetic analysis of the freshwater crayfish *Cherax tenuimanus*. Electrophoresis 18:1660-1665

Jarne P and Lagoda PJL (1996) Microsatellites, from molecules to populations and back. Trends Ecol Evol 11:424-429

Jekielek J and Strobeck C (1999) Characterization of polymorphic brown lemur (*Eulemur fulvus*) microsatellite loci and their amplification in the family Lemuridae. Mol Ecol 8:901-903

Kandpal RP, Kandpal G and Weissman SM (1994) Construction of libraries enriched for sequence repeats and jumping clones, and hybridization selection for region specific markers. Proc Natl Acad Sci USA 91:88-92

Karagyozov L, Kalcheva ID and Chapman VM (1993) Construction of random small-insert genomic libraries highly enriched for simple sequence repeats. Nucleic Acids Res 21:3911-3912

Kijas JMH, Fowler JCS, Garbett CA and Thomas MR (1994) Enrichment of microsatellites from the citrus genome using biotinylated oligonucleotide sequences bound to streptavidin-coated magnetic particles. BioTechniques 16:656-662

Lagercrantz U, Ellegren H and Andersson L (1993) The abundance of various polymorphic microsatellite motifs differs between plants and vertebrates. Nucleic Acids Res 21:1111-1115

Levinson G and Gutman GA (1987) Slipped-strand mispairing: a major mechanism for DNA sequence evolution. Mol Biol Evol 4:203-221

Litt M and Luty JA (1989) A hypervariable microsatellite revealed by in vitro amplification of a dinucleotide repeat within the cardiac muscle actin gene. Am J Hum Genet 44:397-401

Meglécz E and Solignac M (1998) Microsatellite loci for *Parnassius mnemosyne* (Lepidoptera). Hereditas 128:179-180

Moore SS, Sargeant LL, King TJ, Matticks S, Georges M and Hetzel DJS (1991) The conservation of dinucleotide microsatellites among mammalian genomes allows the use of heterologous PCR primer pairs in closely related species. Genomics 10:654-660

Moore SS, Whan V, Davis GP, Byrne K, Hetzel DJS and Preston N (1999) The development and application of genetic markers for the Kuruma prawn *Penaeus japonicus*. Aquaculture 173: 19-32

Nève G and Meglécz E (2000), Microsatellite frequencies in different taxa. Trends Ecol Evol 15:376-377

Ostrander EA, Jong PM, Rine J and Duyk G (1992) Construction of small-insert genomic DNA libraries highly enriched for microsatellite repeat sequences. Proc Natl Acad Sci USA 89: 3419-3423

Palumbi SR and Benzie J (1991) Large mitochondrial DNA differences between morphologically similar penaeid shrimp. Mol Mar Biol Biotechnol 1:27-34

Queller DC, Strassmann JE and Hughes CR (1993) Microsatellites and kinship. Trends Ecol Evol 8:285-289

Refseth UH, Fangan BM and Jakobsen KS (1997) Hybridization capture of microsatellites directly from genomic DNA. Electrophoresis 18:1519-1523

Sambrook J, Fritsch EF and Maniatis T (1989) Molecular Cloning: A Laboratory Manual. 2nd ed. Cold Spring Harbor Laboratory Press, New York

Schlötterer C (1998) Microsatellites. In: Hoelzel AR (ed) Molecular Genetic Analysis of Populations. 2nd ed. (pp 237-261) Oxford University Press, New York

Schlötterer C, Amos B and Tautz D (1991) Conservation of polymorphic simple sequence loci in cetacean species. Nature 354:63-65

Strassmann JE, Solis CR, Peters JM and Queller DC (1996) Strategies for finding and using highly polymorphic DNA microsatellite loci for studies of genetic relatedness and pedigrees. In: Ferraris JD and Palumbi SR (eds) Molecular Zoology (pp 163-180) Wiley-Liss, New York

Sunnucks P (2000) Efficient genetic markers for population biology. Trends Ecol Evol 15:199-203

Tam YK and Kornfield I (1996) Characterization of microsatellite markers in *Homarus* (Crustacea, Decapoda). Mol Mar Biol Biotechnol 5:230-238

Tassanakajon A, Tiptawonnukul A, Supungul P and Rimphanitchayakit V (1998) Isolation and characterization of microsatellite markers in the black tiger prawn *Penaeus monodon*. Mol Mar Biol Biotechnol 7:55-61

Tautz D (1989) Hypervariability of simple sequences as a general source for polymorphic DNA markers. Nucleic Acids Res 17:6463-6471

Tautz D and Renz M (1984) Simple sequences are ubiquitous repetitive components of eukaryotic genomes. Nucleic Acids Res 12:4127-4138

Tóth G, Gáspári Z and Jurka J (2000) Microsatellites in different eukaryotic genomes: survey and analysis. Genome Res 10:967-981

Urbani N, Sévigny J-M, Sainte-Marie B, Zadworny D and Kuhnlein U (1998) Identification of microsatellite markers in the snow crab *Chionoecetes opilio*. Mol Ecol 7:357-358.

Valsecchi E and Amos W (1996) Microsatellite markers for the study of cetacean populations. Mol Ecol 5:151-156

Van Belkum A, Scherer S, Van Alphen L and Verbrugh H (1998) Short sequence DNA repeats in prokaryotic genomes. Microbiol Mol Biol Rev 62:275-293

Vonau V, Ohresser M, Bierne N, Delsert C, Beuzart I, Bédier E and Bonhomme F (1999) Three polymorphic microsatellites in the shrimp *Penaeus stylirostris*. Anim Genet 30:234-235

Weber JL (1990) Informativeness of human (dC-dA)$_n$•(dG-dT)$_n$ polymorphisms. Genomics 7:524-530

Weber JL and May PE (1989) Abundant class of human DNA polymorphisms which can be typed using the polymerase chain reaction. Am J Hum Genet 44:388-396

Whan VA, Wilson KJ and Moore SS (2000) Two polymorphic microsatellite markers from novel *Penaeus monodon* ESTs. Anim Genet 31:143-144

Wierdl M, Dominska M and Petes TD (1997) Microsatellite instability in yeast: dependence on the length of the microsatellite. Genetics 146:769-779

Wimberger P, Burr J, Gray A, Lopez A and Bentzen P (1999) Isolation and characterization of twelve microsatellite loci for rockfish (*Sebastes*). Mar Biotechnol 1:311-315

Winker K, Glenn TC and Graves GR (1999) Dinucleotide microsatellite loci in a migratory wood warbler (Parulidae: *Limnothlypis swainsonii*) and amplification among other songbirds. Mol Ecol 8:1553-1556

Wolfus GM, Garcia DK and Alcivar-Warren A (1997) Application of the microsatellite technique for analyzing genetic diversity in shrimp breeding programs. Aquaculture 152:35-47

Xu Z, Dhar AK, Wyrzykowski J and Alcivar-Warren A (1999) Identification of abundant and informative microsatellites from shrimp (*Penaeus monodon*) genome. Anim Genet 30:150-156

Zhu Y, Queller DC and Strassmann JE (2000) A phylogenetic perspective on sequence evolution in microsatellite loci. J Mol Evol 50:324-338

A SCENARIO FOR THE EVOLUTION
OF THE RHIZOCEPHALA

Henrik Glenner and Jens T. Høeg

HG, Department of Evolution, Zoological Institute, University of Copenhagen
Universitetsparken 15, DK-2100 Copenhagen, Denmark

hglenner@zi.ku.dk

JTH, Department of Zoomorphology, Zoological Institute, University of Copenhagen
Universitetsparken 15, DK-2100 Copenhagen, Denmark

ABSTRACT

New insight into the life cycle leads us to present a scenario for the evolution of the Rhizocephala. We base it on a homology between the rhizocephalan kentrogon and the first juvenile stage of conventional barnacles and on the discovery of the vermigon, the earliest internal parasitic phase. We suggest that the urrhizocephalan was a filter feeder attached to the gills of decapod hosts. The stylet of the kentrogon evolved from the peduncle of the epibiont filter feeder. The vermigon is ontogenetically a part of the stylet, which suggests that the base of the peduncle evolved into the internal parasite. Indirect establishment could have evolved if parasites accidentally detached from their attachment to the overlying cuticle and drifted elsewhere within the host. Those that reached the ventral blood sinus in the abdomen and emerged in the abdominal brood chamber would have a selective advantage if they received the same treatment from the host as a brooded egg mass. This scenario could easily lead to the consequent development of full-fledged host control and castration by the parasitic barnacle.

I. INTRODUCTION

Parasitic barnacles or Rhizocephala is a parasitic group of crustaceans that exclusively infest other crustaceans. The morphology of the adult parasite is extremely reduced. It consists of a sac-formed external reproductive organ, the externa, situated on the abdomen of the host, and an internal trophic organ, the root-system or the interna, which takes up nutrition from the hemolymph of the host. Infection will in most cases have severe consequences for the host. Both infected males and female hosts will change morphology. Males become feminized and females often hyper-feminized and, in spite of the fact that the parasite castrates its host, both sexes will behave as if they were females carrying eggs. In addition the emergence of an externa arrest the host in a permanent inter-moult stage. All modifications that ensure the available energy resources of the host are directed solely into the production of parasite larvae.

In his admirable account on the biology of the rhizocephalan barnacle *Sacculina carcini* (Thompson), Delage (1884) described the kentrogon, the infective stage following the settlement of the cypris on the prospective host crab. He considered it to be a larva unique to the Rhizocephala with no equivalent in other Crustacea. This view has persisted until recently. Delage also discovered what we know as the indirect establishment of parasitism in the Rhizocephala. The kentrogon injects the parasitic material into the host at the very site where the cyprid settled. The parasite thereafter passes through a prolonged endoparasitic phase until a reproductive body (externa) emerges in the brood chamber under the abdomen of the host, far removed from the initial site of infection.

Kluwer Academic/Plenum Publishers

E. Escobar-Briones & F. Alvarez Eds.
MODERN APPROACHES TO THE STUDY OF CRUSTACEA
PP. 301-310

Not surprisingly, there have been very few attempts to construct scenarios for the evolution of the Rhizocephala. Unlike most other parasitic Crustacea, such as many groups of parasitic copepods and isopods, there exist no morphological intermediates between the highly specialized Rhizocephala with their extremely reduced morphology and the conventional filter feeding Cirripedia of the orders Thoracica and Acrothoracica.

The rhizocephalan morphology
Cirripedes of the orders Thoracica and Acrothoracica are highly unusual Crustacea in being permanently sessile as adults. Although this entails numerous specializations in their life cycle and morphology, the larvae and the adults retain numerous crustacean traits. In contrast, the Rhizocephala could hardly be classified as Crustacea, much less as Cirripedia, using adult morphology alone. Only their larvae, which are typical cirripede nauplii and cyprids, disclose their unquestionable affinity with the Cirripedia.

Rhizocephalan nauplii and cyprids are morphologically similar to those found in other cirripedes (Glenner 1998) and until and including settlement of the cyprid on the prospective host crab, development is exactly as in their non-parasitic relatives. Similar to rhizocephalans, many thoracican species settle on live animals including other crustaceans where they live as epibionts. It is only during the metamorphosis following attachment that events in the Rhizocephala diverge remarkably from those seen in the Thoracica. Rather than metamorphosing into a juvenile, filter-feeding barnacle, the rhizocephalan cypris initiates the process which culminates with the infection of the host animal by the injection of parasitic material. In the rhizocephalan suborder Kentrogonida, the cyprid metamorphoses (moults) into the highly specialized kentrogon, which is without visible segmentation and carries no appendages other than the antennules which are cemented to the exterior of the host. The kentrogon initiates an incomplete moult resulting in the formation of an injection stylet, which subsequently penetrates the host and acts as a guide tube, or syringe, for the injection of the parasitic material. The nature of the latter has until recently remained unknown, but it had been implicitly expected that it would be nothing more than one or a few undifferentiated cells (Høeg 1985). This meant that the rhizocephalan life cycle would include a phase not covered by a cuticular layer, something that would be unique in all the Arthropoda.

A breakthrough was made when Glenner and Høeg (1995) discovered that the kentrogon actually injects not naked cells, but a cuticle-covered worm-shaped instar. This so-called *vermigon* originates by the moult of the kentrogon that also forms the stylet. Once inside the host it remains covered by cuticle and grows directly into the root system and the anlage of the externa (Glenner et al. 2000, Bresciani and Høeg 2001, Glenner 2001).

The homology of the kentrogon
Delage considered the kentrogon instar to be unique for the Rhizocephala. His view has persisted until very recently and led to the expectation that the ancestral rhizocephalan lacked a kentrogon and accomplished infection by other means. This put focus on the rhizocephalan suborder Akentrogonida, which contains a number of species that have either been claimed to or have been directly shown to lack a kentrogon stage (Høeg 1990, Høeg and Rybakov 1992). The papers of Bocquet-Vedrine (1961, 1972) concerning the barnacle-infesting akentrogonid *Chthamalophilus delagei* were especially influential in forging this view.

A critical evaluation nevertheless shows that the interpretation of the kentrogon as a unique (autapomorphic) feature relies on emphasizing differences and neglecting homologous similarities. Glenner and Høeg (1994) compared the events following cypris settlement in thoracican and rhizocephalan cirripedes. They emphasized that the metamorphic moult following settlement, which results in a cuticle clad first postcyprid instar is a plesiomorphic ground pattern for all Cirripedia. They therefore homologized both cypris metamorphosis and the resulting instar throughout all Cirripedia. This established a homology between the kentrogon and the first juvenile barnacle in the more conventional cirripedes, although the special morphology of the kentrogon remained apomorphic.

Controversial as this may sound, the first postcyprid instars of the Acrothoracica may actually diverge to an equal extent from those found in the Thoracica (Anderson 1994). Glenner and Høeg's (1994) comparison furthermore entailed that the presence of a kentrogon becomes a rhizocephalan plesiomorphy whereas the absence, such as shown for various Akentrogonida, becomes apomorphic, and could have evolved convergently. The akentrogonid Rhizocephala differ from other Cirripedia because the regular series of moults and instars stops with the cyprid

(Høeg 1990). For this reason alone they are poor candidates for illuminating the nature of the ancestral rhizocephalan. The reinterpretation of the kentrogon by Glenner and Høeg (1994) and the discovery of the vermigon instar open up entirely new scenarios of rhizocephalan evolution. Comparisons can now be made between the Kentrogonida and conventional barnacles when trying to reconstruct the biology and morphology of the ancestral rhizocephalan.

II. A SCENARIO FOR THE URRHIZOCEPHALAN

Filter-feeding ancestor

The Cirripedia consist of three orders; the Thoracica (the conventional stalked and acorn barnacles), the Acrothoracica (burrowing barnacles), and the Rhizocephala. Recent morphological and molecular evidence suggests that the Rhizocephala and the Thoracica are sister groups that form a monophyletic group, whose sister group is the Acrothoracica (Jensen et al. 1994 a,b, Spears et al. 1994, Høeg 1995, Mizrahi et al. 1998, Kolbasov et al. 1999, Perl-Treves et al. 2000, Harris et al. 2000). Since both the Thoracica and the Acrothoracica are filter feeders that use specialized thoracic appendages (cirri) to capture food, this feeding mode is undoubtedly homologous between the two orders and strongly suggests that the ancestral rhizocephalan obtained food in a similar manner.

Ancestral settlement in the branchial chamber

For any sessile organism a crucial factor in the choice of a settling site is the availability of food. For larger filter-feeding barnacles these needs are normally fulfilled in exposed areas, where the chances of being in contact with large volumes of water are optimal. This is also true of thoracican barnacles that are epibiotic on decapods. A typical settling site for these species is the dorsal surface of the carapace of the host. Some minute barnacles belonging to the thoracican genus *Octolasmis* settle in the branchial chamber of brachyuran crabs. Here, protected from predators, the barnacles are constantly supplied with food particles from the water that flows over the gills (Fig. 1). Such a switch from an exposed external surface of the host to a more predator protected habitat, is also thought to have happened in the transition from ectoparasitism to endoparasitism in other parasitic taxa (Euzet and Combes 1998). While cladograms based on molecular

sequences have shown that the Rhizocephala is not nested within the Thoracica (Spears et al. 1994, Perl-Treves et al. 2000, Harris et al. 2000), *Octolasmis* can nevertheless serve as a very interesting model in speculations concerning the origin of the parasitic barnacles. The branchial chamber would appear to be an almost ideal settling site for a true parasite. The infesting stage can be carried passively into the chamber by the gill currents where an extremely thin cuticle functions as a gas exchange organ and facilitates the penetration of the parasite. The extensive blood supply in the gills would also assure an ample supply of nutrients to the parasite.

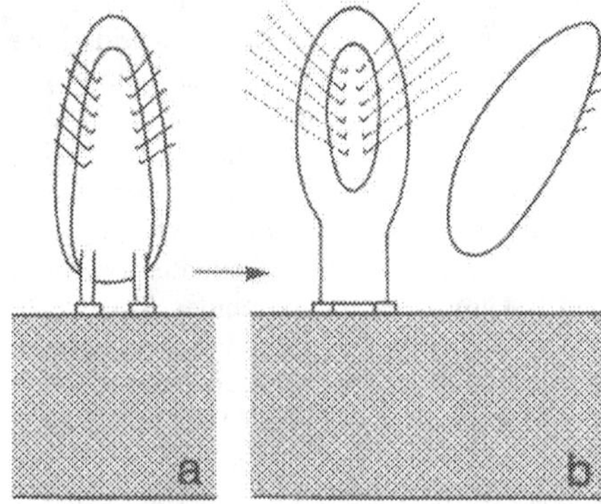

Figure 1. *Metamorphosis in a conventional thoracican cirripede. The cypris cements itself to the substratum (a). The ensuing metamorphic moult, wherein the cypris cuticle is shed, results in a juvenile, filter-feeding barnacle (b).*

Settling site in extant Rhizocephala

For the reasons mentioned above it is easy to imagine the branchial chamber of a decapod as the ancestral settling site for the filter-feeding and potentially parasitic rhizocephalan ancestor. Infection of the host has been examined in 10 kentrogonid species of rhizocephalans and in five of these attachment of the female cyprids takes place in the gill chamber, e.g., *Sacculina pisae, Drepanorchis neglecta, Lernaeodiscus porcellanae, Loxothylacus panopaei, Heterosaccus lunatus*. Of the remaining five species *Sacculina carcini, Loxothylacus texanus, Peltogaster paguri*, and *Peltogasterella sulcata* settle on thin arthrodial membranes, while *Septosaccus rodriguezii* can settle both in the branchial chamber and on arthrodial membranes (Delage 1884, Veillet 1964, Ritchie and Høeg 1981, Glenner and Høeg 1995, Walker 1999). This means that all three families of the Kentrogonida (Sacculinidae, Lernaeodiscidae and Peltogastridae) contain species that settle in the branchial chamber suggesting that this is also the ancestral settling site for all kentrogonid Rhizocephala.

Host location

Prior to the actual infection process (penetration of the host and injection of the vermigon), the most critical part of the life cycle is undoubtedly to locate a suitable host and attach successfully. Much research has been done on the settlement cues of barnacle larvae, but the vast majority of studies have dealt with balanomorph barnacles settling on inanimate substrata. Little information is available on cirripedes settling as epibionts or parasites. All cirripede cyprids have an impressive array of sensory organs including both chemoreceptrors and mechanoreceptors. A special type of chemoreceptors, the lattice organs, is situated in the carapace cuticle and could function in substrate location at a distance while the cyprid is still pelagic. The majority of sensory organs are situated on the third and fourth segments of the antennules. Rhizocephalan cyprids deviate in that the antennules carry aesthetascs and the armature differs between male and female cyprids. This could possibly be linked to the different substrates used by the cyprids: Females settle directly on suitable host crabs often at a certain stage at its lifecycle (Glenner and Werner 1998) while males settle exclusively on virgin external parasites, never directly on the crab. But there is at present no experimental data to support such a statement.

Another point is to which extent rhizocephalan cyprids are substrate specific. The little available data seems to show that some variation exists (Thresher et al. 2000). While cyprids of some species will settle on a range of hosts including some where the parasite is never found in nature, other species seem to be more hosts specific. Obviously the entire field of substrate location, identification and host specificity is wide open to study.

The evolution of host-penetration

An epibiont settled on a crustacean host faces two major threats to its existence: removal by the grooming appendages of the host and the inevitable death following a host moult. Clearly there would be a selective advantage to an ancestral rhizocephalan with a device that penetrated the integument of the host.

In extant thoracican barnacles the cuticle of the peduncle is not shed during ecdysis, a phenomenon that probably differs from the ancestral trait. Formation of a more elongated peduncle underneath the cuticle of the old one could have enabled penetration of the integument of the host (Fig. 2). Although still not a parasite, the epiotic barnacle would have been able to attach to the new post-ecdysial cuticle of the host or even to anchor itself to the subcuticular tissues. Similar to the formation of the peduncle of a thoracican barnacle, the stylet of the rhizocephalan kentrogon is an unpaired structure formed by a moult in the antennulary end of the body

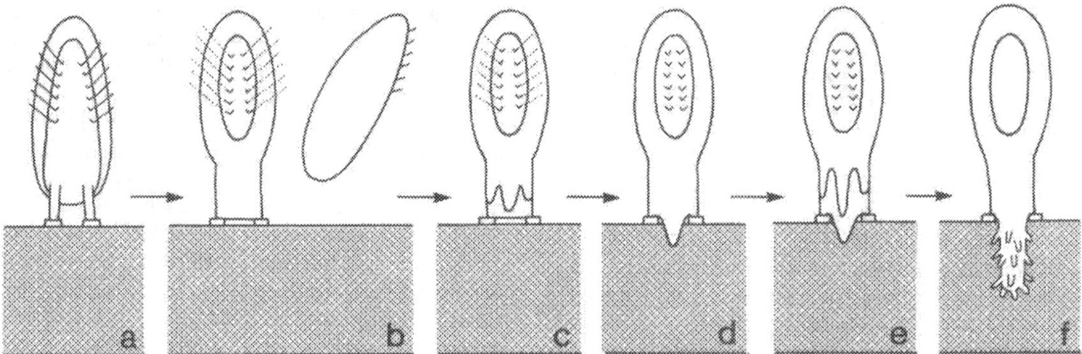

Figure 2. *Development in an ectoparasitic thoracican cirripede. Following the metamorphic moult as in Fig. 1., the peduncle of the juvenile barnacle penetrates the integument of the host (c-e); the cirri either fail to become functional or degenerate altogether. The penetrated peduncle develops into a root-like structure that absorbs nutrients from the host. This series is exemplified by the extant thoracican Anelasma sp. Which is thought to have evolved independent from the stem line of the Rhizocephala.*

(Delage 1884, Høeg 1985, Glenner 2001). We therefore suggest that the stylet of the kentrogon represents the vestige of such a penetrating peduncle in the urrhizocephalan.

The evolution of a root system

Once the peduncle had penetrated the integument of the host, the available hemolymph became a potential source of nutrition. The evolutionary transition towards parasitism would have been furthered by the development of a mechanism by which the barnacle could actively absorb nutrients in the hemolymph through or across the integument of the peduncle (Fig. 2).

In thoracican barnacles such a feeding mode has evolved independently at least twice, e.g., in *Anelasma* parasitizing sharks and in *Rhizolepas* parasitizing polychaetes. Both *Anelasma* and *Rhizolepas* have peduncles that penetrate the integument of their hosts and they gain nourishment via extensions (rootlets) from the peduncle that penetrates some distance into the host tissues. The total dependence on the parasitic mode of life is in both cases illustrated by reductions in the feeding appendages and the alimentary canal. *Anelasma* retains mouth appendages and the gut appears to be functional, but the biramous cirri have lost all setae and the organism derives nourishment from its rootlets (Johnstone and Frost 1927, Hickling 1964). *Rhizolepas* has retrogressed even further. It has lost both the mouth and the mouth appendages and depends wholly on the rootlets. The gut is closed at both ends and rudimentary thoracopods are uniramous (Day 1939).

The systematic position of *Anelasma* and *Rhizolepas* remains unknown. Both are pedunculated barnacles devoid of shell plates, but it is unlikely that they are more closely related to the Rhizocephala than to the Thoracica. They nevertheless illustrate a parallel to the root system of the Rhizocephala and both *Anelasma* and *Rhizolepas* have been used as models to show how Rhizocephala could have evolved from a pedunculated filter-feeding ancestor (Newman et al. 1969, Petriconi 1977).

Reduction of the capitulum

Evolution of a nutrient absorbing system of rootlets on the peduncle would expectedly result in the reduction of all structures associated with thoracic filter feeding, such as cirri, mouthparts, mouth and alimentary tract (Fig 2). The fact that some of these structures still remain in *Anelasma* and *Rhizolepas* could testify to the relatively recent evolution of parasitism in these genera. In the urrhizocephalan, the capitulum could become entirely devoted to reproduction just as is the case with the externa of extant Rhizocephala. Only adult individuals reproduce, so in the period between settlement and sexual maturity the capitulum would be in constant risk of damage or removal by the cleaning devices of the host. Selective pressure that postponed the development of the capitulum might have resulted in the parasite growing so deeply into the gill tissue of the host that it effectively became internal and no longer simply a holdfast at the original settling site (Fig. 3). A parallel can be found in the extant non-parasitic whale barnacles, which partially insert into the cetacean integumental substrate.

When the virtually internal parasite matured, reproduction could now have been accomplished by forming a capitulum with its reproductive apparatus on the exterior surface of the gills. The delayed development of the capitulum would probably have been a gradual process eventually leading to an internally preformed capitulum that could be rapidly evaginated just as in extant Rhizocephala. At this stage it would be appropriate to adopt the rhizocephalan terms "interna" for the trophic system of rootlets and "externa" for the now highly modified, and rapidly emerging "capitulum". A homology between the rhizocephalan externa and the thoracican capitulum is well supported, since both structures contain the ovary, the nervous system, and a brood chamber communicating with the exterior through an orifice that is enclosed in a muscular mantle (Høeg and Lützen 1995).

The evolution of internal migration

The events until this stage of development have been reasonably easy to comprehend if the parasite remained at the same site where the larva initially attached. The remarkable discovery of Delage (1884) was that the rhizocephalan cypris attaches and infects at a site far removed from where the externa eventually develops. This has since been found to hold true for all three families of the Kentrogonida. Such development poses the key question of how the ancestral parasitic barnacle that developed *in situ* became over an evolutionary time period a parasitic castrator that infects the host at one site (such as the gills) and matures into an adult elsewhere (underneath the abdomen of the host).

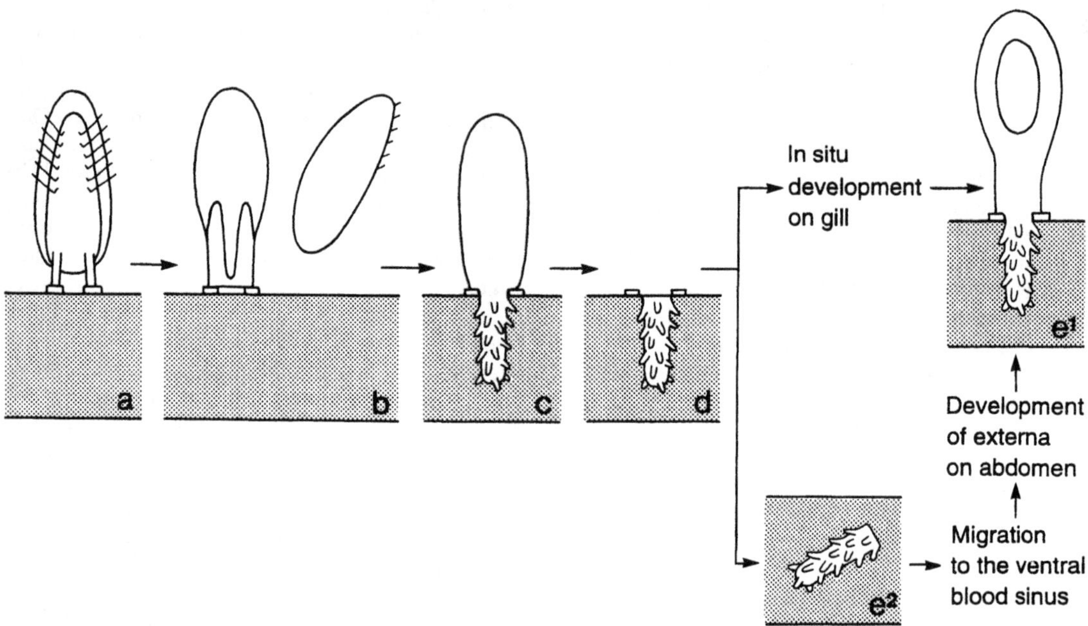

Figure 3. *Hypothetical ancestral ontogeny of the Rhizocephala. The barnacle settles in the host's gill chamber and develops a nutrient absorbing root as in Fig. 2. But the capitulum of the body remains minute or does not develop at all (c-d) which circumvents removal by host grooming behaviour and minimizes chances of being lost when the host moults. The capitulum (now essentially a rhizocephalan externa) develops only when needed for reproduction (e1); this series is comparable to that of extant species of the Chthamalophilidae. We speculate that the root in the ancestral organism lost contact with the integument and flowed away in the hemocoel, where it developed a capitulum in other parts of the host.*

An obvious mode of internal migration would be that the very young parasite grows by a continuous system of rootlets from the site of initial infection to the site where the externa eventually emerges. Newman (1982) suggested that a step towards this migration pattern would be the evolution of colonialism as seen in akentrogonids of the family Thompsoniidae. In this scenario, the root system grows out externae at a continually increasing distance from the site of infection. Eventually one of the externae might emerge underneath the abdomen of the host. The kentrogonid level of organization with a single or a few externae in the host abdominal chamber could then have been reached by the subsequent reduction of externa development in other positions. There are two observations that speak against this theory. Firstly, the colonial Thompsoniidae externae do not emerge gradually, rather they appear simultaneously all over the body (Lützen and Jespersen 1992, Høeg and Lützen 1995). Secondly, no species of the Kentrogonida has been reported to grow roots initiating from the original settling site (Høeg 1985). Instead, Glenner et al. (2000) and Glenner (2001) show that the recently injected parasitic material is a cuticle-covered instar, the vermigon, which breaks from the stylet and is transported in the hemolymph until it reaches the site where it will start to grow into the adult parasite (Fig. 3).

We suggest that free migration via the hemolymph is the ancestral mode of reaching the abdomen of the host. An interna at the evolutionary stage depicted in Figure 3 could occasionally have lost contact with the integument of the gills, e.g., during a host moult. The small "interna" could have drifted through the heart to the central ventral blood sinus just as is observed extant Rhizocephala (Glenner et al. 2000). Here the internal parasite could grow and develop an externa. Most such incidents would have resulted in parasite death, but if an individual succeeded in emerging through the cuticle of the host, the "externa" would be well situated for being the object of brooding behavior by the host. This would occur irrespectively of whether the parasite had already developed some form of host hormonal control as in extant Rhizocephala. Experiments have demonstrated that even neutral, inanimate objects of externa size situated in the brood chamber will elicit some brood caring behaviour in female crabs (DeVries et al. 1989). In addition, the abdominal cavity is a relatively big and protected environment that would allow an externa to grow to a size larger than would be possible in the branchial chamber.

Once a migration pattern as outlined above had been established, natural selection would be expected to reduce the time the internae remained in the gills. Eventually the internal parasite (in the form of a vermigon) would be injected directly by a kentrogon into the hemolymph of the gills.

Our proposed scheme does not take into account any possible influence of immune mechanisms in the host. If some host organs, such as the gonads, are immunologically more permissive to parasite growth this could have affected where the interna first began to grow. Data is virtually absent in this respect but our histological sections at least do not indicate any difference between the different host tissues where internas were found. Aside from the almost automatic egg-mimicry effect gained by emergence in the abdominal brood chamber, extant rhizocephalans start to affect their hosts and being recognized as "self" already during the internal phase. This we suggest has evolved only after the life cycle arrived at the stage, where the interna always developed and emerged in the brood chamber. The details whereby the parasite affects the neuroendocrine system of the host and takes over control was reviewed by Høeg (1995) but the entire field is clearly in need of a modern molecularly based approach.

Abnormally situated externae

In May-June 1986 one of us (JTH) surveyed a population of *Carcinus maenas* infested with *Sacculina carcini* at Roscoff in Britanny, France and found evidence in support of our theory. At this time of the year, most of the external parasites are juveniles, i.e., small externae before their first brood release (Høeg and Lützen 1995). Among 112 such small, juvenile externae, three (2.7 %) were situated

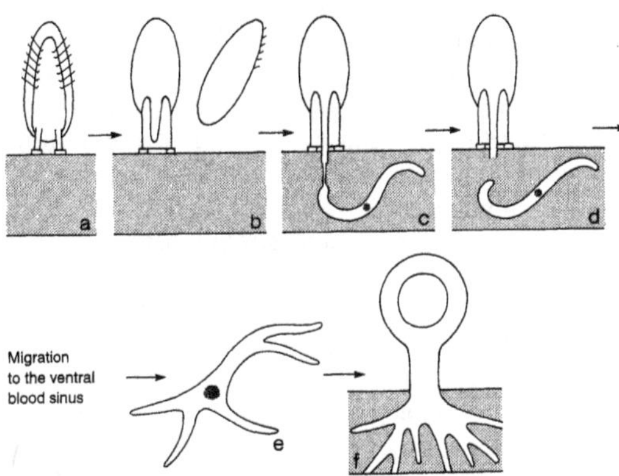

Figure 4. *Ontogeny of extant kentrogonid Rhizocephala. The cyprid settles in a site remote from where the externa will later emerge. The metamorphic moult of the cypris results in an external stage, the kentrogon, which is the first postcyprid instar and thus homologous to the juvenile barnacle in (b) of Figs 1-3. The kentrogon undergoes another moult but the kentrogon cuticle is not shed. The new instar consists of a vermiform part, coiled up inside the kentrogon and an external cuticular stylet, derived from the peduncle in Figs. 1-3. The stylet pierces the integument of the host. Following this, the vermigon part is evaginated through the stylet, whereafter it breaks free leaving the stylet (part of 2nd postcypris instar) and the kentrogon cuticle (1st postcypris instar) behind on the surface. The vermigon travels through the hemocoel to the ventral abdominal sinus, where it grows and eventually develops the reproductive body (externa).*

not underneath the abdomen in the brood chamber but abnormally on the dorsal (external) abdominal surface. These three individuals had a morphology exactly as those that emerged "correctly" underneath the abdomen and could even receive males and continue growth (Høeg and Ritchie 1985, Høeg 1987). No adult externa has ever been observed on the dorsal side of the abdomen and the reason is quite clear. Such parasites would almost certainly be damaged when the crab burrows if not simply torn to pieces during combative encounters with other crabs. These observations show, however, that even among extant kentrogonid Rhizocephala a sizable percentage of the externae emerge in a place other than the normal site. We see this variation in the site of emergence of externae as a trait upon which natural selection could have acted and as indicative of how the indirect establishment of the external parasite could have evolved.

The evolution of host-control

Once the abdominal brood chamber had become the normal site for emergence (Fig. 4), there would be a selective advantage to any parasite that could elaborate its "egg mimicry" into controlling host development, reproduction, and behaviour (Ritchie and Høeg 1981). Since both host and parasite are crustaceans, the ancestral parasite probably had the capability of expressing hormones that could influence host development and reproduction. Hence natural selection would only have to work on the timing of expression of such genes that affected feminisation, size at maturity, and inhibition of gonad maturation rather than developing homology of amino acid sequences. Høeg (1995) reviewed these aspects in detail. Eventually the hosts of extant kentrogonid Rhizocephala could be considered almost robot-like supplying the parasite's reproductive requirements or perhaps more aptly a host phenotype with a parasite genotype (O'Brien 1999).

III. CONCLUSIONS

Because we cannot point to obvious ways in which our ideas could be tested and falsified, what we have presented here is a scenario, not a hypothesis, on rhizocephalan evolution. The reason is that we find it unlikely that further studies of the metamorphosis and ontogeny of the Rhizocephala Kentrogonida will lead to much new insight concerning rhizocephalan origins, although they would undoubtedly shed light on the intrinsic relationships in the group. The few Akentrogonida, where host infection has been studied, use antennulary penetration with the cypris being the terminal instar, an ontogeny unique in Cirripedia (Høeg 1990, Glenner and Høeg 1994). Other Akentrogonida such as the elusive families Duplorbidae and Chthamalophilidae may yet hold surprises but this remains for the future.

A more promising avenue may lie in renewed studies of metamorphosis and early ontogeny in conventional cirripedes. Unfortunately the only thoracican species where metamorphosis has been studied in any detail belong to the highly derived balanomorph barnacles (Walley 1969, Glenner and Høeg 1998, Takenaka et al. 1993). We therefore see an urgent need for studies of metamorphosis in pedunculated barnacles and also in the Acrothoracica to elucidate the ancestral type of ontogeny in the now highly divergent cirripede orders.

ACKNOWLEDGEMENTS

We both wish to provide our sincere thanks to the organizers of the very successful 2000 Summer Meeting of The Crustacean Society in Puerto Vallarta, Mexico. HG gratefully acknowledges a grant no. 980076/20-1234 from the Carlsberg Foundation and JTH grants nos. 9701589 and 9901769 from the Danish Research Agency for financial support.

REFERENCES

Anderson DT (1994) [available 1993] Barnacles - structure, function, development and evolution. Chapman and Hall, London

Bocquet-Védrine J (1961) Monographie de *Chthamalophilus delagei* J. Bocquet-Védrine, Rhizocéphale parasite de *Chthamalus stellatus* (Poli). Cah Biol Mar 2:455-593

Bocquet-Védrine J (1972) Les Rhizocéphales. Cah Biol Mar 13:615-626

Bresciani J and Høeg JT (2001) Comparative ultrastructure of the root system in rhizocephalan barnacles (Crustacea: Cirripedia: Rhizocephala). J Morphol 9-42

Day JH (1939) A new cirripede parasite- *Rhizolepas annelidicola*, nov. gen. et sp. Proc Linn Soc Lond 151:64-78

Delage Y (1884) Évolution de la sacculine (*Sacculina carcini* Thomps.) crustacé endoparasite d l'ordre nouveau des kentrogonides. Archs Zool exp gén (ser 2) 2:417-736

DeVries MC, Rittschof D and Forward RB Jr (1989) Response by rhizocephalan-parasitized crabs to analogues of crab larval-release pheromones. J Crust Biol 9:517-524

Euzet L and Combes C (1998) The selection of habitats among the monogenea. Int J Parasitol 28:1645-1653

Glenner H (1998) Functional morphology of the cirripede cypris: a comparative approach. In: Thompson MF, Nagabhushanam R (eds) Barnacles. The Biofoulers (pp 161-187) Regency Publications, New Delhi, India

Glenner H (2001) Larval metamorphosis, injection and earliest internal development of the rhizocephalan *Loxothylacus panopaei* (Gissler). Crustacea: Cirripedia: Rhizocephala: Sacculinidae. J Morphol 43-75

Glenner H and Høeg JT (1994) Metamorphosis in the Cirripedia Rhizocephala and the homology of the kentrogon and trichogon. Zool Scr 23:161-173

Glenner H and Høeg JT (1995) A new motile, multicellular stage involved in host invasion by parasitic barnacles (Rhizocephala). Nature 377:147-150

Glenner H and Høeg JT (1998) Fate of the cypris and adult adductor muscles during metamorphosis of *Balanus amphitrite* (Cirripedia; Thoracica). J Crust Biol 18: 463-470

Glenner H and Werner M (1998) Increased susceptibility of recently moulted *Carcinus maenas* (L.) to attack by the parasitic barnacle *Sacculina carcini* Thompson 1836. J. Exp Mar Biol Ecol 228:29-33

Glenner H, Høeg JT, O'Brien JJ and Sherman TD (2000) Invasive vermigon stage in the parasitic barnacles *Loxothylacus texanus* and *L. panopaei* (Sacculinidae): closing of the rhizocephalan life cycle. Mar Biol 136:249-257

Harris DJ, Maxson LS, Braithwaite LF and Crandall KA (2000) Phylogeny of the thoracican barnacles based on 18S rDNA sequences. J Crust Biol 20:393-398

Hickling CF (1964) On the small deep-sea shark *Etmopterus spinax* L., and its cirripede parasite *Anelasma squalicola* (Loven). J Linn Soc (Zool) 45:17-24

Høeg JT (1985) Cypris settlement, kentrogon formation and host invasion in the parasitic barnacle *Lernaeodiscus porcellanae* (Müller) (Crustacea: Cirripedia: Rhizocephala). Acta Zool (Stockholm) 66:1-45

Høeg JT (1987) Male cypris metamorphosis and a new male larval form, the trichogon, in the parasitic barnacle *Sacculina carcini* (Crustacea: Cirripedia: Rhizocephala). Phil Trans R Soc Lond 317B:47-63

Høeg JT (1990) "Akentrogonid" host invasion and an entirely new type of life cycle in the rhizocephalan parasite *Clistosaccus paguri* (Thecostraca: Cirripedia). J Crust Biol 10: 37-52

Høeg JT (1995) The biology and life cycle of the Cirripedia Rhizocephala. J mar biol Ass UK 75:517-550

Høeg JT and Lützen J (1995) Life cycle and reproduction in the Cirripedia Rhizocephala. Oceanogr Mar Biol Annu Rev 33:427-485

Høeg JT and Ritchie LE (1985) Male cypris settlement and its effects on juvenile development in *Lernaeodiscus porcellanae* Müller (Crustacea: Cirripedia: Rhizocephala). J Exp Mar Biol Ecol 87:1-11

Høeg JT and Rybakov AV (1992) Revision of the Rhizocephala Akentrogonida (Cirripedia), with a list of all the species and a key to the identification of families. J Crust Biol 12: 600-609

Jensen PG, Moyse J, Høeg JT and Al-Yahya H (1994a) Comparative SEM studies of lattice organs: Putative sensory structures on the carapace of larvae from Ascothoracida and Cirripedia (Crustacea: Maxillopoda: Thecostraca). Acta Zool (Stockholm) 75:125-142

Jensen PG, Høeg JT, Bower S and Rybakov AV (1994b) Scanning electron microscopy of lattice organs in cyprids of the Rhizocephala Akentrogonida (Crustacea: Cirripedia). Can J Zool 72:1018-1026

Johnstone JAS and Frost WE (1927) *Anelasma squalicola* (Lovén): Its general morphology. Proc Trans Liverpool Biol Soc 41:29-91

Kolbasov G, Høeg JT and Elfimov AS (1999) Scanning electron microscopy of acrothoracican cypris larvae (Crustacea, Thecostraca, Cirripedia, Acrothoracica, Lithoglyptidae). Contrib Zool 68:143-160

Lützen J and Jespersen Å (1992) A study of the morphology and biology of *Thompsonia littoralis* (Crustacea: Cirripedia: Rhizocephala). Acta Zool (Stockholm) 73:1-23

Mizrahi I, Achituv Y, Katcoff DJ and Perl-Treves R (1998) Phylogenetic position of *Ibla* (Cirripedia: Thoracica) based on 18S rDNA sequence analysis. J Crust Biol 18:363--368

Newman WA (1982) Cirripedia. In: Abele LG (ed) The Biology of the Crustacea . Vol. 1. Systematics, the Fossil Record, and Biogeography (pp 197-221) Academic Press. London, New York

Newman WA, Zullo VA and Withers TH (1969) Cirripedia. In: Moore RC (ed) Treatise on Invertebrate Paleontology. Arthopoda 4(1) (pp R206-R296) Geol Soc Amer, Univ Kansas, Lawrence, Kansas

O'Brien JJ (1999) Parasites and reproduction. In: Encyclopedia of reproduction. Vol. 3 (pp 638-646) Academic Press

Perl-Treves R, Mizrahi L, Katcoff DJ and Achituv Y (2000) Elucidation of the phylogenetic relationship of three thecostracans, *Verruca*, *Paralepas*, and *Dendrogaster* based on 18S rDNA sequence. J Crust Biol 20:385-392

Petriconi V (1977) Wege zum Parasitismus - aufgezeigt an der Gruppe der Rankenfusskrebse. Natur und Museum 107:353-361

Ritchie LE and Høeg JT (1981) The life history of *Lernaeodiscus porcellanae* (Cirripedia Rhizocephala) and co-evolution with its porcellanid host. J Crust Biol 1:334-347

Spears T, Abele LG and Applegate MA (1994) A phylogenetic study of cirripeds and their relatives (Crustacea: Thecostraca). J Crust Biol 14:641-656

Takenaka M, Suzuki A, Yamamoto T, Yamamoto M and Yoshida M (1993) Remodeling of the nauplius eye into the adult ocelli during metamorphosis of the barnacle, *Balanus amphitrite hawaiiensis*. Dev Growth Differ 35:245-255

Thresher RE, Werner M, Høeg JH, Svane I, Glenner H, Murphy N and Wittwer W (2000) Developing the options for managing marine pests: specificity trials on rhizocephlan parasitic castrators against the European shore crab, *Carcinus meanas*, and related species. J Exp Mar Biol Ecol 254:37-51

Veillet A (1964) La métamorphose des cypris femelles des rhizocéphales. Zool Meded 39:573-576

Walker G (1999) The cypris larvae of the rhizocephalan barnacle *Heterosaccus lunatus* with particular reference to antennular morphology. Acta Zool (Stockholm) 80:209-217

Walley LJ (1969) Studies on the larval structure and metamorphosis of *Balanus balanoides* (L.). Phil Trans R Soc Lond 256B:237-280

MORPHOLOGICAL VARIATION IN THE CRAYFISH *CAMBARELLUS (CAMBARELLUS) MONTEZUMAE* (DECAPODA: CAMBARIDAE)

Yolanda Rojas, Fernando Alvarez and José Luis Villalobos

Colección Nacional de Crustáceos, Instituto de Biología, Universidad Nacional Autónoma de México, Apartado Postal 70-153, México 04510, D.F., México

ABSTRACT

The morphological variation exhibited by populations of *Cambarellus (Cambarellus) montezumae* in central and western Mexico is analyzed using somatic and genitalic characters. Five populations (Tequisquiapan, Queretaro; Chapultepec and Xochimilco, DF; Atlangatepec Dam, Tlaxcala; and Aljojuca Crater Lake, Puebla) were analyzed through seven morphological (shape and position of the caudal process, mesial processes, and central projection of the gonopod, shape and position of the *annulus ventralis* and postannular sclerite) and nine morphometric variables (total and cephalothorax length, areola length and width, rostrum length, chela length and height, palm length and movable finger or dactyl length). The results show that both morphological and morphometric characters can be useful to distinguish among populations and that there are two new species within this species complex.

I. INTRODUCTION

The species of the genus *Cambarellus*, family Cambaridae, are distributed from northern Florida and southern Illinois to Texas, and along Mexico's central plateau from the crater lakes in eastern Puebla to the Pacific slope (Hobbs 1989).

The genus includes 17 species classified in three subgenera. The subgenus *Cambarellus*, endemic of Mexico, groups 9 species in the states, from north to south, of: Chihuahua, Coahuila, Nuevo León, Sinaloa, Nayarit, Jalisco, Colima, Michoacán, Guanajuato, Querétaro, Tlaxcala, Morelos, México, Distrito Federal (DF), Hidalgo and Puebla.

Within *Cambarellus (Cambarellus)*, *C. montezumae* is without question the most widespread species, distributed along central Mexico, from the State of Puebla in the east, to Jalisco in the west. Although the morphological variation among populations of *C. montezumae* has long been recognized (e.g., Villalobos (1943, 1955) who described the variation in relative size, shape, and ornamentation of the rostrum among different populations), recent studies have revealed that this variation can be considerable, mainly among the populations distributed in the central plateau and the Pacific slope. With the purpose of documenting the extent of this variation and establishing the possible existence of new species, we embarked on the revision of five populations from Queretaro, DF, Tlaxcala, and Puebla (Fig. 1). These studies have shown that both reproductive (first pleopod of male form I and annulus ventralis in females) and somatic (rostrum, acumen, and chelipeds) structures can be sources of useful characters for discriminating among populations.

Kluwer Academic/Plenum Publishers

E. Escobar-Briones & F. Alvarez Eds.
MODERN APPROACHES TO THE STUDY OF CRUSTACEA
PP. 311-317

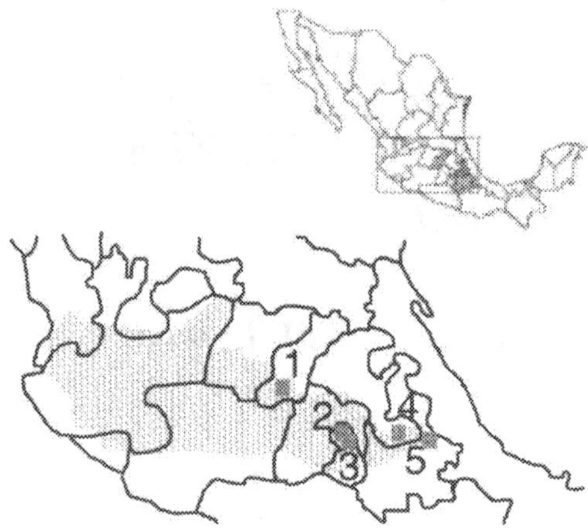

Figure 1. *Distribution range of **Cambarellus (Cambarellus)** **montezumae** in central Mexico. Vertical hatching represents the distribution area, black squares represent the collection sites used in this study.*

II. MATERIALS AND METHODS

All the organisms used in this study are deposited in the Colección Nacional de Crustáceos (CNCR) of the Instituto de Biología, UNAM. Approximately 60 individuals of each population were considered from the following sites (from west to east): Tequisquiapan, Queretaro; Chapultepec and Xochimilco, DF; Atlangatepec Dam, Tlaxcala; and Aljojuca Crater Lake, Puebla. Morphological and morphometric analyses of each organism were conducted. The morphological analysis considered the following characters: shape and position of the caudal process, mesial processes, and central projection of the gonopod, shape and position of the *annulus ventralis* and postannular sclerite. Drawings, photographs, and scanning electron microscope (SEM) micrographs were produced to illustrate the morphological variation.

The morphometric analysis was based on nine variables: total (TL) and cephalothorax length (CL), areola length (AL) and width (AW), rostrum length (RL), chela length (CHL) and width (CHW), palm length (PL), and movable finger or dactyl length (DL). The statistical analysis consisted of one way analysis of variance (ANOVA), using the five populations as treatments. In those cases where significant differences were found, Tukey's HSD test was used to make all possible individual comparisons among populations. Total length

was analyzed alone, while the rest of the variables were used as proportions in order to avoid obtaining significant differences due to size. The proportions examined are the following: RL/CL, AL/AW, AL/CL, CHL/ CHW, CHL/CL, PL/DL, and DL/CHL. All the new variables, resulting from calculating the proportions, were tested for normality. Males and females were analyzed separately in those cases where the chelae was involved due to the sexual dimorphism exhibited.

III. RESULTS

The variation observed among the populations of *Cambarellus montezumae* studied is described below, in two sections. First, the morphological analysis, which includes a brief description of the male form I and the female. The populations are presented from west to east and the characters analyzed are shown in figures 2 to 4. Second, the results of the statistical analysis and its graphical representation are presented in figure 5. Tables 1 through 3 contain: the areola length/ areola width ratios, the ANOVA results, and a summary and comparison of the most significant morphological characters of the five populations studied, respectively.

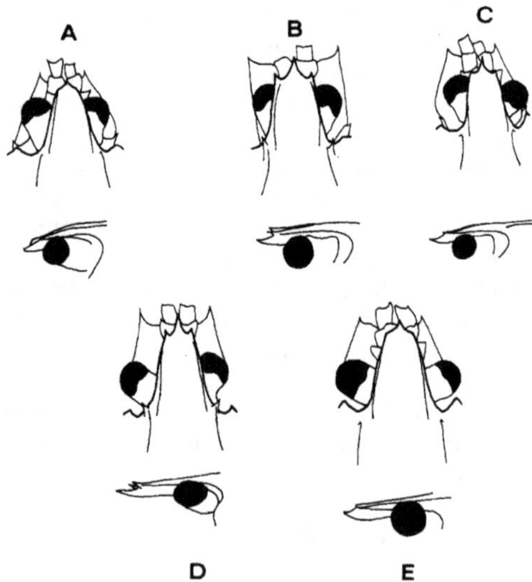

Figure 2. *Dorsal and lateral views of the rostrum of **Cambarellus (Cambarellus) montezumae**: A, Tequisquiapan, Queretaro; B, Chapultepec Lake, D.F.; C, Xochimilco, D.F.; D, Atlangatepec, Tlaxcala; E, Aljojuca, Puebla.*

Morphological analysis
Tequisquiapan, Querétaro

Rostrum without marginal spines, dorsal surface flat, smooth; rostral borders convergent, subrostral borders evident in dorsal view. Postorbital borders without spines. Acumen triangular, wide (Fig. 2A). Areola length/areola width ratio 3.4 (Table 1). Chelae of first pair of pereiopods of male form I slender, smooth, subcylindrical, with scattered setae along the cutting surface; cutting edge sharp, divided into small indentations (Fig. 3A). In mesial view, first pleopod of male form I symmetrical, caudal process subsetiform, slender, caudodistally oriented; basal portion of caudal process next to central projection; central projection wide basally, tapering distally (Fig. 4A). Mesial process wide, distal third grooved, ending in rounded tip, twice as long as wide (Fig. 4B). *Annulus ventralis* bell-shaped, wide and transversally positioned. Lateral sinus close to the base of *annulus*, with straight borders, cutting the lateral and caudal surfaces. Basal lateral projections cephalically oriented. Postannular sclerite wide basally, tapering distally, ending in rounded tip.

Table 1. *Areola length/areola width ratio and areola length/ cephalothorax length ratio for the five populations studied.*

Population	Sample size	Ratio	
		Lar/AAr	(Lar/LC)x100
Tequisquiapan	21	3.0-3.6 (3.4)	30.3-33.3 (32.2)
Chapultepec	50	2.8-3.5 (3.3)	30.3-34.4 (32.8)
Xochimilco	67	2.9-4.3 (3.5)	30.2-35.2 (32.7)
Atlangatepec	52	3.0-3.7 (3.3)	31.6-34.3 (32.6)
Aljojuca	60	3.2-4.4 (3.7)	31.6-35.6 (33.4)

Chapultepec, DF (type locality)

Rostrum with marginal spines, dorsal surface grooved, rostral borders subparallel, subrostral borders evident in dorsal view. Postorbital borders ending anteriorly in spines. Acumen wide at base, becoming sharp distally (Fig. 2B). Areola length/areola width ratio 3.3 (Table 1). Chelae of first pair of pereiopods of male form I smooth, slender, subcylindrical, with scattered setae on the cutting surfaces of fingers (Fig. 3B). In mesial view, first pleopod of male form I symmetrical. Caudal process caudomesially oriented, wide, curved apically. Central projection wide basally, tapering distally (Fig. 4C). Mesial process, moderately wide, mid portion grooved, distolaterally oriented, almost twice as long as wide (Fig. 4D). *Annulus ventralis* bell-shaped,

with slight torsion in one branch. Lateral sinus in central portion of lateral branch, with undulated borders, cutting transversally lateral and caudal surfaces. Basal lateral projections short. Postannular sclerite triangular.

Xochimilco, DF

Rostrum with marginal tubercles, dorsal surface almost flat; rostral borders subparallel, converging anteriorly; subrostral borders not visible in dorsal view. Postorbital borders ending in tubercles anteriorly. Acumen wide at base tapering distally, lateral borders with setae (Fig. 2C). Areola length/areola width ratio 3.5 (Table 1). Chela of first pair of pereiopods of male form I smooth, slender, subcylindrical, with scattered setae along cutting edges of fingers (Fig. 3C). In mesial view, first pleopod of male form I symmetrical, caudal process subsetiform, slender, caudodistally oriented, slightly shorter than central projection; central projection wide at base tapering apically (Fig. 4E). Mesial process moderately wide, tubular with apical opening, ending in rounded tip, its length almost twice its basal width (Fig. 4F). *Annulus ventralis* bell-shaped, with torsion in one branch. Lateral sinus in central portion of lateral branch, with undulated borders, cutting tranversely lateral and caudal surfaces. Lateral projections short. Postannular sclerite conical.

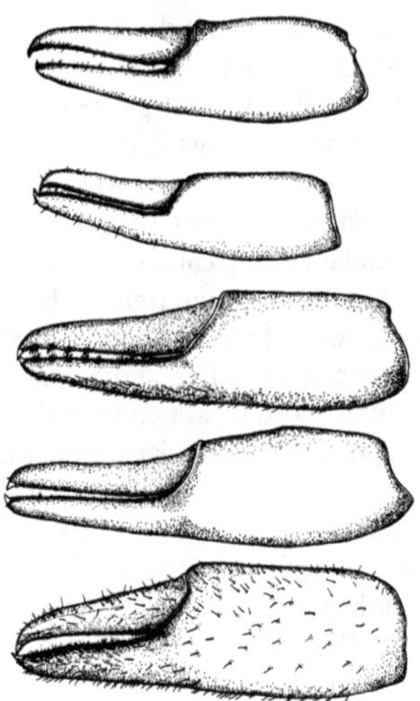

Figure 3. *Lateral view of the chela of* **Cambarellus** *(***Cambarellus***) **montezumae***: A, Tequisquiapan, Queretaro; B, Chapultepec Lake, D.F.; C, Xochimilco, D.F.; D, Atlangatepec, Tlaxcala; E, Aljojuca, Puebla.*

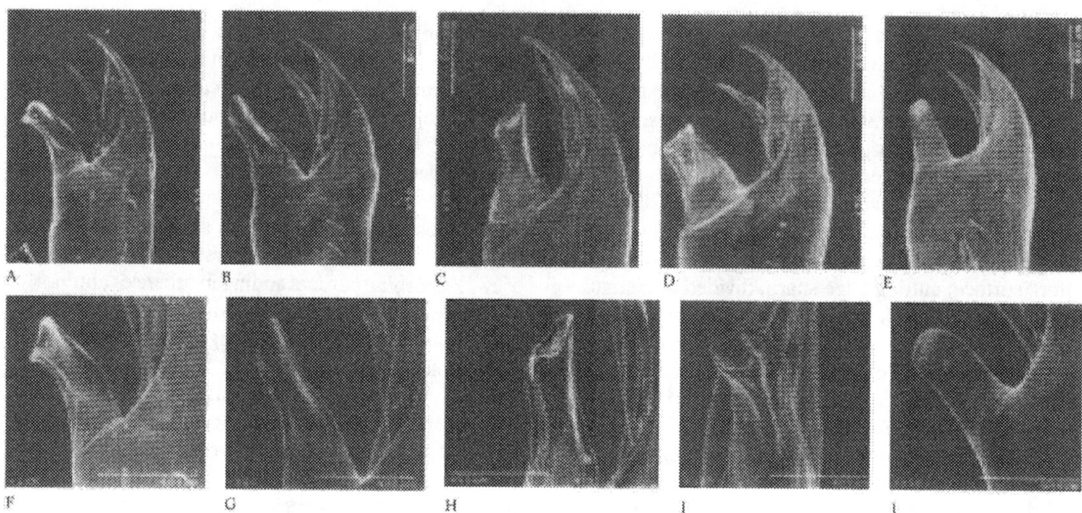

Figure 4. *Mesial view and detail of mesial process of the male form I first pleopod A-B, Tequisquiapan, Queretaro; C-D Chapultepec Lake, D.F.; E-F, Xochimilco, D.F.; G-H, Atlangatepec, Tlaxcala; I-J, Aljojuca, Puebla.*

Atlangatepec, Tlaxcala

Rostrum with marginal spines, dorsal surface grooved, smooth; rostral borders subparallel, subrostral borders evident in dorsal view. Postorbital borders ending in spines laterally oriented. Acumen triangular, long, wide (Fig. 2D). Areola length/areola width ratio 3.3 (Table 1). Chelae of first pair of pereiopods of male form I smooth, slender, subcylindrical, with scattered setae on cutting edge (Fig. 3D). In mesial view, first pleopod of male form I symmetrical. Caudal process wide, caudomesially oriented, curved apically. Central projection wide, tapering apically (Fig. 4G). Mesial process short, wide, tubular, distal third grooved; in caudal view apical aperture visible, ending in rounded tip (Fig. 4H). *Annulus ventralis* bell-shaped, with slight torsion in one of the branches. Lateral sinus with ondulated borders, cutting lateral and caudal surfaces. Basal lateral projections short, separated from the *annulus* by a deep and wide notch. Postannular sclerite triangular.

Aljojuca, Puebla

Rostrum without marginal spines, dorsal surface flat; rostral borders convergent; subrostral borders evident in dorsal view. Postorbital borders visible in dorsal view and ending anteriorly in small tubercles. Acumen short, triangular (Fig. 2E). Areola length/areola width ratio 3.7 (Table 1). Chelae of first pair of pereiopods of male form I strong, surface of palm and fingers covered with short setae (Fig. 3E). In mesial view, first pleopod of male form I symmetrical. Caudal process subsetiform, slender, caudodistally oriented; basal portion separated from mesial process, slightly shorter than other two processes; central projection wide at base tapering apically (Fig. 4I). Mesial process moderately wide, short, flattened, without groove in the central portion (Fig. 4J). *Annulus ventralis* bell-shaped, in transversal position. Sinus in distal portion of lateral branch, with undulated borders, cutting lateral and caudal surfaces. Lateral projections short. Postannular sclerite pyramidal.

Table 2. *Results from the ANOVA and Tukey's HSD test. The five populations studied were taken as treatments (1, Tequisquiapan; 2, Chapultepec; 3, Xochimilco; 4, Atlangatepec; 5, Aljojuca).*

	ANOVA		Tukey's HSD
Variable	F – value	Probability	
TL	33.829	0.0001	1 = 2, 3 = 5
RL/CL	53.490	0.001	1 = 5, 2 = 4
AL/AW	27.580	0.0001	1 = 3 = 4
AL/CL	5.866	0.0001	1 = 2 = 3 = 4, 4 = 5
CHL/CHH♂	50.957	0.0001	1 = 2, 4 = 5
CHL/CHH♀	29.944	0.0001	1 = 2 = 3, 4 = 5, 1 = 5
CHL/CL♂	14.100	0.0001	1 = 2 = 4 = 5
CHL/CL♀	21.965	0.0001	1 = 2, 3 = 4, 4 = 5
PL/DL♂	8.231	0.0001	1 = 2 = 3 = 4
PL/DL♀	13.680	0.0001	1 = 2 = 3 = 4
DL/CHL♂	2.408	0.054	1 = 2 = 3 = 4 = 5
DL/CHL♀	3.876	0.0056	1 = 2, 1 = 3 = 4 = 5

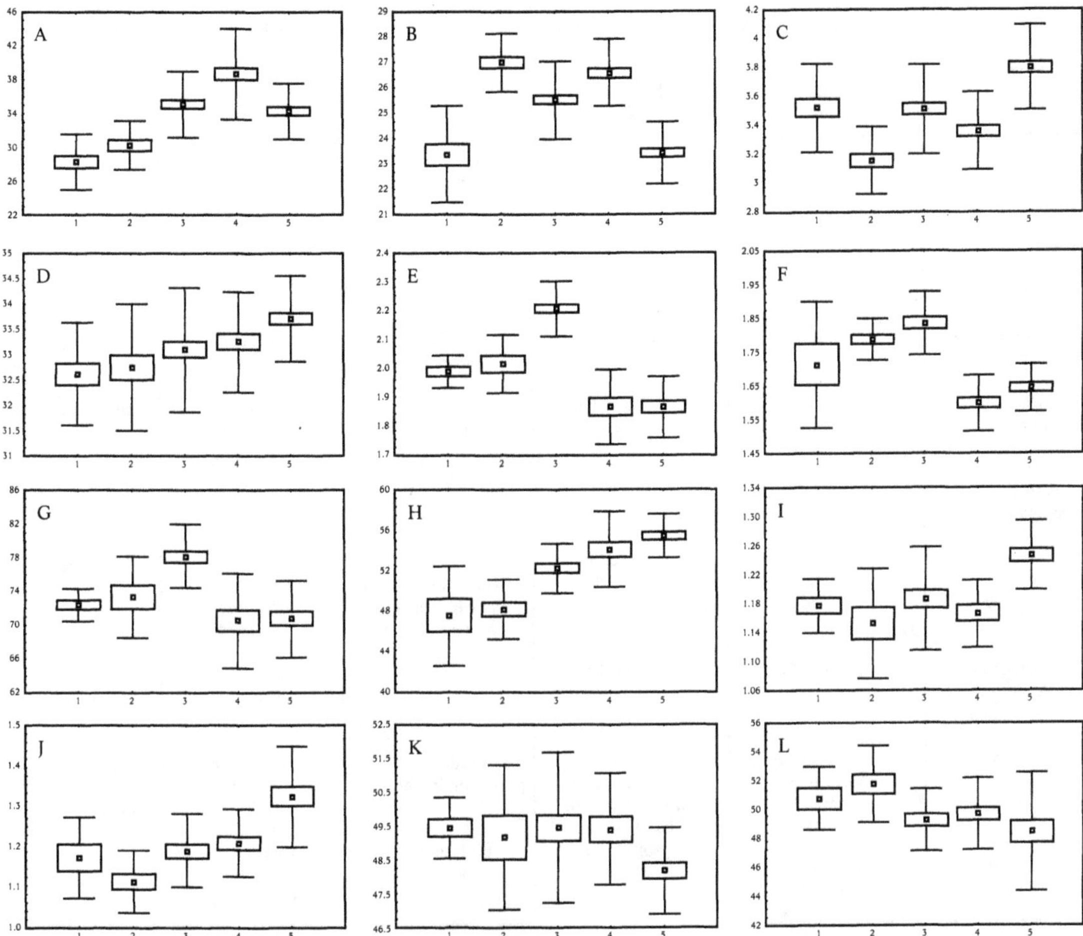

Figure 5. *Graphs of the mean, standard error (box), and standard deviation (error bars), of selected structures and ratios obtained from the five populations of **Cambarellus (Cambarellus) montezumae** examined in this study: A, total length; B, rostrum length/ cephalothorax length; C, areola length/areola width; D, areola length/cephalothorax length; E, chela length/chela width (males); F, chela length/chela width /females); G, chela length/cephalothorax length (males); H, chela length/cephalothorax length (females); I, palm length/dactyl length (males); J, palm length/dactyl length (females); K, dactyl length/chela length (males); L, dactyl length/chela length (females). In the horizontal axis each number corresponds to one of the five populations studied as follows: 1, Tequisquiapan, Queretaro; 2, Chapultepec, DF; 3, Xochimilco, DF; 4, Atlangatepec, Tlaxcala; 5, Aljojuca, Puebla.*

Morphometric analysis

The nine variables and proportions analyzed showed significant differences among populations, except for the DL/CHL proportion for males, where no differences were detected (Table 2). The differences among populations do not follow a defined pattern, the results show that the differences and similarities among populations change depending on the charac-

ters analyzed, these differences are shown in figure 6. Some salient differences obtained are: 1) the organisms from Atlangatepec attain a larger size than the rest (Fig. 5A), 2) Tequisquiapan and Aljojuca have significantly larger rostral lengths (Fig. 5B), 3) the Xochimilco population differs from the rest in rostrum length relative to cephalothorax length (Fig. 5B), 4) areola length presents a clinal variation from west to

Table 3. *Comparison of characters of the five populations of **Cambarellus (Cambarellus) montezumae** examined.*

	Tequisquiapan, Queretaro	Chapultepec, DF	Xochimilco, DF	Atlangatepec, Tlaxcala	Aljojuca, Puebla
Marginal spines of rostrum	Absent	Spines or tubercles	Modified as tubercles	Present	Absent
Rostral borders	Convergent	Subparallel	Subparallel	Subparallel	Convergent
Rostral surface	Flat	Grooved	Grooved	Concave	Flat
Postorbital spines	Absent	Spines or tubercles	Modified as tubercles	Present	Modified as small tubercles
Chela	Slender and smooth	Slender and smooth	Slender and smooth	Slender and smooth	Robust, covered with short setae
Palm length/dactyl length ratio	1.18 ♂ 1.17 ♀	1.15 ♂ 1.08 ♀	1.19 ♂ 1.19 ♀	1.17 ♂ 1.21 ♀	1.25 ♂ 1.32 ♀
First pleopod of male form I, mesial process	Robust, mid section grooved	Mid section grooved	Moderately wide, cylindrical with apical opening	Wide, cylindrical, distal third grooved	Moderately wide
First pleopod of male form I, caudal process	Caudodistally oriented	Caudomesially oriented	Caudodistally oriented	Caudomesially oriented	Caudodistally oriented
Annulus ventralis	Without torsion, basal lateral projections short, cephalically oriented	Moderate torsion in one of the branches, basal lateral projections short	Moderate torsion in one of the branches, basal lateral projections short	Moderate torsion in one of the branches, basal lateral projections with a notch	Without torsion, basal lateral projections short

east with Aljojuca presenting longer areolas (Fig. 5D), 5) the Aljojuca population is significantly different from the rest in the length/width proportion of the areola and in the palm length/dactyl length proportion (Figs. 5C, I, J).

IV. DISCUSSION

Cambarellus (Cambarellus) montezumae is a widespread species distributed throughout Mexico's central plateau, from the crater lakes in the eastern limit of the State of Puebla (19° 05' N, 97° 33' W) to Lake Chapala in the State of Jalisco (20° 18' N, 103° 42' W). Without a doubt *C. montezumae* (sens. lat.) is the species of freshwa-

ter decapod with the largest geographical distribution range in Mexico. Although with a common origin, many of the extant populations are now effectively isolated and some of them occur only in very small reservoirs. The distribution pattern and the degree of isolation may be two of the factors that have promoted the extensive morphological variation found among populations.

The morphological variation found in the pleopod of the male form I of *C. montezumae* had never before been documented. Both within and among population variation was found. The highest degree of within population variation was found in the two populations located in the Valley of Mexico: Chapultepec and Xochimilco. This result contrasts with the generalized perception that the characters found in the male's first pleopods are stable and reliable for discriminating between species (Hobbs and Villalobos 1964, Chambers et al. 1980). In the case of these two populations the effect of man-mediated introductions and possibly pollution can not be ruled out as causes of the observed variation. In contrast, the other three populations studied exhibit very little variation in the gonopod morphology. Based on the gonopod morphology, mainly on the morphology of the mesial process, the populations from Atlangatepec and Aljojuca could be regarded as two new species, whose descriptions will be published elsewhere.

Somatic characters such as rostrum, chelae, and areola, may be more variable than characters obtained from the genitalia (Chambers et al. 1979, Hobbs 1991, Taylor 1997); however, they can be useful in supplying complementary information. In this study, it was found through the statistical analysis that the rostrum, areola, and chela can be used to discriminate among populations. These analyses also showed that there is a considerable amount of variation among populations that are still now considered to belong to the same species (Table 3). Further, these results also suggest that as more populations are included in the analysis, more variations will be found. Only five populations from the eastern section of the range of *C. montezumae* (sens. lat.) were considered in this study (Fig. 1), leaving to the west probably more than 15 other populations that need to be examined.

The results shown serve to illustrate what has become a common problem in the taxonomy of freshwater species, especially cambarids, with ample distribution ranges. In previous studies (Rojas 1998, Rojas et al. 1999, 2000) we have detected the existence of new forms within the range of well known species, which were thought to present a uniform morphology throughout their distribution area. In fact, within the Mexican cambarids there are still several species complexes (e.g., *Procambarus* (*Austrocambarus*) *mexicanus*, *P.* (*A.*) *acanthophorus*, *P.* (*A.*) *llamasi*) that are probably composed of more than one species. The pace at which these populations are studied should increase to allow for the planning of appropriate conservation policies before they start disappearing.

REFERENCES

Chambers CL, Payne JF and Kennedy ML (1979) Geographic variation in the dwarf crayfish *Cambarellus puer* Hobbs (Decapoda: Cambaridae). Crustaceana 36:39-55

Chambers CL, Payne JF and Kennedy ML (1980) Geographic variation in the first pleopod of the form I male crayfish *Cambarellus puer* Hobbs (Decapoda: Cambaridae). Crustaceana 38:169-177

Hobbs HH Jr and Villalobos A (1964) Los cambarinos de Cuba. An Inst Biol, Univ Nal Autón Méx 84:307-366

Hobbs HH III (1991) Decapoda. In: Thorp JH and Covich AP (eds) Ecology and Classification of North American Freshwater Invertebrates, Academic Press Inc

Rojas Y (1998) Revisión taxonómica de ocho especies del género *Procambarus* (Crustacea: Decapoda: Cambaridae) del centro de Veracruz, México. Tesis Profesional, Facultad de Ciencias, Univ Nal Autón Méx, 158 p

Rojas Y, Alvarez F and Villalobos JL (1999) A new species of crayfish of the genus *Procambarus* (Crustacea: Decapoda: Cambaridae) from Veracruz, Mexico. Proc Biol Soc Wash 112:396-404

Rojas Y, Alvarez F and Villalobos JL (2000) A new species of crayfish (Crustacea: Decapoda: Cambaridae) from Lake Catemaco, Veracruz, Mexico. Proc Biol Soc Wash 113:792-798

Taylor CA (1997) Taxonomic status of members of the subgenus *Erebicambarus*, genus *Cambarus* (Decapoda: Cambaridae), east of the Mississippi River. J Crust Biol 17:352-360

Villalobos A (1943) Estudios de los Cambarinos Mexicanos I. Observaciones sobre *Cambarellus montezumae* (Saussure) y algunas de sus formas con descripción de una subespecie nueva. An Inst Biol, Univ Nal Autón Méx 15:587-611

Villalobos A (1955) Cambarinos de la Fauna Mexicana (Crustacea, Decapoda). Tesis Doctoral, Facultad de Ciencias, Univ Nal Autón Méx 290 p

PHYLOGENETIC RELATIONSHIPS IN SOME SPECIES OF THE GENUS *MACROBRACHIUM* BASED ON NUCLEOTIDE SEQUENCES OF THE MITOCHONDRIAL GENE CYTOCHROME OXIDASE I

Guido Pereira, Hilda De Stefano, Joseph Staton and Brian Farrell

GP, Universidad Central de Venezuela, Instituto de Zoología Tropical,
Apartado 47058, Caracas 1041-A, Venezuela
HDS, Universidad Nacional Experimental Simón Rodríguez, Instituto de Estudios Científicos
y Tecnológicos, Centro de Estudios Biomédicos y Veterinarios,
Apartado 47925 Caracas 1041-A, Venezuela
JS, BF, Harvard University, Museum of Comparative Zoology, 26 Oxford Street,
Cambridge, Massachusetts 02138, USA

ABSTRACT

A phylogeny for four species of the freshwater prawn genus *Macrobrachium* (*M. carcinus, M. olfersii, M. acanthurus* and *M. rosenbergii*) based on the partial nucleotide sequence of the mitochondrial cytochrome c oxidase subunit I gene is presented. The results are consistent with previous studies based on morphological data.

I. INTRODUCTION

The genus *Macrobrachium* comprises a group of approximately 250 species distributed in the tropics worldwide, they are very common in freshwater rivers and one species, *Macrobrachium rosenbergii* (De Man) is cultivated worldwide. The alpha taxonomy of the genus as well as that of the whole subfamily Palaemoninae has remained without major changes after the monographic studies of Holthuis (1950, 1951, 1952), who combined several subgenera within the single genus *Macrobrachium*. Regarding the phylogenetic relationships within the genus, some authors have recognized several species groups, but no formal action has been taken to reorganize the genus so that it reflects phylogenetic relationships. Pereira (1997), based on previous work (Pereira 1989) showed a cladogram of the subfamily Palaemoninae in which the most obvious conclusion was that the subfamily shows different degrees of paraphyly at the subfamily and generic levels (Fig. 1). However, further corroboration is necessary in order to improve this phylogenetic hypothesis towards a more stable model. Molecular systematics may provide further evidence independent of previous morphological analysis that would be of great help in order to improve our understanding of phylogenetic relationships among the group.

The objective of the present study is to determine the partial nucleotide sequence of the mitochondrial cytochrome c oxidase subunit I gene (COI) of representative species of *Macrobrachium* to begin a genetic database of palaemonid shrimps and to establish phylogenetic hypothesis to compare it with previously published phylogenies based on morphological characters.

II. MATERIALS AND METHODS

Four species of freshwater shrimps in the family Palaemonidae (*Macrobrachium carcinus* (L), *M. olfersii* (Wiegmann), *M. acanthurus* (Wiegmann) and *M. rosenbergii* (De Man)) and one species in the family Atyidae (*Potimirim potimirim* (Müller)) were selected. Additionally, five taxa whose sequences were obtained from

E. Escobar-Briones & F. Alvarez Eds.
MODERN APPROACHES TO THE STUDY OF CRUSTACEA
PP. 319-322

GenBank were included (*Litopenaeus vannamei* (Boone), *Artemia franciscana* (L), *Daphnia pulex* (Leidig), *Anopheles gambie* (Giles) and *Drosophila yakuba*).

The laboratory methods used in this study followed Hillis et al. (1996). DNA was extracted from muscle tissue and visualized in agarose gels. mtDNA fragments were amplified by the polymerase chain reaction (PCR) using tested primers for COI (Table 1)

Table 1. *Sequence information for oligonucleotide primers (Simon et al. 1994, Folmer et al.1994).*

A2963	5' AGGTAGTTCTTCATTATAIGAATGTTC 3'
A2771	5' GGATAA/GTCAGAA/GTAACGTCGA/TGG/TGGTATA/C 3'
S1718	5' GGAGGATTTGGAAATTGATTAGTTCC 3'
S1991	5' GTAATTAATATACGACCTAAAGG 3'
LCO1490	5' GGTCAACAAATCATAAAGATATTG 3'
HCO2198	5' TAAACTTCAGGGTGACCAAAAAATCA 3'

with the following thermal cycle protocol (40 cycles): 30 sec at 94 °C (DNA denaturing); 30 sec at 50 °C (primer annealing); 60 sec at 72 °C (primer extension). The product of this reaction was purified after being run on an agarose gel (QIAquick columns, Qiagen, Valencia, CA, USA) and visualized in agarose gel. Sequencing was carried out using DNA automatic sequencing machine (ABI PRISM 370A). The sequences were edited, translated to amino acids of the COI, aligned and translated back to DNA sequences. DNA sequences for related species were obtained fron GenBank. Phylograms were generated using a distance method (Saitou and Nei 1987) and maximum parsimony (MP). Several outgroups were used simultaneously in order to use the overall outgroup parsimony (Madison et al. 1984, Pereira 1997). Phylogenetic trees were generated with the computer program PAUP (Swofford 1998), using the Branch and Bound algorithm excluding 3rd position.

Figure 1. *Partial view of the morphological based cladogram after Pereira (1997). (*)refers to species of **Macrobrachium** used in the analysis*

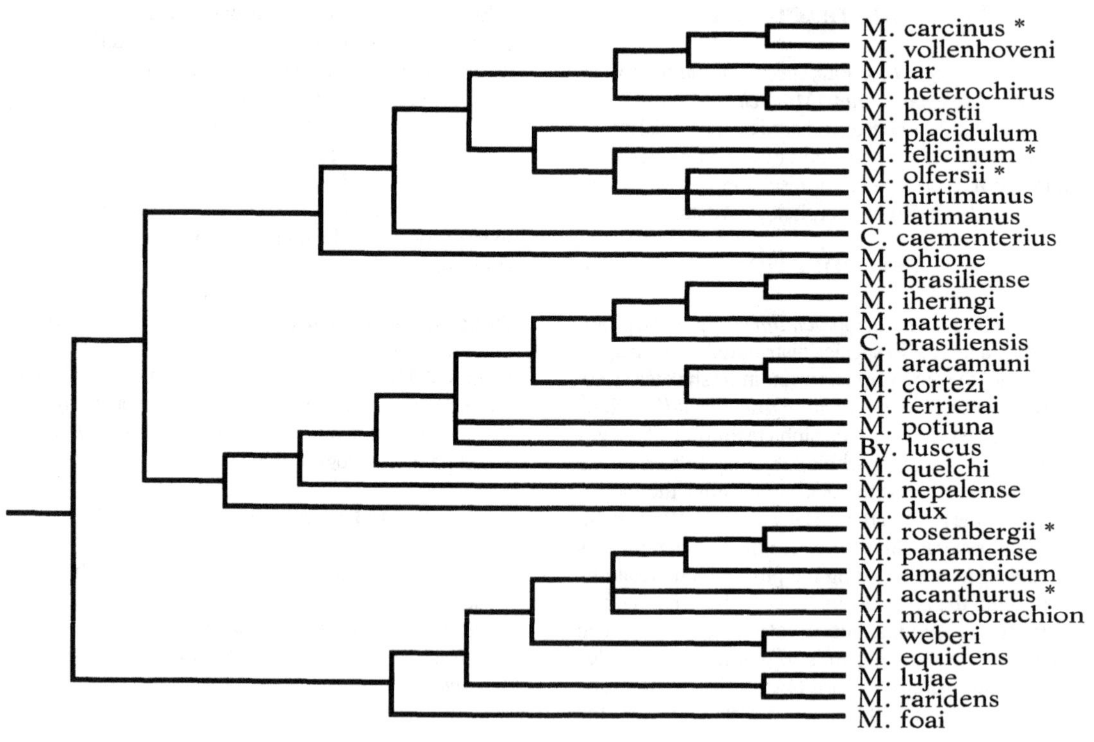

Table 2. *Sequence and alignment of 651 nucleotide for the Cytochrome C Oxidase Subunit I for five species used in the analysis.*

Potimirim potimirim

?????????????????????????????GAATAGTAGGAACTGCCCTAAGTCTCTTAATTCGAGCTGAACTAGGTCAGCCAGGCAGACTAATTGGGA
ATGATCAAATTTACAACGTAWTTGTTACAGCCCACGCCTTCGTTATAATCTTCTTCATAGTTATACCGATTATAATTGGTGGATTCGGCAACT
GACTGGTCCCACTAATGCTAGGAGCCCCAGATATGGCTTTTCCCCGGATAAACAACATAAGATTCTGACTTTTACCACCCTCCCTAACTCTT
TTACTATCCAGAGGAATAGTAGAAAGGGGCGTGGGCACAGGATGAACTGTGTACCCACCCCTTGCTAGAGGAATTGCTCATGCCGGGGCT
TCTGTCGACCTTGGTATTTTTTCACTCCACTTAGCAGGAGTGTCTTCTATCCTAGGGGCCGCTAACTTCATATCTACAGTAATCAATATACGA
AGAACAGGAATACTAATAGACCGGATACCCCTCTTTGTGTGATCTGTCTTTATTACTGCCATTTTATTACTTCTCTCTCTCCCGGTACTAGCC
GGAGCTATTACTATACTTTCAACCGACCGTAATTTAAATACCTCATTCTTTGACCCCGCAGGAAGTGGTGACC????????????????????

Macrobrachium acanthurus

CTCATCTTGATCTTCGGAKCTCGAGCTGGCATATTAAGCACATCTCTTTAAGACTTAATCCGGGCTGAACTAGGGCAGCCAGGCAGACTAAT
GAGAAATTATCAGATCTACAACGTCATTGTCACTGCCCACGCATTCGTGATAATCTTCTTCATGGTTATACCCATCATAATTGGAGGATTCGGTA
ACTGACTACTGCCCCTAATACTAGGTGCCCCCGACATGGCCTTCCCCCGAATAAACAACATGAGATTCTGACTTTTACCCCCATCACTCACAC
TCCTTTTATCGAGAGGTATGGTAGAAAGCGGGGTAGGCACGGGGTGAACTGTATACCCCCCACTAGCAGCCGGGACTGCACATGCAGGAGC
ATCAGTAGACCTCGGGATTTTCTCATTACACCTCGCAGGTGTTTCTTCCATCCTAGGAGCGGTCAATTTTATTACCACAGTAATTAACATGCG
GTCACCGGGCATAACCATGGACCGACTCCCCCTATTTGTATGGGCGGTATTCCTGACAGCCATTCTTCTTTTTATTTCCCTTCCTGTATTAG
CCGGAGCAATTACCATGTTACTAACAGACCGAAACCTAAATACTTCCTTCTTTGACCCTGTTGGGGGAGGAGACCCA????????????????

Macrobrachium olfersii

CTATCTTGGATCTTCGGAGCTTGAGCAGGTATAGTCGGCACATCCCTAAGACTCTTAATTCGAGCTGAATTAGGTCAACCCGGGAGACWGAT
TGGGAATGACCAAATCTACAACGTCATWGTCACCGSWCACGCTTTCGTAATAATTTTCTTCATGGTAATGCCTATCATAATTGGAGGATTCGG
TAATTGACTAGTCCCTCTAATACTAGGAGCACCTGATATGGCCTTCCCCCGAATGAATAACATAAGATTCTGACTCCTGCCCCCCTCTCTAAAT
CTTCTTCTATCCAGAGGAATGGTAGAAAGAGGTGTAGGAACCGGGTGAACCGTGTATCCCCCCCTAGCTGCAGGAACCGCCCACGCCGGAG
CCTCAGTTGACCTCGGTATCTTTTCCCTCCACTTAGCCGGAGTCTCATCAATCCTAGGAGCCGTAAATTTCATTACAACTGTAGTTAACATGC
GATCTCCTGGAATAACTATAGACCGACTACCCCTATTCGTTTGAGCGGTATTATTAACCGCCATTATTGTCGTTCTTTCCCTCCCAGTATTAGC
TGGAGCTATCACTATGGTTATAACTGACCGAAACCTAAATACATCCTTCTTTGACCCTGCCGGGGGTGGAGACCCAATTTTAT??????????

Macrobrachium carcinus

?????????????????????????????????????GCACATCTCTGANACTCTTAATTCGTGCTGAGTTAGGACAACCGGGCAGACTAATCGGAAA
TGATCAAATCTACAACGTTATTGTCACAGCTCACGCGTTCGTAATAATTTTCTTCATGGTAATGCCAATTATAATTGGAGGCTTTGGATATTG
GCTAGTCCCCCTTATACTAGGAGCCCCAGACATGGCCTTCCCGCGAATAAATAATATAAGATTCTGGCTCTTACCTCCCTCTCTAACTCTCCT
ACTATCTAGAGGAATAGTGGAAAGAGGGGGTGGGGACAGGATGAACTGTGTACCCCCCTCTAGCAGCAGGAACTGCTCACGCGGGACCTG
CAGTAGACCTTGGAATCTTTTCCCTTCACCTTGCCGGTGTCTCATCTATCCTGGGTGCCGTCAATTTCATCACCACTGTAATTAACATGCGA
TCACCAGGAATAACCATAGACCGGCTACCCCTATTAGTGTGAGTCGTNTTGTTAACAGCAATTTTATTTCTATTATCCCTGCCCGTGTTAGCG
GGAGCCATCACTATATTATTGACAGACCGAAACTCAAATACTGANNNTTTYGACCCAGCAGGNGGAGGCGACCTCAAT??????????????

Macrobrachium rosenbergii

CCCATCTTGGACYTCGGAGCGTGAGCAGGCATGGTAGGTACGTCACTAAGACTCTTAATTCGAGCAGAATTAGGGCAGCCGGGCAGACTGA
TCGGAAATGACCAAATCTACAACGTAATTGTCACTGCCCACGCATTCGTAATAATTTTTTTCATGGTTATACCGATCATAATTGGTGGTTTCGGT
AATTGACTAGTACCCCTAATATTAGGGGCCCCAGACATAGCATTCCCACGCATAAACAACATAAGATTCTGACTCCTACCCCCATCTCTAACACT
TCTTCTCTCCAGAGGAATAGTAGAAAGAGGGGGTTGGCACAGGATGAACTGTTTATCCACCACTAGCGGCCGGTACCGCCCACGCCGGGGCA
TCGGTAGATCTAGGTATTTTTTCCCTCCACCTAGCAGGAGTTTCTTCAATCTTAGGGGCTGTCAACTTTATTACCACAGTGATTAACATACGAG
CCCCAGGAATAACTATAGATCGACTGCCCCTATTCGTATGAGCCGTATTTCTAACAGCCATCCTGCTTCTTCTCTCATTACCAGTTTTAGCCGG
AGCCATTACCATACTCTTAACTGATCGAAACCTAAATACATCCTTTTTCGACCCAGCGGGAGGAGGGGACCCTATTCTCTACCGACA????

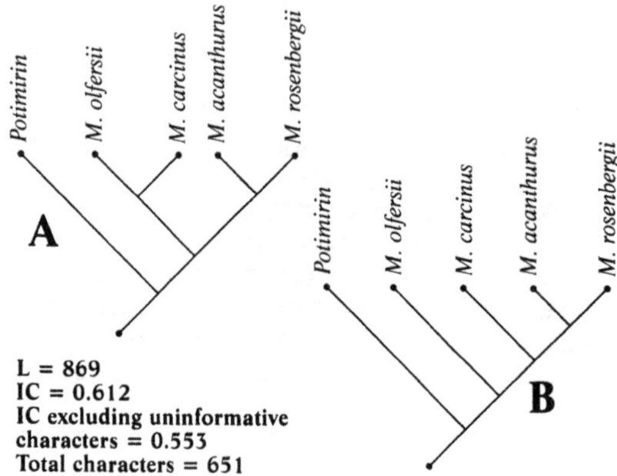

Figure 2. *A: distance based tree (Neighbor-Joining). B: maximum parsimony tree.*

III. RESULTS

We sequenced 651 DNA nucleotide bases at the 5' end of the COI mitochondrial gene (Table 2), for four species of freshwater shrimps in the family Palaemonidae and one species in the Family Atyidae. The results of the distance and phylogenetic analyses are shown using neighbor-joining (Fig. 2A) and maximum parsimony (Fig. 2B). Maximum parsimony resulted in a single tree of 191 steps and a consistency index (CI) of 0.6. The distance based (DB) tree (Fig. 2A) agrees more closely with morphological-based hypotheses. Regarding outgroups, insects cluster together and outside of crustaceans, showing the progressive splitting of *Artemia* and Cladocera, both within the Class Branchiopoda and considered a primitive group; then follow the Decapoda, with *Litopenaeus* appearing first, traditionally considered a primitive genus. Next the representatives of the two caridean families, first *Potimirim potimirim* from the family Atyidae and then all the members of the Palaemonidae, *M. olfersii* and *M. carcinus* in a single cluster and *M. acanthurus* and *M. rosenbergii* in another.

The MP tree (Fig. 2B) does not provide resolution at higher levels, however results within caridean species are almost identical using both methods. When compared with previous phylogenetic trees based on morphological data (Fig. 1), the results are fully consistent. They agree in that Atyidae is the sister group of the Palaemonidae, *M. carcinus* and *M. olfersii* are more closely related to each other as well as the pair *M. acanthurus-M. rosenbergii*. This result agrees with

Pereira (1997), however more species should be included in order to have a more robust hypothesis. We hope that this work will motivate further research on molecular systematics of this important group.

REFERENCES

Folmer OM, Black W, Hoeh R, Lutz R, and Vrijenhoek R (1994) DNA primers for amplification of mitochondrial cytochrome c oxidase subunit I from diverse metazoan invertebrates. Mol Mar Biol Biotech 3:294-299

Hillis MH, Moritz C, and Mable BK (1996) Molecular Systematics. Second edition, Sinauer Associates 655 p

Holthuis LB (1950) The Decapoda of the Siboga-Expedition. Part X. The Palaemonidae collected by the Siboga and Snellius expeditions with remarks on other species I. Subfamily Palaemoninae. Siboga Expeditie, Monographie 39, 268 p

Holthuis LB (1951) A general revision of the Palaemonidae (Crustacea, Decapoda, Natantia) of the Americas. I. The Subfamilies Euryrhynchinae and Pontoniinae. Allan Hancock Foundation Publications, Occasional Papers 11, 332 p

Holthuis LB (1952) A general revision of the Palaemonidae (Crustacea, Decapoda, Natantia) of the Americas. II. The subfamily Palaemonidae. Allan Hancock Foundation Publications of the University of Southern California, Ocasional Paper 12, 396 p

Maddison WP, Donoghue MJ and Maddison DR (1984) Outgroup analysis and parsimony. Syst Zool 33:83-03

Pereira G (1989) Cladistics, Taxonomy, Biogeography and the evolutionary history of the shrimp Family Palaemonidae (Crustacea, Decapoda, Caridea). Ph.D. Thesis, University of Maryland, College Park, USA, 450 p

Pereira G (1997) A cladistic analysis of the freshwater shrimps of the family Palaemonidae (Crustacea, Decapoda, Caridea). Acta Biol Venezuel 17, 68 p

Saitou N and Nei M (1987) The neighbor-joining method: a new method for reconstructing phylogenetic trees. Mol Biol Evol 6:514-525

Simon C, Frati F, Beckenbach A, Crespi B, Liu H and Flook P (1994) Evolution, weighting and phylogenetic utility of mitochondrial gene sequences and a compilation of polymerase chain reaction primers. Ann Entom Soc Am 87:651-701

Swofford D (1998) PAUP*, Phylogenetic analysis using parsimony (*and other methods). Version 4a, Sinauer Associates, Sunderland, Massachussets

THE FIRST TWO LARVAL STAGES OF TWO FRESHWATER SHRIMPS GENUS *MACROBRACHIUM* (DECAPODA, PALAEMONIDAE) REARED IN THE LABORATORY, WITH A DISCUSSION ON THE SIGNIFICANCE OF ABBREVIATED DEVELOPMENT FOR THE RADIATION OF THE GENUS

Guido Pereira and José V. García

Universidad Central de Venezuela, Instituto de Zoología Tropical, Facultad de Ciencias,
Apartado 47058, Caracas 1041-A, Venezuela
gpereira@strix.ciens.ucv.ve

ABSTRACT

The description of the first two larval stages of *Macrobrachium cortezi* Rodríguez, 1982 and *M. rodriguezi*, Pereira, 1986, are presented for the first time. The origin and significance of abbreviated development in *Macrobrachium* is discussed.

I. INTRODUCTION

The shrimps of the genus *Macrobrachium* represent a group of crustaceans that have successfully radiated into the freshwater and estuarine environments. Within the genus, most species need some saline concentration to complete their larval development, while others are totally independent of salinity and develop far from estuaries. Estuarine species possess relatively small and numerous eggs that develop through as many as 13 larval stages, before reaching the juvenile stage. Inland freshwater species produce relatively few large eggs and have abbreviated larval development. Around 17 species of *Macrobrachium* presenting few large eggs occur in South America. So far, the larval development of *M. brasiliense* (Vega, 1984), *M. iheringi* (Bueno, 1979), *M. jelskii* (Gamba, 1980), *M. nattereri* (Magalhães, 1989), *M. petronioi* (Melo and Garcia, 1999), *M. potiuna* (Bueno, 1981) and *M. reyesi* (Pereira and García, 1995) have been described. These species present from three to six stages before the juvenile stage. For *M. borelli*, only the first larval stage has been described (Boschi, 1961). In the remainder nine species, the larval development is unknown. The present paper provides the description and illustrations of the first and second larval stages of *M. cortezi* Rodríguez, 1982 and *M. rodriguezi*, Pereira, 1986, and a discussion on the significance of abbreviated development in understanding the radiation of *Macrobrachium*.

II. MATERIALS AND METHODS

Ovigerous females of *M. cortezi* were collected in Puerto Ayacucho, Amazonas State, Venezuela (5° 45' N, 67° 25' W). Females of *M. rodriguezi* were collected in San Tomé, Anzoategui State, Venezuela (9° 0' N, 64° 0' W). Shrimps were collected with hand net, transported to the laboratory, and placed in aquaria at 27 °C. Shrimps were fed daily with fish food and newly hatched *Artemia* nauplii. Females carrying eggs with eye-spots were transferred to 4-l beakers with aerated freshwater. The larvae were reared in separated 100-ml beakers, and fed with newly hatched *Artemia* nauplii and micronized Tetra® fish food. The larvae were

![] Kluwer Academic/Plenum Publishers

E. Escobar-Briones & F. Alvarez Eds.
MODERN APPROACHES TO THE STUDY OF CRUSTACEA
PP. 323-328

checked daily and the water changed. Samples of 2-3 animals were fixed in 70 % ethanol solution for dissections and measurements. Drawings were made with stereoscopic and compound microscopes with camera lucida. Measurements were made using an ocular micrometer. Total body length and carapace length were measured.

III. RESULTS

Ovigerous females of *M. cortezi* carry between 18 - 21 red colored oval eggs (2.09 ± 0.11 mm average largest diameter); while those of *M. rodriguezi* carry between 7 - 11 gray colored oval eggs (1.66 ± 0.07 mm average largest diameter). Description of those more important characteristics of the first and second larval stages of *M. cortezi* and *M. rodriguezi* are presented in Table 1, as well as in figures 1 to 5.

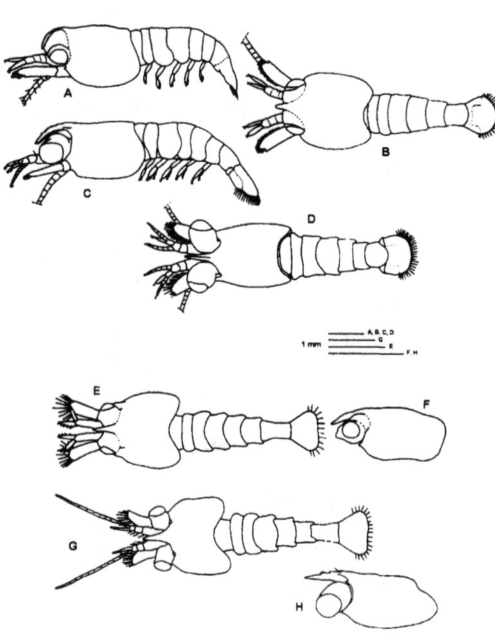

Figure 1. *Macrobrachium cortezi*, stage I. A, lateral view; B, dorsal view. Stage II. C, lateral view; D, dorsal view. *Macrobrachium rodriguezi*, stage I. E, dorsal view; F, lateral view of the carapace. Stage II. G, dorsal view; H, lateral view of the carapace.

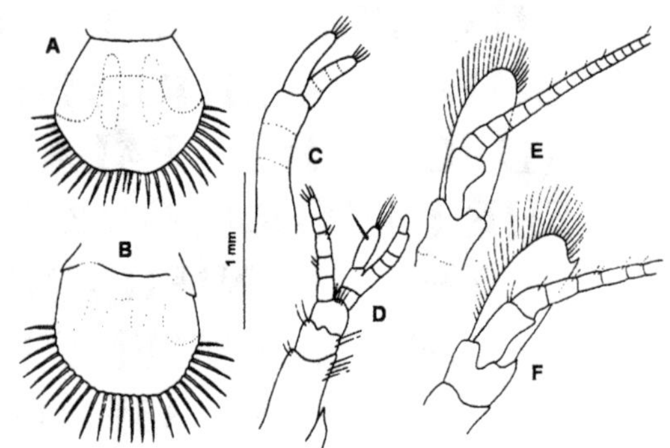

Figure 2. *Macrobrachium cortezi*, stage I. A, telson; C, antennula; E, antena. Stage II. B, telson; D, antennula; F, antena.

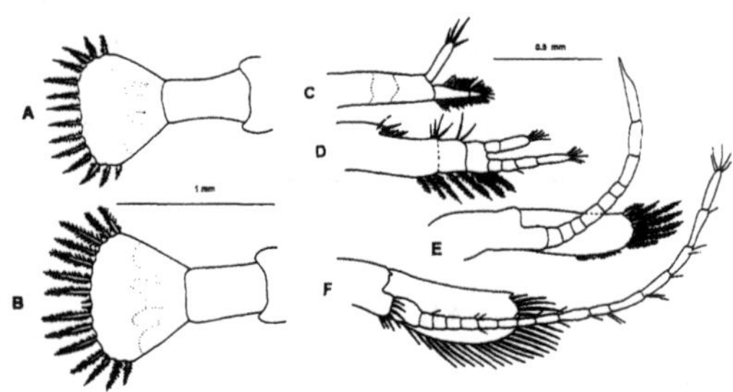

Figure 3. Macrobrachium rodriguezi, stage I. A, telson. Stage II. B, telson. Stage I. C, antennula; E, antena. Stage II. D, antennula; F, antena.

IV. DISCUSSION

A summary of characteristics of the first larval stage of 10 species of *Macrobrachium* from South America is presented in Table 2. We can group them into three categories following Jalihal et al. (1993), on the basis of development presented by the first larval stage. One formed by *M. nattereri*, *M. petronioi* and *M. borellii*; a second by *M. brasiliense*, *M. jelskii*, *M. iheringi*, and *M. rodriguezi* and a third group with *M. cortezi*, *M. potiuna*, and *M. reyesi*, corresponding the last group to species that hatch in a more advanced stage.

Colonization of inland freshwater habitats, has been interpreted in different ways by several authors. Boschi (1961) found that *M. borellii*, an inland freshwater species, hatched in an advanced developmental stage and suggested, based on geological evidence, that the origin of this species in freshwater could be ancient. Jalihal et al. (1993) suggest a gradual modifica-

Table 1. *Description of the characteristics of the first and second larval stages of* **M. cortezi** *and* **M. rodriguezi**.

Characteristic	Macrobrachium cortezi (Fig. 1,2,3)		Macrobrachium rodriguezi (Fig. 1,4,5)	
	Stage I	**Stage II**	**Stage I**	**Stage II**
TOTAL LENGTH	6 mm	6.3 mm	3.9 mm	4.7 mm
ROSTRUM				
-shape	curved downward	curved downward	slightly curved	straight
-teeth on upper border	absent	six teeth	absent	3 teeth
	(Fig. 1 A)	(Fig. 1 C)	(Fig. 1 F)	(Fig. 1 H)
CARAPACE				
-length	2.4 mm	2.7 mm	1.3 mm	1.5 mm
-spines	2 spines	2 spines	1 spine	2 spines
	(Fig. 1 A, B)	(Fig. 1 C, D)	(Fig. 1 E, F)	(Fig. 1 G, H)
TELSON				
-shape	rounded	rounded	triangular	triangular
-# setae	28	25	14	16
	(Fig. 2 A)	(Fig. 2 B)	(Fig. 3 A)	(Fig. 3 B)
ANTENNULAE				
# segments on				
-inner flagellum	unsegmented	5 segments	unsegmented	2 segments
-outer flagellum	unsegmented	unsegmented	unsegmented	4 segments
	(Fig. 2 C)	4 segments	(Fig. 3 C)	(Fig. 3 D)
		(Fig. 2 D)		
ANTENA				
-# segments on flagellum	multisegmented	multisegmented	9 segments	16
	(Fig. 2 E)	(Fig. 2 F)	(Fig. 3 E)	(Fig. 3 F)
MANDIBLE				
-palp	absent	absent	absent	absent
-# of molar processes	3 rounded lobes	2 rounded lobes	2 serrate rounded lobes	2 processes, upper
	(Fig. 4 A)	(Fig. 4 M)	(Fig. 5 A)	serrate, lower smooth
				(Fig. 5 M)
FIRST MAXILLA				
-# of endites	3	3	3	3
-surface of endites	central lobe dentate	smooth	all endites with spines	upper and lower 1 spine,
	(Fig. 4 B)	(Fig. 4 N)	(Fig. 5 B)	central smooth
				(Fig. 5N)
THIRD MAXILLIPED				
-# of segments on endopod	unsegmented	3	5	4
- setae on tip of endopod	absent	present	present	absent
	(Fig. 4 F)	(Fig. 4 R)	(Fig. 5 F)	(Fig. 5 R)
CHELIPEDS				
-exopods	absent	absent	present	present
-shape of cutting edge	smooth	smooth	smooth	smooth
	(Fig. 4 G, H)	(Fig. 4 S, T)	(Fig. 5 G, H)	(Fig. 5 S, T)
PEREIOPODS				
-exopods	absent	absent	present	only on the third
-# of segments	7	7	7	7
	(Fig. 4 I, J, K)	(Fig. 4 U, V, W)	(Fig. 5 I, J, K)	(Fig. U, V, W)
PLEOPODS				
-shape	birramous	birramous	birramous	birramous
-appendix interna	present	present	present	present
-setae	present	present	absent	present
	(Fig. 4 L)	(Fig. 4 X)	(Fig. 5 L)	(Fig. 5 X)

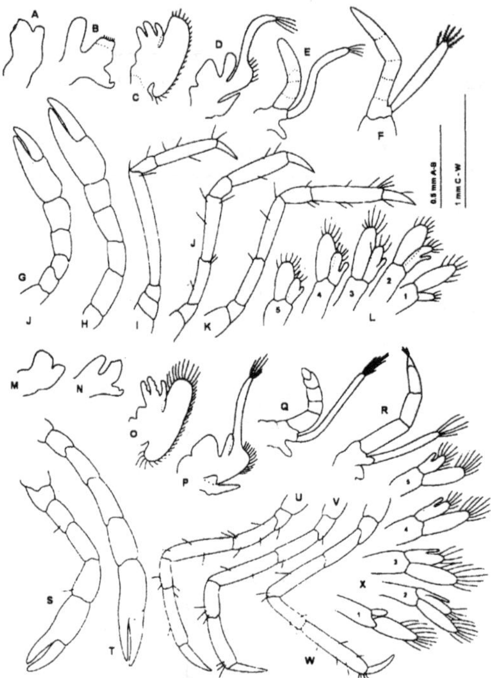

Figure 4. *Macrobrachium cortezi*, *stage I. A, mandible; B, first maxilla; C, second maxilla; D, first maxilliped; E, second maxilliped; F, third maxilliped; G, first cheliped; H, second cheliped; I, J, K, third, fourth and fifth pereiopods; L, pleopods. Stage II. M, mandible; N, first maxilla; O, second maxilla; P, first maxilliped; Q, second maxilliped; R, third maxilliped; S, first cheliped; T, second cheliped; U, V, W, third fourth and fifth pereiopods; X, pleopods.*

tion because of a freshwaterization process. Magalhães and Walker (1988) considered abbreviated development as a multiple convergent phenomenon overriding phylogenetic relationships even above the generic level. Pereira (1989) argued that, since primitive palaemonids such as the Euryrhynchidae and *Troglocubanus* possess abbreviated development, this feature could be considered primitive. Figure 6 shows a segment of the phylogenetic tree obtained by Pereira (1997) that includes the species treated in Table 2. The only one species not included in the tree was *M. petronioi*, but should fall near *M. nattereri* (following comments on morphological similarity suggested by Melo et al. 1988). Those species considered in Table 2 are distributed in the tree with no obvious pattern, this is not consistent with the idea of a gradual and transitional directed process from

estuarine to inland freshwater habitats as suggested by Jalihal et al. (1993). In our view, there is no reason to believe that a scenario in which estuarine species are colonizing inland freshwater is prevailing as oppose to another in which inland freshwater ancient species are colonizing estuarine environments. The following two examples illustrate the point, Mashiko (1999) found a mechanism of polypatric evolution of egg size among 28 local populations of *M. nipponense* during recent colonization in inland waters of Japan, indicating that populations with intermediate-egg sizes probably evolved independently from estuarine small-egg populations. López and Pereira (1996, 1998) found *Eurhyrynchus amazoniensis* and *Palaemonetes carteri*,

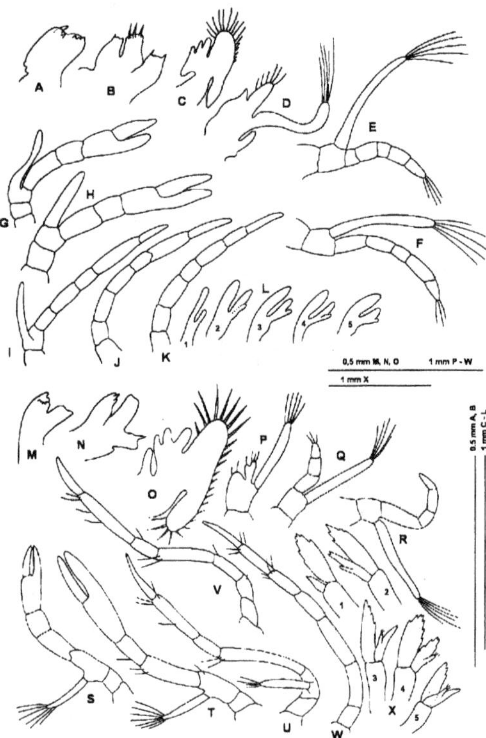

Figure 5. *Macrobrachium rodriguezi*, *stage I. A, mandible; B, first maxilla; C, second maxilla; D, first maxilliped; E, second maxilliped; F, third maxilliped; G, first cheliped; H, second cheliped; I, J, K, third, fourth and fifth pereiopods; L, pleopods. Stage II. M, mandible; N, first maxilla; O, second maxilla; P, first maxilliped; Q, second maxilliped; R, third maxilliped; S, first cheliped; T, second cheliped; U, V, W, third, fourth and fifth pereiopods; X, pleopods.*

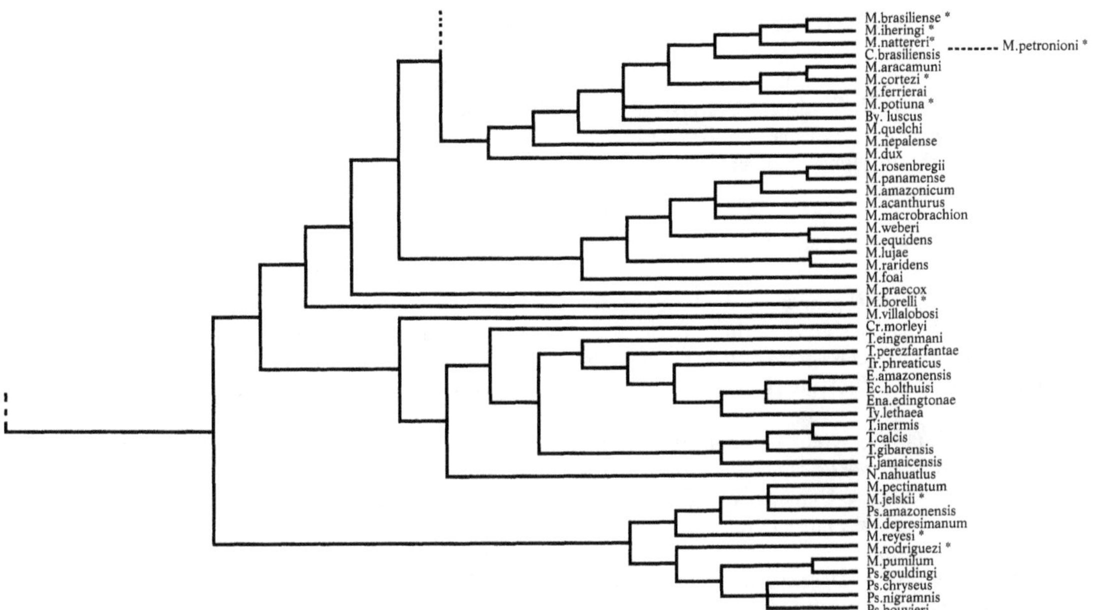

Figure 6. *Partial view of the phylogenetic tree obtained by Pereira (1997). Asterisk denotes the species treated in Table 2. For explanations see discussions on the text.*

Table 2. *Characteristics of the first larval stage in shrimps of the genus* **Macrobrachium** *in South America.*

Characteristic	M. nattereri	M. petronioi	M. borellii	M. brasiliense	M. jelskii	M. iheringi	M. rodriguezi	M. potiuna	M. cortezi	M. reyesi
Carapace										
- Anterolateral border	unarmed	unarmed	unarmed	1 spine	1 spine	1 spine	1 spine	1 spine	2 spines	2 spines
Antenna										
- Segments on flagellum	unsegmented	unsegmented	7	10	9	8	9	Multisegmented (21)	multisegmented (32)	multisegmented 20 - 25
- Setae on scale border	11	12	11	12	20	12	7	12	23	
Pereiopods										
- Quelipeds	unirramous	unirramous	unirramous	unirramous	birramous	birramous	birramous	unirramous	unirramous	unirramous
- Pereiopod 3	unirramous	unirramous	unirramous	unirramous	unirramous	birramous	birramous	unirramous	unirramous	unirramous
- Pereiópods 4-5	unirramous	unirramous	unirramous	unirramous	unirramous	4 birramous 5 unirramous	unirramous	unirramous	unirramous	unirramous
Pleopods										
- Exopod	without seate	without setae	without setae	without setae	without setae	without setae	without setae	without setae	with setae	with setae
- Appendix interna (2-5)	absent	present without setae	present without setae	present without setae	absent	present without setae	present without setae	present without setae	present without setae	present with setae
Telson										
- Number of setae	14	30-36	14	14	18	18	14	30	28	17
Reference	Magalhães, 1989	Melo and Garcia 1999	Boshi, 1961	Vega, 1984	Gamba, 1980	Bueno, 1979	Present work	Bueno, 1981	Present work	Pereira and García, 1995

typical inland freshwater species, living in the Orinoco Delta, under the tide influence. So, an alternative scenario for the radiation of *Macrobrachium*, would have first to consider primitive inland freshwater species as well as modern estuarine species. Also, an adaptation process in both ways, estuarine species colonizing lentic or running inland freshwaters and inland species colonizing estuarine environments should be taken into account. It is possible that migrations in both directions have been going on since the origin of the family in the early Cretaceous time.

ACKNOWLEDGEMENTS

Financial support was provided by Consejo de Desarrollo Científico y Humanístico (Universidad Central de Venezuela) Research Project No 03-31-2763-98. We also tanks to Henry Egáñez for helping in collection of shrimps.

REFERENCES

Boschi E (1961) Sobre el primer estadío larval de dos especies de camarones de agua dulce (Crustacea, Palaemonidae). In: Actas y trabajos del 1er Congreso Sudamericano de Zoología. La Plata, Argentina 2:69-77

Bueno SL de S (1979) Notas preliminares sobre o desenvolvimento larval de *Macrobrachium iheringi* (Ortman, 1897) (Crustacea, Decapoda, Palaemonidae), obtido em laboratório. Ciência e Cultura 32:486-488

Bueno SL de S (1981) Desenvolvimiento larval de *Macrobrachium potiuna* (Müller, 1880) e *Macrobrachium iheringi* (Ortman, 1897) (Crustacea, Decapoda, Palaemonidae). M Sc Thesis. Universidade de São Paulo, Brasil

Gamba AL (1980) Desarrollo abreviado del camarón de agua dulce *Macrobrachium jelskii* (Miers, 1877). (pp 169-89) In: Biología, ecología y cultivo de organismos acuáticos. Simposia 16, USB, Editorial Equinoccio, Caracas

Jalihal DR, Sankolli KN and Shenoy S (1993) Evolution of larval developmental patterns and the process of freshwaterization in the prawn genus *Macrobrachium* Bate, 1868 (Decapoda, Palaemonidae). Crustaceana 65:365-376

López B and Pereira G (1996) Inventario de los crustáceos decápodos de las zonas alta y media del delta del Orinoco, Venezuela. Act Biol Ven 16:45-64

López B and Pereira G (1998) Actualización del inventario de crustáceos decápodos del Delta del Orinoco. In López JL, Saavedra I and Dubois M (eds.) El Río Orinoco. Aprovechamiento sustentable (pp. 76—85) Memorias de las jornadas de investigación sobre el Río Orinoco. Facultad de Ingeniería, Universidad Central de Venezuela, Caracas

Magalhães C (1989) The larval development of palaemonid shrimp from the Amazon Region reared in the laboratory. VI. Abbreviated larval development of *Macrobrachium nattereri* (Heller, 1862) (Crustacea, Decapoda). Amazoniana 10:379-392

Magalhães C and Walker I (1988) Larval developemnt and ecological distribution of central Amazonian palaemonid shrimps (Decapoda, Caridea). Crustaceana 55:279-292

Mashiko K (1999) Polypatric evolution of egg size among local populations of the palaemonid prawn *Macrobrachium nipponense* (De Haan, 1849) during colonization of inland waters. Crustacean Issues 12:527-532

Melo G and Garcia AL (1999) Postembryonic development of *Macrobrachium petronioi* (Caridea: Palaemonidae) in the laboratory. J Crust Biol 19:622-642

Melo GAS, Lobao VI and Fernandes WM (1988) Resdescrição de *Macrobrachium birai* Lobão, Melo & Fernandes e de *Macrobrachium petronioi* Melo, Lobão & Fernandes (Crustacea: Decapoda), palemonídeos da região sul do Estado de São Paulo, Brasil. Bol Inst Pes São Paulo 15:87-89

Pereira G (1989) Cladistics, taxonomy, biogeography and the evolutionary history of the Palaemonidae (Crustaca, Decapoda, Palaemonidae). Ph D Dissertation, University of Maryland, College Park, Maryland, USA

Pereira G (1997) A cladistic analysis of the freshwater shrimps of the family Palaemonidae (Crustacea, Decapoda, Caridea). Act Biol Ven 17:1-69

Pereira G and García JV (1995) Larval development of *Macrobrachium reyesi* Pereira (Decapoda: Palaemonidae), with a discussion on the origin of abbreviated development in palaemonids. J Crust Biol 15:117-133

Vega LA (1984) Desenvolvimento larval de *Macrobrachium heterochirus* (Wiegman, 1836), *Macrobrachium amazonicum* (Heller, 1862) e *Macrobrachium brasiliense* (Heller, 1862) (Crustacea, Decapoda, Palaemonidae) em laboratório. Ph D Dissertation, Universidade de São Paulo, Brasil

A NEW SPECIES OF *LEIDYA* CORNALIA AND PANCERI, 1861, AND THE FIRST RECORD OF THE GENUS *LOBOCEPON* NOBILI, 1905, BOTH FROM THE EASTERN PACIFIC OCEAN, WITH A REVIEW OF THE PARASITES OF GRAPSID CRABS WORLDWIDE (ISOPODA, BOPYRIDAE, IONINAE)

John C. Markham

Arch Cape Marine Laboratory, Arch Cape, Oregon 97102-0105, USA

jmarkham@seasurf.com

ABSTRACT

Leidya infelix, new species, is described as a parasite of the grapsid crab *Pachygrapsus crassipes* Randall from the west coast of Baja California, Mexico. Another male belonging to the genus *Leidya* from the same host and nearly the same locality may belong to the same species, but that is not certain. A lone immature female bopyrid assignable to the genus *Lobocepon* is recorded as a parasite of *Grapsus grapsus* (L.) from the Pacific coast of Nicaragua; although probably a representative of a new species, it was not suitable for description. Included are reviews of the genera *Leidya* Cornalia and Panceri, 1861, and *Lobocepon* Nobili, 1905, and a tabulation of all species of bopyrid isopods reported to infest grapsid crabs worldwide.

I. INTRODUCTION

The first species of bopyrid isopod described as a parasite of a brachyuran crab, and the first recorded from either coast of the Americas, was recorded as a parasite of the fiddler crab *Gelasimus pugilator* (Bosc) [now *Uca pugilator* (Bosc)] at Atlantic City, New Jersey, USA. Leidy (1855) described the parasite as *Cepon distortus*, considering it congeneric with *Kepon typus* Duvernoy (1841) from an unknown host on the Indian Ocean island of Mauritius but using an emended spelling of that name. Shortly afterward, Cornalia and Panceri (1861) created the new genus *Leidya* and reassigned Leidy's species to it as the type-species, *Leidya distorta*, but did not examine any further material or add to the original description, which was seriously deficient in some crucial details. Despite the common occurrence of *Uca pugilator* along the carcinologically well-studied Atlantic coast of the United States, for over a century there were no further published records of *Leidya distorta*. The inadequate description and drawings presented by Leidy (1855) were repeated without further information by Bonnier (1900), Richardson (1905), Fowler (1912), Hay and Shore (1918), Miner (1950) and as recently as Schultz (1969).

Meanwhile, Pearse (1951), in a description little better than that of Leidy (1855), described the species *Leidya bimini* as a parasite of the grapsid crab *Pachygrapsus transversus* (Gibbes). Unfortunately, he failed to consider previous published citations of infestation of this widespread species. The first of these was by Müller (1871), who reported a *"Bopyrus"* infesting it in Brazil; Giard and Bonnier (1887), with no description, gave this parasite the name *Grapsicepon fritzii*, which was thus a nomen nudum. The next mention of

E. Escobar-Briones & F. Alvarez Eds.
MODERN APPROACHES TO THE STUDY OF CRUSTACEA
PP. 329-338

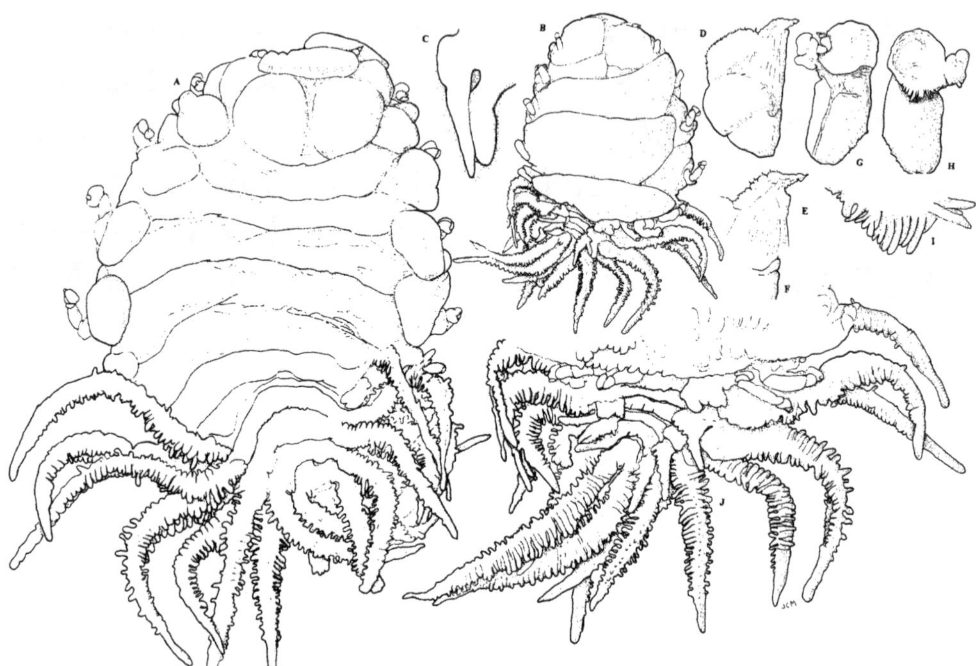

Figure 1. *Leidya infelix*, n. sp., holotype female. A. Dorsal view. B. Ventral view. C. Right side of barbula. D. Right maxilliped. E. Palp of same. F. Plectron of same. G. Right oostegite 1, external view. H. Same, internal view. I. Same, detail of internal ridge. J. Pleon in ventral view. Scale: 2.1 mm for B, G, H; 1.0 mm for A, D, J; 0.5 mm for C, E, F, I.

infestation of *Pachygrapsus transversus* was that of Verrill (1908) in Bermuda. Afterwards, Bourdon (1968) reported it in Bermuda, Brazil, Jamaica and the Bahamas: both of the latter authors called its parasite *Leidya distorta*. Only when Bourdon and Bowman (1970) thoroughly redescribed the parasites of several species of *Uca* and of *Pachygrapsus transversus* in the western Atlantic did it become clear that there are two valid and distinguishable species of *Leidya* there, *L. distorta* infesting *Uca* spp. and *L. bimini* from *Pachygrapsus* and some other grapsid crabs.

Elsewhere, mentions of the genus *Leidya* have been almost as confusing. Pearse (1930) described *L. sesarmae* as a parasite of *Sesarma dehaani* (Milne Edwards) [now called *Chiromantes dehaani*] and *L. ucae* infesting *Uca forcipata* (Adams and White), both from Fujien, China. Unfortunately, both descriptions were severely deficient, the types have not been examined since, and no one has reported collecting either species again. In the eastern Pacific, a parasite of *Pachygrapsus crassipes* Randall has been recorded from the coasts of California and adjacent Mexico on three

different occasions (Baker 1912, Hilton 1917, Hoard 1937); it was never described but at least twice tentatively considered to belong to the genus *Grapsicepon*, while the possibility existed that it was yet another species of *Leidya*. It is noteworthy that *Pachygrapsus transversus*, the host of *L. bimini* in the western Atlantic, also occurs along the west coast of Mexico, though it seems not to be known to harbor any bopyrid parasites there.

Recently, material of infested *Pachygrapsus crassipes* from western Mexico has become available for examination, and its parasite has proved to belong to an undescribed species of *Leidya*. At the same time, a parasite infesting *Grapsus grapsus* (L.) on the Pacific coast of Nicaragua, which had long been present but overlooked in the collection of the Smithsonian Institution, became available for study and proved to belong to the genus *Lobocepon*. The first of these species is herein described, while the latter, only an immature female, is inadequate for description. These findings provided opportunity for reconsideration of the genera *Leidya* and *Lobocepon* and remarks on world-wide records of bopyrid parasites infesting grapsid crabs.

II. RESULTS

Genus *Leidya* Cornalia and Panceri, 1861
Type-species, by original designation, *Cepon distortus*
Leidy, 1855
Leidya infelix, new species
Figs. 1-2

"[P]arasitic isopod."—Baker, 1912: 102 [Laguna Beach, California; infesting *Pachygrapsus crassipes* Randall].—Schmitt, 1921: 271. *Grapsicephon* [sic].—Hilton, 1917: 25-26 [Same locality and host; study of nervous system anatomy].

"[U]nidentified parasite similar to *Leidya* or *Grapsiceon.*"—Hoard, 1937: 105-106 [Seal Beach, Laguna Beach and La Jolla, California; same host; brief description]. "Unidentified genus."—Markham, 1992: 3.

Material examined

Infesting *Pachygrapsus crassipes* Randall. San Quintín, Baja California, México, 30°28'N, 115°58'W, intertidal, 23 May 1999, C. Schmidt, collector, Sta. PC83: 1♀, holotype, USNM 230506; 1♂, allotype, USNM 230507.

Description of holotype female, fig.1, fig. 2A, B. Length 7.9 mm, maximal width 5.6 mm, head length

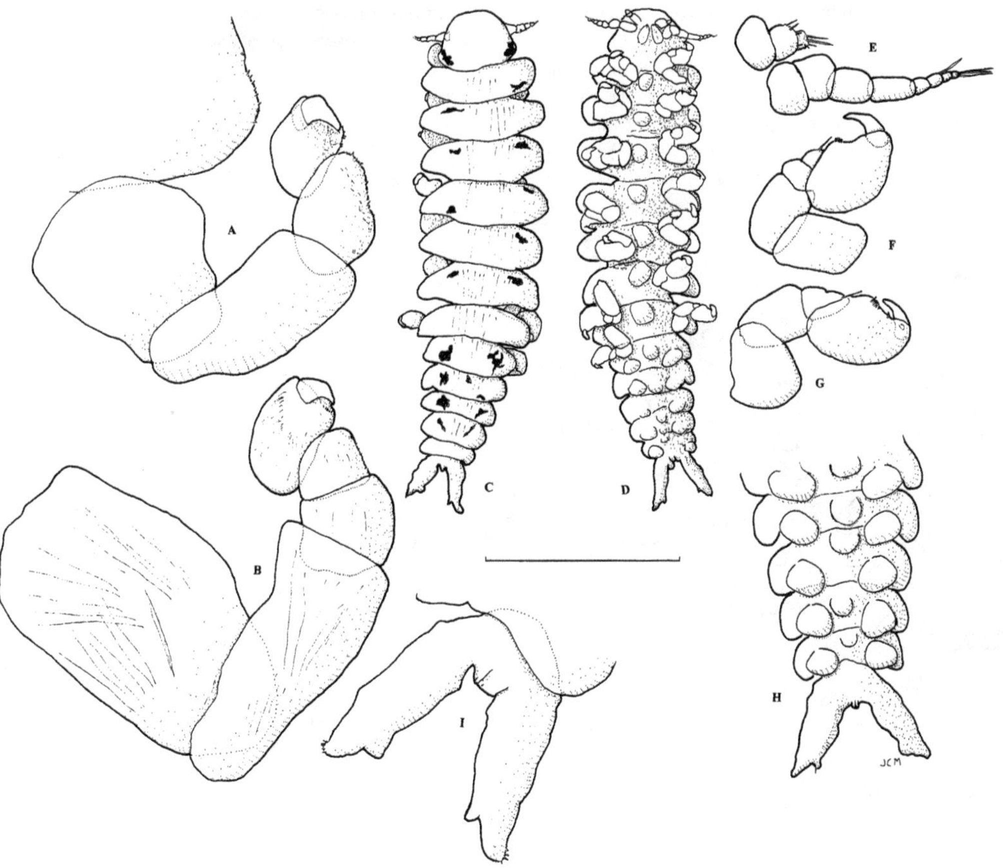

Figure 2. *Leidya infelix*, n. sp. A, B. *Holotype female. C-I. Allotype male. A. Right pereopod 1. B. Right pereopod 7. C. Dorsal view. D. Ventral view. E. Left antennae. F. Left pereopod 1. G. Left pereopod 7. H. Pleon in ventral view. I. End of pleon in ventral view. Scale: 0.4 mm for A-D; 0.2 mm for H; 0.1 mm for E-G, I.*

1.8 mm, pleon length 2.2 mm. Body distortion 25°. Body, exclusive of pleopods, broadly suboval, all body regions and segments distinct (Figs. 1A, B).

Head completely embedded in pereon, its dorsal surface divided into 2 large suboval lobes; prominent frontal lamina extending across middle half of head only. Barbula (Fig. 1C) with 2 long slender flaps on each side. Maxilliped (Fig. 1D) incompletely segmented, fringed with setae anterolaterally, bluntly pointed posteriorly; produced into large medially-directed nonarticulating palp placed anteromedially (Fig. 1E); plectron (Fig. 1F) small, rounded, not extended medially.

Pereon broadest across pereomere 4, both lateral margins almost completely covered by globose coxal plates; no midventral projections. Oostegites completely enclosing brood pouch. Oostegite 1 (Fig. 1G-I) about 2½ times as long as wide, its sides nearly parallel, bluntly rounded anteriorly and slightly pointed posteriorly; both segments of same width, anterior one less than half as long as posterior one; internal ridge (Fig. 1I) of many small lanceolate projections. Oostegites 3-5 extending

clear across brood pouch, last pair covering anterior part of pleon. Pereopods (Fig. 2A, B) all extending laterally, slightly larger posteriorly, especially in bases and ischia; carpi and meri of anterior pereopods fused, others separate.

Pleon (Fig. 1A, B, J) of 6 distinct pleomeres, its sides enclosed by long lanceolate lateral plates and exopodites all of about same size and extending far out and with similar deeply digitate margins; 5 pairs of smaller foliate endopodites covering half of ventral surface of pleon, narrower but longer posteriorly, their margins crenulate. Uniramous uropods similar in structure to lateral plates and exopodites but somewhat larger, extending far posteriorly. Anal cone minute, filling margin between uropods.

Description of allotype male, Fig. 2C-I. Length 1.04 mm, maximal width 0.26 mm, head length 0.12 mm, pleon length 0.34 mm. Sides of pereon nearly parallel, head and pleon both narrower (Fig. 2C, D). Head subtrapezoidal, truncate and narrowest anteriorly, broadest just in front of posterior margin. Large

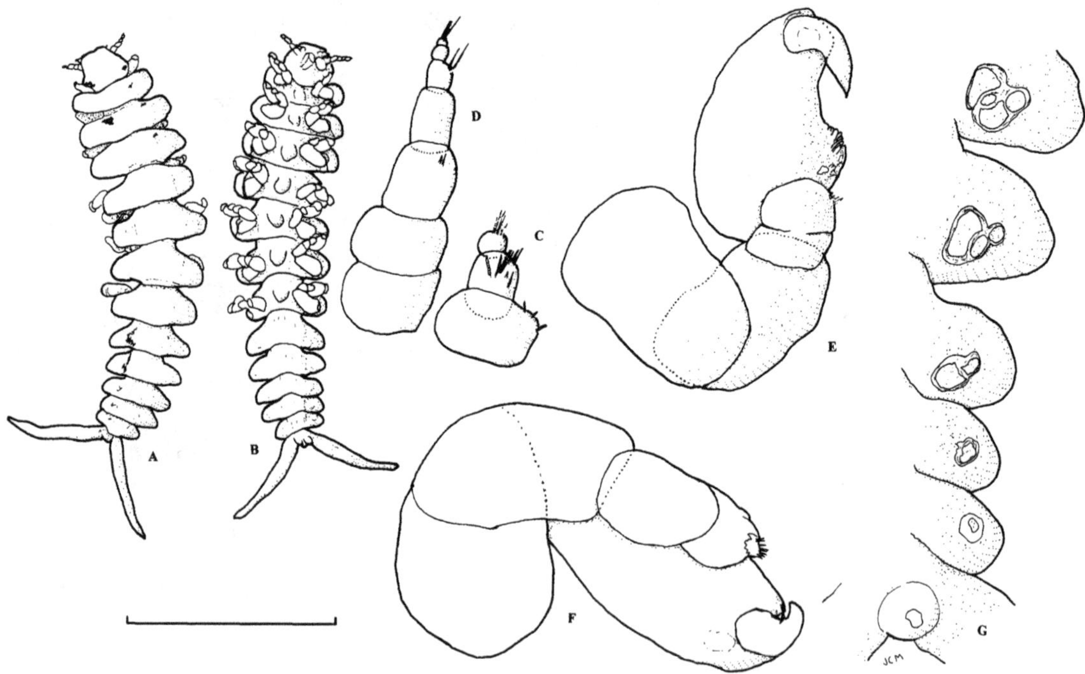

Figure 3. Leidya *sp., male. A. Dorsal view. B. Ventral view. C. Right antenna 1. D. Right antenna 2. E. Right pereopod 1. F. Left pereopod 7. G. Left side of pleon in ventral view. Scale: 1.15 mm for A, B; 0.20 mm for G; 0.10 mm for C-F.*

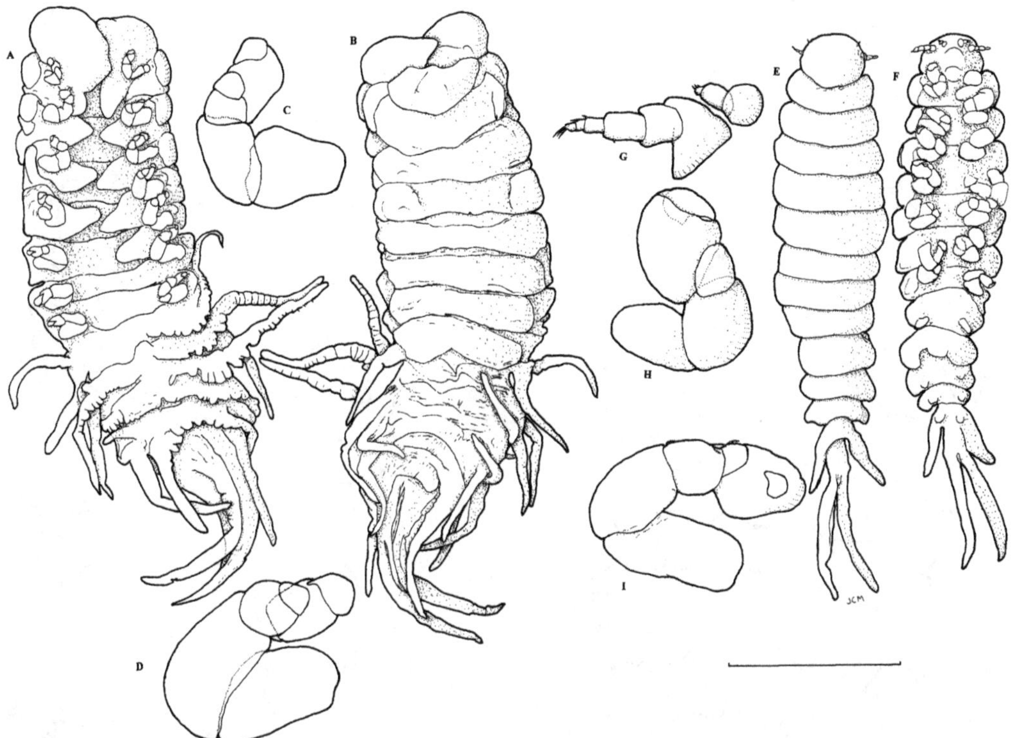

Figure 4. *Leidya ucae Pearse, 1930. A- D. Holotype female. E-I. Allotype male. A, Ventral view, B. Dorsal view. C. Left pereopod 1. D. Left pereopod 7. E. Dorsal view. F. Ventral view. G. Right antennae. H. Right pereopod 1. I. Left pereopod 6. Scale: 2.3 mm for A, B; 1.0 mm for E, F; 0.4 mm for C, D; 0.2 mm for G-I.*

irregularly shaped dark eyespots near posterolateral corners of head. Antennae (Fig. 2E) well-developed, both terminally setose, first of 3 articles, second of 8 articles and extending far laterally from side of head; sparse setae also on middle article of antenna 1 and on antepenultimate article of antenna 2.

Pereon with pereomeres distinctly separated by anterolateral notches, all of nearly same length and width. Dark dorsal splotches near sides of most pereomeres. Prominent conical midventral tubercle on each pereomere. Pereopods (Fig. 2F, G) all equally developed, some extending slightly beyond body margin, all articles distinct in anterior pereopods, meri and carpi fused in posterior ones.

Pleon long, tapering gradually posteriorly, of 6 distinct pleomeres. Irregularly shaped dark dorsal splotches near sides of most pleomeres. Five pairs of flaplike uniramous pleopods and 5 extended conical midventral pleonal projections (Fig. 2H). Final pleomere (Fig. 2I) produced into extended divergent uropods, each with posteromedial process and tuft of tiny setae on posterolateral margin.

Etymology

The specific name *infelix*, "ill-fated" or "unfortunate," has been selected to denote the long inability of this species to be described, despite its having been collected originally almost 90 years ago.

Remarks

Because the two species of *Leidya* reported from China are either poorly known or described from immature specimens only, the comparisons here deal with the two western Atlantic species, as redescribed by Bourdon and Bowman (1970). The new species, *L. infelix*, is assigned to the genus *Leidya* by the following criteria. The

333

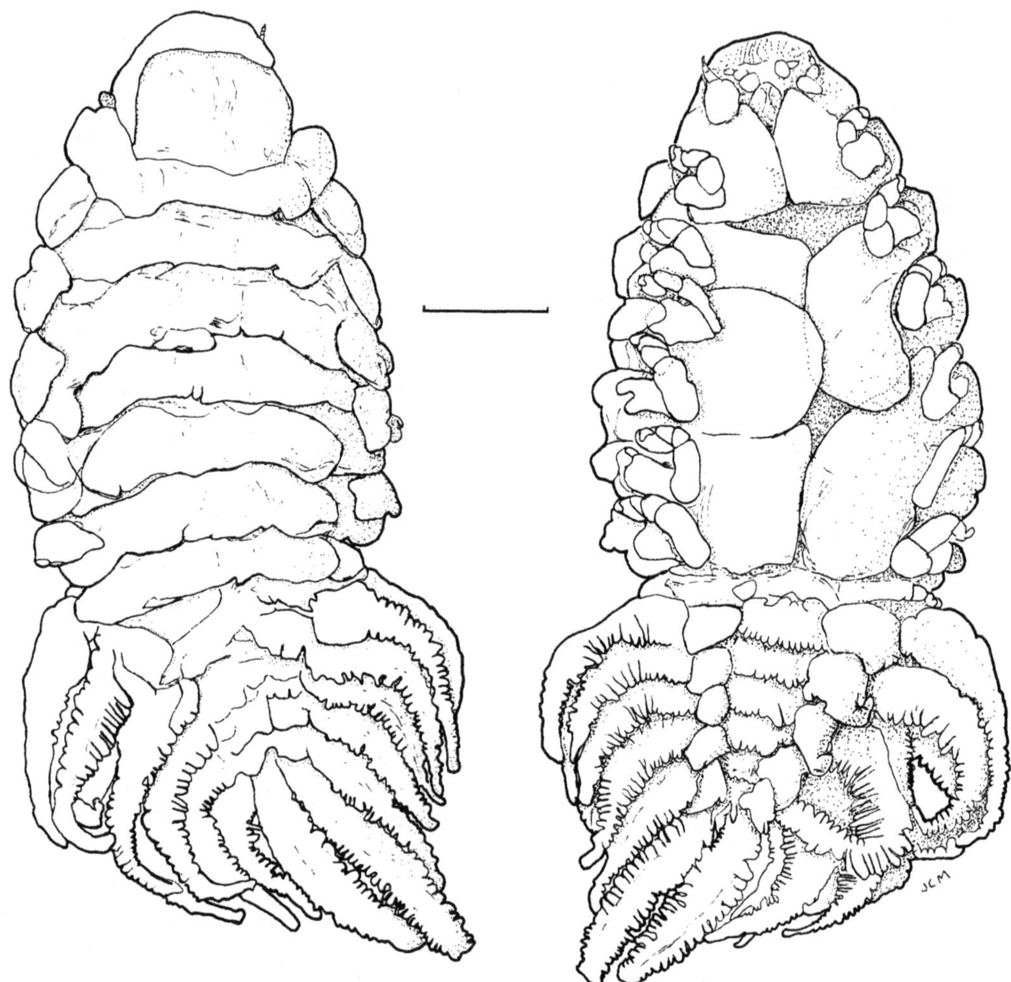

Figure 5. ***Lobocepon*** *sp., immature female, in dorsal and ventral views. Scale: 1.0 mm.*

female has only moderate body distortion; a conspicuous frontal lamina; a dorsally bilobate head; a prominent nonarticulating maxilliped palp; two long slender projections on each side of the barbula; well-developed coxal plates; 5 pairs of similar long, slender lanceolate lateral plates and pleopodal exopodites and larger but similar uniramous uropods, all with digitate margins; and 5 pairs of reduced pleopodal endopodites. It lacks the middorsal pereonal bosses of *L. distorta* and *L. bimini*; its first oostegite, similar to that of *L. distorta*, is much more elaborate than that of *L. bimini*; its pereopods are relatively larger and

their dactyli better developed than those of either species; and its pleonal appendages are proportionately considerably longer and its endopodites more ornamented than those of either species. The male of *L. infelix* has the long slender body, narrow head, deeply separated body segments, midventral pereonal and pleopodal tubercles and extended divergent pleopods typical of *Leidya*. Its antennae and pereopods are proportionately larger than those of either other species, and its pleopods are tuberculiform and extended rather than sessile, while its uropods are much shorter.

Table 1. *Bopyrid parasites of species of family Grapsidae.*

Host species	Parasite species	Locality	Reference
Armases ricordi (H. Milne Edwards)	*Leidya bimini* Pearse, 1951	S. Florida, USA	Markham, 1972
Chiromantes dehaani (H. Milne Edwards)	*L. sesarmae* Pearse, 1930	Fuzhou, China	Pearse, 1930
	Megacepon choprai George, 1946	Okayama, Japan	Shiino, 1958
Cyclograpsus integer (H. Milne Edwards)	*Leidya bimini* (?)	Jamaica	Hartnoll, 1965
Episesarma mederi (H. Milne Edwards)	*Megacepon choprai*	Samut Sakhan, Thailand	Markham, 1982
Goetice depressus (de Haan)	*Allokepon goetici* (Shiino, 1934)	Seto, Mie &Ryukyus, Japan	Shiino, 1934,1939, 1958
		Hong Kong	Markham, 1982
Grapsus albolineatus Lamarck	*Kepon typus* Duvernoy, 1841	Mauritius	
G. grapsus (Linnaeus)	*Lobocepon* sp.	W. Nicaragua	This paper
G. tenuicrustatus (Herbst)	*L. grapsi* Nobili, 1905	New Guinea	Nobili, 1905
		Taiwan	Shiino, 1936
Metaplax distincta H. Milne Edwards	*Kepon orientalis* Markham, 1985	Phuket, Thailand	Markham, 1985
M. elegans de Man	*K. orientalis*	Phuket, Thailand	Markham. 1985
Metapograpsus messor Forskål	*Grapsicepon messoris* (Kossmann, 1877)	Red Sea	Kossmann, 1877
M. oceanicus Jacquinot & Lucas	*Megacepon (?)* sp.	Sumbawa, Indonesia	Bourdon & Stock, 1979
Muradium tetragonum (Fabricius)	*Megacepon choprai*	Madras, India	George, 1946
Pachygrapsus crassipes Randall	*Leidya infelix*, n. sp.	Calif., USA; B. Calif, Mex	This paper
P. transversus (Gibbes)	*L. bimini*	Bermuda; Bahamas; Florida; Caribbean; Panama; Brazil	Bourdon & Bowman, 1970
Perisesarma maipoensis (Soh)	*Megacepon choprai*	Hong Kong	Markham, 1990
Planes minutus (Linnaeus)	*Grapsicepon edwardsi* Bonnier, *1900*	Sargasso Sea	Markham, 1988
Varuna litterata (Fabricius)	*Megacepon* (?) *pleopodatus* Bourdon, 1981	Sumba, Indonesia	Bourdon, 1981

Leidya sp.
Fig. 3

Material examined

Infesting *Pachygrapsus crassipes* Randall. Intertidal, Estero de Ensenada, Baja California, México, 31°53'N, 116°35'W, intertidal, 2 December 1998, M. Torchin, coll.: 1♂, USNM 000000.

Remarks

Because there was no female present in this collection, I cannot assign this specimen to species, even though it came from the same host and nearly the same locality as *L. infelix*, n. sp. As Bourdon and Bowman (1970) demonstrated in their study of *L. distorta*, there is much variation within at least that species, so it is possible that the present specimen is actually assignable to *L. infelix*. The absence of an accompanying female here and the necessity to base the description above on only the allotype male make it impossible to be certain.

It is noteworthy that all of the characters by which the male of *L. infelix* differs from those of the western Atlantic species are absent from this male. As in the males of both Atlantic species, its pleopods are reduced to scroll-like scars, and its uropods are greatly extended. According to Bourdon and Bowman (1970), the only character by which the Atlantic species' males differ from each other is that the male of *L. bimini* has midventral tubercles on pleomeres 1-5, while in *L. distorta* they occur on only pleomeres 1-2. The present male differs from both of those species only in having no such tubercles on any pleomere, in contrast with all three species.

Leidya ucae Pearse, 1930
Fig. 4.

Leidya ucae Pearse, 1930: 12-13; pl. I, figs. 2, 4-7; pl. II, figs. 11, 12 [Type-locality Mui Hua, Foochow [=Fuzhou], Fukien [=Fujian Province], China, infesting *Uca forcipata* (Adams & White)].—Markham, 1980: 629.—Bourdon, 1981: 106.—Markham, 1982: 361, 365.

Material examined

Infesting *Uca forcipata* (Adams & White), Mui Hua, Fuzhou, Fujian Province, China, 10 April 1930, A. S. Pearse, coll.: 1 ♀, holotype, 1♂, allotype, USNM 66737.

Remarks

Reexamination and reillustration of the type-specimens of *L. ucae*, still the only material of this species known, confirm that the holotype is a very immature female, even though the allotype male appears essentially mature. The characters of the female are easier to assess than from the original description, but its immaturity means that it is not possible to make meaningful comparisons with other species.

Genus *Lobocepon* Nobili, 1905
Type-species, by monotypy, *Lobocepon grapsi*,
Nobili, 1905
Lobocepon sp.
Fig. 5

Material examined

Infesting *Grapsus grapsus* (Linnaeus). Masachapa (Pacific coast), Nicaragua, 11°47'N, 86°31'W, intertidal, 4 December 1955, W. Lemons, collector. 1E, immature, USNM 000000.

Remarks

Because the single female is clearly immature and unaccompanied by a male, it is unsuitable for description. Further, its assignment to the genus *Lobocepon* is somewhat doubtful, though it appears the most likely placement for this species. The only prior records of a species of *Lobocepon* are of the type-species, *L. grapsi*, reported as a parasite of *Grapsus grapsus* (evidently a reference which should have been to *Grapsus tenuicrustatus* (Herbst, 1783), because it is far outside the known range of the species now recognized as *G. grapsus*) at Tami Island, New Guinea (Nobili, 1905) and Kotosho, Taiwan (Shiino, 1936). Both Nobili (1905) and Shiino (1936) presented good descriptions of *L. grapsi*, on the basis of which I consider the present material probably to be congeneric with it. Like the female of *L. grapsi*, the present female has a well-developed frontal lamina; a faintly bilobate head; distinctly separated pereomeres and pleomeres; five pairs of biramous pleomeres, their endopodites clumped together and extended laterally, with finely serrate margins, their endopodites suboval, small, dorsomedially placed and with smooth margins; and its uniramous uropods filling the posterior gap between the exopodites and similar to them in shape and ornamentation but slightly broader than any pleopodal exopodites. The most conspicuous differences are the much broader body and the presence of large prominent tuberculate coxal plates on the first 4 pereomeres of *L. grapsi*, but these characters probably become manifest only with maturity, so it is not unexpected for the present female to lack them.

Even though its name has been cited previously as a host of bopyrid isopod, this is probably the first actual report of bopyrid infestation of *Grapsus grapsus* (Linnaeus), as presently recognized. See the discussion above about the use of the name for *Grapsus tenuicrustatus* (Herbst, 1783). Another citation of a species of *Grapsus* is probably also erroneous. Giard (1906) reported that *G. strigosus* Herbst was the long-unknown host of *Kepon typus* in Mauritius, but that reference should probably have been to *Grapsus albolineatus* Lamarck, 1818.

III. ADDENDUM

Summary of parasites of Grapsidae

This discussion of parasites of grapsid crabs of the eastern Pacific Ocean has provided a suitable opportunity to summarize published records of bopyrid parasites of crabs in the Family Grapsidae worldwide. Because some of the host species were originally recorded under names now considered incorrect or subsequently reassigned to more recently established genera, it is also timely to update those names. The parasites are summarized by host in Table 1. To date, 18 host species in 13 genera are known. Their parasites belong to 11 species in 6 genera all in the subfamily Ioninae. Most parasite species are reported from single host species, but *Leidya bimini* infests 2 and probably 3 species in as many genera, and *Megacepon choprai* infests 4 host species, all in different genera.

ACKNOWELDGEMENTS

I thank all those people who responded to my requests for parasites of *Pachygrapsus crassipes*, especially Armand Kuris, Chris Schmidt and Mark Torchin. Marilyn Schotte located and lent the type-specimens of *Leidya ucae* and the material of *Lobocepon* sp. My requests for the current names and records of distribution of grapsid crabs brought essential and welcome responses from Ana Costa, Michel Hendrickx, Peter Hogarth and Christoph Schubart among others.

REFERENCES

Baker CF (1912) Notes on the Crustacea of Laguna Beach. Ann Rept Laguna Mar Lab 1: 100-117

Bonnier J (1900) Contribution à l'étude des épicarides. Les Bopyridae. Trav Inst Zool Lille et Lab Zool Marit Wimereux 8:1-476

Bourdon R (1968) Les Bopyridae des mers Europeénnes. Mém Mus Natl Hist Nat Paris, Nouv Sér (A) 50:77-424

Bourdon R (1981) Sur une nouvelle espèce de Céponien (Isopode: Bopyride), parasite du crabe Varuna litterata (Fabricius). Verhandl Naturf Basel 91:101-107

Bourdon R and Bowman TE (1970) Western Atlantic species of the parasitic genus *Leidya* (Epicaridea: Bopyridae). Proc Biol Soc Wash 83:409-424

Bourdon R and Stock JH (1979) On some Indo-West Pacific Bopyridae (Isopoda, Epicaridea) in the collections of the Zoologisch Museum, Amsterdam. Beaufortia 28:205-218

Cornalia E, and Panceri P (1861) Osservazioni zoologische ed anatomische spora un nuovo genere di isopodo sedentari (Gyge branchialis). Mem Reale Accad Sci Torino (2) 19:85-118

Duvernoy GL (1841) Sur un nouveau genre de l'orde des Crustacés isopodes et sur l'espèce type de ce genre, le Képone type. Ann Sci Nat (sér 2) 15:10-22

Fowler HW (1912) The Crustacea of New Jersey. Ann Rept New Jersey State Mus 1911 (part II):29-650

George PC (1946) Megacepon choprai gen. et sp. nov., a bopyrid isopod from the gill chamber of Sesarma tetragonum (Fabr.). Rec Indian Mus 44:385-390

Giard A (1906) Sur le Grapsicepon typus Duvernoy, parasite de Grapsus strigosus Herbst. Compt Rend Soc Biol Paris 61:704-706

Giard A and Bonnier J (1887) Contributions à l'étude des bopyriens. Trav Inst Zool Lille et Lab Marit Wimereux 5:1-272

Hartnoll RG (1965) Notes on the marine grapsid crabs of Jamaica. Proc Linn Soc London 176:113-147

Hay WP and Shore CA (1918) The decapod crustaceans of Beaufort, N. C., and the surrounding region. Bull U S Bur Fish 35:369-475

Hilton WA (1917) The central nervous system of the parasitic isopod, *Grapsicephon* [sic]. J Paras 4:25-26

Hoard R (1937) Preliminary report on parasitic isopod. Pomona Coll J Entomol Zool 29: 105-106

Kossmann R (1877) Malacostraca. Zoologische Ergebnisse einer im Auftrage der Königlichen Academie der Wissenschaften zu Berlin ausgeführten Reise in die Küstengebieten des Rothen Meeres III:1-140

Leidy J (1855) Contributions towards a knowledge of the marine invertebrate fauna of the coasts of Rhode Island and New Jersey. J Acad Nat Sci 3:135-152

Markham JC (1972) Extension of range, a new host record and color notes for the parasitic isopod *Leidya bimini* Pearse, 1951 (Isopoda, Bopyridae). Crustaceana 3:190-192

Markham JC (1982) Bopyrid isopods parasitic on decapod crustaceans in Hong Kong and southern China. (pp 325-391) In: Morton BS and Tseng CK (eds) Proc First Internatl Mar Biol Workshop: The Marine Flora and Fauna of Hong Kong and Southern China, Vol. 1, Hong Kong University Press, Hong Kong

Markham JC (1985) Additions to the bopyrid isopod fauna of Thailand. Zool Verhand 224:1-63

Markham JC (1988) Descriptions and revisions of some species of Isopoda Bopyridae of the north western Atlantic Ocean. Zool Verhand 246:1-63

Markham JC (1990) Further notes on the Isopoda Bopyridae of Hong Kong. (pp 555-568) In: Morton BS (ed) Proc Second Internatl Mar Biol Workshop: The Marine Flora and Fauna of Hong Kong and southern China, Hong Kong, 1986, Vol. 2, Hong Kong University Press, Hong Kong

Markham JC (1992) The Isopoda Bopyridae of the eastern Pacific – missing or just hiding? Proc San Diego Soc Nat Hist 17:1-4

Müller F (1871) Bruchstücke zur Naturgeschichte der Bopyriden. Jenaische Zeitsch Medezin Naturw (Leipzig) 6:53-73

Nobili G (1905) Decapodi e isopodi della Nuova Guinea Tedesca raccolti dal Sign. L. Biro. Ann Mus Nationalis Hungarici 3:480-507

Pearse AS (1930) Parasites of Fukien crabs. Fukien Christian Univ Nat Hist Soc, Proc (Foochow) 3:10-18

Pearse AS (1951) Parasitic Crustacea from Bimini, Bahamas. Proc U S Natl Mus 101:341-372

Richardson H (1905) A monograph on the isopods of North America. Bull U S Natl Mus 54:1-727

Schmitt WL (1921) The marine decapod Crustacea of California with special reference to the decapod Crustacea collected by the United States Bureau of Fisheries steamer "Albatross" in connection with the biological survey of San Francisco Bay during the years 1912-1913. Univ California Publs Zool 23:1-470

Schultz GA (1969) How to know the marine isopod crustaceans. Wm. C. Brown Co., Dubuque, Iowa

Shiino SM (1934) Bopyrids from Tanabe Bay II. Mem Coll Sci, Kyoto Univ (Ser B) 9: 257-287

Shiino SM (1936) Bopyrids from Shimoda and other districts. Recs Oceanogr Works Japan 8:161-176

Shiino SM (1939) Bopyrids from Kyûsyû and Ryûkyû. Recs Oceanogr Works Japan 10: 79-99

Shiino SM (1958) Note on the bopyrid fauna of Japan. Rep Fac Fish, Pref Univ Mie 3: 29-73

Verrill AE (1908) Decapod Crustacea of Bermuda, I. Brachyura and Anomura. Their distribution, variations and habits. Trans Connecticut Acad Sci 13:299-474

SUBJECT INDEX

SYSTEMATIC INDEX

Printed in the USA
CPSIA information can be obtained
at www.ICGtesting.com
LVHW081920261123
764947LV00008B/587